Modern Carbonylation Methods

Edited by
László Kollár

Related Titles

Demchenko, A.

Handbook of Chemical Glycosylation

Advances in Stereoselectivity and Therapeutic Relevance

2008

ISBN: 978-3-527-31780-6

Dodziuk, H. (ed.)

Strained Hydrocarbons

2008

ISBN: 978-3-527-31767-7

Dyker, G. (ed.)

Handbook of C-H Transformations

Applications in Organic Synthesis

2005

ISBN: 978-3-527-31074-6

Tolman, W. B. (ed.)

Activation of Small Molecules

Organometallic and Bioinorganic Perspectives

Hardcover

ISBN: 978-3-527-31312-9

Dyker, G. (ed.)

Handbook of C-H Transformations

Applications in Organic Synthesis

2005

ISBN: 978-3-527-31074-6

Modern Carbonylation Methods

Edited by
László Kollár

WILEY-VCH Verlag GmbH & Co. KGaA

The Editor

Prof. László Kollár
University of Pécs
Department of Inorganic Chemistry
Ifjúság u. 6
7624 Pécs
Hungary

Library of Congress Card No.: applied for

British Library Cataloguing-in-Publication Data
A catalogue record for this book is available from the British Library.

Bibliographic information published by the Deutsche Nationalbibliothek
Die Deutsche Nationalbibliothek lists this publication in the Deutsche Nationalbibliografie; detailed bibliographic data are available in the Internet at <http://dnb.d-nb.de>.

Composition Thomson Digital, Noida, India
Printing Strauss GmbH, Mörlenbach
Binding Litges & Dopf GmbH, Heppenheim
Cover Design Grafik-Design Schulz, Fußgönheim

Printed in the Federal Republic of Germany
Printed on acid-free paper

ISBN: 978-3-527-31896-4

Contents

Modern Carbonylation Methods. Edited by László Kollár

ISBN: 978-3-527-31896-4

Preface

The organotransition metal chemistry, after an unbelievable expansion in the last decades, reached the stage of general application in synthetic organic chemistry. The recognition of the carbon–metal bonding properties and the mechanistic understanding of the basic catalytic reactions, as well as the definition of the scope and limitations, have rendered many of the transition metal-catalyzed reactions, among them carbonylation reactions involving the use of carbon monoxide as reactant, the most efficient solution to practical problems. Many general treatises and reviews, as well as increasing number of papers, demonstrate the increasing role of transition metal-catalyzed carbonylations in the field of organometallic synthesis.

Overcoming the fear of the novel type of organometallic reactants and a myth of using "expensive" transition metal complexes in a different way than "classical" organic reagents, some of these systems are used routinely as a tool for the introduction of C = O functionalities of various skeletons of practical importance. The enhanced selectivity, the well-defined mechanism, and the applicability of standard techniques are the main features that make these homogeneous catalytic reactions attractive.

Nowadays, the gap between "organometallic chemists" and those seeking to develop the results ("synthetic chemists") has been substantially narrowed: most organometallic reagents are used as "tools" in organic synthesis. Both the main-group organometallic chemistry and the transition metal organic chemistry have become indispensable and significant part of the curriculum of the students of chemistry.

As widely known, the earliest step towards organometallic chemistry was done 180 years ago by the synthesis and characterization of the Zeise's salt (1827). Although several milestones in the development of organometallic chemistry, such as the discovery of biner carbonyls (Mond, 1890), were marked by catalytic significance, the fundamental findings like these remained sporadic in the nineteenth century. At the beginning of the twentieth century, inorganic chemistry was overshadowed by developments in organic and physical chemistry, the developments in both of which laid the foundations for the subdisciplines of coordination

Modern Carbonylation Methods. Edited by László Kollár

ISBN: 978-3-527-31896-4

chemistry and organometallic chemistry. The achievements in both fields characterized the chemistry of the second half of the last century.

As far as the application of carbon monoxide as small "building block" is concerned, its history started with the cobalt-catalyzed alkene hydroformylation developed by Roelen in 1939. This seminal work is generally considered as the start of homogeneous catalysis as well. Since then fundamental work of the highest standard has been carried out in homogeneous catalysis featured by several Nobel laureates.

Nowadays, the use of carbon monoxide as a carbonyl source of aldehydes, ketones, carboxylic acids, and their derivatives in various transition metal-catalyzed reactions has become probably the most widespread methodology for homogeneous catalytic reactions. The unbelievably rich chemistry of homogeneous catalysis like carbon–carbon bond forming reactions or hydrogenations of alkenes, alkines, ketones, oximes or azomethines have also provided products with unprecedented structures. Several known compounds were synthesized by the new synthetic "tools" in acceptable yields, whereas the formation of side products were substantially reduced, that is, the novel methodologies met the environmental requirements. However, since the earliest history of homogeneous catalysis, the highest volume homogeneous catalytic reactions of industrial importance have been the production of *n*-butanal and acetic acid by the rhodium-catalyzed carbonylation of propene and methanol, respectively.

The scope of the book is largely confined to the most recent developments in carbonylation chemistry. Since this book of special focus is not intended to go into the fine details of homogeneous catalysis as well as its historical background, only the most recent achievements of carbonylation chemistry are discussed. I believe there is no need to elaborate further on the earlier findings in view of the excellent textbooks already available. During the last decade, several novel synthetic reactions involving carbon monoxide have been discovered, as well as new methods such as biphasic carbonylation or application of ionic liquids have been developed. It is our purpose to provide a perspective of this formative period through the contributions of the experts on special topics of carbon monoxide activation by transition metals.

Since I am sure that there are several details to be criticized and commented on, I would appreciate if the readers of this book would send their remarks both to the corresponding subchapter authors and to me preferably via e-mail (kollar@ttk.pte.hu).

László Kollár

List of Contributors

Antonio Arcadi
Universita dell'Aquila
Dipartimento di Quimica, Ingegniera
Chimica e Materiali
Via Vetoio snc – Coppito Due
67100 L'Aquila
Italy

Carmen Claver
Universitat Rovira i Virgili
Facultat de Química
c/ Marcel.li Domingo s/n
43007 Tarragona
Spain

Crestina S. Consorti
Laboratory of Molecular Catalysis
Institute of Chemistry, UFRGS
Av. Bento Gonçalves,
9500 Porto Alegre, 91501-970 RS
Brasil

Montserrat Diéguez
Universitat Rovira i Virgili
Facultat de Química
c/ Marcel.li Domingo s/n
43007 Tarragona
Spain

Simon Doherty
Newcastle University
School of Natural Sciences
Chemistry Bedson Building
Newcastle upon Tyne, NE1 7RU
United Kingdom

Jairton Dupont
Laboratory of Molecular Catalysis
Institute of Chemistry, UFRGS
Av. Bento Gonçalves
9500 Porto Alegre
91501-970 RS
Brasil

Zoraida Freixa
Universiteit van Amsterdam
Van't hoff Institute for Molecular
Sciences
Nieuwe Achtergracht 166
1018 WV Amsterdam
The Netherlands

Cyril Godard
Universitat Rovira i Virgili
Facultat de Química
c/ Marcel.li Domingo s/n
43007 Tarragona
Spain

Modern Carbonylation Methods. Edited by László Kollár

ISBN: 978-3-527-31896-4

Tamás Kégl
Pannon University
Research Group for Petrochemistry of the Hungarian Academy of Sciences
Egyetem u. 8.
H-8200 Veszprém
Hungary

Julian G. Knight
Newcastle University
School, of Natural Sciences
Chemistry Bedson Building
Newcastle upon Tyne, NE1 7RU
United Kingdom

László Kollár
University of Pécs
Department of Inorganic Chemistry
Ifjúság u. 6
7624 Pécs
Hungary

Mats Larhed
Uppsala University
Department for Medical Chemistry
Boksz 574
SE-75123 Uppsala
Sweden

Piet W.N.M. van Leeuwen
Universiteit van Amsterdam
Van't hoff Institute for Molecular Sciences
Nieuwe Achtergracht 166,
1018 WV Amsterdam
The Netherlands

David Morales-Morales
Universidad Nacional Autonoma de Mexico
Instituto Química
Circuito Exterior S/N, Ciudad Universitaria
Coyocan C.P. 04510 Mexico DF
Mexico

Akihiro Nomoto
Osaka Prefecture University
Department of Applied Chemistry
Gakuen-cho, Nakaku, Sakai
Osaka, 599-8531
Japan

Akiya Ogawa
Osaka Prefecture University
Department of Applied Chemistry
Gakuen-cho, Nakaku, Sakai
Osaka, 599-8531
Japan

Oscar Pàmies
Universitat Rovira i Virgili
Facultat de Química
c/ Marcel.li Domingo s/n
43007 Tarragona
Spain

Elisabetta Rossi
Università di Milano
Istituto di Chimica Organica 'A. Marchesisni'
Via Venezian 21
20133 Milano
Italy

Aurora Ruiz
Universitat Rovira i Virgili
Facultat de Química
c/ Marcel.li Domingo s/n
43007 Tarragona
Spain

Detlef Selent
Leibniz-Institut für Katalyse e.V.
Albert Einstein Str. 29a
D-18059 Rostock
Germany

Rita Skoda-Földes
Pannon University
Institute of Chemistry
Department Organic Chemistry
H-8200 Veszprém, Egyetem u. 8.
Hungary

Catherine H. Smyth
Newcastle University
School of Natural Sciences
Chemistry Bedson Building
Newcastle upon Tyne, NE1 7RU
United Kingdom

Ferenc Ungváry
Pannon University
Department for Organic Chemistry
H-8200 Veszprém, Egyetem u. 8.
Hungary

Neszta Ungvári
Pannon University
Department for Organic Chemistry
H-8200 Veszprém, Egyetem u. 8.
Hungary

Johan Wannberg
Uppsala University
Department for Medical Chemistry
Boksz 574, SE-75123 Uppsala
Sweden

1
Bite Angle Effects of Diphosphines in Carbonylation Reactions

Piet W.N.M. van Leeuwen, Zoraida Freixa

1.1
Introduction

The first two wide bite angle diphosphines, BISBI [1] and Xantphos [2], were introduced with the aim of improving the selectivity for linear aldehyde in the rhodium-catalyzed hydroformylation reaction. For designing Xantphos and related ligands, molecular mechanics methods were used. The concept of the natural bite angle β_n, that is, the ligand backbone preferred bite angle, was introduced by Casey and Whiteker [3], and β_n can be easily obtained by using molecular mechanics calculations. This angle gives the relative magnitudes of bite angles of the bidentate ligands, but it does not predict the angles for X-ray structures for two reasons. First, because the parameter for phosphorus–metal–phosphorus bending, the metal preferred bite angle, is set to zero in these calculations. Second, while parameters for the organic part of the molecules are highly accurate, the parameters involving the metal (for bond stretch and dihedral bending) are inaccurate and variable, but this need not distort the relative order of the ligands. The effect on hydroformylation was fairly well predicted and so was the favorable effect on metal-catalyzed hydrocyanation [4]. The bite angle effect on the activity or selectivity has been studied and reviewed for many catalytic reactions [5–9]. Initially for palladium-catalyzed reactions the results seemed rather capricious, but today these reactions are understood reasonably well [10].

For our study of the effect of (wide) bite angle diphosphines on catalytic reactions, a distinction between two different effects, both related to the bite angle of diphosphine ligands, can be made [5]:

- The first one, which we have called the steric bite angle effect, is related to the steric interactions (ligand–ligand or ligand–substrate) generated when the bite angle is modified by changing the backbone and keeping the substituents at the phosphorus donor atom the same. The resulting steric interactions can change the energies of the transition and the catalyst resting states. In rhodium-catalyzed hydroformylation reactions steric effects dominate [11], although an electronic bite angle effect was observed in one instance [12].

Modern Carbonylation Methods. Edited by László Kollár

ISBN: 978-3-527-31896-4

- The second one, the electronic bite angle effect, is associated with electronic changes in the catalytic center when changing the bite angle [9]. It can be described as an orbital effect, because the bite angle determines metal hybridization and as a consequence metal orbital energies and reactivity. This effect can also manifest itself as a stabilization or destabilization of the initial, final, or transition state of a reaction. The reductive elimination occurring in hydrocyanation and cross-coupling catalysis is an example of an electronic bite angle effect.

1.2 Rhodium-Catalyzed Hydroformylation

1.2.1 Introduction

The hydroformylation of alkenes is one of the most extensively applied homogeneous catalytic processes in industry. More than 9 million tons of aldehydes and alcohols are produced annually [13]. Many efforts have been devoted in the last few years to the development of systems with improved regioselectivity toward the formation of the industrially more important linear aldehyde. Both phosphine- and phosphite-based systems giving high regioselectivities to linear aldehyde for the hydroformylation of terminal and internal alkenes have been reported [1,2,14–16] (Scheme 1.1).

The generally accepted mechanism for the rhodium triphenylphosphine catalyzed hydroformylation reaction as proposed by Heck and Breslow [17] is shown in Scheme 1.2. The catalytically active species is a trigonal bipyramidal hydrido rhodium complex, which usually contains two phosphorus donor ligands. In early mechanistic studies [18], it was already demonstrated that this catalyst exists with two isomeric structures, depending on the coordination of the triphenylphosphine ligands, namely, equatorial–equatorial (ee) and equatorial–apical (ea) in an 85/15 ratio. Ever since the first rhodium–phosphine system was developed, a lot of research has been devoted to the development of more active and selective systems. In 1987, Devon *et al.* at Texas Eastman [1] patented the BISBI–rhodium catalyst, which gave excellent selectivity toward the linear aldehyde compared to other diphosphine ligands previously studied [19]. To rationalize this result, Casey and Whiteker [15] studied the relationship between selectivity and bite angle for different diphosphine ligands. They found a good correlation between the bite angle of the diphosphines and the

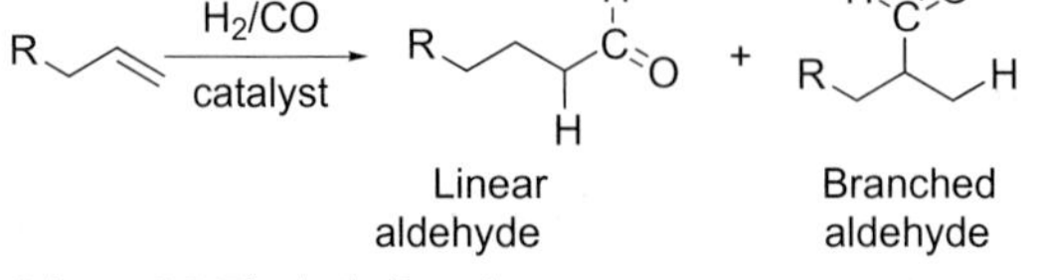

H_2/CO, catalyst

Linear aldehyde + Branched aldehyde

Scheme 1.1 The hydroformylation reaction.

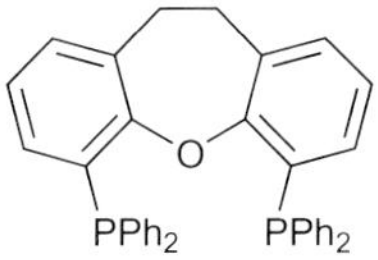

1, Homoxantphos

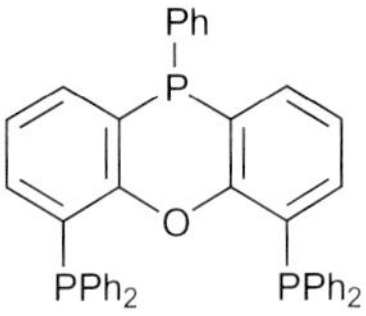

2, Phosxantphos

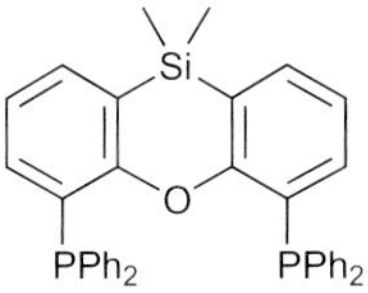

3, Sixantphos

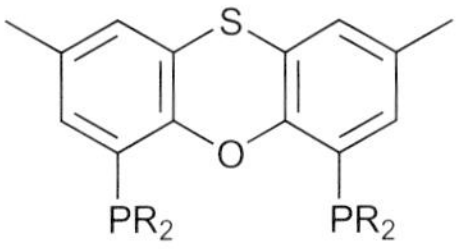

4d, R = Ph, Thixantphos

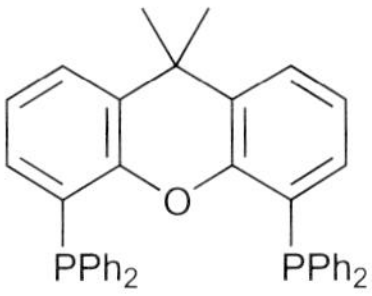

5, Xantphos

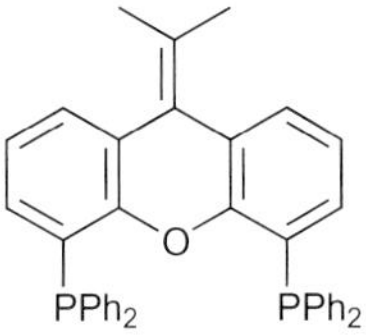

6, Isopropxantphos

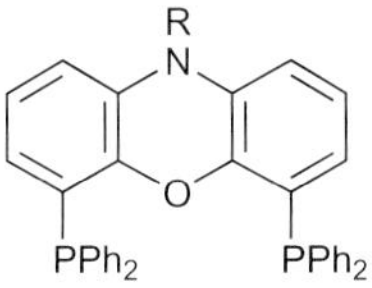

7, R = Bn, Benzylnixantphos

8, R = H, Nixantphos

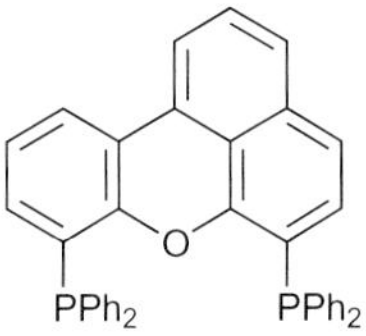

9, Benzoxantphos

regioselectivity. The high regioselectivity observed with BISBI was attributed to the preferential coordination mode, ee, in the catalytically active [RhH(diphosphine)$(CO)_2$] species, due to BISBI's natural bite angle close to 120°.

In the last decade, van Leeuwen *et al.* synthesized a series of xantphos-type diphosphines possessing backbones related to xanthene and having natural bite angles ranging from 102° to 123° [2]. These ligands, designed to ensure the bite angle is the only factor that has a significant variation within the series (the differences in electronic properties are minimal), have been applied to study the bite angle effect on the coordination mode, selectivity, and activity in hydroformylation (Scheme 1.2).

1.2.2
Steric Bite Angle Effect and Regioselectivity

In the first publication on the xantphos series [2], a regular increase in the selectivity to linear product in 1-octene hydroformylation while increasing the bite angle was

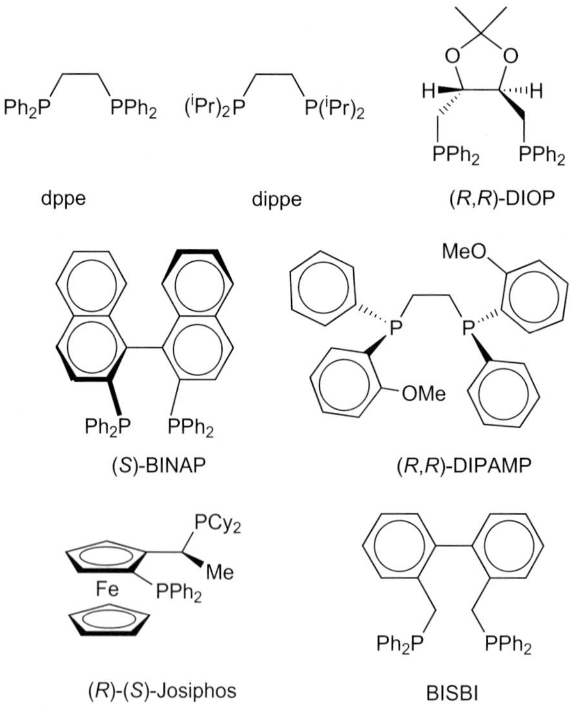

dppe dippe (*R*,*R*)-DIOP

(*S*)-BINAP (*R*,*R*)-DIPAMP

(*R*)-(*S*)-Josiphos BISBI

reported. The suggestion of a shift in the ee : ea equilibrium in the rhodium hydride resting state toward the ee isomer, considered to be the more selective one, was the tentative explanation. Later studies [16,20,21] (Table 1.1) showed that, even though there is a clear bite angle–selectivity correlation when a wide range of angles

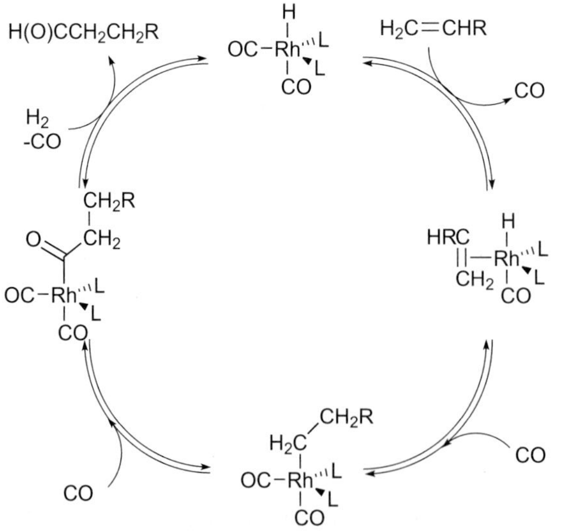

Scheme 1.2 Simplified catalytic cycle for the hydroformylation reaction.

Table 1.1 1-Octene hydroformylation using xantphos ligands (**1–10**).[a]

Ligand	β_n (°)[b]	l/b ratio[c]	% linear aldehyde[c]	% isomer[c]	tof[c,d]	ee : ea ratio
1	102.0	8.5	88.2	1.4	37	3:7
2	107.9	14.6	89.7	4.2	74	7:3
3	108.5	34.6	94.3	3.0	81	6:4
4d	109.6	50.0	93.2	4.9	110	7:3
5	111.4	52.2	94.5	3.6	187	7:3
6	113.2	49.8	94.3	3.8	162	8:2
7	114.1	50.6	94.3	3.9	154	7:3
8	114.2	69.4	94.9	3.7	160	8:2
9	120.6	50.2	96.5	1.6	343	6:4
10	123.1	66.9	88.7	10.0	1560	>10:1

[a]Conditions: $CO/H_2 = 1$, $P(CO/H_2) = 20$ bar, ligand/Rh = 5, substrate/Rh = 637, [Rh] = 1.00 mM, number of experiments = 3. In none of the experiments, hydrogenation was observed.
[b]Natural bite angles taken from Ref. [20].
[c]Linear-to-branched ratio and turnover frequency were determined at 20% alkene conversion.
[d]Turnover frequency = (moles of aldehyde)(moles of Rh)$^{-1}$ h^{-1}.

is considered, the ee/ea equilibrium in the hydride precursor is not the factor governing the regioselectivity when a smaller range of bite angles is considered. The RhH(diphosphine)$(CO)_2$ species itself is not involved in the step that determines the selectivity, but the selectivity is determined in the alkene coordination to RhH(diphosphine)(CO) or in the hydride migration step. A plausible explanation of the bite angle effect is that in these steps, an augmentation of the steric congestion around the metal center is produced when the bite angle is increased. This favors the sterically less demanding transition state of the possible ones, driving the reaction toward the linear product. Later [11], this was quantified by means of an integrated molecular orbital/molecular mechanics method, using the two limiting examples in the bite angle in the xantphos series, homoxantphos **1** and benzoxantphos **9**.

1.2.3 Electronic Bite Angle Effect and Activity

While the effect of the bite angle on selectivity in 1-octene hydroformylation (and styrene as well [2]) seems to be steric, the existence of a relationship between activity and bite angle in the hydroformylation reaction that can be easily deduced from the experiments done within the xantphos ligands family, might well have an electronic origin. An increase in the rate was found with the increasing bite angle (**1–9**), but ligand **10**, having the widest bite angle, showed a sharp increase in the rate of reaction (see Table 1.1).

The rate of dissociation of CO was studied separately via ^{13}CO exchange in a rapid scan IR spectroscopy study under pressure [16]. In this study, no influence of the natural bite angle on the rate of formation of the unsaturated (diphosphine)Rh(CO)H complexes was found for ligands **2**, **4**, and **6**. Ligand **10** shows a sharp increase in CO t-Bu t-Bu O P P O O 10

dissociation rate (seven times that of the other ligands). As steric effects on CO coordination are supposed to be small, this was explained by assuming a larger stabilization of the four-coordinate intermediate for ligand **10** with a wider bite angle and more electron-withdrawing character.

In a series of electronically distinct but sterically equal ligands **4**, it was found that the overall selectivity for linear aldehyde was constant, whereas the linear branched ratio and the rate increased concomitantly with the ee/ea ratio in the hydrido isomers (Table 1.2) [20]. The higher l/b ratio was because of an increase in the 2-octene formation – the "escape" route for the formed branched alkylrhodium intermediate.

It is possible that increasing the bite angle increases the activation energy for alkene coordination on steric grounds. What kind of electronic effect the widening of the bite angle has on the activation energy for alkene coordination depends on the dominant type of the alkene bonding; if electron donation from alkene to rhodium dominates, alkene coordination will be enhanced by wide bite angles.

In summary, a wider bite angle increases the concentration of unsaturated (diphosphine)Rh(CO)H and, other effects being absent or insignificant, the overall effect will result in an acceleration of the hydroformylation reaction.

When the backbone of a ligand allows both ee and ea coordination, the basicity of the phosphine has a pronounced effect on the chelation mode [22]. One of the first

Table 1.2 1-Octene hydroformylation using ligands **4a–g**.[a]

Ligand	R′	ee : ea ratio	l/b ratio[b]	% linear aldehyde[b]	% isomer[b]	tof[b,c]
4a	$N(CH_3)_2$	47 : 53	44.6	93.1	4.8	28
4b	OCH_3	59 : 41	36.9	92.1	5.3	45
4c	CH_3	66 : 34	44.4	93.2	4.7	78
4d	H	72 : 28	50.0	93.2	4.9	110
4e	F	79 : 21	51.5	92.5	5.7	75
4f	Cl	85 : 15	67.5	91.7	6.9	66
4g	CF_3	92 : 8	86.5	92.1	6.8	158

[a] Data taken from Ref. [20]. R = p-C_6H_4R'. Conditions: $CO/H_2 = 1$, $P(CO/H_2) = 20$ bar, ligand/Rh = 5, substrate/Rh = 637, [Rh] = 1.00 mM, number of experiments = 3.
[b] Linear-to-branched ratio and turnover frequency were determined at 20% alkene conversion.
[c] Turnover frequency = (moles of aldehyde)(moles of Rh)$^{-1}$ h^{-1}.

systematic studies using diphosphines was by Unruh [23] who used substituted dppf. Both rate and selectivity increase when the χ-value of the ligands increase. There are two possible reasons: electron's preference for linear alkyl complex formation when the π-back-donation to the phosphine increases; or, alternatively, EWD ligands enhance the formation of ee isomer as was observed later in the xantphos complexes [20]. This can be explained by the general preference of electron-withdrawing ligands for the equatorial positions in trigonal bipyramidal complexes. The loss of CO is faster for complexes containing ligands with higher χ-values. As mentioned above, a stronger complexation of the alkene donor ligand may be expected for more electron-deficient rhodium complexes. Thus, higher rates can be explained because in most phosphine-based systems the step involving replacement of CO by alkene contributes to the overall rate. The reaction rate is first order in alkene concentration and -1 in CO in many catalyst systems.

The introduction of electron-withdrawing substituents on the aryl rings of the bis-equatorial chelate of (BISBI)RhH$(CO)_2$ leads to an increase in linear aldehyde selectivity as well as the rate. This must be an electronic effect on the l/b ratio since BISBI containing phenyl substituents coordinates already purely in the bis-equatorial fashion [15].

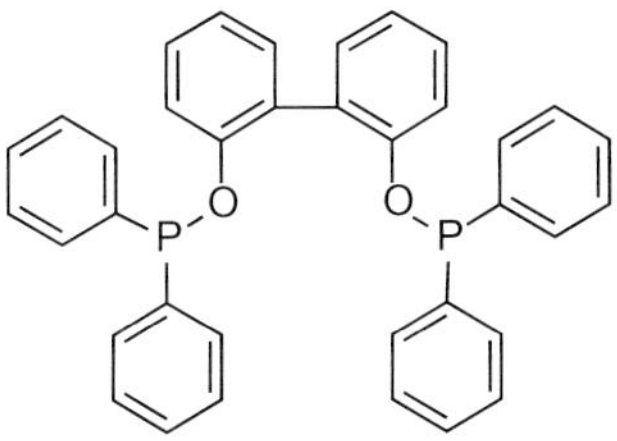

11 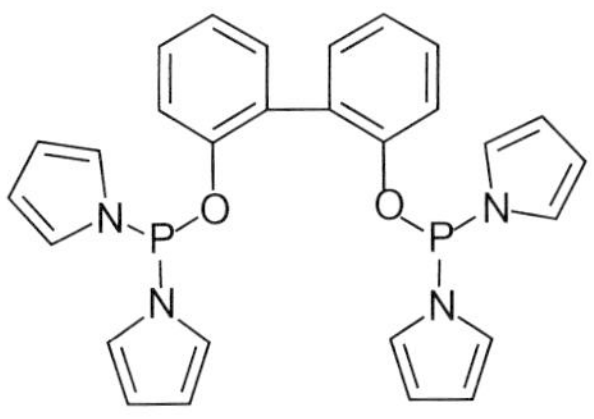 **12**

A similar electronic effect has been observed for ligands **11** and **12**. Both coordinate exclusively in the ee mode in rhodium hydrido dicarbonyl; but for the electron-withdrawing ligand **11**, a moderate l/b ratio of 6 was found while that for the electron-poor ligand **12** was as high as 100. Increased l/b ratios at higher χ-values are relatively general for ligand effects in hydroformylation, but in the last cases they cannot be assigned to an electronic bite angle effect and they must represent an electronic effect per se, which is not fully understood yet [12].

1.2.4
Isotope Effects [24]

The above studies still left open the possibility of two steps that could be rate determining: alkene coordination or insertion in the rhodium hydride bond. To this end, the rate-determining step in the hydroformylation of 1-octene, catalyzed by the rhodium–xantphos catalyst system, was determined using a combination of experimentally determined $^{1}H/^{2}H$ and $^{12}C/^{13}C$ kinetic isotope effects and a theoretical approach. From the relative rates of hydroformylation and deuterioformylation of 1-octene, a small $^{1}H/^{2}H$ isotope effect of 1.2 was determined on the hydride moiety of the rhodium catalyst. $^{12}C/^{13}C$ isotope effects for the olefinic carbon atoms of 1-octene were determined at natural abundance. Both quantum mechanics/molecular mechanics (QM/MM) and full quantum mechanics calculations were carried out on the key catalytic steps, using "real-world" ligand systems. The combination of kinetic isotope effects determination and theoretical studies suggest that the barrier for hydride migration has a slightly higher free energy than that of the alkene insertion under these conditions. Dissociation of CO constitutes the main part of the overall energy barrier, as is quite common in catalysis.

1.3
Platinum-Catalyzed Alkene Hydroformylation

Phosphine platinum complexes give active hydroformylation catalysts and both terminal and internal alkenes can be hydroformylated by selectively employing platinum–diphosphine complexes, often activated by an excess of tin chloride as the cocatalyst [25,26]. The combination of platinum chloride and tin(II) chloride leads to the formation of the trichlorostannate anion, which presumably acts as a weak coordinating anion, as tin-free catalyst systems have also been reported [27]. The group of Vogt found that the preformation of the catalyst also proved to be effective with only one equivalent of the tin source [28].

These systems have mainly been applied to asymmetric hydroformylation [29], although their strength in normal alkene hydroformylation rests in their high selectivity for linear aldehyde.

In platinum/tin-catalyzed hydroformylation, widening of the natural bite angle of the diphosphine ligands has proven to be favorable for the catalytic performance [21,25]. The synthesis of the (mixed) group 15 derivatives of the di-*t*-butyl-xantphos backbone, including the arsine-analogues of xantphos **13**, has been explored. Xantarsine and xantphosarsine ligands **14** and **15** constituted the first efficient arsine modified platinum/tin catalysts for selective hydroformylation of terminal alkenes [30].

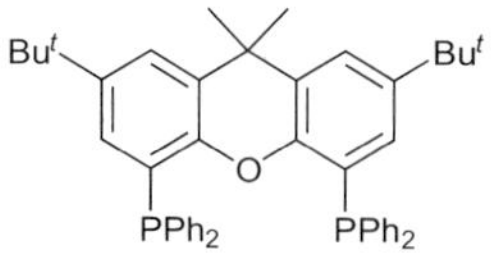

Bu^t Bu^t PPh_2 PPh_2 **13**

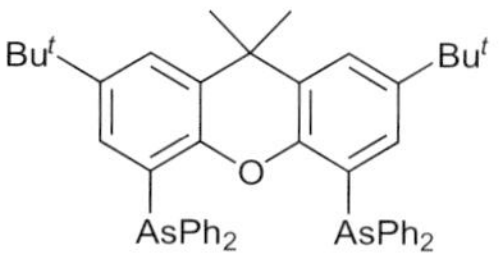

Bu^t Bu^t $AsPh_2$ $AsPh_2$ **14**

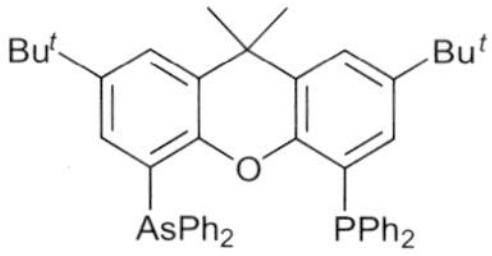

Bu^t Bu^t $AsPh_2$ PPh_2 **15**

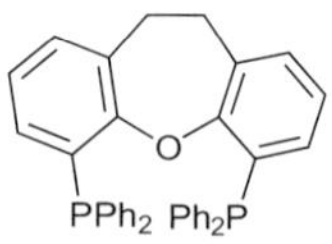

PPh_2 Ph_2P **16**

The calculated natural bite angles of ligands **13**, **14**, **15**, and **16** are 110°, 113°, 111°, and 102°, respectively. Ligands **13–16** were tested in the platinum/tin-catalyzed hydroformylation (Table 1.3). In the hydroformylation of 1-octene, the arsine-based ligands **14** and **15** proved to give more efficient catalysts than the parent xantphos ligand **13**. The

Table 1.3 Platinum/tin-catalyzed hydroformylation of 1-octene at 60 °C.[a]

Ligand	l/b ratio[b]	% *n*-nonanal[b]	% isomerization[b]	tof [b,c]
13	230	95	4.5	18
14	>250	92	8.0	210
15	200	96	3.1	350
16	>250	88	12	720

[a]Reactions were carried out in a 180-ml stainless steel autoclave in dichloromethane at 60 °C under 40 bar of CO/H_2 (1 : 1), catalyst precursor $[Pt(cod)Cl_2]$, [Pt] = 2.5 mM, Pt : $SnCl_2$: P : 1-octene = 1 : 2 : 4 : 255.
[b]Determined by GC with decane as the internal standard.
[c]Averaged turnover frequencies (tof) were calculated as (moles of aldehyde) (moles of Pt)$^{-1}$ h^{-1} at 20–30% conversion.

xantarsine ligand **14** is only slightly less selective than xantphos **13**, but it is more than 10 times as active. The xantphosarsine ligand **15** is even 20 times as active as xantphos **13**, while displaying the same excellent selectivity for linear aldehyde formation.

Comparison of the activities of the xantphos ligands **13** and **16** revealed a dramatic effect of the natural bite angle. Narrowing the natural bite angle from 110° to 102° results in a 40-fold higher hydroformylation rate. The high selectivities of ligands **13**, **14**, and **15** compared to xantphos **16** can be ascribed to the wider natural bite angles of the former ligands. Widening of the bite angle of the ligand will increase the steric congestion around the platinum center resulting in more selective formation of the sterically less hindered linear aldehydes.

1.4 Palladium-Catalyzed CO/Ethene Copolymerization

1.4.1 Polyketone Formation

One of the most astonishing manifestations of the dependence of a catalytic reaction on the bite angle of chelating diphosphines is the subtle balance between CO/alkene copolymerization and alkoxycarbonylation of alkenes (Scheme 1.3) [5,6]. Ethene–propene–CO polymers were produced commercially for a short while, oligomers have been studied as starting materials for resinlike materials, and methyl propanoate has been commercialized by Lucite and it is the starting material for making methyl methacrylate. In fact, methyl propanoate (product of the methoxycarbonylation of ethene) is the smallest possible product of the CO/ethene copolymerization using

$$n\ H_2C{=}CH_2 + n\ CO + MeOH \xrightarrow{\text{cat.}} H(CH_2CH_2CO)_nOMe$$

n = 1 methoxycarbonylation
n > 1 oligo- or polymerization

Scheme 1.3 Scheme of alkoxycarbonylation and CO/ethene copolymerization.

methanol as the chain transfer agent. It is formed when chain transfer occurs immediately after the insertion of just two monomers. Consequently, the selectivity control between copolymerization and alkoxycarbonylation implies a tuning between chain propagation and chain transfer rates, which can be directed by modifications in the ligands.

Sen reported in the early 1980s that certain Pd(II)–PPh_3 complexes containing weak coordinating anions (i.e., $[Pd(CH_3CN)_4][BF_4]_2{\cdot}nPPh_3$, $n = 1$–3) produce oligomers (25 °C, 4–1.5 bar) [31]. Weakly coordinating anions improve the productivity probably because they create easily accessible coordination sites; this also explains the lower activity obtained when a large excess of PPh_3 was used.

In the same year, Drent (Shell), when studying the alkoxycarbonylation reaction in methanol by using palladium complexes similar to those used by Sen, discovered that replacing the excess of PPh_3 by a stoichiometric amount of diphosphine generates catalysts for the polymerization reaction that are orders of magnitude faster [32]. Using these complexes, $PdX_2(L{\cap}L)$ ($L{\cap}L$ being a bidentate phosphorus or nitrogen ligand chelating in a *cis* fashion, X a weak coordinating anion), perfectly alternating CO/ethene copolymer was produced with only ppm quantities of residual catalyst. Suitable ligands were coordinating diphosphines (i.e., dppe, dppp, and dppb). The number of carbon atoms in the backbone showed to have a dramatic influence on the activity and selectivity (see Table 1.4) [33].

The change of selectivity from alkoxycarbonylation to oligomerization or polymerization when changing from monophosphines to chelating diphosphines was first rationalized in terms of a bite angle effect [33]. With monophosphines, a *trans* orientation of the phosphine ligands is more stable for the acyl or alkyl species. Therefore, immediately after an insertion, a *cis–trans* isomerization occurs. The new species formed opposes further insertions and chain growth. Thus, the acyl–palladium species will eventually terminate by alcoholysis of the Pd–acyl bond,

Table 1.4 Palladium-catalyzed CO/C_2H_4 copolymerization: the effect of variation of the chain length of bidentate phosphines.[a]

Ligand $Ph_2P(CH_2)_mPPh_2$	β_n[b] (°)	Product[c] $H(CH_2CH_2CO)_nOCH_3$	Reaction rate[d] ($g\,g^{-1}$ Pd h^{-1})
$m = 1$	72	$\tilde{n} = 2$	1
2	85	100	1000
3	91	180	6000
4	98	45	2300
5		6	1800
6		2	5

[a] Data taken from Ref. [33]. Solvent CH_3OH, catalyst $Pd(CH_3CN)_2(OTs)_2$, and diphosphine; $C_2H_4/CO = 1$; temperature = 84 °C; pressure = 45 bar.
[b] Natural bite angles taken from Ref. [9].
[c] The average degree of polymerization ($\tilde{n}$) determined by end group analysis from ^{13}C NMR spectra, except for the low molecular weight products, where a combination of GC and NMR was used.
[d] The rate was the highest measured during the reaction period (1–5 h).

which was initially thought to take place in the *trans* species [34]. When *cis* and *trans* isomers occur in equilibrium, this is reflected in a tendency to form oligomers.

However, when diphosphines are used, in which the phosphorus donor atoms are always *cis* to one another (all the ligands assayed were *cis* coordinating), the growing chain and monomer are in *cis* positions as well – a prerequisite for insertion reactions. As a result, diphosphines with natural bite angles close to 90° (dppp) stabilize the transition state for insertion reactions (chain growth), explaining also the higher activity and polymer selectivity of dppp when compared to monophosphines. The trend for the bidentates in Table 1.4 together with those of other series of diphosphine ligands [35] will be discussed below. Later, this explanation for the difference between mono- and diphosphines has been reconsidered.

1.4.2
Chain Transfer Mechanisms (Initiation–Termination)

In the earliest publications [33], it was proposed that the initiation step in both hydroxycarbonylation and polymerization reactions involved the reaction of the alcohol with the palladium complex to give the catalytically active palladium–methoxy complexes. After chain growing reactions, the termination mechanism was supposed to proceed via protonolysis of the alkyl–palladium complex to give the keto-ester (KE) product (methyl propanoate or polymer) and regenerate the active catalyst (Scheme 1.4). In addition, hydrido palladium species are smoothly formed from palladium(II) salts in methanol (not shown).

However, a GC analysis of the oligomeric fractions obtained when using dppb (1,4-diphenylphosphinobutane) showed that although the ketone/ester end group ratio was close to 1, together with the KE polyketone, products containing diketo (KK) or diester (EE) end groups were also obtained. The appearance of these palindrome products cannot be explained via the catalytic cycle mentioned before. If only one chain transfer mechanism is active (one termination releasing the polymer and regenerating the initiation active species) via methanol reaction with the palladium-chain compound, it is not possible to obtain KK or EE products in the absence of oxidants or reductants. To explain the formation of these products, the two different

INITIATION:

$$[L_2PdX_2] + MeOH \longrightarrow [L_2PdOMeX] + HX$$

PROPAGATION:

$$[L_2PdOMeX] + CO \longrightarrow [L_2Pd(CO)OMe]X$$

$$[L_2Pd(CO)OMe]X \longrightarrow [L_2PdCOOMeX]$$

$$[L_2PdCOOMeX] + C_2H_4 \longrightarrow [L_2Pd(C_2H_4)COOMe]X$$

$$[L_2Pd(C_2H_4)COOMe]X \longrightarrow [L_2PdC_2H_4COOMeX]$$

TERMINATION:

$$[L_2PdC_2H_4COOMeX] + MeOH \longrightarrow [L_2PdOMeX] + MeOC(O)CH_2CH_3$$

Scheme 1.4 Mechanism initially proposed for ethene hydroxycarbonylation.

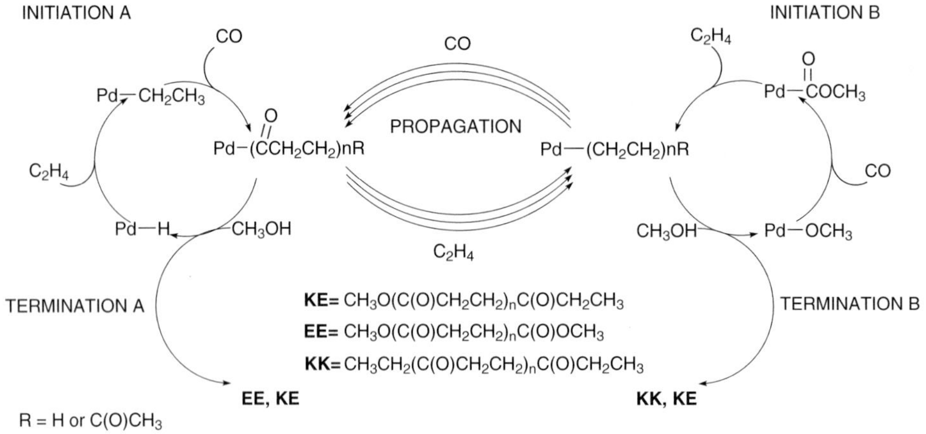

INITIATION A

INITIATION B

CO

CO

C_2H_4

$Pd-CH_2CH_3$

$Pd-C(O)OCH_3$

PROPAGATION

$Pd-(C(O)CH_2CH_2)nR$

$Pd-(CH_2CH_2)nR$

C_2H_4

CO

$Pd-H$

CH_3OH

CH_3OH

$Pd-OCH_3$

C_2H_4

TERMINATION A

TERMINATION B

KE= $CH_3O(C(O)CH_2CH_2)_nC(O)CH_2CH_3$

EE= $CH_3O(C(O)CH_2CH_2)_nC(O)OCH_3$

KK= $CH_3CH_2(C(O)CH_2CH_2)_nC(O)CH_2CH_3$

EE, KE

KK, KE

R = H or $C(O)CH_3$

Scheme 1.5 Proposed catalytic cycle for CO/ethene polymerization.

chain-transfer mechanisms A and B (corresponding to Pd-hydride and the Pd-methoxy initiation species) must occur simultaneously (Scheme 1.5) [33]. When other ligands are used at lower temperatures and in the absence of reagents that can convert hydride species into methoxy species or vice versa, predominantly KE oligomers have been obtained.

The effect of the bite angle on the termination reaction has been the focus of recent studies. The mechanism B, involving the enolate formation [36], is only slightly sensitive to changes in the bidentate ligand (dppe, dppp, and dppf) and the reaction is slightly faster for ligands with a wider bite angle [36].

The effect of the bite angle on termination reaction A has also been studied recently on model acyl–palladium compounds containing a variety of bidentate phosphine ligands [37]. The reaction turned out to be extremely sensitive to the steric properties of the ligand and therefore to the bite angle also. From the results obtained so far, it was concluded that the dppp backbone (bite angle close to 90°) is decisive in obtaining high molecular weight polymers in CO/ethene copolymerization owing to electronic bite angle effects. We will return to this after we have introduced the catalysts that give methyl propanoate as the product. The use of bdompp (1,3-bis(di-(*o*-methoxyphenyl)phosphino)propane) gave a polymer of higher molecular weight than the one obtained with dppp [32,38]. This is not easily explained as, on the one hand, the ligand is somewhat more bulky than dppp, but the hemilabile methoxy groups, on the other hand, may participate in the coordination sphere of palladium, the effect of which is unknown. At this point of time, it seemed fair to conclude that [38] "Nowadays the catalyst selected for the manufacture of Carilon polymer at commercial scale is $Pd(bdompp)X_2$. This catalyst is not only more active than the most prominent member of the first generation of CO–ethene copolymerization catalysts (dppp), but also produces co- and terpolymers with a considerably higher molecular weight." In conclusion, the ligand chosen by the industry in the early 1990s showed a synergism between the electronic and steric bite angle effect.

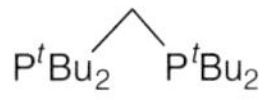

P^tBu_2 P^tBu_2 **17** — P^tBu_2 N P^tBu_2 **18**

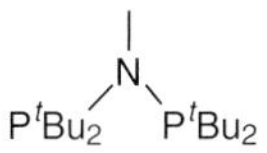

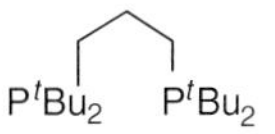

P^tBu_2 P^tBu_2 dtbpp **19** — P^tBu_2 P^tBu_2 dtbpx **20** — **21**

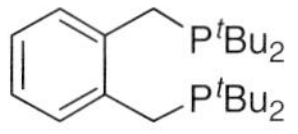

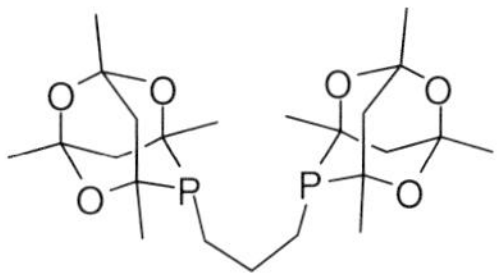

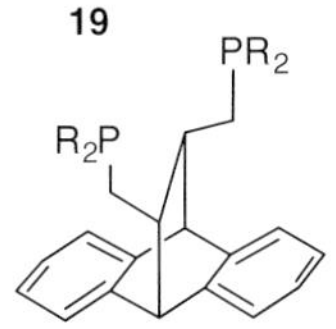

22 PR_2 = PPh_2, Me_4-phospholyl — **23** — Ph_2P O PPh_2 DPEphos **24**

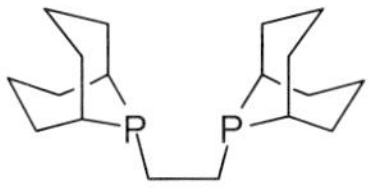

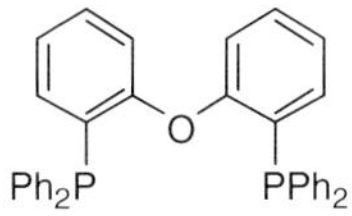

Not only the use of substituents at the phenyl groups at phosphorus, but also the introduction of groups attached to the backbone has been exploited. A study on dppp modified ligands showed that the introduction of alkyl substituents on the 2 position of the propane chain did not improve catalyst performance. In contrast, the productivity increased remarkably when methyl groups were introduced on 1 and 3 positions of the diphosphine ligand, the effect depending on the configuration of the stereogenic centers generated [39]. The same effect has been observed for 2,3-substituted dppb derivatives [40]. Thus, a slight increase of the steric constraints leads to a faster catalyst, while the length of the backbone and the bite angle remains the same. Some mechanistic studies have also been developed to determine the origin of these steric effects [40]. Although a conclusive explanation is still lacking, it seems clear that beta chelates (with the oxygen of the acyl group occupying a coordination site) are resting states in the catalytic cycle [41]. A possible explanation could involve that the opening of the beta chelates via a five-coordinate transition state constitutes the rate-limiting step or a step close to it. This step could be strongly influenced by the steric environment of the metal center.

For a long time, it had been generally accepted that diphosphine ligands containing only one carbon atom in the backbone (dppm derivatives) do not generate active catalysts for CO/ethene oligomerization or polymerization (Table 1.4). An important observation is that ligands **17** and **18** with still larger substituents at phosphorus but a backbone that consists of just one atom gave highly active catalysts producing polymer. Recently, it has been reported that several dppm derivatives with various types of bulky groups on the phosphorus atoms form active catalysts for polyketone synthesis, whereas the dppm ligand under the same reaction conditions shows lower productivities [42]. Thus, neither the bite angle nor the flexibility of the backbone is a prerequisite for making a polymer; instead it would seem that a certain steric bulk around the palladium site is required that tunes the reactivity for insertion and termination reactions and also reduces the amount of catalyst residing in one of the inactive resting states. The latter is usually neglected, but it surely is of great importance in palladium catalysis.

1.4.3
Methyl Propanoate Formation

By introducing steric modifications on the ligand maintaining the propane backbone, it is possible to radically change the selectivity. Drent reported that the use of dtbpp, **19**, a bidentate in which tertiary butyl groups replace the phenyl groups in dppp, changes the selectivity of the catalyst completely from polyketone to methyl propanoate [43]. Both the selectivity and the rate were further improved by slightly enlarging the backbone of the catalyst with the use of a xylene moiety [44]. In recent years, a whole range of bulky bidentate phosphine ligands have been added to the initial two examples all giving methyl propanoate, or mixtures with oligomers, with moderate-to-high rates (**21** [45], **22** [46], **23** [47], and **24** [37]). For the fast catalyst systems such as those based on **20**, it has been proven that the catalytic cycle follows the hydride mechanism A (Scheme 1.5).

Another example in which backbone substitution affords the effect of increasing steric bulk is provided by octamethyl-dppf, carrying eight methyl substituents at the ferrocene rings [48]. While dppf gives oligomers in the methoxycarbonylation/polyketone reaction (rate 5000 mol mol^{-1} h^{-1}, at 85 °C), octamethyl-dppf gave methyl propanoate. Octamethyl-dppf is sterically more crowded albeit not via substitution directly at the phosphorus atoms, as appears from the P–Pd–P angle, which is 101° as compared to 96° for dppf in the dicationic palladium diaqua adducts. Indeed, octamethyl-dppf gives methyl propanoate in the palladium-catalyzed reaction with ethene, CO, and CH_3OH, albeit at a modest rate (600 mol mol^{-1} h^{-1}, at 85 °C).

DPEphos, **24**, gave methyl propanoate at a rate of 2000 mol mol^{-1} h^{-1}, at 80 °C and 20 bar, and an additional 10% of the lowest oligomer [37]. Surprisingly, xantphos, **5**, gave hardly any activity in this reaction. Xantphos can form *trans* complexes, which remain inactive in this type of catalytic reactions [7,49] – not unexpectedly, as insertion reactions require a *cis* disposition of the migrating group and unsaturated fragment.

When the *t*-butyl groups were replaced by the smaller *i*-propyl groups in **19** [50] or **20** [44], both systems produced oligomers at high rates instead of methyl propanoate. When the 1,3-propanediyl bridge in $(t\text{-Bu})_2P(CH_2)_3P(t\text{-Bu})_2$ was replaced by a 1,2-ethanediyl bridge, the accessibility of the catalyst for ethene increased so that in the reaction of ethene, CO, CH_3OH, and H_2 pentan-3-one was formed at extremely high rates instead of methyl propanoate, the product of the more bulky ligand.

From the data of the last decade emerges a new picture of the effect of the steric bulk. Clearly, starting from dppe, continuing with dppp and then on with still larger ligands, the overall rate of polymer production increases, which can be attributed to the destabilization of resting states preceding the insertion of ethene, the rate-determining step. However, at a certain point this relationship is broken, perhaps by hampering the coordination of ethene altogether to the intermediate palladium–acyl species. Instead, methanol coordination and reductive elimination take place (see Section 1.4.4). Most interestingly, the termination reaction not only becomes faster in the bulky catalysts relative to propagation, but, also, in absolute terms it becomes orders of magnitude faster [37] with increasing steric bulk of the ligand. The first insertion of ethene in the methyl propanoate forming reaction is much less sensitive to changes in the ligand bulk, because this takes place at a palladium hydrido species, as has been proven for a few bidentate ligands [51] and monodentate phosphine catalysts [34,52].

Previously, the formation of methyl propanoate has been associated with *trans* complexes generated by monodentate ligands. Indeed, *trans* acyl complexes are the resting states of these catalytic systems [6,34]. Following this explanation, an "arm-off" mechanism for the strained bidentates such as **19–24** could be imagined replacing the phosphine *trans* to acyl also with a solvent molecule. Recent measurements have shown that this is not the case and that the alcoholysis reaction requires *cis* orientation for the acyl group and the alcohol, and thus a *cis*-diphosphine [37]. The decisive factor is the steric hindrance exerted by the ligand: the larger the steric bulk, the faster the ester formation. For monodentate ligands such as PPh_3, the *trans* complexes undergo an isomerization to a *cis* complex, which behaves effectively as a complex containing a bulky bidentate, and then the sequence of reactions is terminated by alcoholysis; however, the catalyst is slower than a bulky *cis* bidentate as the complex resides mainly in the nonactive *trans* configuration. Thus, we arrive at the conclusion that in both the polymerization and the methoxycarbonylation reaction, all data point to merely steric causes.

1.4.4
Theoretical Support

A recent DFTstudy gave further insight into the effects of ligands on various insertion and elimination steps [53]. Both chain propagation and methanolysis termination mechanisms catalyzed by palladium complexes containing electron-donating diphosphine ligands were studied. The rate-determining step in the chain propagation mechanism is the insertion of ethene into the palladium–ethanoyl bond, yielding a β-chelate complex of the formed keto group. For the methanolysis pathways, the formation of a 14-electron intermediate is crucial, because unlike the formation of CO an ethene complexes, which occur through associative ligand exchange processes, methanol coordination requires a vacant site. The calculations show that the most likely methanolysis pathway involves a proton transfer/reductive elimination mechanism, in which the solvent acts as a proton transfer agent.

Both increasing the bite angle and increasing the steric bulk of the diphosphine ligand stabilize the 14-electron η^2-acyl complex. Increasing the steric bulk of the ligand strongly disfavors the formation of ethene complex and consequently increases the barrier for ethene insertion, as expected.

For all methanolysis pathways considered in this study, increasing the bite angle of the diphosphine ligand *increases* the rate of methanolysis. This is attributed to the involvement of electron-rich intermediates and/or transition states in all three methanolysis pathways.

The steric bulk of the diphosphine ligand hardly affects the barriers for methanolysis via the stepwise and concerted reductive elimination pathways. Based on these observations, it was postulated that the high activity and chemoselectivity in the methoxycarbonylation of ethene observed for *t*-butyl substituted wide bite angle diphosphine ligands are determined by a combination of electronic and steric effects induced by the diphosphine ligand. The high electron density at the metal center induced by the s-donating diphosphine ligand stabilizes 14-electron η^2-acyl intermediate, while suppressing the formation of 18-electron intermediates such as ethene complex. Methanol coordination is hardly affected by the steric bulk of the ligand, although increasing steric bulk of the ligand enhances the direct nucleophilic attack of methanol on η^2-acyl complex. Furthermore, for all methanolysis pathways considered, the barrier for the formation of the ester product decreases when the bite angle of the diphosphine ligand increases. This is attributed to the formation of zero-valent palladium complexes, which are stabilized by wide bite angle ligands [10].

1.5
Rhodium-Catalyzed Methanol Carbonylation: the Ligand-Modified Monsanto Process

Acetic acid is a bulk commodity whose actual annual global production reaches 8 million tons, led by two companies, Celanese (using "Monsanto" technology) and BP Chemicals, with processes based on both rhodium and iridium [54]. Methanol carbonylation, from an economic point of view, presents a clear advantage over all the former processes developed for the industrial production of acetic acid. Methanol and CO are relatively cheap feedstocks that can be obtained from different raw materials, which makes the whole process independent of petroleum prices. Methanol can be obtained from syngas (CO and H_2 mixture) coming from petroleum, natural gas, or even from coal or biomass. These latest technologies are not yet onstream, but industries are considering them as powerful candidates for the future and pilot plants operate since the mid-1990s [55].

The process presents a 100% atom economy with all the atoms in the reactants going into the product (see Equation 1.1), and compared to previous methods benefits of a reduced waste and easier (and cheaper) product separation.

$$CH_3OH + CO \rightarrow CH_3COOH \quad \Delta G^\circ = -75\,\text{kJ}\,\text{mol}^{-1}, \ \Delta H^\circ = -135\,\text{kJ}\,\text{mol}^{-1}. \tag{1.1}$$

From the early beginning, a lot of effort was devoted to elucidate the reaction mechanism, and the key steps of the catalytic cycle are nowadays well established (see Figure 1.1). The rate-limiting step is the oxidative addition of CH_3I to the Rh(I) species, as kinetic studies showed first-order dependence on CH_3I and Rh precursor and zero order on CO and CH_3OH.

Although the rhodium-catalyzed processes currently onstream are based on the original rhodium iodide Monsanto system (no ligands added), in the last 20 years many reports deal with the use of phosphines (and other ligands) to tune the catalyst properties. A general approach to facilitate the oxidative addition of CH_3I to the active rhodium species, and therefore to enhance the rate of the overall process, is to increase the nucleophilicity of the metal center. This is accomplished by substituting CO and I^- for stronger donor ligands. In most of the cases, either chelating or two donating ligands are used, and a neutral [Rh(L)(CO)I] (L = chelating or two monodentate ligands) active species is generated. Ligand modified systems are aimed at

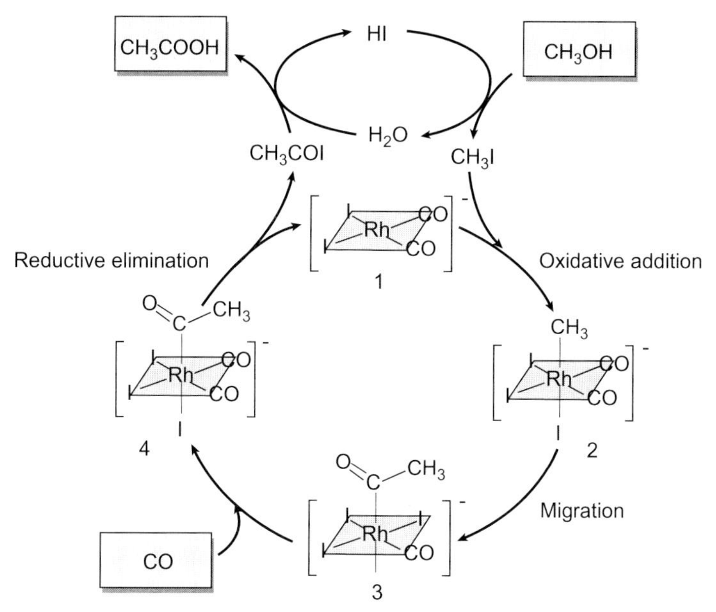

Figure 1.1 Monsanto catalytic cycle.

increasing not only the activity but also the selectivity and stability of the metal in the areas of low CO pressure (one of the main drawbacks of the process).

Even though several chelating diphosphines have been considered as likely ligands for this reaction (using the chelate effect to stabilize the ligand under the harsh conditions of the process), their performance has not traditionally been analyzed in terms of bite angle.

Only the recent patent literature [56,57] refers explicitly to the beneficial use of wide bite angle ($\beta > 105°$) and rigid (flexibility range <40°) diphosphines for this reaction (i.e., xantphos **5** and BISBI). They claim the positive effect of these ligands as related to the inhibition of side reactions leading to acetaldehyde and the corresponding undesired hydrogenation products (namely ethanol and its derivatives ethylmethylether and ethyl acetate). Acetaldehyde formation implies the existence of a complex containing a hydride and an acyl group in mutually *cis* positions. Owing to the postulated geometry of the rhodium–acyl species ([Rh(P∩P)$(COCH_3)I_2]^-$ with the acyl occupying the apical position of a pentacoordinated square base pyramid (sbp), AcH reductive elimination requires a concerted movement of ligands prior to or after H_2 addition, as postulated for the related reductive methanol carbonylation reaction [58]. Although detailed calculations are still missing, the argument used is that a rigid ligand backbone should prevent the complex to easily change from the original sbp conformation avoiding hydrogenolysis of the metal–acyl species.

Using bite angle considerations this should be interpreted as a bite angle effect, where rigid wide bite angle diphosphines produce a destabilizing effect on the transition states required for the undesired side reaction to take place. Based on the

Table 1.5 Methanol carbonylation using chelating diphosphines.[a]

					Selectivity		
Ligand	β (°C)	Flexibility range	Reaction time (min)	CH_3OH conversion (%)	HAcO and derivatives	EtOH derivatives/AcH	CH_4
dppp	91	81–112	30	38.8	28.1	1.2/42.9	26.9
dppp,[b]	91	81–112	120	16.8	20.0	42.7/15.3	21.9
Xantphos	111[c]	97–133[d]	21	29.2	38.3	0/0.3	60.9
Xantphos[b]	111[c]	97–133[d]	17	31.1	35.7	2.6/0.5	60.7
BISBI[b]	112[e]	92–155[e]	51	38.0	37.6	1.7/1.2	59.3

Reaction conditions: $[Rh(acac)(CO)_2]$ (0.65 g, 2.5 mmol), ligand 2.8 mmol, 80 g CH_3OH, 14 g CH_3I, 140 °C, 67–70 bar syngas (H_2 : CO = 2).
[a] Data extracted from Ref. [56].
[b] $RuCl_3$ approximately 10 mmol as additive.
[c] Data extracted from Ref. [67].
[d] Data extracted from Ref. [16].
[e] Data extracted from Ref. [15].

same considerations, the coplanar terdentate coordination of xantphos derivatives (P–O–P) in the corresponding acyl intermediates are also postulated as responsible for the superior performance of these derivatives when compared to other wide bite angle diphosphines (Table 1.5).

Another extreme bite angle effect was observed when *trans*-spanning diphosphines were used. When SPANphos ligands where studied in CH_3OH carbonylation, kinetic experiments performed on the commonly accepted rate-limiting step, the CH_3I oxidative addition to the complex *trans*-[Rh(P∩P)(CO)Cl] (P∩P = SPANphos derivatives), a complete inhibition of the reaction was observed [59]. Halogen exchange happened instead leading to *trans*-[Rh(P∩P)(CO)I]. The same behavior has been observed for other diphosphanes that do not easily isomerize from *trans* to *cis* positions [60] and for ligands that even coordinating in a *cis* manner sterically block one of the axial sites of the metal [61].

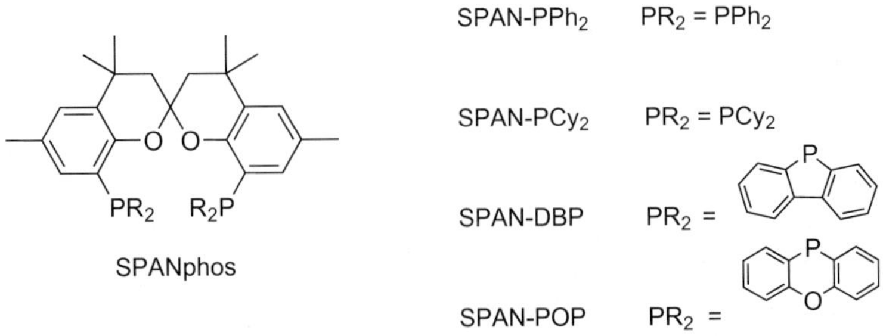

O O
PR$_2$ R$_2$P
SPANphos

SPAN-PPh$_2$ PR$_2$ = PPh$_2$
SPAN-PCy$_2$ PR$_2$ = PCy$_2$
SPAN-DBP PR$_2$ =
SPAN-POP PR$_2$ =

It is generally accepted that oxidative addition follows a two-step S_N2 mechanism: nucleophilic attack by the metal on the methyl carbon to displace iodide, presumably

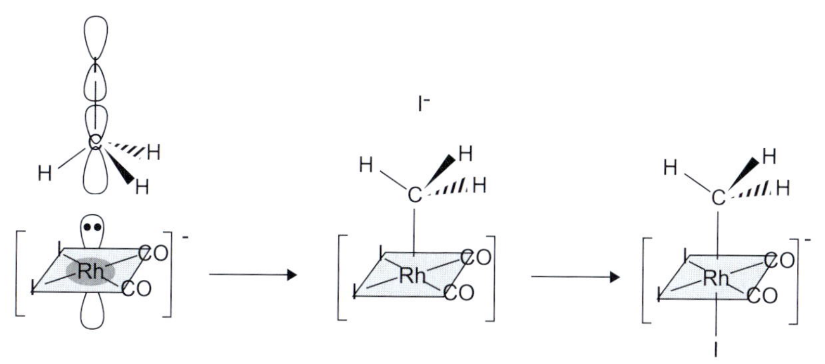

Figure 1.2 S_N2 mechanism for CH_3I oxidative addition.

with inversion of configuration at the carbon atom, and a subsequent iodide coordination to the five-coordinate rhodium complex to give the alkyl complex 2 (Figure 1.2) [62]. When considering the bare rhodium (Monsanto) system, the product of this reaction has been fully characterized spectroscopically [63]. Although there is general agreement on the mechanism, the theoretical calculations with respect to the geometry of the TS are still controversial [64], but the main interaction in the Rh–C bond formation seems to take place between a full metal d_{z^2} orbital and an empty C–I σ^* orbital.

The complete inhibition of CH_3I oxidative addition observed in the case of *trans* chelating diphosphines complexes has been attributed to the fact that one of the axial sites of the rhodium center is blocked by the ligand backbone, avoiding the CH_3I addition to proceed (Figure 1.3).

Surprisingly, when SPANphos derivatives [59] and other *trans* chelating diphosphines [65] were tested as ligands for rhodium-catalyzed methanol carbonylation, they showed to be one of the most active systems reported until now for this reaction. The observed activities (taking into account the complete absence CH_3I oxidative addition to the presynthesized mononuclear *trans* species) have been attributed to the formation of active dinuclear species. Kinetic studies on the CH_3I oxidative addition to presynthesized SPANphos dinuclear species of the form $[Rh_2(SPANphos)(\mu\text{-}Cl)_2(CO)_2]$ showed fast reaction rates ($k_1 = 0.025\ s^{-1}\ M^{-1}$), but the mechanism operating under catalytic conditions is being investigated [66].

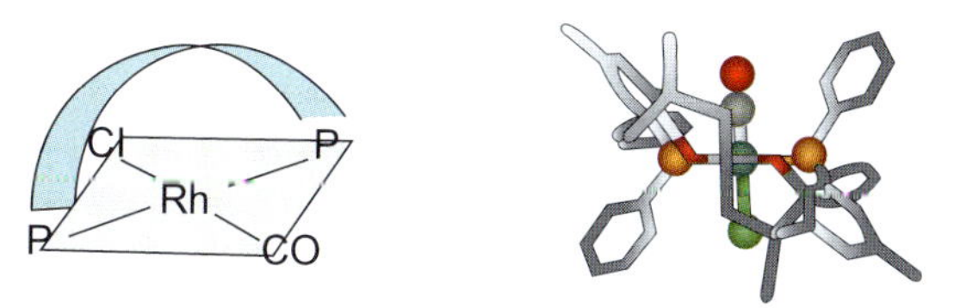

Figure 1.3 [Rh(SPANphos)(CO)Cl] complex (X-ray structure).

References

1. Devon, T.J., Phillips, G.W., Puckette, T.A., Stavinoha, J.L. and Vanderbilt, J.J. (1987) *Chelate Ligands for Low Pressure Hydroformylation Catalyst and Process Employing Same*, Eastman Kodak Company.
2. Kranenburg, M., van der Burgt, Y.E.M., Kamer, P.C.J. and van Leeuwen, P.W.N.M. (1995) New diphosphine ligands based on heterocyclic aromatics inducing very high regioselectivity in rhodium-catalyzed hydroformylation: effect of the bite angle. *Organometallics*, **14**, 3081.
3. Casey, C.P. and Whiteker, G.T. (1990) The natural bite angle of chelating dipho-sphines. *Israel Journal of Chemistry*, **30**, 299.
4. Goertz, W., Kamer, P.C.J., van Leeuwen, P. W.N.M. and Vogt, D. (1997) Application of chelating diphosphine ligands in the nickel-catalysed hydrocyanation of alk-1-enes and omega-unsaturated fatty acid esters. *Journal of the Chemical Society, Chemical Communications*, 1521.
5. Freixa, Z. and van Leeuwen, P.W.N.M. (2003) Bite angle effects in diphosphine metal catalysts: steric or electronic? *Journal of the Chemical Society, Dalton Transactions*, 1980.
6. van Leeuwen, P.W.N.M., Kamer, P.C.J., Reek, J.N.H. and Dierkes, P. (2000) Ligand bite angle effects in metal-catalyzed C–C bond formation. *Chemical Reviews*, **100**, 2741.
7. Kamer, P.C.J., van Leeuwen, P.W.N.M. and Reek, J.N.H. (2001) Wide bite angle diphosphines: xantphos ligands in transition metal complexes and catalysis. *Accounts of Chemical Research*, **34**, 895.
8. Kamer, P.C.J., Reek, J.N.H. and van Leeuwen, P.W.N.M. (1998) Designing ligands with the right bite. *Chemtech*, **28**, 27.
9. Dierkes, P. and van Leeuwen, P.W.N.M. (1999) The bite angle makes the difference: a practical parameter for diphosphine ligands. *Journal of the Chemical Society, Dalton Transactions*, 1519.
10. Zuidema, E., van Leeuwen, P.W.N.M. and Bo, C. (2005) Reductive elimination of organic molecules from palladium–diphosphine complexes. *Organometallics*, **24**, 3703.
11. Carbó, J.J., Maseras, F., Bo, C. and van Leeuwen, P.W.N.M. (2001) Unraveling the origin of regioselectivity in rhodium diphosphine catalyzed hydroformylation. A DFT QM/MM study. *Journal of the American Chemical Society*, **123**, 7630.
12. van der Slot, S.C., Duran, J., Luten, J., Kamer, P.C.J. and van Leeuwen, P.W.N.M. (2002) Rhodium-catalyzed hydroformylation and deuterioformylation with pyrrolyl-based phosphorus amidite ligands: influence of electronic ligand properties. *Organometallics*, **21**, 3873.
13. Pacciello, R. (2007) private communication.
14. van Rooy, A., Goubitz, K., Fraanje, J., Kamer, P.C.J., van Leeuwen, P.W.N.M., Veldman, N. and Spek, A.L. (1996) Bulky diphosphite-modified rhodium catalysts: hydroformylation and characterization. *Organometallics*, **15**, 835; Broussard, M.E., Juma, B., Train, S.G., Peng, W.J., Laneman, S.A. and Stanley, G.G. (1993) A bimetallic hydroformylation catalyst: high regioselectivity and reactivity through homobimetallic cooperativity. *Science*, **260**, 1784; van der Veen, L.A., Kamer, P.C.J. and van Leeuwen, P.W.N.M. (1999) Hydroformylation of internal olefins to linear aldehydes with novel rhodium catalysts. *Angewandte Chemie-International Edition in English*, **38**, 336.
15. Casey, C.P., Whiteker, G.T., Melville, M.G., Petrovich, L.M., Gavney, J.A., Jr, and Powell, D.R. (1992) Diphosphines with natural bite angles near 120° increase selectivity for *n*-aldehyde formation in rhodium-catalyzed hydroformylation. *Journal of the American Chemical Society*, **114**, 5535.
16. van der Veen, L.A., Keeven, P.H., Schoemaker, G.C., Reek, J.N.H., Kamer, P.C.J., van Leeuwen, P.W.N.M., Lutz, M. and Spek, A.L. (2000) Origin of the bite

angle effect on rhodium diphosphine catalyzed hydroformylation. *Organometallics*, **19**, 872.

17 Heck, R.F. and Breslow, D.S. (1961) Reaction of cobalt hydrotetracarbonyl with olefins. *Journal of the American Chemical Society*, **83**, 4023; Heck, R.F. and Breslow, D.S. (1962) Acylcobalt carbonyls and their triphenylphosphine complexes. *Journal of the American Chemical Society*, **84**, 2499.

18 Brown, J.M. and Kent, A.G. (1987) Structural characterisation in solution of intermediates in rhodium-catalysed hydroformylation and their interconversion pathways. *Journal of the Chemical Society, Perkin Transactions 2*, 1597.

19 Hughes, O.R. and Unruh, J.D. (1981) Hydroformylation catalyzed by rhodium complexes with diphosphine ligands. *Journal of Molecular Catalysis*, **12**, 71; Sanger, A.R. (1977) Hydroformylation of 1-hexene catalysed by complexes of rhodium(I) with di- or tritertiary phosphines. *Journal of Molecular Catalysis*, **3**, 221; Sanger, A.R. and Schallig, L.R. (1977) The structures and hydroformylation catalytic activities of polyphosphine complexes of rhodium(I), and related complexes immobilised on polymer supports. *Journal of Molecular Catalysis*, **3**, 101; Pittman, C.U. and Hirao, A. (1978) Hydroformylation catalyzed by *cis*-chelated rhodium complexes – extension to polymer-anchored *cis*-chelated rhodium catalysts. *The Journal of Organic Chemistry*, **43**, 640.

20 van der Veen, L.A., Boele, M.D.K., Bregman, F.R., Kamer, P.C.J., van Leeuwen, P.W.N.M., Goubitz, K., Fraanje, J., Schenk, H. and Bo, C. (1998) Electronic effect on rhodium diphosphine catalyzed hydroformylation: the bite angle effect reconsidered. *Journal of the American Chemical Society*, **120**, 11616.

21 van der Veen, L.A., Kamer, P.C.J. and van Leeuwen, P.W.N.M. (1999) New phosphacyclic diphosphines for rhodium catalyzed hydroformylation. *Organometallics*, **18**, 4765.

22 Casey, C.P., Paulsen, E.L., Beuttenmueller, E.W., Proft, B.R., Petrovich, L.M., Matter, B. A. and Powell, D.R. (1997) Electron withdrawing substituents on equatorial and apical phosphines have opposite effects on the regioselectivity of rhodium catalyzed hydroformylation. *Journal of the American Chemical Society*, **119**, 11817; Casey, C.P., Paulsen, E.L., Beuttenmueller, E.W., Proft, B.R., Matter, B.A. and Powell, D.R. (1999) Electronically dissymmetric DIPHOS derivatives give higher *n*:*i* regioselectivity in rhodium-catalyzed hydroformylation than either of their symmetric counterparts. *Journal of the American Chemical Society*, **121**, 63; Herrmann, W.A., Kohlpaintner, C.W., Herdtweck, E. and Kiprof, P. (1991) Structure and metal coordination of the diphosphane 2,2′-bis((diphenylphosphino) methyl)-1,1′-biphenyl (BISBI). *Inorganic Chemistry*, **30**, 4271.

23 Unruh, J.D. and Christenson, J.R. (1982) A study of the mechanism of rhodium phosphine catalyzed hydroformylation: use of 1,1′-bis(biarylphosphino) ferrocene ligands. *Journal of Molecular Catalysis*, **14**, 19.

24 Zuidema, E., Escorihuela, L., Eichelsheim, T., Carbó, J.J., Bo, C., Kamer, P.C.J. and van Leeuwen, P.W.N.M. (2007) The rate-determining step in the rhodium-xantphos catalyzed hydroformylation of 1-octene, *Chem. Eur. J. B*, accepted.

25 Kawabata, Y., Hayashi, T. and Ogata, I. (1979) Platinum–diphosphine–tin systems as active and selective hydroformylation catalysts. *Journal of the Chemical Society, Chemical Communications*, 462.

26 Hayashi, T., Kawabata, Y., Isoyama, T. and Ogata, I. (1981) Platinum chloride–diphosphine–tin(II) halide systems as active and selective hydroformylation catalysts. *Bulletin of the Chemical Society of Japan*, **54**, 3438; Schwager, I. and Knifton, J.F. (1976) Homogeneous olefin hydroformylation catalyzed by ligand stabilized platinum(II)-group-IVb metal halide complexes. *Journal of Catalysis*, **45**, 256; Ancillotti, F., Lami, M. and Marchionna, M. (1990) Hydroformylation of *Z*-2-butene with the $PtCl_2(cod)/SnCl_2/L$ catalytic system.

3. The effect of the phosphorus ligand L. *Journal of Molecular Catalysis*, **63**, 15.

27 Tang, S.C. and Kim, L. (1982) Homogeneous hydroformylation of internal olefins by platinum tin cationic complexes. *Journal of Molecular Catalysis*, **14** (1), 231; van Leeuwen, P.W.N.M., Roobeek, C.F., Wife, R.L. and Frijns, J.H.G. (1986) Platinum hydroformylation catalysts containing diphenylphosphine oxide ligands. *Journal of the Chemical Society, Chemical Communications*, (1), 31; Botteghi, C., Paganelli, S., Matteoli, U., Scrivanti, A., Ciorciaro, R. and Venanzi, L.M. (1990) A new catalytic system for the hydroformylation of styrene using alkene complexes of platinum(0). *Helvetica Chimica Acta*, **73**, 284.

28 Meessen, P., Vogt, D. and Keim, W. (1998) High regioselective hydroformylation of internal, functionalized olefins applying Pt–Sn complexes with large bite angle diphosphines. *Journal of Organometallic Chemistry*, **551**, 165; van Duren, R., Cornelissen, L.L.J.M., Van der Vlugt, J.I., Huijbers, J.P.L., Mills, A.M., Spek, A.L., Muller, C. and Vogt, D. (2006) Chiral (diphosphonite)platinum complexes in asymmetric hydroformylation. *Helvetica Chimica Acta*, **89**, 1547.

29 Agbossou, F., Carpentier, J.F. and Mortreux, A. (1995) Asymmetric hydroformylation. *Chemical Reviews*, **95**, 2485.

30 van der Veen, L.A., Keeven, P.K., Kamer, P.C. J. and van Leeuwen, P.W.N.M. (2000) Wide bite angle amines, arsine and phosphine ligands in rhodium and platinum/tin catalysed hydroformylation. *Journal of the Chemical Society, Dalton Transactions*, 2105.

31 Sen, A. and Lai, T.W. (1982) Novel palladium(II) catalyzed copolymerization of carbon monoxide with olefins. *Journal of the American Chemical Society*, **104**, 3520; Lai, T.W. and Sen, A. (1984) Palladium(II)-catalyzed copolymerization of carbon monoxide with ethylene. Direct evidence for a single mode of chain growth. *Organometallics*, **3**, 866.

32 Drent, E. (1984) Process for the preparation of polyketones. EP 0121965 (to Shell), *Chemical Abstracts*, 1985, **102**, 464223.

33 Drent, E., Broekhoven, J.A.M. and Doyle, M.J. (1991) Efficient palladium catalyst for the copolymerization of carbon monoxide with olefins to produce perfectly alternating polyketones. *Journal of Organometallic Chemistry*, **417**, 235.

34 del Rio, I., Claver, C. and van Leeuwen, P.W.N.M. (2001) On the mechanism of the hydroxycarbonylation of styrene with palladium systems. *European Journal of Inorganic Chemistry*, (11), 2719; Verspui, G., Moiseev, I.I. and Sheldon, R.A. (1999) Reaction intermediates in the Pd/tppts-catalyzed aqueous phase hydrocarboxylation of olefins monitored by NMR spectroscopy (tppts = $P(C_6H_4\text{-}m\text{-}SO_3Na)_3$). *Journal of Organometallic Chemistry*, **586** (11), 196.

35 Doherty, S., Eastham, G.R., Tooze, R.P., Scanlan, T.H., Williams, D., Elsegood, M.R.J. and Clegg, W. (1999) Palladium complexes of C2-, C3- and C4-bridged bis(phospholyl) ligands: remarkably active catalysts for the copolymerization of ethylene and carbon monoxide. *Organometallics*, **18**, 3558.

36 Zuideveld, M.A., Kamer, P.C.J., van Leeuwen, P.W.N.M., Klusener, P.A.A., Stil, H.A. and Roobeek, C.F. (1998) Chain-transfer mechanisms of the alternating copolymerization of carbon monoxide and ethene catalyzed by palladium(II) complexes: rearrangement to highly reactive enolates. *Journal of the American Chemical Society*, **120**, 7977.

37 van Leeuwen, P.W.N.M., Zuideveld, M.A., Swennenhuis, B.H.G., Freixa, Z., Kamer, P.C.J., Goubitz, K., Fraanje, J., Lutz, M. and Spek, A.L. (2003) Alcoholysis of acylpalladium(II) complexes relevant to the alternating copolymerization of ethene and carbon monoxide and the alkoxycarbonylation of alkenes: the importance of *cis*-coordinating diphosphines. *Journal of the American Chemical Society*, **125**, 5523.

38 Mul, W.P., Dirkzwager, H., Broekhuis, A.A., Heeres, H.J., van der Linden, A.J. and Orpen, A.G. (2002) Highly active, recyclable catalyst for the manufacture of viscous, low molecular weight, CO–ethene–propene based polyketone, base component for a new class of resins. *Inorganica Chimica Acta*, **327**, 147.

39 Bianchini, C., Lee, H.M., Meli, A., Moneti, S., Vizza, F., Fontani, M. and Zanello, P. (1999) Copolymerization of carbon monoxide with ethene catalyzed by palladium(II) complexes of 1,3-bis (diphenylphosphino)propane ligands bearing different substituents on the carbon backbone. *Macromolecules*, **32**, 4183.

40 Bianchini, C., Lee, H.M., Meli, A., Oberhauser, W., Peruzzini, M. and Vizza, F. (2002) Ligand and solvent effects in the alternating copolymerization of carbon monoxide and olefins by palladium–diphosphine catalysis. *Organometallics*, **21**, 16.

41 Mul, W.P., Oosterbeek, H., Beitel, G.A., Kramer, G.-J. and Drent, E. (2000) *in situ* monitoring of a heterogeneous palladium-based polyketone catalyst, *Angewandte Chemie-International Edition*, **39**, 1848.

42 Dossett, S.J., Gillon, A., Orpen, A.G., Fleming, J.S., Pringle, P.G., Wass, D.F. and Jones, M.D. (2001) Steric activation of chelate catalysts: efficient polyketone catalysts based on four-membered palladium(II) diphosphine chelates. *Journal of the Chemical Society, Chemical Communications*, 699.

43 Drent, E. and Kragtwijk, E. (1992) Process for the carbonylation of olefin. EP 0495548 (to Shell).

44 Clegg, W., Eastham, G.R., Elsegood, M.R.J., Tooze, R.P., Wang, X.L. and Whiston, K. (1999) Highly active and selective catalysts for the production of methyl propanoate via the methoxycarbonylation of ethane. *Journal of the Chemical Society, Chemical Communications*, 1877; Tooze, R.P., Eastham, G.R., Whiston, K. and Wang, X.L. (1996) Process for the carbonylation of ethylene and catalyst system for use therein. WO 09619434 (to ICI PCL).

45 Pugh, R.I., Drent, E. and Pringle, P.G. (2001) Tandem isomerisation–carbonylation catalysis: highly active palladium(II) catalysts for the selective methoxycarbonylation of internal alkenes to linear esters. *Journal of the Chemical Society, Chemical Communications*, 1476.

46 Doherty, S., Robins, E.G., Knight, J.G., Newman, C.R., Rhodes, B., Champkin, P.A. and Clegg, W. (2001) Selectivity for the methoxycarbonylation of ethylene versus CO–ethylene copolymerization with catalysts based on C-4-bridged bidentate phosphines and phospholes. *Journal of Organometallic Chemistry*, **640**, 182.

47 Drent, E., van Kragtwijk, E. and Pello, D.H. L. (1992) Carbonylation of olefins. EP 495547 (to Shell).

48 Gusev, O.V., Kalsin, A.M., Peterleitner, M. G., Petrovskii, P.V., Lyssenko, K.A., Akhmedov, N.G., Bianchini, C., Meli, A. and Oberhauser, W. (2002) Palladium(II) complexes with 1,1′-bis (diphenylphosphino) ferrocenes $[Fe(\eta^5)\text{-}C_5R_4PPh_2)_2]^{n+}$ (dppf, $R = H$, $n = 0$; dppomf, $R = Me$, $n = 0$; dppomf$^+$, $R = Me$, $n = 1$). Synthesis, characterization, and catalytic activity in ethene methoxycarbonylation. *Organometallics*, **21**, 3637.

49 Zuideveld, M.A., Swennenhuis, B.H.G., Boele, M.D.K., Guari, Y., van Strijdonck, G.P.F., Reek, J.N.H., Kamer, P.C.J., Goubitz, K., Fraanje, J., Lutz, M., Spek, A.L. and van Leeuwen, P.W.N.M. (2002) The coordination behaviour of large natural bite angle diphosphine ligands towards methyl and 4-cyanophenylpalladium(II) complexes. *Journal of the Chemical Society, Dalton Transactions*, 2308; Yin, J.J. and Buchwald, S.L. (2002) Pd-catalyzed intermolecular amidation of aryl halides: the discovery that xantphos can be *trans*-chelating in a palladium complex. *Journal of the American Chemical Society*, **124**, 6043.

50 Pugh, R.I. and Drent, E. (2002) Methoxycarbonylation versus hydroacylation of ethene: dramatic influence of the ligand in cationic palladium catalysis. *Advanced Synthesis & Catalysis*, **344**, 837.

51 Eastham, G.R., Heaton, B.T., Iggo, J.A., Tooze, R.P., Whyman, R. and Zacchini, S. (2000) Synthesis and spectroscopic characterisation of all the intermediates in the Pd-catalysed methoxycarbonylation of ethane. *Journal of the Chemical Society, Chemical Communications*, 609; Clegg, W., Eastham, G.R., Elsegood, M.R.J., Heaton, B.T., Iggo, J.A., Tooze, R.P., Whyman, R. and Zacchini, S. (2002) Characterization and dynamics of [Pd(L-L)H (solv)](+), $[Pd(L\text{-}L)(CH_2CH_3)]^+$, and $[Pd(L\text{-}L)(C(O)Et)(THF)]^+$ (L-L = 1,2-$(CH_2PBu_2t)_2C_6H_4$): key intermediates in the catalytic methoxycarbonylation of ethene to methylpropionate. *Organometallics*, **21**, 1832; Liu, J.K., Heaton, B.T., Iggo, J.A., Whyman, R., Bickley, J.F. and Steiner, A. (2006) The mechanism of the hydroalkoxycarbonylation of ethene and alkene–CO copolymerization catalyzed by Pd-II-diphosphine cations. *Chemistry – A European Journal*, **12**, 4417; Liu, J.K., Heaton, B.T., Iggo, J.A. and Whyman, R. (2004) The complete delineation of the initiation, propagation, and termination steps of the carbomethoxy cycle for the carboalkoxylation of ethene by Pd–diphosphane catalysts. *Angewandte Chemie-International Edition in English*, **43**, 90.

52 Tooze, R.P., Whiston, K., Malyan, A.P., Taylor, M.J. and Wilson, N.W. (2000) Evidence for the hydride mechanism in the methoxycarbonylation of ethene catalysed by palladium–triphenylphosphine complexes. *Journal of the Chemical Society, Dalton Transactions*, 3441.

53 Zuidema, E., Bo, C. and van Leeuwen, P.W. N.M. (2007) Ester versus polyketone formation in the palladium–diphosphine catalyzed carbonylation of ethene. *Journal of the American Chemical Society*, **129**, 3989.

54 Smejkal, Q., Linke, D. and Baerns, M. (2005) Energetic and economic evaluation of the production of acetic acid via ethane oxidation. *Chemical Engineering and Processing*, **44**, 421.

55 http://www.greener-industry.org/(2007).

56 Gaemers, S. and Sunley, G.J. (2004) Carbonylation process using metal–polydentate ligands catalysis. WO 2004101487 (to BP Chemicals Limited).

57 Gaemers, S. and Sunley, G.J. (2004) Carbonylation process using metal–tridentate ligand catalysts. WO 2004101488 (to BP Chemicals Limited).

58 Moloy, K.G. and Wegman, R.W. (1989) Rhodium-catalyzed reductive carbonylation of methanol. *Organometallics*, **8**, 2883; Moloy, K.G. and Petersen, R.A. (1995) Structural characterization of intermediates in the rhodium-catalyzed reductive carbonylation of methanol: $Rh(COCH_3)(I_2)$(dppp) and $[Rh(H)(\mu\text{-}I)(dppp)]_2$. *Organometallics*, **14**, 2931.

59 Freixa, Z., Kamer, P.C.J., Lutz, M., Spek, A. L. and van Leeuwen, P.W.N.M. (2005) Activity of SPANphos rhodium dimers in methanol carbonylation. *Angewandte Chemie-International Edition*, **44**, 4385.

60 Müller, C., Freixa, Z., Spek, A.L., Lutz, M., Vogt, D. and van Leeuwen, P.W.N.M. (2007) Wide bite angle diphosphines: design, synthesis and coordination properties. *Organometallics*, submitted.

61 Broussier, R., Laly, M., Perron, P., Gautheron, B., Nifant'ev, I.E., Howard, J.A.K., Kuz'mina, L.G. and Kalck, P. (1999) Neutral and cationic 1,4-diphospha-2-rhoda-3-thia-4-ferrocenophane. *Journal of Organometallic Chemistry*, **587**, 104; Martin, H.C., James, N. H., Aitken, J., Gaunt, J.A., Adams, H. and Haynes, A. (2003) Oxidative addition of MeI to a rhodium(I) *N*-heterocyclic carbene complex. A kinetic study. *Organometallics*, **22**, 4451.

62 Griffin, T.R., Cook, D.B., Haynes, A., Pearson, J.M., Monti, D. and Morris, G.E. (1996) Theoretical and experimental evidence for S_N2 transition states in oxidative addition of methyl iodide to *cis*-$[M(CO)_2I_2]^-$ (M = Rh, Ir). *Journal of the American Chemical Society*, **118**, 3029; Chauby, V., Daran, J.-C., Berre, C.S.-L., Malbosc, F., Kalck, P., Gonzalez, O.D., Haslam, C.E. and Haynes, A. (2002) A mechanistic investigation of oxidative addition of methyl iodide to [TpRh(CO)L]. *Inorganic Chemistry*, **41**, 3280.

63 Forster, D. (1976) Mechanism of a rhodium-complex-catalyzed carbonylation of methanol to acetic acid. *Journal of the American Chemical Society*, **98**, 846; Adamson, G.W., Daly, J.J. and Forster, D. (1974) Reaction of iodocarbonylrhodium ions with methyl-iodide – structure of rhodium acetyl complex – $[Me_3PhN^+]_2[Rh_2I_6(MeCO)_2(CO)_2]^{2-}$. *Journal of Organometallic Chemistry*, **71**, C17; Haynes, A., Mann, B.E., Gulliver, D.J., Morris, D.E. and Maitlis, P.M. (1991) Direct observation of $MeRh(CO)_2I_3^-$: the key intermediate in rhodium catalyzed methanol carbonylation. *Journal of the American Chemical Society*, **113**, 8567; Haynes, A., Mann, B.E., Morris, G.E. and Maitlis, P.M. (1993) Mechanistic studies on rhodium-catalyzed carbonylation reactions – spectroscopic detection and reactivity of a key intermediate, $[MeRh(CO)_2I_3]^-$. *Journal of the American Chemical Society*, **115**, 4093.

64 Cheong, M. and Ziegler, T. (2005) Density functional study of the oxidative addition step in the carbonylation of methanol catalyzed by $[M(CO)_2I_2]^-$ (M = Rh, Ir). *Organometallics*, **24**, 3053; Feliz, M., Freixa, Z., van Leeuwen, P.W.N.M. and Bo, C. (2005) Revisiting the methyl iodide oxidative addition to rhodium complexes: a DFT study of the activation parameters. *Organometallics*, **24**, 5718; Kinnunen, T. and Laasonen, K. (2001) The oxidative addition and migratory 1,1-insertion in the Monsanto and Cativa processes. A density functional study of the catalytic carbonylation of methanol. *Journal of Molecular Structure: Theochem*, **542**, 273.

65 Thomas, C.M., Mafua, R., Therrien, B., Rusanov, E., Stoeckli-Evans, H. and Süss-Fink, G. (2002) New diphosphine ligands containing ethyleneglycol and amino alcohol spacers for the rhodium-catalyzed carbonylation of methanol. *Chemistry – A European Journal*, **8**, 3343; Thomas, C.M. and Süss-Fink, G. (2003) Ligand effects in the rhodium-catalyzed carbonylation of methanol. *Coordination Chemistry Reviews*, **243**, 125; Burger, S., Therrien, B. and Süss-Fink, G. (2005) Square-planar carbonylchlororhodium(I) complexes containing *trans*-spanning diphosphine ligands as catalysts for the carbonylation of methanol. *Helvetica Chimica Acta*, **88**, 478.

66 Wells, J., Williams, G.L., Haynes, A., Freixa, Z. and van Leeuwen, P.W.N.M., unpublished results.

67 Kranenburg, M., Kamer, P.C.J., van Leeuwen, P.W.N.M., Vogt, D. and Keim, W. (1995) Effect of the bite angle of diphosphines ligands on activity and selectivity in the nickel-catalysed hydrocyanation of styrene. *Journal of the Chemical Society, Chemical Communications*, 2177.

2
Reactivity of Pincer Complexes Toward Carbon Monoxide

David Morales-Morales

Carbon monoxide is an ubiquitous molecule in organometallic chemistry and an important feedstock in multiple catalytic processes both at the laboratory and industrial levels [1]. This compound has been used in diagnostics to bind different ligands in a given complex, to stabilize molecules that are otherwise difficult or almost impossible to isolate, and, in general, as a regular substrate for the initial examination of the reactivity of new organometallic compounds. The chemistry of pincer complexes, however, has undergone tremendous development in the past few decades [2], owing partly to their high thermal stability; these compounds play an important role in the activation of unreactive bonds, particularly aliphatic C–H bonds [3]. Additionally, the reaction of pincer complexes with carbon monoxide has been fundamental to the better understanding of the chemical nature of these compounds and has allowed further modification of the donor properties of pincer ligands on the basis of the analysis of metal-carbonyl derivatives, some of which have been identified as kinetic resting pools in different catalytic processes and transformations mediated by pincer complexes. Thus, this chapter will look at the reactivity of different transition metal pincer complexes with carbon monoxide, their products, and the relevance of these species in a given process.

Carbonyl pincer complexes that are obtained not from the direct reaction of pincer species with carbon monoxide but from the reaction of pincer ligands with metal-carbonyl as starting materials and those that are obtained from decarbonylative processes of organic substrates are out of the scope of this chapter and have not been discussed.

2.1
Reactivity of CO with Pincer Complexes of the Group 10 (Ni, Pd, Pt)

2.1.1
Nickel

Moulton and Shaw were the first to report direct reactions of carbon monoxide with transition metal pincer complexes in their seminal work in 1976 [4]. In this paper, the

Modern Carbonylation Methods. Edited by László Kollár

ISBN: 978-3-527-31896-4

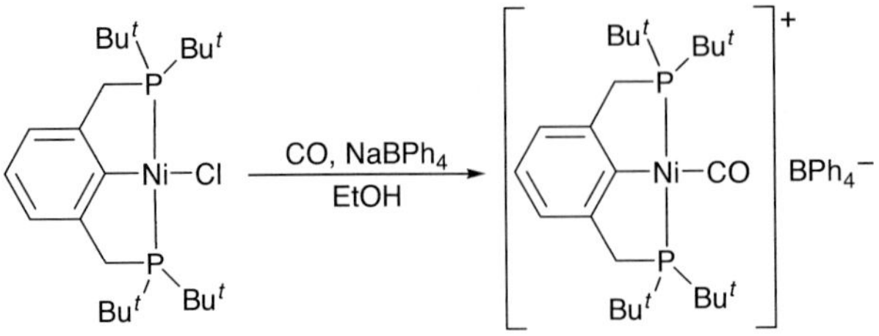

Bu^t Bu^t P Ni–Cl P Bu^t Bu^t → CO, $NaBPh_4$ / EtOH → [Bu^t Bu^t P Ni–CO P Bu^t Bu^t]$^+$ BPh_4^-

(1) (2)

Scheme 2.1

authors describe the synthesis of (PCP) pincer Ni(II), Ir(III), and Rh(I) carbonyl derivatives. In the case of (PCP)Ni, the reaction of compound (**1**) with CO gas in an ethanolic solution in the presence of $NaBPh_4$ affords complex (**2**) as a yellow microcrystalline powder according to Scheme 2.1.

However, new reactions of carbon monoxide with nickel pincer complexes were not reported until 1980. Thus, Sacco and coworkers [5] were able to identify pincer carbonyl species important in water gas shift reaction using (PNP) pincer complexes of type (**3**) as catalysts (Scheme 2.2).

These species readily react with carbon monoxide in water–ethanol solution at room temperature and atmospheric pressure to give (PNP)Ni(0)–carbonyl complexes of type (**4**) (Scheme 2.3).

If the reaction is carried out in the presence of bases such as NaOH or NEt_3, the reaction proceeds more rapidly to afford the carbonylated neutral species (**5**) (Scheme 2.4).

However, the propensity of CO to react with other organic molecules led Sacco and coworkers [6] to study the reactivity of complex (**3a**) with different aryl amines $ArNH_2$ (Ar=R=C_6H_5, C_6H_4-4-CH_3, C_6H_4-2-CH_3, C_6H_4-4-Cl, C_6H_4-4-OCH_3, and C_6H_4-4-NO_2) under a carbon monoxide atmosphere to afford carbamoyl complexes of type (**6**) (Scheme 2.5).

Sacco and coworkers found that the basicity of the nucleophile was strong enough to displace the equilibrium of the reaction completely to products when the pK_a of the amine was higher than 4. Thus, when using the much less basic *p*-nitroaniline ($H_2NC_6H_4$-4-NO_2), pK_a = 1, the carbamoyl complex was formed in good yield only in

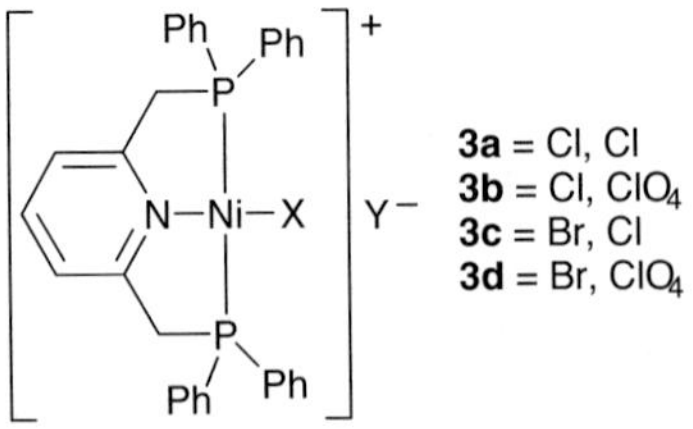

[Ph Ph P N–Ni–X P Ph Ph]$^+$ Y^-

3a = Cl, Cl
3b = Cl, ClO_4
3c = Br, Cl
3d = Br, ClO_4

(3)

Scheme 2.2

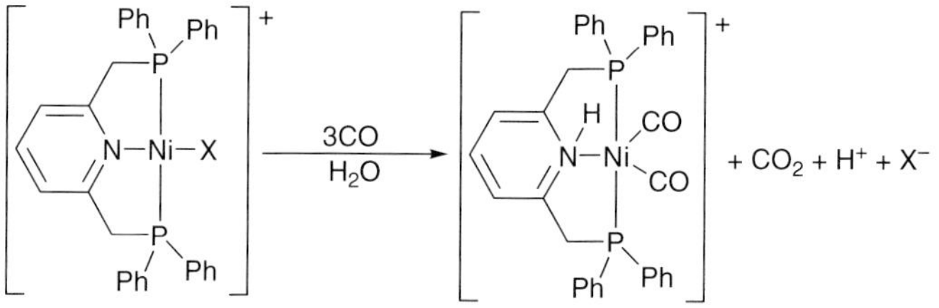

Scheme 2.3

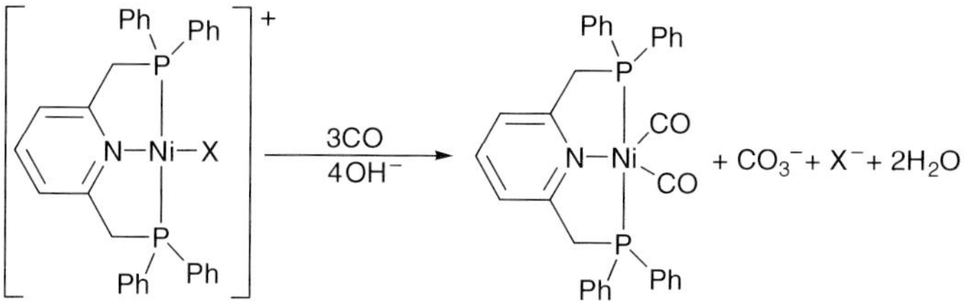

Scheme 2.4

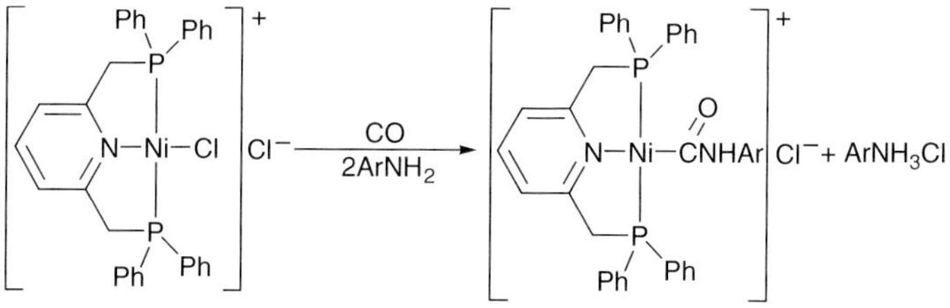

$Ar = C_6H_5$ (**6a**), C_6H_4-4-CH_3 (**6b**), C_6H_4-2-CH_3 (**6c**), C_6H_4-4-Cl (**6d**), C_6H_4-4-OCH_3 (**6e**)

Scheme 2.5

the presence of a base such as sodium or potassium carbonate or triethylamine. In contrast, a strict control of the stoichiometry was necessary when more basic amines such as benzylamine or *n*-butylamine ($pK_a = 9$–10) were used, since the formed carbamoyl complex reacted further with the excess amine. In fact, the phenylamine carbamoyl derivative (**6a**) reacted with NEt_3 or K_2CO_3 in CH_2Cl_2 to afford phenylisocyanate. However, most of the isocyanate formed remained bound to the nickel center, giving a complex formulated as (PNP)Ni(PhNCO).

In contrast, the aryl urethanes were obtained easily and with better yields by treating the carbamoyl complexes in ethanol with anhydrous bases such as NEt_3, K_2CO_3, or CaO, having complex (**5**) as resting pool for this reaction.

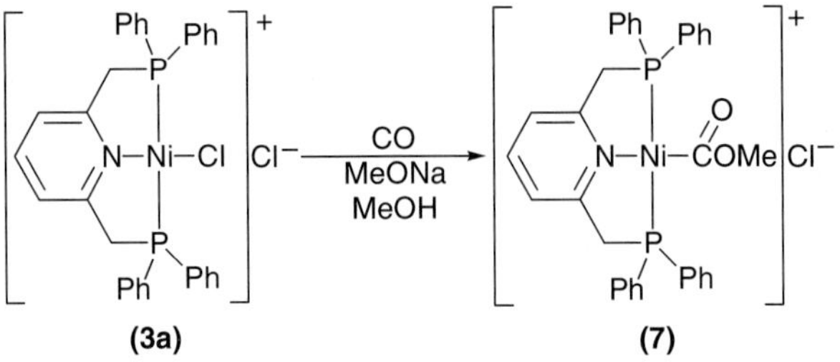

(3a) → CO, MeONa, MeOH → (7)

Scheme 2.6

Additionally, in this work, the methoxycarbonyl complex (**7**) was also synthesized from complex (**3a**) according to Scheme 2.6.

The ionic nature of compound (**7**) was determined on the basis of its molar conductivity values ($56\,\Omega^{-1}\,cm^{-2}\,mol^{-1}$) in CH_3NO_2.

2.1.2
Palladium and Platinum

In general, the chemistry of pincer derivatives of palladium and platinum is very closely related, affording in many cases analogous species and reactivity. Thus, for the sake of discussion, the chemistry of both metals is included in this section and we will be switching from one metal to another unless a specific reaction or product refers to one metal solely.

As is the case for pincer nickel complexes, their palladium and platinum counterparts exhibit very similar reactivity. Hence, similar carbamoyl (**10**) and alkoxycarbonyl (**11**) products were obtained by Sacco and coworkers in 1985 [7] (Scheme 2.7), when the Pd(II) (**8**) and Pt(II) (**9**) (PNP) pincer derivatives reacted under similar conditions as their Ni analogues (see Schemes 2.5 and 2.6).

Interestingly, when the reaction of compounds (**8**) and (**9**) with methoxide is carried out in a 1 : 2 molar ratio in methanol, complexes of the type (**12**) with a single methoxycarbonyl ligand and no chloride ion are obtained (Scheme 2.8) [8]. These compounds are the result of a nucleophilic attack by a methoxide ion over one of the methylenic groups of the (PNP) ligand to lose a proton, thus forming a carbanion (Scheme 2.8).

A similar compound of formula (PNP)PtCl (**13**) was obtained [8] from the reaction of (**9**) and the methoxide ion in methanol under nitrogen (Scheme 2.9). In both cases, the structures of the complexes were unequivocally confirmed by single-crystal X-ray diffraction analysis.

An interesting feature of the chemistry of monomeric organoplatinum(II) hydroxides and alkoxides is that CO inserts into the Pt–O bond more readily than into the Pt–C bond. Thus in 1993, Bennett and coworkers in an effort to examine the structures and properties of Pt(II)-hydroxycarbonyls with *trans*-arranged phosphines explored the reaction of the monomeric pincer Pt(II)-hydroxo species (**14**) with CO [9]. Hence, treating the (PCP)Pt(OH) pincer complex (**14**) with CO in benzene under ambient conditions caused the formation of the hydroxycarbonyl complex (**15**) as a

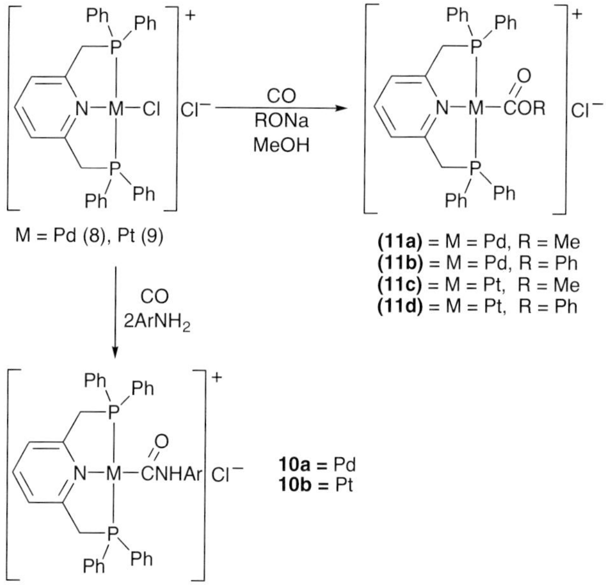
CO
RONa
MeOH
M = Pd (8), Pt (9)
(11a) = M = Pd, R = Me
(11b) = M = Pd, R = Ph
(11c) = M = Pt, R = Me
(11d) = M = Pt, R = Ph
CO
2ArNH2
10a = Pd
10b = Pt

Scheme 2.7

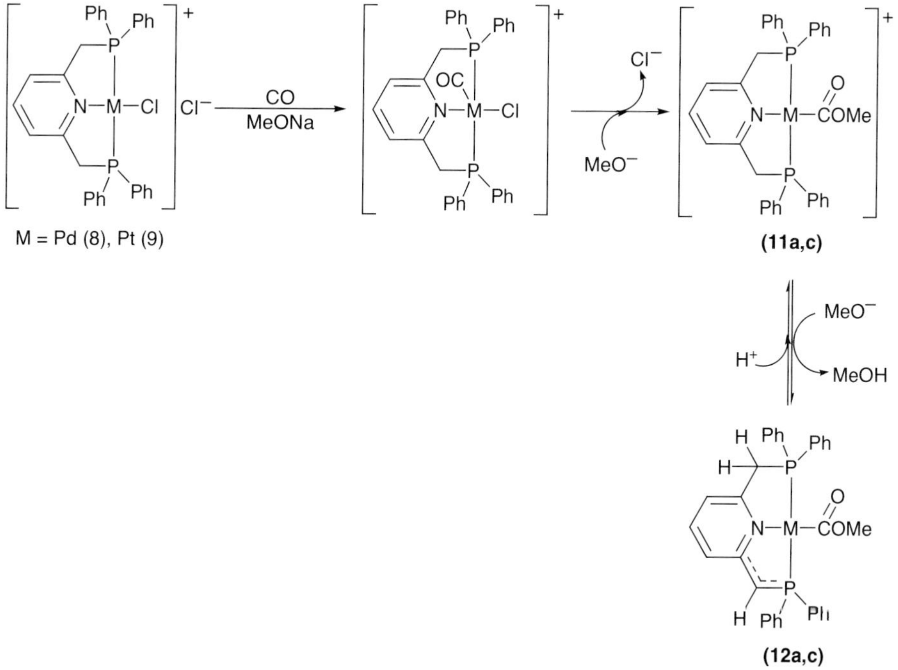
CO
MeONa
Cl−
MeO−
M = Pd (8), Pt (9)
(11a,c)
MeO−
H+
MeOH
(12a,c)

Scheme 2.8

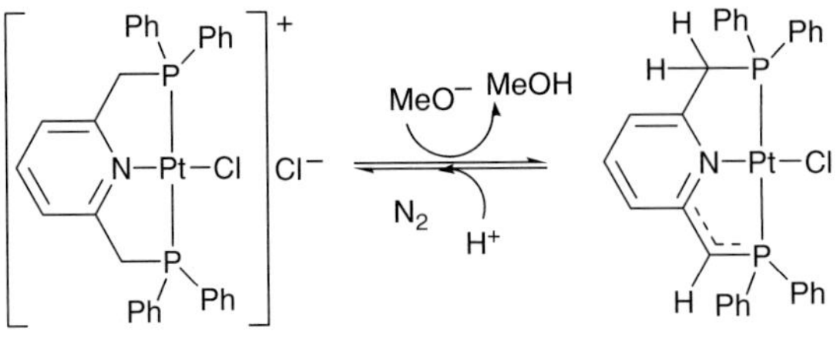

(9) (13)

Scheme 2.9

colorless microcrystalline precipitate (Scheme 2.10). The identity of this compound was unequivocally determined by single-crystal X-ray diffraction analysis. These results show that the complex exhibits strong O–H...O hydrogen bonded units, in the form of discrete dimeric units in the solid state. However, this association disappears in solution (CH_2Cl_2) and thus complex (**15**) exists mainly as a monomer in solution (Scheme 2.10).

As it has been shown so far, many of the species presented are the result of research focused in the most interesting water gas shift reaction. A different approach with similar consequences is the electrochemical reduction of CO_2 catalyzed by (PCP)Pd (II) pincer complexes. Of particular interest is the study by DuBois in 1994 [10] in which several different pincer complexes were evaluated as potential catalyst for the electrochemical reduction of CO_2. Thus, because CO is the main product of the CO_2 reduction electrochemical process, the effect of CO on complex (**16**) was investigated.

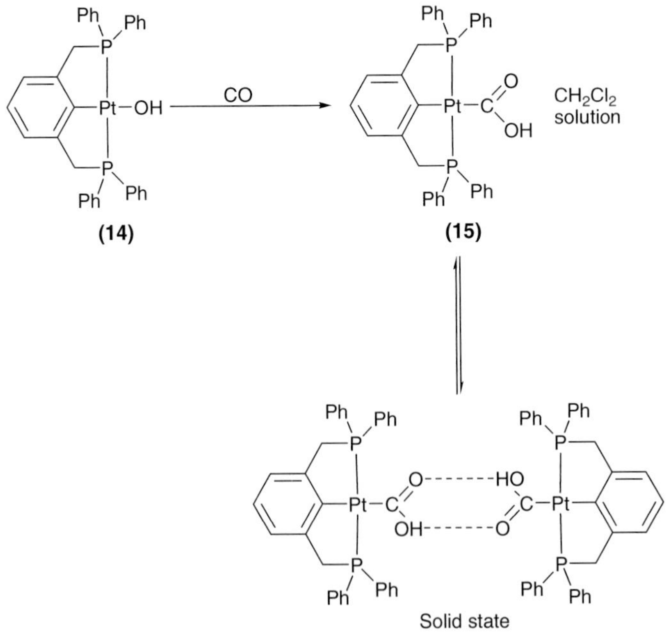

(14) (15)

CH_2Cl_2 solution

Solid state

Scheme 2.10

Hence, complex (**16**) reacts with CO in noncoordinating solvents, such as dimethylformamide (DMF) or acetone but not in acetonitrile to afford the carbonyl complex (**18**). The spectroscopic data support a rapid reversible binding of CO from (**17**) to (**18**) (Scheme 2.11).

Thus, most of the information collected for this system, including catalytic and electrochemical experiments, led the authors to conclude that the PCP pincer system (**16**) is just the second best of the series tested in this study and that the rate-determining step for the electrochemical reduction of CO_2 using this compound is probably the formation of an intermediate (**19**) such as that shown in Scheme 2.12. However, most of the efforts to prepare this species (**19**) independently were unsuccessful.

A similar approach by Park and coworkers has been reported for the synthesis of (**21**), an analogous species to complex (**18**), which essentially reports the same results as those of DuBois [10] (see Scheme 2.11), with the only variant that the pincer starting material contains either ammonia (**20**) [11] or a dimethyl amino group (**22**) [12] (Scheme 2.13). Surprisingly, complex (**22**) does not lead to the formation of either the carbamoyl pincer derivatives or free dimethylformamide as it has been observed before for similar compounds (see Scheme 2.7).

Given the already mentioned importance that ^{-}OH, CO, and CO_2 derivatives have had as potential intermediates in the water gas shift reaction and the relevance that late transition metal complexes containing metal–oxygen covalent bonds have in

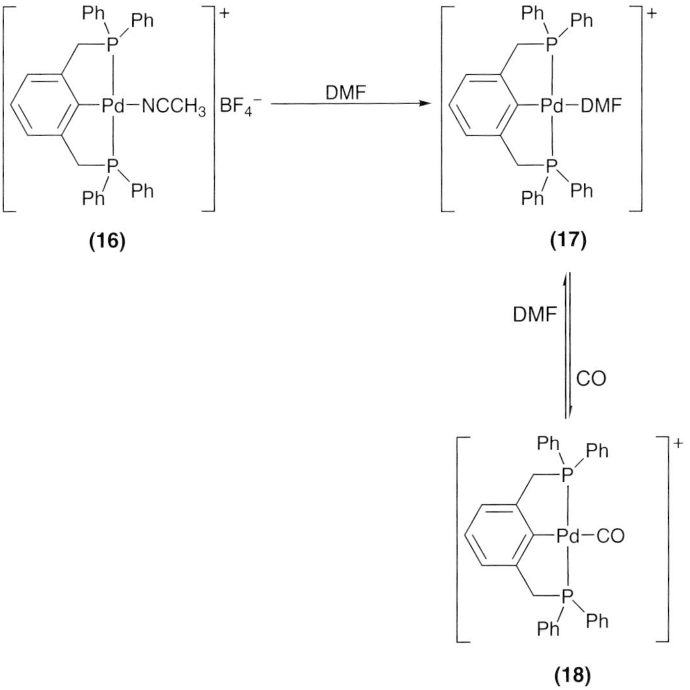

Ph Ph P Pd–NCCH3 BF4− P Ph Ph (**16**) DMF → Ph Ph P Pd–DMF P Ph Ph (**17**) DMF ⇌ CO Ph Ph P Pd–CO P Ph Ph (**18**)

Scheme 2.11

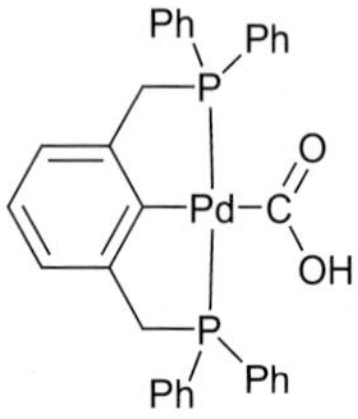

(19)

Scheme 2.12

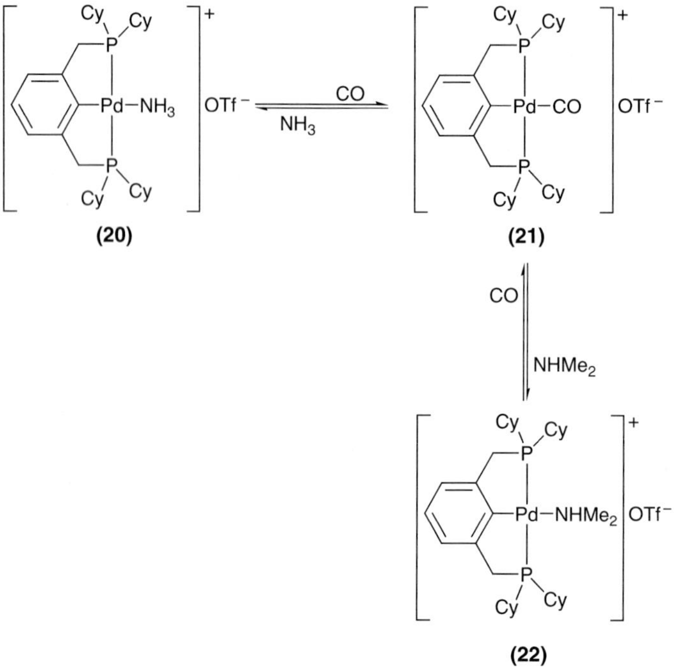

(20) CO / NH_3 **(21)** CO / $NHMe_2$ **(22)**

Scheme 2.13

biological systems, Campora and coworkers in 2004 reported [13] the study of the reactivity of the (PCP)Pd–hydroxy complexes (**23a**) and (**23b**) toward carbon monoxide. Thus, under the conditions depicted in Scheme 2.14, the corresponding binuclear CO_2 complexes (**24a**) and (**24b**) were obtained. From which, the palladium derivative (**24b**) was fully characterized by single-crystal X-ray diffraction analysis.

Monitoring of C_6D_6 solutions of (**23b**) at different pressures of CO by $^{31}P\{^1H\}$ NMR revealed the formation of (**24b**) and the intermediate hydroxylcarbonyl complex (**25b**), with an almost complete conversion (about 90%) into (**25b**) at a pressure of 3.5 atm of CO. This process is reversible and hence (**24b**) is regenerated when the excess CO is removed (Scheme 2.15).

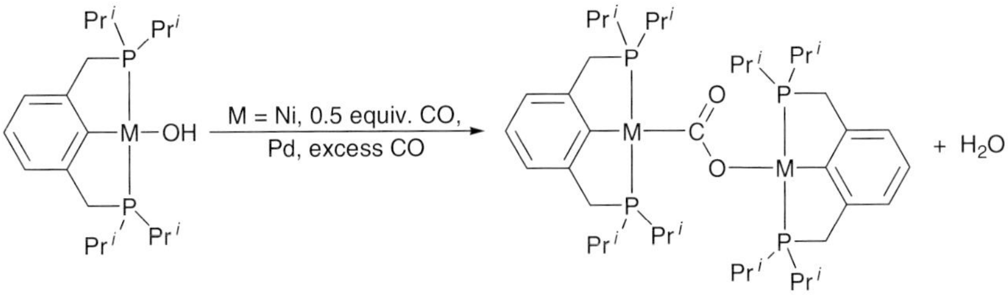

M = Ni (**23a**), Pd (**23b**)

M = Ni (**24a**), Pd (**24b**)

Scheme 2.14

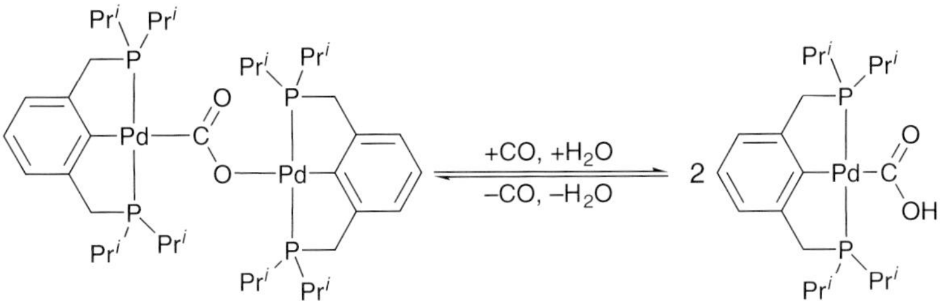

(24b)

(25b)

Scheme 2.15

Analogous reactions like those performed by Campora and coworkers [13] have recently been reported by Wendt and Johansson (2007) [14] with the Bu^t(PCP)Pd(OH) derivative (**26**). Thus, the reaction of hydroxide species (**26**) with carbon monoxide affords the hydroxycarbonyl complex (**27**). Single-crystal X-ray diffraction analysis of this species shows fairly strong hydrogen-bonded dimers formed from symmetry equivalent molecules. Additionally, when benzene solutions of (**27**) are heated, the complex decomposes by losing CO_2, giving place to the hydride compound (PCP) PdH (**28**) unequivocally characterized by single-crystal X-ray diffraction techniques. This process is not reversible since reaction of (**28**) with CO_2 produces the formate complex (**29**) instead, not the hydroxycarbonyl derivative (**27**). The reactivity observed for complex (**27**) contrasts with that of the isopropyl derivative (**25b**), which was not isolated but just detected in solution under carbon monoxide pressure. Moreover, for (**27**) there is no evidence of a monomer—dimer equilibrium in solution (Scheme 2.15) [13] and attempts to synthesize the dimeric analogue by reacting (**26**) with (**27**) were unsuccessful. The difference in behavior was attributed to the higher steric hindrance conferred by the *tert*-butyl substituents.

A unique case of the chemistry of pincer-type complexes was the synthesis of the first carbonyl hemilabile pincer complex (PCN)Pt(CO) reported by Milstein and coworkers [15]. Hence, the carbonyl derivatives (**32**) and (**33**) were synthesized from the direct reaction of CH_2Cl_2 solutions of the corresponding cationic complexes (**30**) and (**31**) with a slight excess of CO. Here, the weakly coordinated BF_4^- counteranion is easily displaced by the better coordinating ligand CO (Scheme 2.17).

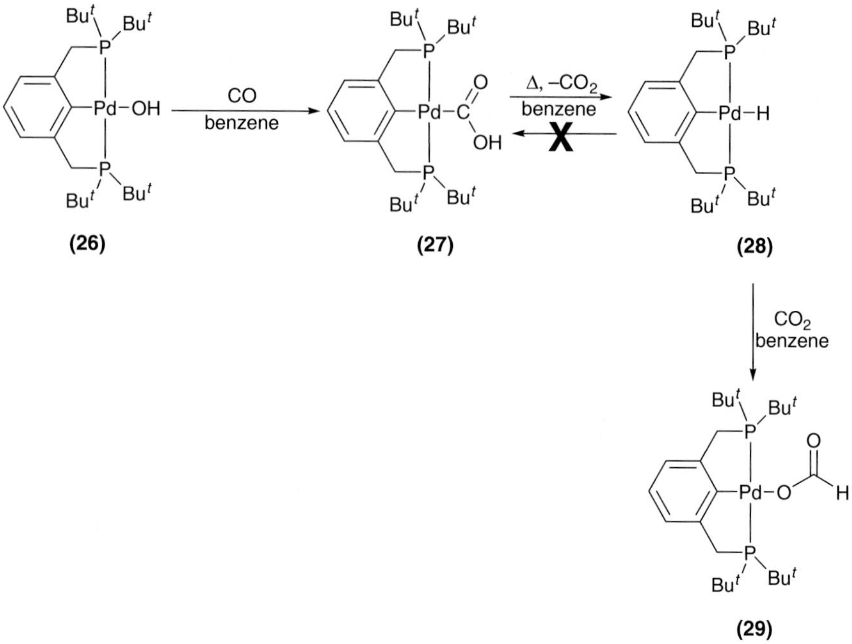

Scheme 2.16

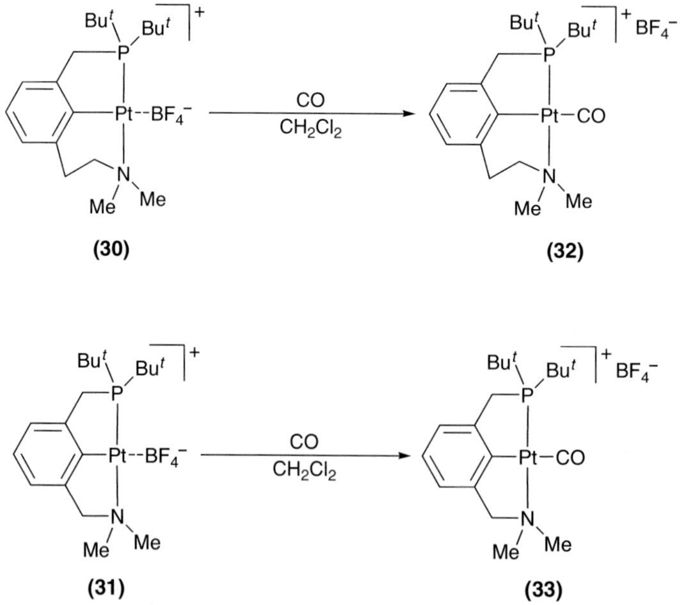

Scheme 2.17

It is noteworthy that further reaction of species (**32**) in acetone with 5 atm of H_2 gas in a Fischer Porter tube at 65 °C affords the novel trimeric cluster complex (**34**) (Scheme 2.18). In this compound, the Pt atoms are connected by metal–metal bonds and by three bridging carbonyl ligands, forming a rigid cluster (Scheme 2.18). The aromatic rings of the ligands underwent demetalation and the amine "arms" were detached from the metal center to leave the phosphine groups as the only remaining coordinated part of the PCN ligand. Significantly, reaction of (PCN)Pt(H) complex (**35**) at 5 atm of CO under identical conditions resulted in the same cluster (**34**) (Scheme 2.18).

One striking difference to the reactivity of complex (**32**) is that the "normal arm" PCN-based carbonyl complex (**33**) does not react with dihydrogen under the same conditions and remains unchanged. This result clearly reflects the influence of the amine "arm" length on its hemilability properties and, as a consequence, on the alternative reactivity patterns.

Finally, an interesting reaction induced by carbon monoxide leading to the formation of a platinum *tris*-carbene pincer complex (**37**) is produced by the reaction at room temperature in pentane or hexanes of the Pt–olefin–carbene complex (**36**) with CO (1 atm) [16]. This compound contains a tridentate pincer ligand in which all three donors are carbenes (Scheme 2.19). The structure of complex (**37**) was fully determined by single-crystal X-ray diffraction analysis.

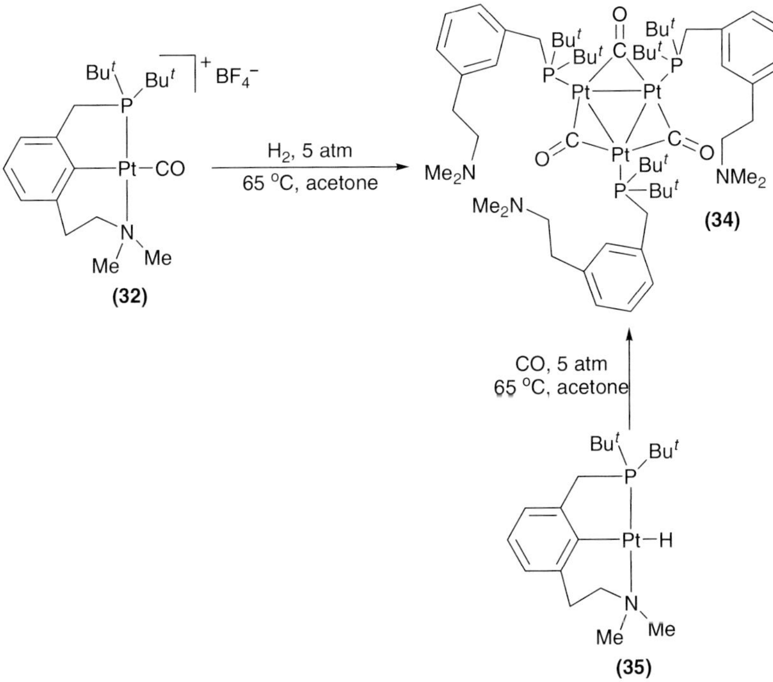

Scheme 2.18

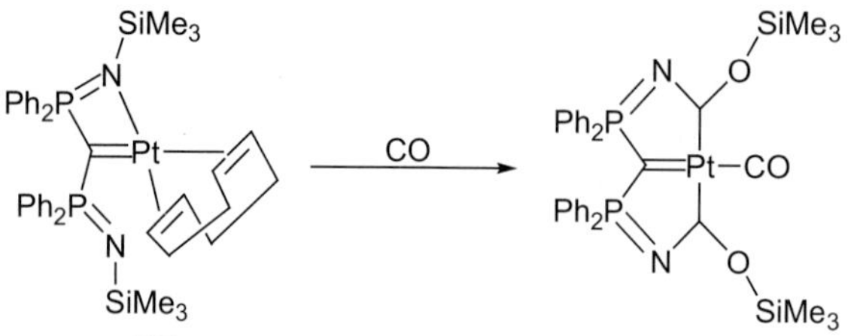

(36) (37)

Scheme 2.19

It is assumed that compound (**37**) is produced by an analogous mechanism to that postulated for the synthesis of Fischer carbenes from metal–carbonyl or metal–isonitrile complexes according to Scheme 2.20.

2.2
Reactivity of CO with Pincer Complexes of the Group 9 (Rh and Ir)

2.2.1
Rhodium

The chemistry of pincer complexes of the group 9 has been dominated by the study of the derivatives with rhodium and iridium. To date, the reactivity of cobalt pincer compounds with carbon monoxide has not been reported, mainly due to the fact that the number of pincer complexes derived from cobalt is reduced [17]. Thus, the interest in the Rh and Ir PCP pincer complexes is mainly owing to the fact that they have been proved efficient in the catalytic dehydrogenation of aliphatic C–H bonds; this is particularly true in the case of the iridium PCP pincer complexes [3].

Thus, as is the case for the first pincer derivatives with nickel, Moulton and Shaw also reported the first PCP pincer rhodium and iridium derivatives and studied their reactivity toward carbon monoxide [4]. Hence, the (PCP) rhodium derivative (**38**) affords complex (**40**), as a product of the decomposition of the unstable hydrido chloro carbonyl intermediate (**39**) via a hydrodechlorination process, although this compound is obtained impure. To favor the hydrodechlorination process, an alternative route was attempted employing EtONa as base. This approach afforded exclusively compound (**40**) (Scheme 2.21). It is noteworthy that analogous reactions with the iridium analogous to complex (**38**) only afford impure samples of the carbonyl species analogous to complex (**40**).

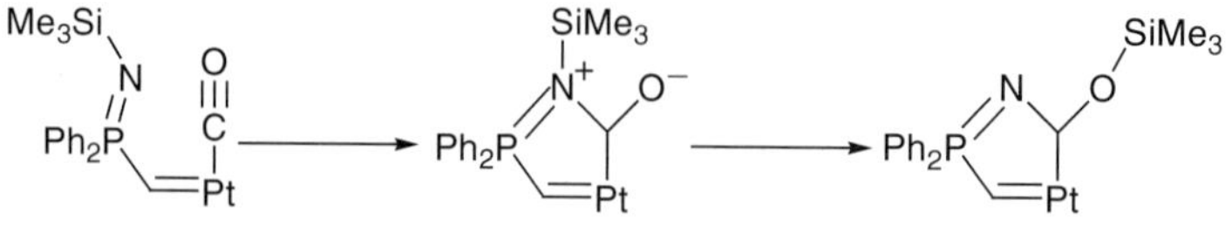

Scheme 2.20

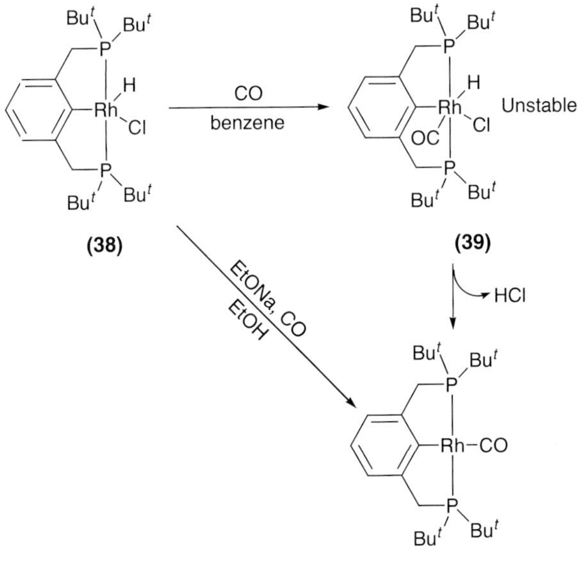

Scheme 2.21

Sacco and coworkers [18] also made an important contribution to the chemistry of rhodium carbonyl pincer complexes. Thus, reactions of the Rh derivative (PNP)Rh(I) (**41**) were performed with carbon monoxide to yield the corresponding carbonyl complex (**42**) (Scheme 2.22).

Interestingly, analogous reaction of the (PNP)Rh(I)-O_2 adduct (**43**) affords complex (**44**) in which the carbon monoxide promotes the O—O bond cleavage, transferring, most likely in a concerted manner, the resulting oxygen atoms to both CO and one of the phosphorus atoms in the (PNP) pincer ligand according to Scheme 2.23.

However, Kaska and coworkers, in 1983 [19], retook the work carried by Shaw and coworkers and carried out similar reactions (Scheme 2.21) of the (PCP)Rh(III) pincer complex (**38**) with carbon monoxide in the presence of thc very basic compound $NaN(Si(CH_3)_3)_2$ obtaining the same species (**40**).

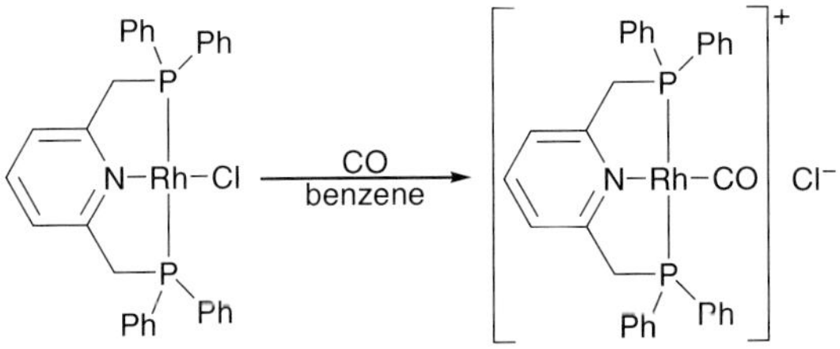

Scheme 2.22

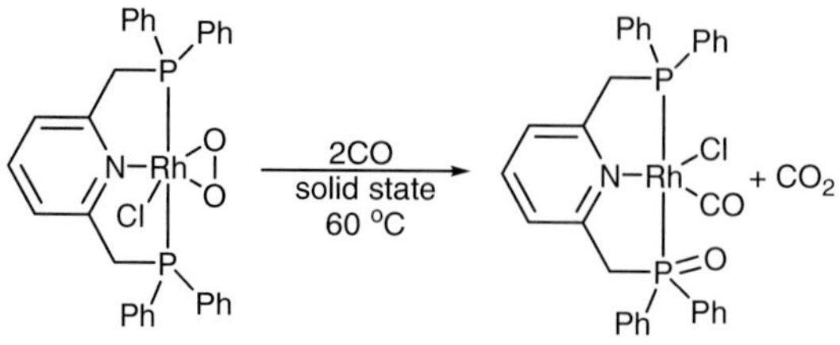

Scheme 2.23

Later, in the study of the reactivity of the (PCP)Rh(I) complex (**38**), Kaska and coworkers [20] realized that the reaction of this compound with CO_2 was, in fact, a CO_2 reducing process to water and CO in what amounted to be a reversal of the water gas shift reaction. In fact, the reaction of the hydrido–hydroxy complex (**45**) with CO afforded complex (**40**) quantitatively (Scheme 2.24).

Quinone methides (QMs) have been postulated as intriguing transients in various biochemical transformations and are of general importance in modern organic chemistry [21]. However, most of these compounds have only been observed in dilute solutions; thus, in an effort to explore the chemistry of this sort of compounds, in 1998, Milstein and coworkers reported [22] the synthesis and reactivity of PCP-type complexes containing QM moiety (**46**). This compound was shown to be remarkably stable toward the reactions with nucleophiles and electrophiles. And to probe the effect of the electron density in the rhodium atom on the stability of the methylene

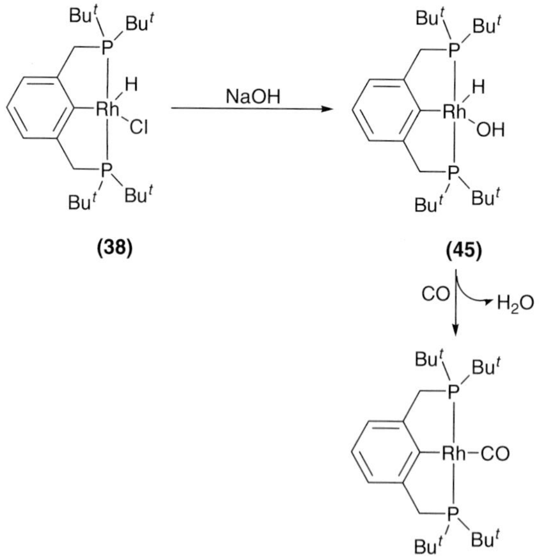

Scheme 2.24

arenium carbocation relative to that of the parent quinone methide complex, the cationic Rh(I) carbonyl complex (**47**) was prepared (Scheme 2.25).

Complex (**47**) could not be protonated with 1 equiv. of HOTf. This mean the protonated rhodium complex has a pK_a value lower than that of triflic acid. Thus, protonation of (**47**) was achieved *in situ* by using 5 equiv. of triflic acid, although the resulting adduct (**48**) could not be isolated in the solid state (Scheme 2.25). Hence, complex (**48**) represents one of the strongest stable acids known. The structure of compound (**47**) was unequivocally identified by single-crystal X-ray diffraction analysis.

Extending the above-mentioned work on methylene arenium complexes (Scheme 2.25), in 1999, Milstein and coworkers disclosed [23] a general method for the synthesis of these highly perturbed aromatic systems involving the loss of aromaticity under mild conditions and a new approach for the synthesis of the closely related σ-arenium metal complexes. These species being stabilized by an unusual agostic interaction of a C–C bond with a metal center. Thus, when the cationic (PCP)Rh (III) complex (**49a**) was reacted with CO in THF or $CDCl_3$ at room temperature, immediate formation of the methyl arenium complex (**50a**) took place as a result of an apparent 1,2-methyl shift (Scheme 2.26). Low-temperature NMR experiments did not show traces of possible intermediates in this reaction, which proceeded smoothly even at −60 °C. Similarly, analogous reaction of the benzyl triflate (PCP) Rh(III) (**49b**) with carbon monoxide resulted in a 1,2 benzyl shift and formation of the benzyl arenium complex (**50b**) (Scheme 2.26). The structure of this compound was unequivocally determined by single-crystal X-ray diffraction analysis.

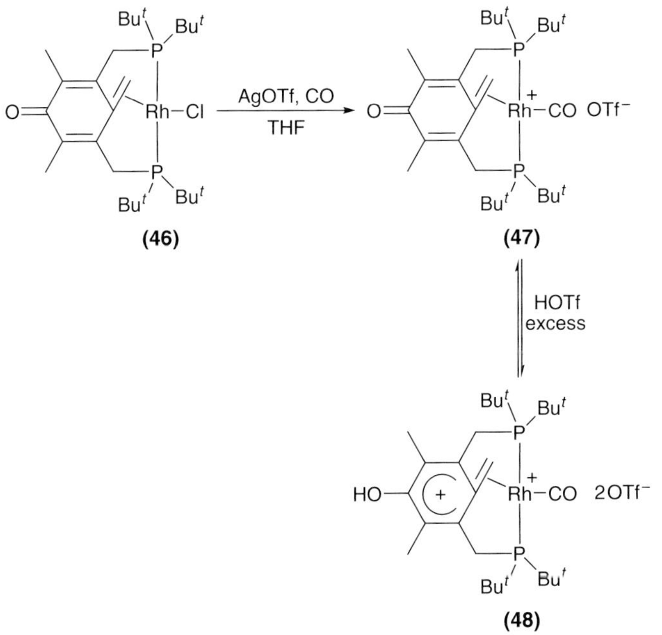

Scheme 2.25

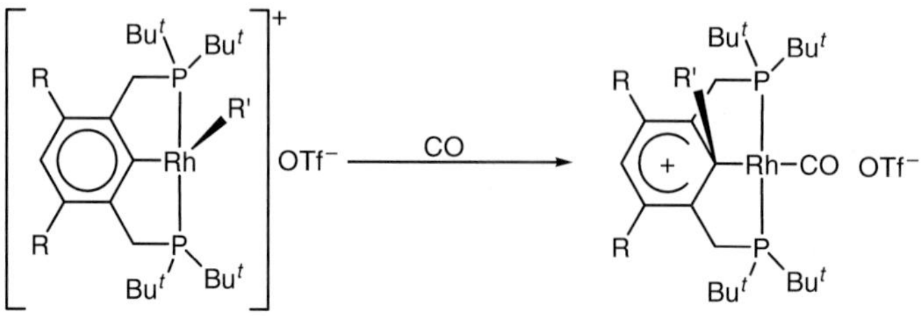

R = R'= Me (**49a**)
R = Me, R'= CH_2Ph (**49b**)
R = R'= H (**51**)

R = R'= Me (**50a**)
R = Me, R'= CH_2Ph (**50b**)
R = R'= H (**52**)

Scheme 2.26

The described apparent 1,2-shift is not limited to alkyl groups. Thus, when the hydrido triflate complex (**51**) was reacted with carbon monoxide in $CDCl_3$, quantitative formation of the corresponding (PCP)Rh(I) complex (**52**) took place (Scheme 2.26). This complex has been described as having only a small contribution of the arenium form due to a very strong η^2 agostic interaction between the ipso C–H bond and the metal center.

Additionally, the reactivity of some of the methylene arenium and σ-arenium complexes toward carbon monoxide was tested. For instance, bubbling of carbon monoxide through a solution of complex (**53**) in CD_2Cl_2 causes the immediate color change from green to red as a result of the quantitative formation of the pentacoordinate CO adduct (**54**) (Scheme 2.21). The later appears to be stable only under a CO atmosphere in solution. Hence, moderate heating of a CH_2Cl_2 solution of (**54**) or drying of (**54**) in vacuum resulted in the loss of carbon monoxide and formation of the precursor (**53**) (Scheme 2.27). However, upon addition of $(CH_3)_3SiOTf$ to a CH_2Cl_2 solution of (**54**), slow Cl^- abstraction took place to afford the tetracoordinated dicationic complex (**55**) (Scheme 2.27).

Another very unusual pincer species was recently (2005) reported by Milstein and coworkers [24], disclosing the synthesis of the first π-accepting PCP pincer ligand bearing pyrrolyl substituents and their corresponding rhodium complexes. Interestingly, complexes (**56a–c**) react with 1 equiv. of carbon monoxide to form the thermally stable, pentacoordinate (PCP)Rh(I) pincer complexes (**57a–c**) (Scheme 2.28). This is in sharp contrast with the usually observed reactivity of CO with (PCP)Rh(I) complexes where generally an ancillary ligand is replaced. Hence, complexes (**57a–c**) can be viewed as arrested intermediates in an associative substitution process involving PCP-type complexes. The structure of one of these complexes (**57b**) was determined fully by single-crystal X-ray diffraction experiments, showing that the pincer framework on these carbonyl complexes bears the phosphine-donor atoms in *cisoid* (rather than in the normally observed *transoid*) conformation. Thus, the unusual stability of these PCP d^8 ML_5 complexes is attributed to the π-accepting nature of the PCP ligand, which disfavors formation of the coordinatively unsaturated ML_4 complexes.

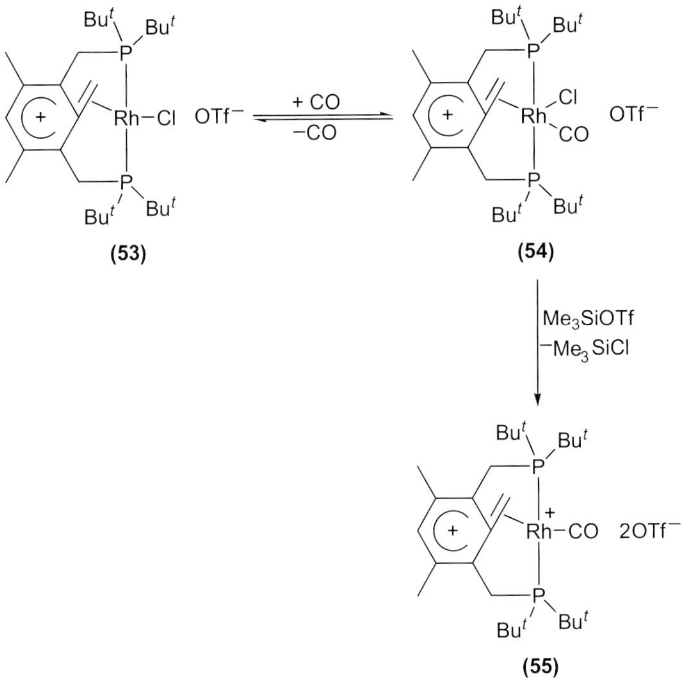
+ CO
−CO
OTf−
2OTf−
Me_3SiOTf
$-Me_3SiCl$
(53)
(54)
(55)

Scheme 2.27

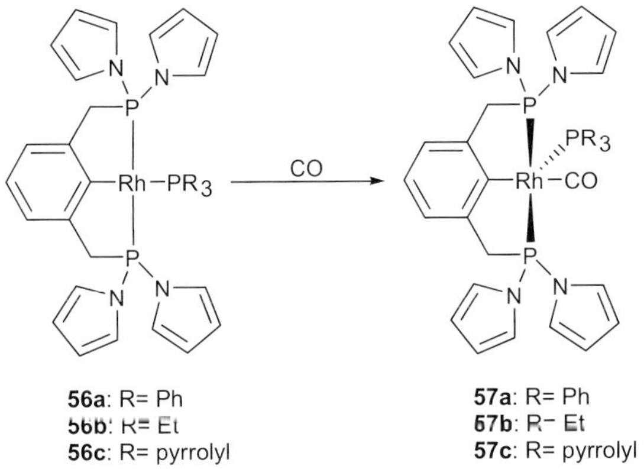
CO
56a: R= Ph
56b: R= Et
56c: R= pyrrolyl
57a: R= Ph
57b: R= Et
57c: R= pyrrolyl

Scheme 2.28

Most recently, Milstein and coworkers [25] with the aim of studying the potential effect in changing the phosphino PCP versus the phosphinite POCOP pincer systems, synthesized the (POCOP)Rh(N_2) complex (**58**) and reacted it with carbon monoxide, observing that the dinitrogen ligand is readily displaced by 1 equiv. of CO to give compound (**59**) (Scheme 2.29). Analysis of the stretching frequencies of the CO ligand showed that the effect of the phosphinite (POCOP) pincer ligand

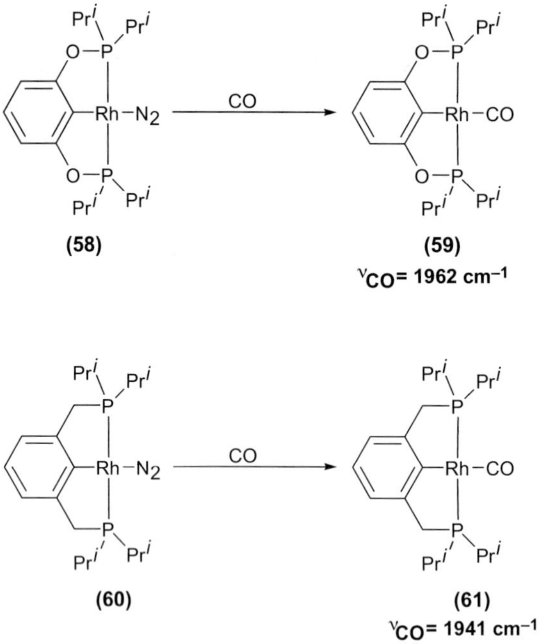

CO

(58) (59)

$\nu_{CO} = 1962\ cm^{-1}$

CO

(60) (61)

$\nu_{CO} = 1941\ cm^{-1}$

Scheme 2.29

compared to the corresponding PCP pincer ligand is only minimal. To have a direct comparison, the compound (**61**) Pr^{i}(PCP)Rh(CO) was also synthesized using an analogous procedure for complex (**59**) (Scheme 2.29).

An additional variation to the pincer ligands was provided by Milstein's group, the modifications this time consisting of the change of the usually employed benzene moiety for a naphthalene fragment [26]. Thus, the nafto-(PCP) Rh dinitrogen complexes (**62a–b**) were reacted in benzene with an equimolar amount of CH_3I at room temperature to afford quantitatively the corresponding dark red, oxidative addition products (**63a–b**) (Scheme 2.30). Compound (**63b**) was further treated with an excess (~100 equiv.) of CO to give (**64b**) and CH_3I as the only products of the reductive elimination process (Scheme 2.30). The formation of CH_3I was confirmed both by ^1H NMR and GC–MS techniques. It is worth noting that this compound can be directly prepared from the reaction of compound (**62b**) with carbon monoxide and that the reverse reaction between complex (**64b**) and methyl iodide does not take place, even when large excess (~100 equiv.) of CH_3I was used for periods as long as 48 h.

Conversely, the treatment of the analogous complex (**63a**) with CO under the same conditions rather resulted in the formation of the carbonyl adducts (**65a–a′**), not in reductive elimination of methyl iodide (Scheme 2.30). However, reaction of (**63a**) with CO at $-50\,^{\circ}C$ resulted in the formation of (**65a**) exclusively. This result indicates that the later compound is the kinetic product. In fact upon warming to room temperature, an equilibrium mixture of 17% (**65a**) and 83% (**65a′**) was obtained. This ratio did not change upon cooling to $-50\,^{\circ}C$.

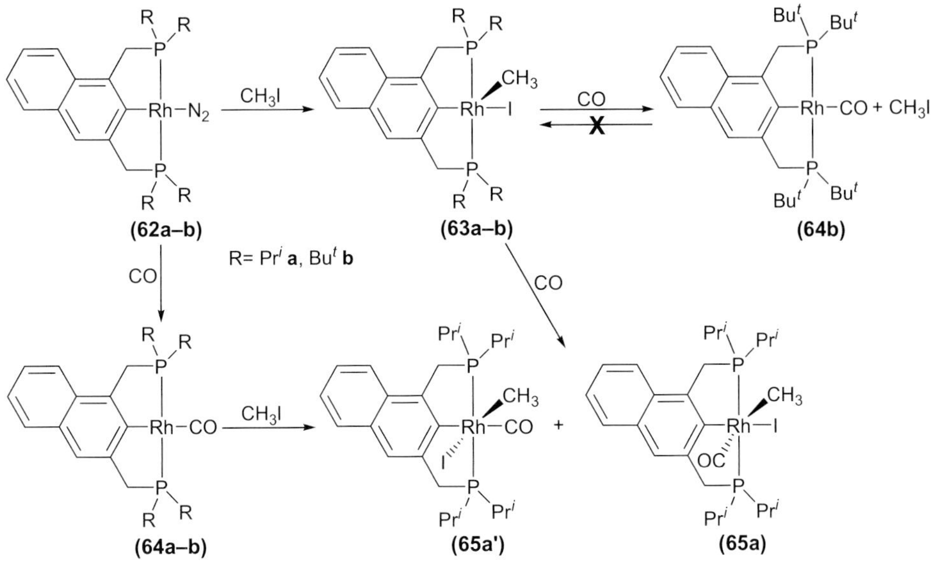

Scheme 2.30

Furthermore, abstraction of the iodide ligand from (**63b**) with $AgBF_4$ yielded the BF_4-coordinated complex (**66**), which upon treatment with CO underwent methyl migration to the ipso carbon, forming the η^2-C–C agostic complex (**67**). Treatment of CH_2Cl_2 or benzene solutions of (**67**) with an equimolar amount (or excess) of $Bu^n{}_4NI$ at room temperature did not result in the formation of CH_3I and (**64b**) (Scheme 2.31), confirming that η^2-C–C agostic intermediates such as (**67**) are not involved in the mechanism.

With the aim of studying the potential activation of molecular oxygen using (PCP)Rh complexes, Milstein and coworkers synthesized [27] the low-valent η^2-O_2 complex (**69**) from the dinitrogen derivative (**68**) (Scheme 2.32). Further reaction of complex (**69**) with excess of carbon monoxide afforded the carbonyl complex (**70**). These results are in sharp contrast with those observed by Sacco and coworkers [18] (see Scheme 2.23) where the oxygen complex (**43**) was capable of oxidizing CO to CO_2.

Interestingly, when complex (**60**) (see Scheme 2.29), structurally similar to compound (**68**), was treated with molecular oxygen, only partial decomposition was observed but no formation of the corresponding (PCP)Rh(O_2). This indicates a strong influence of electron density at the metal center on its ability to bind molecular oxygen.

The easy synthesis and successful application in different organic transformations has made phosphinite PCP pincer complexes very popular [28]. As a result, the synthesis of phosphite PCP pincer complexes is growing. In addition, these ligands can very easily allow the incorporation of chiral motifs in the PCP ligand adding an extra bonus to these already versatile complexes. Although similar compounds have been known since 1999 and applied in metal-mediated organic synthesis [29], to the best of our knowledge, the first report regarding the reactivity of phosphite (PCP)Rh

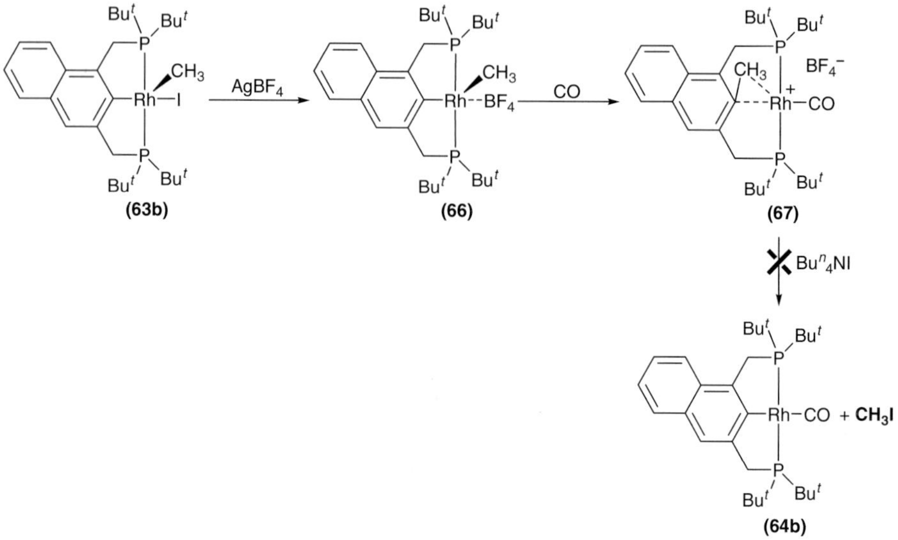

Scheme 2.31

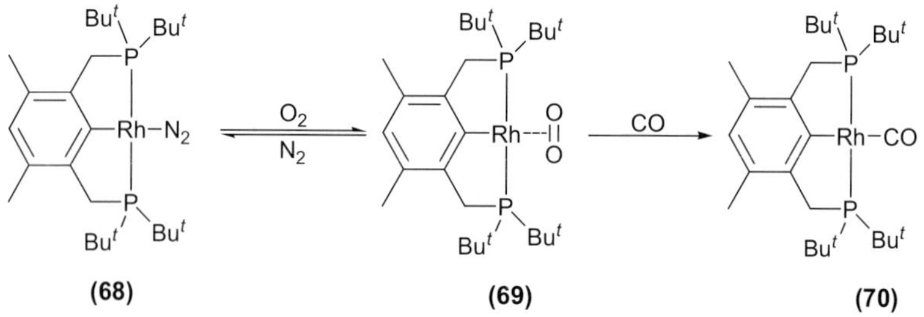

Scheme 2.32

complexes with carbon monoxide has been published only recently in 2007. Thus, Galindo and Pizzano have reported [30] the synthesis of complex (**71a–b**) and examined its reactivity toward CO. It was observed that PPh_3 ligand in (**71a–b**) is labile and easily displaced by CO to afford the carbonyl compounds (**72a–b**) (Scheme 2.33). This is in sharp contrast with the reactivity observed by Milstein when complex (**56**) was reacted with CO to afford the phosphine carbonyl complexes (**57**) with no displacement of the phosphine.

2.2.2
Iridium

Iridium complexes of anionic tridentate pincer ligands are popular catalysts for alkane dehydrogenation. Recently, the reactivity of these complexes has been

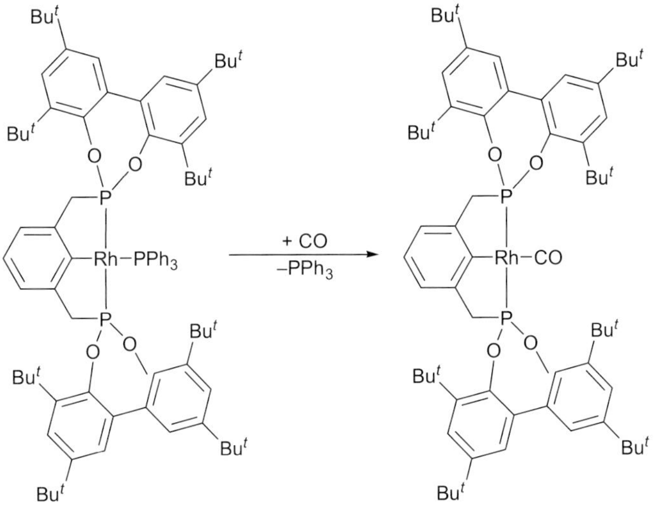

(71a) **(72a)**

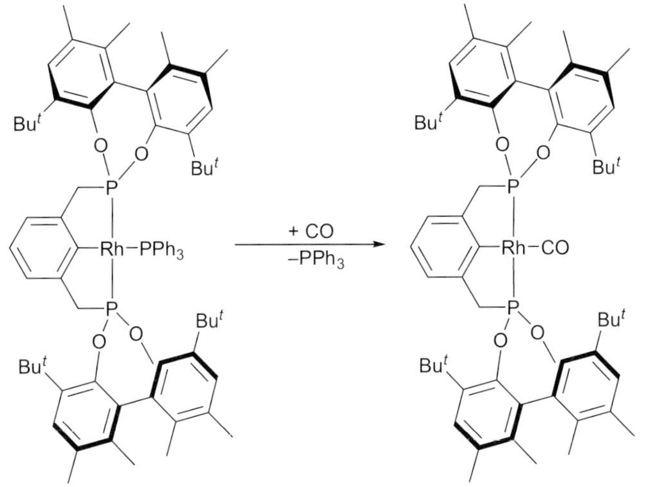

(71b) **(72b)**

Scheme 2.33

extended to the activation of other D–H (D = donor atom) bonds including the dehydrogenation of alcohols [31] and amines [32]. In this respect, authors have gained valuable information from the reactions of some of these complexes with carbon monoxide. In some cases, this sort of reaction has lead to the isolation of reactive intermediates in a given process. The reactivity of iridium pincer complexes with carbon monoxide was first reported by Moulton and Shaw [4]. Hence, analogous

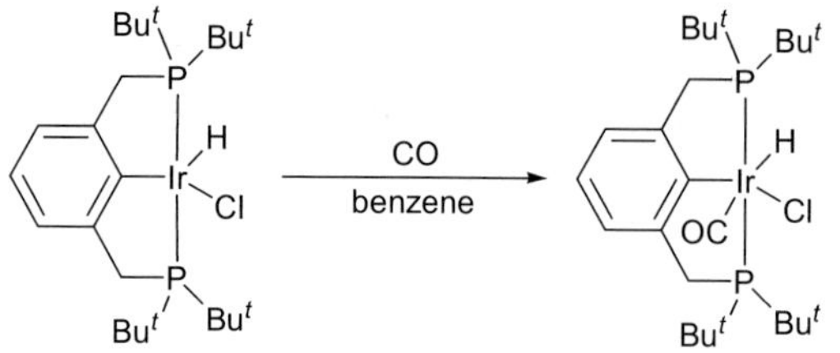

Scheme 2.34

reactions to those reported for (PCP) rhodium complexes with the (PCP)Ir derivative (**73**) afforded the hydrido chloro carbonyl compound (**74**) (Scheme 2.34).

As a ligand for trapping unstable or transient species, carbon monoxide worked well for Goldman and coworkers [33]. They reacted the pentacoordinated (PCP)Ir(III) pincer complex (**75**) with CO in benzene at 6 °C affording compound (**76**) quantitatively (Scheme 2.35). Unlike (**75**), carbonyl compound (**76**) does not undergo arene exchange and is stereochemically rigid at ambient temperature.

It was first theorized by Kaska in 1983 [19] that the formation of PCP double cyclometallated complexes of the type (**77**) could arise from the intramolecular C–H oxidative addition of a *tert*-butyl C–H of the PCP ligand (Scheme 2.36). Although the authors claimed the observation of this sort of species, the evidence shown was not conclusive [19].

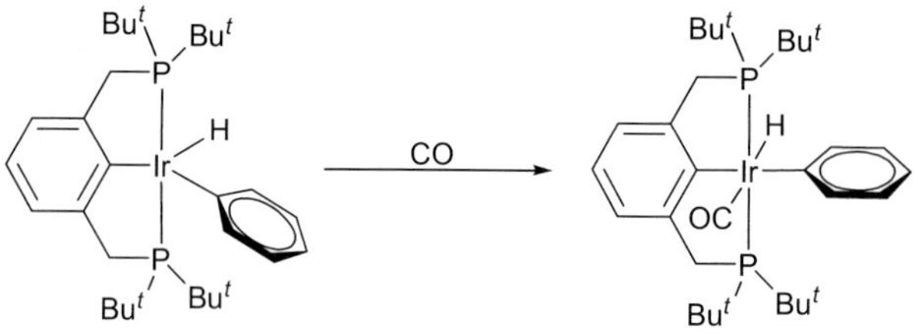

Scheme 2.35

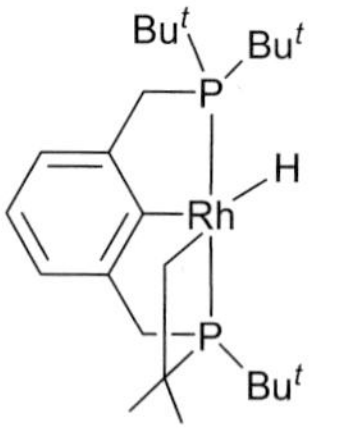

Scheme 2.36

However, in 2002 Kaska and Mayer retook the issue by making variations on the PCP ligand to provide, this time, hard evidence including single-crystal X-ray diffraction studies for the characterization of these PCP pincer double cyclometallated species. Hence, by including a CH_3O-donor group in the benzene moiety of the ligand they were able to isolate complex (**78**). Spectroscopic evidence indicates that in solution species (**78**) is in equilibrium with the most commonly observed compound (**79**) (Scheme 2.37). Complex (**78**) was found both air and thermally stable (up to 200 °C). The reactivity of both these species was tested toward CO affording complexes (**80**) and (**81**), respectively (Scheme 2.37). It is noteworthy that the four-membered ring in (**80**) remains and this species does not interconvert into (**81**) in solution as evidenced from the spectroscopic information exhibited.

In 2003, Morales-Morales and Jensen reported [35] on the reactivity of different (PCP)Ir derivatives with carbon monoxide. Thus, direct reactions of the dihydride species (**82**) (in the presence of TBE), dinitrogen (**83**), and hydrido–hydroxy (**84**) afforded the (PCP)Ir(I) carbonyl complex (**85**) (Scheme 2.38). Worth noting is the reaction of compound (**84**) with CO, which besides complex (**85**) also affords a very small amount of a pale yellow air sensitive (PCP)Ir(I) carboxyl compound (**86**) (Scheme 2.38). The structure of this compound was unequivocally determined by single-crystal X-ray diffraction analysis.

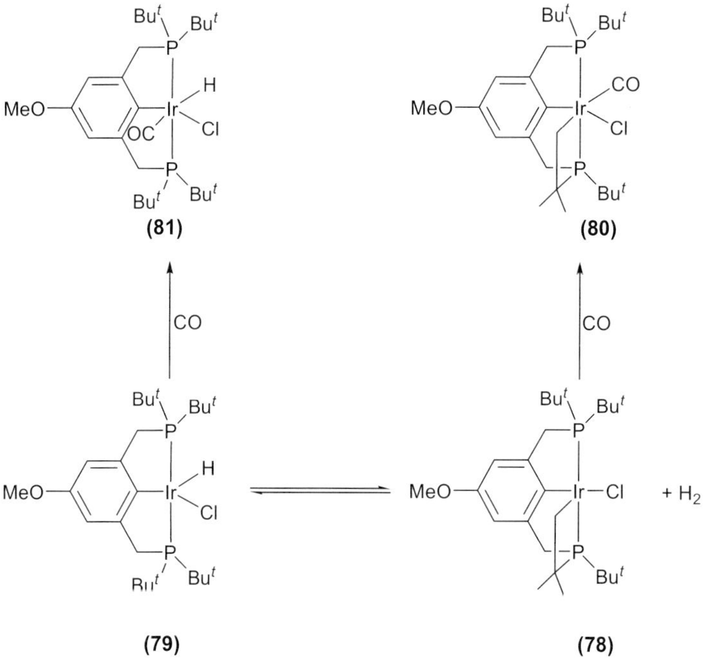

Scheme 2.37

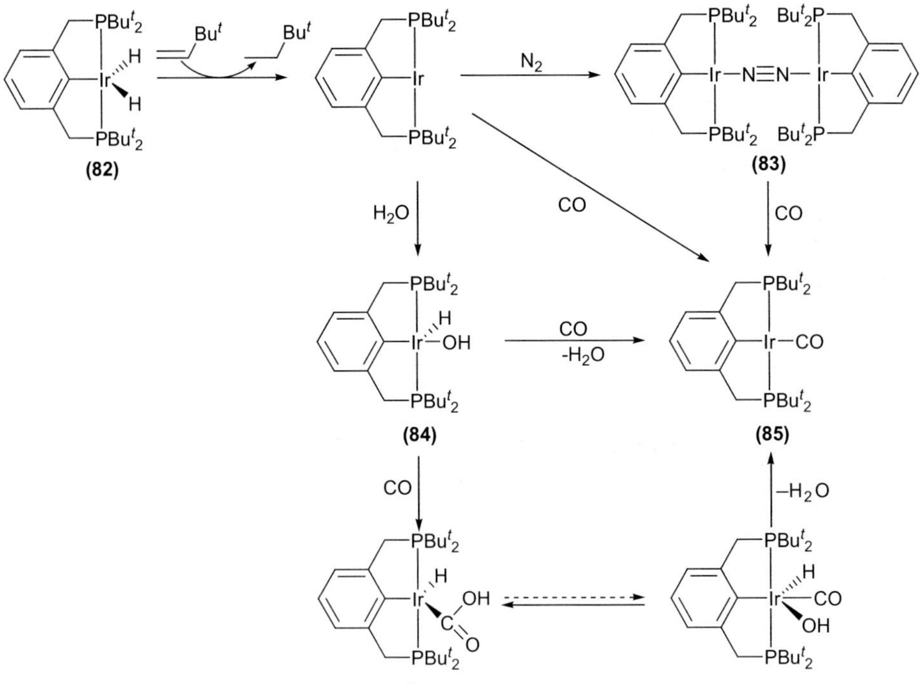

Scheme 2.38

Additionally, compound **(87)**, attained from the reaction of complex **(84)** and CO_2 or from the reaction of **(83)** and wet CO_2, was reacted with carbon monoxide to quickly afford the pale yellow bicarbonate hydride carbonyl species **(88)** (Scheme 2.39).

Isoelectronic, differently charged metal complexes containing the same metal center that undergo C−H reductive elimination/oxidative addition are very rare. Thus, with the aim of comparing the electronic density at the metal center of the very closely related complex **(76)** reported by Goldman and coworkers [33] (see Scheme 2.35), Milstein and coworkers synthesized the corresponding carbonyl derivatives **(90)** (Scheme 2.40) [36], finding, by comparing the absorption frequencies in infrared (1973 cm^{-1} for **(76)** versus 2005 cm^{-1} for **(90)**), that the electronic density is lower at the metal center in complex **(90)** (Scheme 2.41).

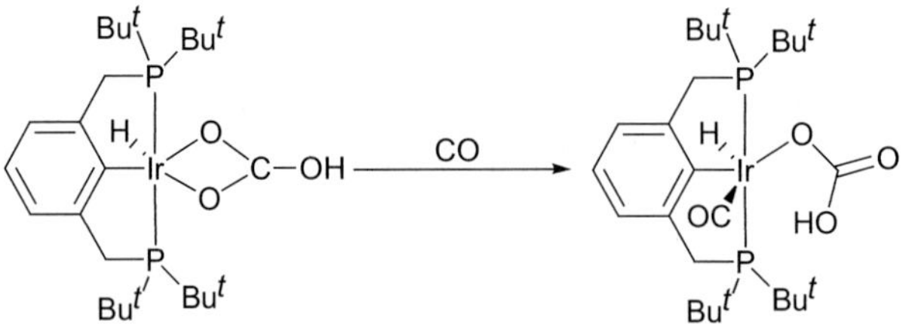

Scheme 2.39

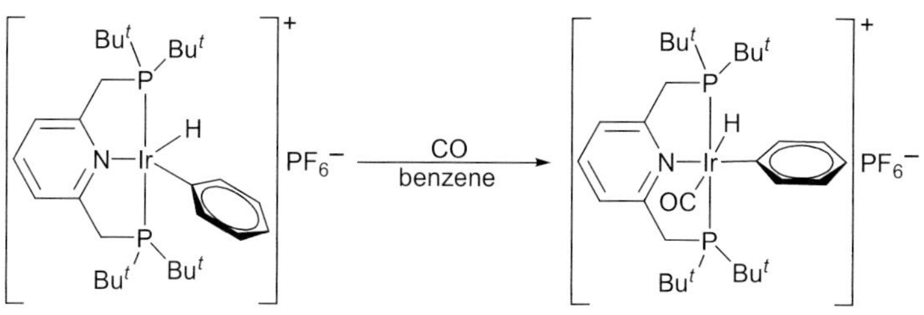

(89) (90)

Scheme 2.40

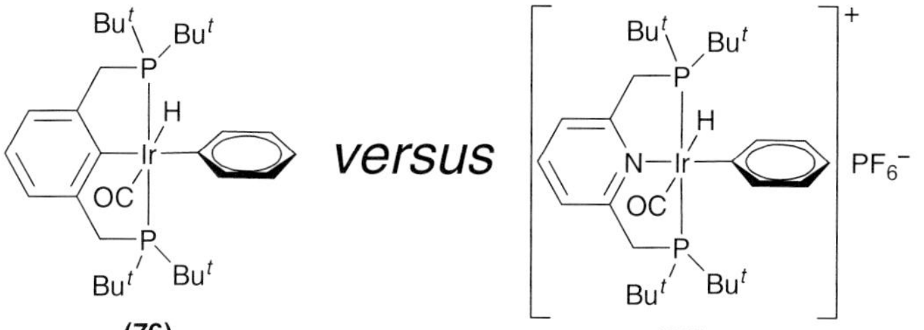

(76) versus (90)

ν_{CO} 1973 cm^{-1} ν_{CO} 2005 cm^{-1}

Scheme 2.41

Aiming at studying the reactivity of ammonia with Ir(III) PCP pincer complexes, Goldman and Hartwig reported [37] the synthesis of the anilide hydride complex (**91**); addition of carbon monoxide to this species produces neither carbamoyl derivative nor free formamide, but gives CO adduct (**92**) that is unstable at room temperature and undergoes reductive elimination to give the known carbonyl (PCP) Ir(I) compound (**85**) (Scheme 2.42).

As it has been observed in other examples (see Schemes 2.29 and 2.41), a sensitive tool for determining the electronic influence of different ligands or substituents in a given system is the ν_{CO} stretching frequency of the corresponding carbonyl complexes. Thus, in 2004, Brookhart and coworkers synthesized [38], in an analogous manner as reported by Morales-Morales and Jensen [35] (Scheme 2.38), a series (*p*-XPCP)Ir(CO) derivatives (**95a–f**) to study the electronic effects of the *p*-substituents

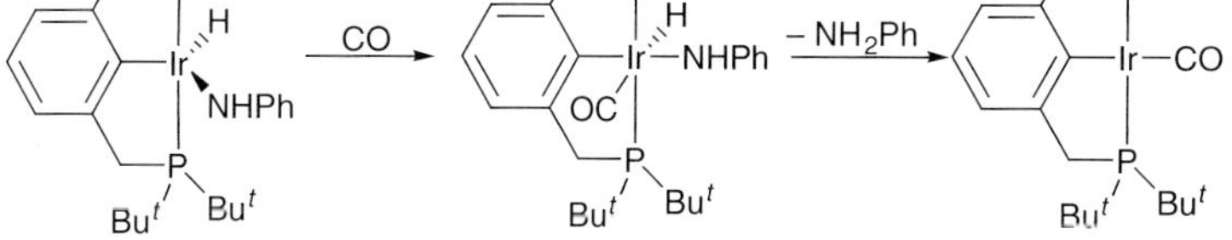

(91) (92) (85)

Scheme 2.42

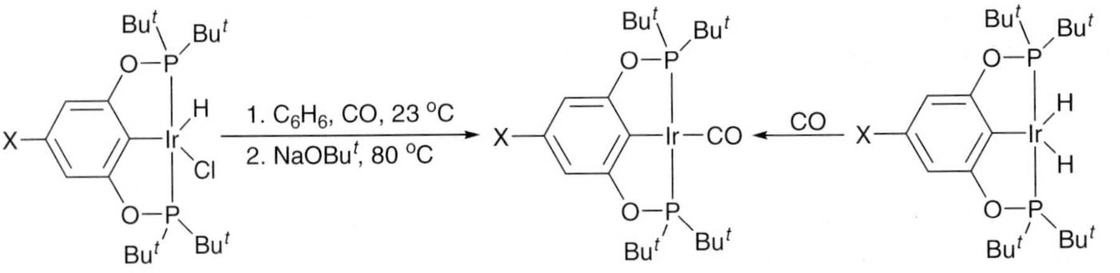

	(95a)	(95b)	(95c)	(95d)	(95e)	(95f)
X	MeO	Me	H	F	C_6F_5	ArF
ν_{CO} (cm^{-1})	1947	1947	1949	1953	1955	1955

Scheme 2.43

in the (*p*-XPCP)Ir fragments (Scheme 2.43). It is worth noting that these studies were carried out to further understand and improve the performance of the parent dihydride species in the catalytic dehydrogenation of aliphatic C—H bonds.

Due to the well-known high reactivity of the (PCP)Ir dihydride species (**82**) [3] and the long standing interest of the scientists in both inorganic and organometallic chemistry of nitroalkanes [39], Goldman and coworkers reported in 2006 [40] that complex (**82**) reacts with CH_3NO_2 to afford compound (**96**) (Scheme 2.44).

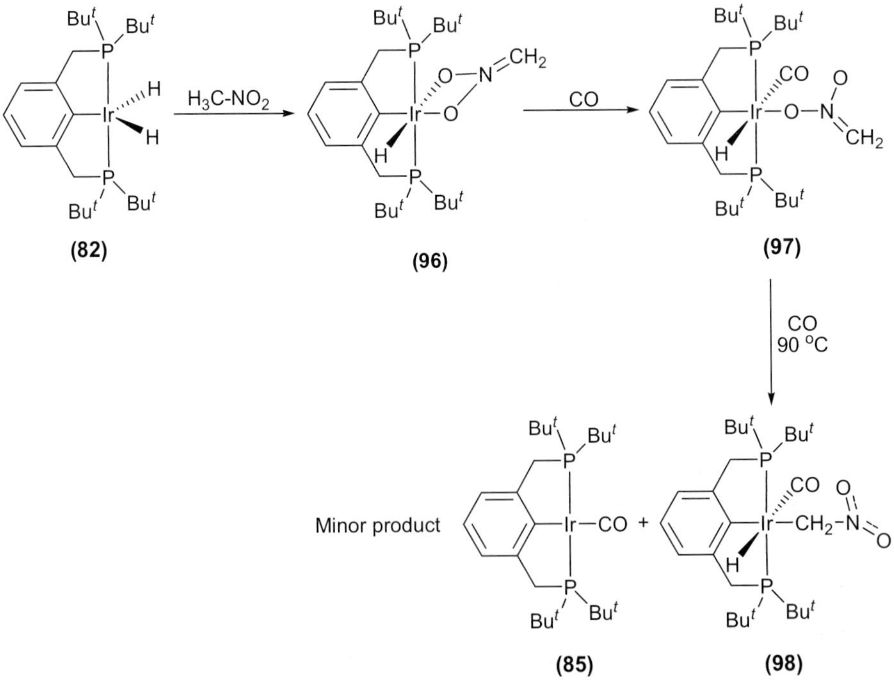

Scheme 2.44

This complex readily reacts with CO (800 Torr) to afford compound (**97**) as a pale yellow solid in an almost quantitative yield (Scheme 2.44). The structure of this compound was unequivocally determined by the common spectroscopic techniques, including a single-crystal X-ray diffraction experiment. Upon heating under CO atmosphere in *p*-xylene solution at 90 °C for 24 h, complex (**97**) was converted into compound (**98**) in 75% yield with the additional formation of some four-coordinate carbonyl species (**85**) (Scheme 2.44). The structure of complex (**98**) was also fully determined by single-crystal X-ray diffraction techniques.

Besides the studies carried out by Goldman and Hartwig in 2003 [37] with the anilide pincer complex (PCP)Ir(H)(NHPh) (**91**) (see Scheme 2.42), Brookhart and coworkers also reported [41] similar results by changing the phosphino PCP ligand for their phosphinite POCOP analogues. Thus, complexes (**99a–b**) were obtained by successive reaction of complex (**93c**) with NaO^tBu, the corresponding aniline, $C_6F_5NH_2$ (**99a**) or 4-$CF_3C_6H_4NH_2$ (**99b**) and CO (Scheme 2.45). Heating an *in situ* generated toluene-d_8 solution of complex (**99b**) at 63 °C resulted in the formation of complex (**95c**) and free 4-(trifluoromethyl)aniline. In contrast, reductive elimination from compound (**99a**) occurred slowly at −78 °C. Additionally, analysis by ^{1}H NMR of a toluene solution of (**99a**) revealed reductive elimination of $C_6F_5NH_2$ to afford complex (**95c**) at 25 °C after 18 h (Scheme 2.45). Thus, the phosphinite POCOP pincer complexes resulted more stable to reductive elimination when compared to their phosphino PCP pincer analogues (see Scheme 2.42).

Most recently and following the studies of the reactivity of pincer iridium species for metal-mediated organic transformations, Goldberg and coworkers reported [42] the stereoselective decarbonylation of methanol, in which an unusual carbonyl *trans*-hydride Ir(III) species (**100**) was obtained (Scheme 2.46).

Thus, to understand how this compound is produced, the reactivity of $(PNP)Ir(H)_2$ complex (**101**) was explored with carbon monoxide. The addition of CO to a CH_2Cl_2 solution of complex (**101**) resulted in the immediate and quantitative formation of the *cis*-dihydride carbonyl complex (**102**). This compound is not stable in solution at room temperature, even under an atmosphere of hydrogen, and undergoes reductive elimination of hydrogen to yield the iridium carbonyl complex (**103**) (Scheme 2.47). This observed reactivity of (**102**) clearly established that this complex is not generated as an intermediate in the formation of (**100**) and that the decarbonylation reaction proceeds stereospecifically.

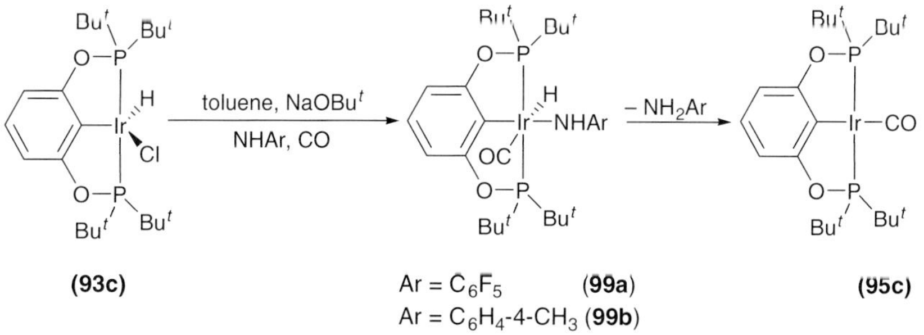

toluene, $NaOBu^t$
NHAr, CO
$-NH_2Ar$
(**93c**)
Ar = C_6F_5 (**99a**)
Ar = C_6H_4-4-CH_3 (**99b**)
(**95c**)

Scheme 2.45

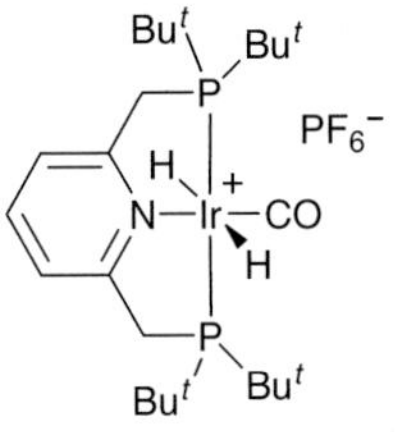

(100)

Scheme 2.46

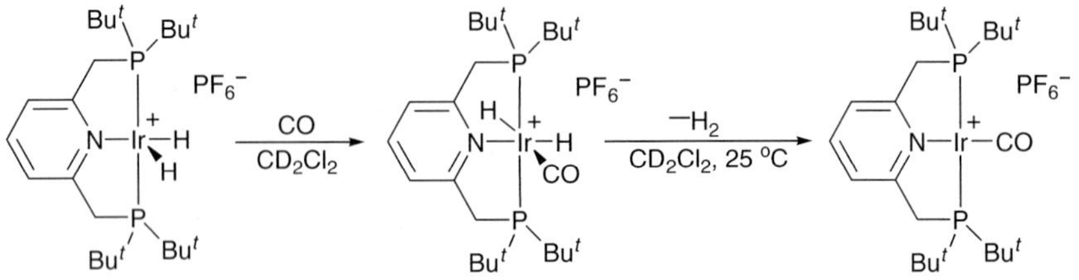

(101) (102) (103)

Scheme 2.47

Worth noting is the fact that almost simultaneously Milstein and coworkers reported [43] the synthesis of complex (**103**). However, their synthesis involved the thermolysis of complex (**90**) (see Scheme 2.40) or the displacement of the cyclooctene ligand from the derivative (**104**) (Scheme 2.48).

An interesting modification of the backbone of the PCP pincer ligands including ferrocene or ruthenocene motifs has been provided recently by Koridze and coworkers [44]. They reported the synthesis of the hydrido chloride complexes (**105a–b**) and their reactivity with carbon monoxide, affording hydrido chloride carbonyl derivatives (**106a–b**) and small amounts of carbonyl compounds (**107a–b**) (Scheme 2.49). It is worth noting that the latter complex can be obtained from the reaction of the dihydride species (**108a–b**) with carbon monoxide. In general, the reactivity of these species is very similar to that shown by other pincer derivatives already discussed in this section. However, the presence of ferrocene fragment in complexes (**105a**) and (**107a**) allows further reactions. In the case of these species, the chemical oxidation by $[Cp_2Fe]^+$ salts affords the corresponding paramagnetic species (**109a**) and (**110a**) (Scheme 2.49).

2.3
Reactivity of CO with Pincer Complexes of the Group 8 (Fe, Ru, Os)

2.3.1
Iron

Reactions of carbon monoxide involving iron pincer complexes were first reported by Chirik and coworkers [45] in 2006. They showed that stirring a pentane slurry of the

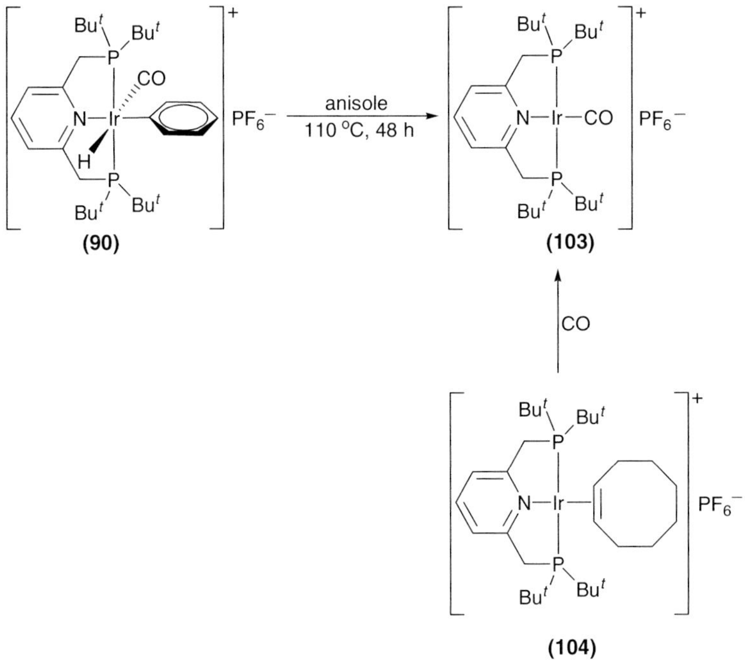

Scheme 2.48

pincer compound (PNP)$Fe^{II}Cl_2$ (**111**) with an excess of 0.5% sodium amalgam in the presence of 4 atm of CO affords a red solid identified as (PNP)$Fe^0(CO)_2$ (**112**) (Scheme 2.50). This species was fully identified by the common spectroscopic techniques and unequivocally identified by single-crystal X-ray diffraction analysis.

Additionally, the reaction of an ethereal solution of complex (**111**) with $NaBEt_3H$ under 1 atm of dinitrogen affords the dihydride dinitrogen species (PNP) $Fe^{II}(H)_2(N_2)$ (**113**) (Scheme 2.51). Treating (**113**) further with 1 equiv. of $PhSiH_3$ results in the displacement of one of the hydride ligands for the silyl group to yield a unique isomer of complex (**114**) (Scheme 2.51). Compound (**114**) was tested as catalyst for the hydrosilation of 1-hexene and cyclohexane and was found totally ineffective. On the contrary, addition of 4 atm of CO to a pentane solution of complex (**113**) induced rapid reductive elimination of H_2 and the concomitant formation of the dicarbonyl complex (**112**) (Scheme 2.51). A similar behavior is observed when the same experiment is performed with compound (**114**) attaining complex (**115**) as a product of the substitution of the dinitrogen ligand with carbon monoxide (Scheme 2.51). It is worth noting that extended thermolysis of the latter species at 95 °C for 1 week under excess of CO produced approximately 50% conversion to complex (**112**) (Scheme 2.51), thus establishing the higher barrier for Si−H reductive coupling and dissociation compared to H_2.

Recently, another study on the synthesis of the novel (PNP) pincer ligands based on phosphoramidites and their corresponding iron complexes (**116a–b**) was reported

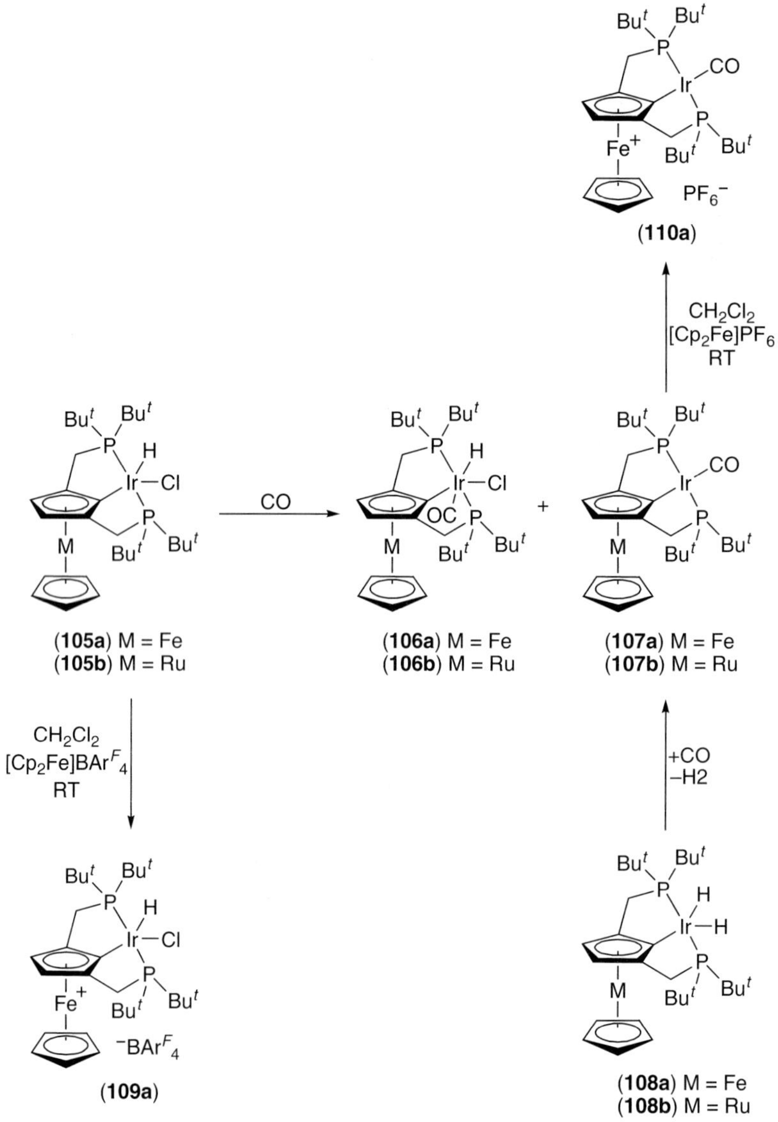

Scheme 2.49

[46]. The reactivity of these compounds toward carbon monoxide was evaluated at room temperature giving place to the corresponding carbonyl species (**117a–b**). The structure of complex (**117b**) was fully determined by single-crystal X-ray diffraction analysis. The species (**117a–b**) were tested as catalyst precursors for the coupling of aromatic aldehydes with (EDA = ethyldiazoacetate). Complex (**117b**) turned out to be the most effective of the series when the reactions were carried out in CH_3NO_2.

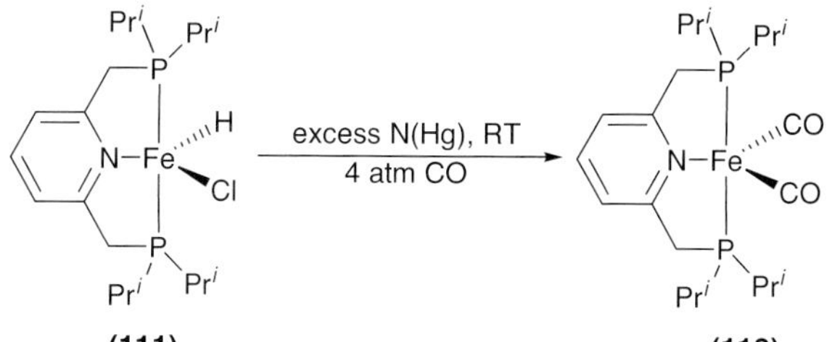

Scheme 2.50

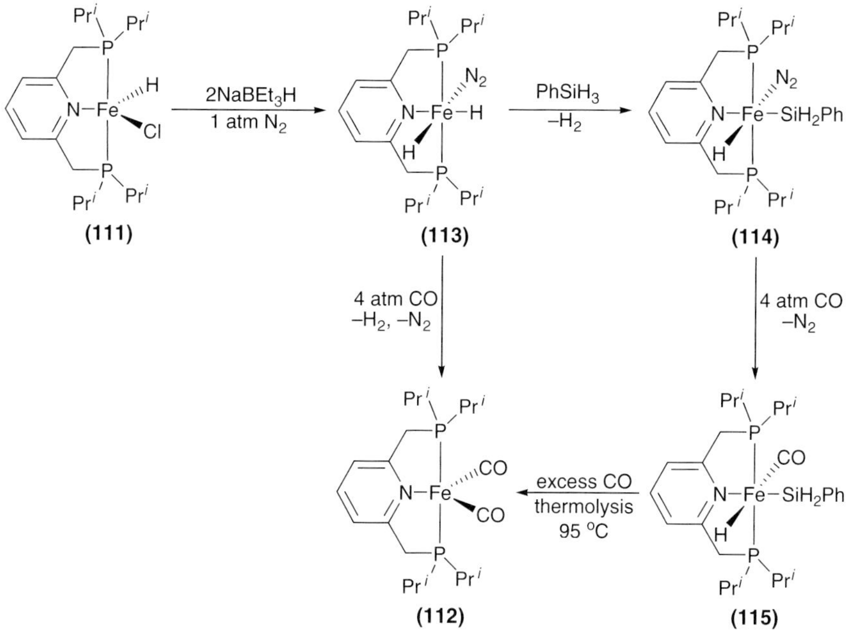

Scheme 2.51

2.3.2
Ruthenium

It was not until 1997 that the first study involving a direct reaction of carbon monoxide with a ruthenium pincer compound was reported. Jia and coworkers reported [47] the synthesis of (PCP)Ru(II) (**118**) and examined its reactivity toward carbon monoxide. They found that the reaction of (**118**) with CO gas leads to the replacement of the PPh_3 ligand, giving place to the white compound (PCP)RuCl$(CO)_2$ (**119**) (Scheme 2.53).

In 2000, Milstein and coworkers reported the synthesis of the first stable metalloquinone [48], a compound in which one of the oxygen atoms of quinone has been replaced by a metal. The procedure for the synthesis of such species consisted first in the formation of (PCP)Ru(II) pincer complexes (**120a**) and (**120b**) (Scheme 2.54) from the cyclometallation reaction of the PCP ligand 3,5-bis(di-*tert*-butylphosphinomethy-

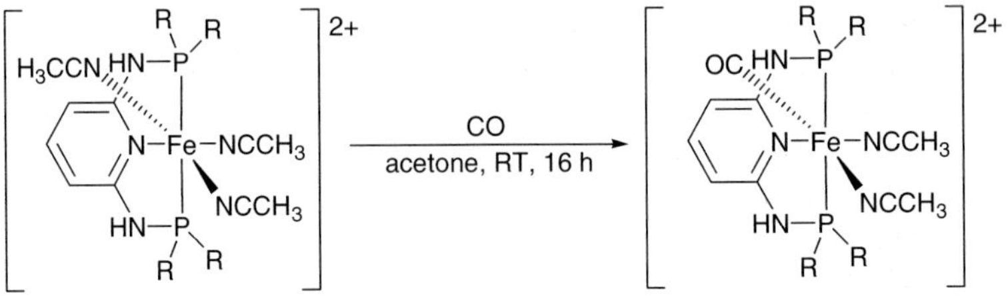

(116a) R = Ph
(116b) R = Pr^i

(117a) R = Ph
(117b) R = Pr^i

Scheme 2.52

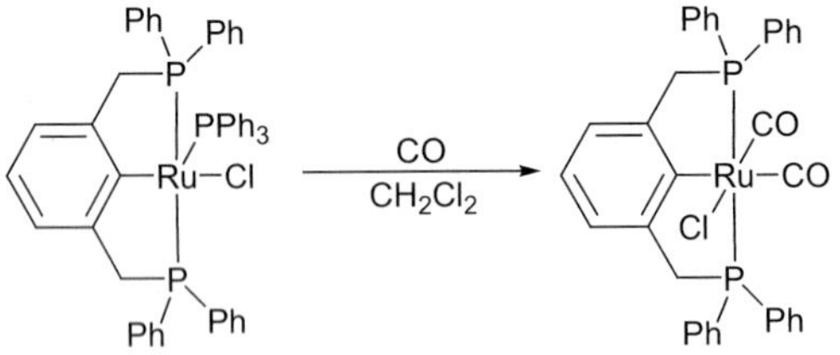

(118)

(119)

Scheme 2.53

lene)phenol with either $Ru(O_2\text{-}CCF_3)_2CO(PPh_3)_2$ or $Ru(DMSO)_4Cl_2$ in the presence of 2 equiv. of NEt_3. Further bubbling of carbon monoxide through solutions of (**120a**) and (**120b**) quantitatively afforded the dicarbonyl complexes (**121a**) and (**121b**), respectively (Scheme 2.54). Finally, deprotonation of the later compounds with equimolar amounts of KOH in THF at 25 °C overnight or with 1,8-bis(dimethylamino)naphthalene proton sponge or 1,4-diazabicyclo[2.2.2]octane (Dabco) upon refluxing in CH_3OH for 4 days produced the metalloquinone (**122**) (Scheme 2.54). In this case, the presence of two strong π-acceptor CO ligands is essential for the charge transfer from the phenoxide moiety to the metal center, stabilizing the Ru(0)-metalloquinone formed. For comparison, addition of $P(CH_3)_3$ to (**120a**) followed by deprotonation results in decomposition.

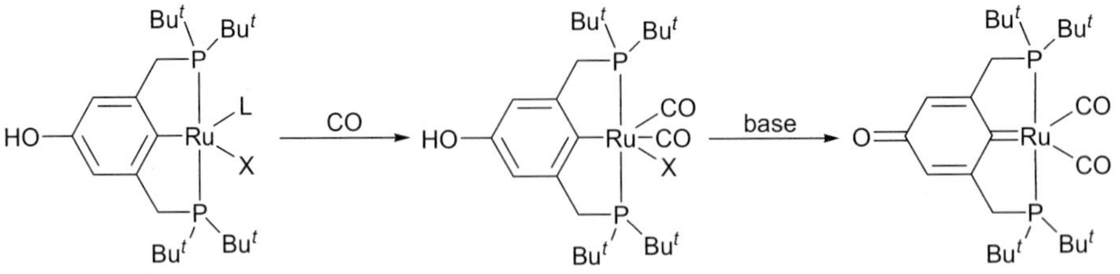

L, X = CO, TFA (**120a**)
L, X = NEt_3, Cl (**120b**)

X = TFA (**121a**)
X = Cl (**121b**)

(**122**)

Scheme 2.54

Furthermore, evidence obtained from the spectroscopic analyses led the authors to conclude that metalloquinone (**122**) appears in two different forms in different solvents: the quinoid form (**122a**), present in less polar solvents such as benzene and THF, and the zwitterionic form (**122b**), present in polar solvents such as methanol and acetone (Scheme 2.55). The latter and the intermediate (**121a**) were characterized unequivocally by single-crystal X-ray diffraction analyses.

Experimental and theoretical observations have shown that C−H bond formation and, by microscopic reversibility, C−H bond activation proceed via η^2-(C−H...M) "agostic" intramolecular or σ-C−H intermolecular intermediates. Thus, studying the trapping of such complexes provides fundamental insight into the nature of the synthetically important hydrocarbons, their activation and utilization. In 2004, Milstein and coworkers [49] synthesized compound (**124**) from the equimolar reaction of $RuCl_3{\cdot}3H_2O$ with ligand (**123**) in methanol at 80 °C in a pressure vessel followed by reaction with CO (1 atm) at room temperature. This complex was fully characterized by X-ray diffraction analysis (Scheme 2.56). Compound (**124**) was further reacted in CH_2Cl_2 with silver salts to abstract the halide and quantitatively afford the cationic complex (**125**). The structure of the latter compound was unequivocally determined by single-crystal X-ray diffraction analysis. Analysis by dynamic NMR of complex (**125**) indicates that the agostic interaction is reversible. To verify this, complex (**125**) was further reacted with H_2 to trap the open coordination site, affording complex (**126**). The addition of NEt_3 to a solution of (**126**) led to the formation of complex (**127**). The same complex is attained when the cationic complex (**125**) is reacted with 1 atm of H_2 in CH_2Cl_2 in the presence of triethylamine (Scheme 2.56).

Soon after Milstein reported in 2004 the synthesis of the (PCP)Ru carbonyl complex [49] (**124**) (Scheme 2.56), Fogg and coworkers [50] reported the analogous complex (**129**) (Scheme 2.57), whose structure was unequivocally determined by single-crystal X-ray diffraction analysis showing this species to be structurally analogous also to that determined for complex (**119**) synthesized by Jia and coworkers in 1997 [47] (Scheme 2.53). Unfortunately, the infrared analysis for complex (**124**) was not performed, thus a potential analysis of the donor properties of the ligands in these three different complexes could not be carried out.

Finally, the most recent report regarding reactivity of ruthenium pincer complexes with carbon monoxide was published by Milstein and coworkers [51]. They informed

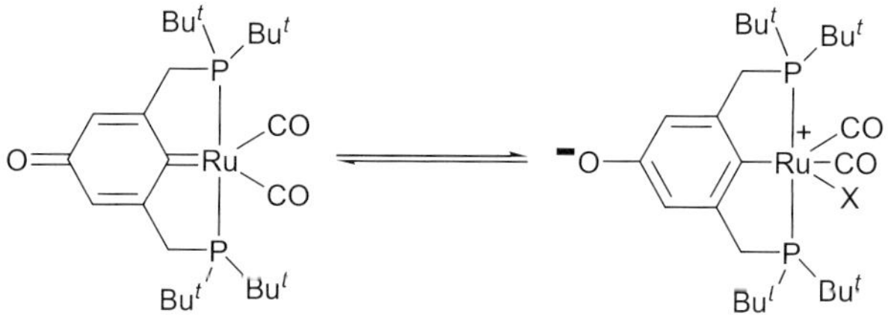

(**122a**) ⇌ (**122b**)

Scheme 2.55

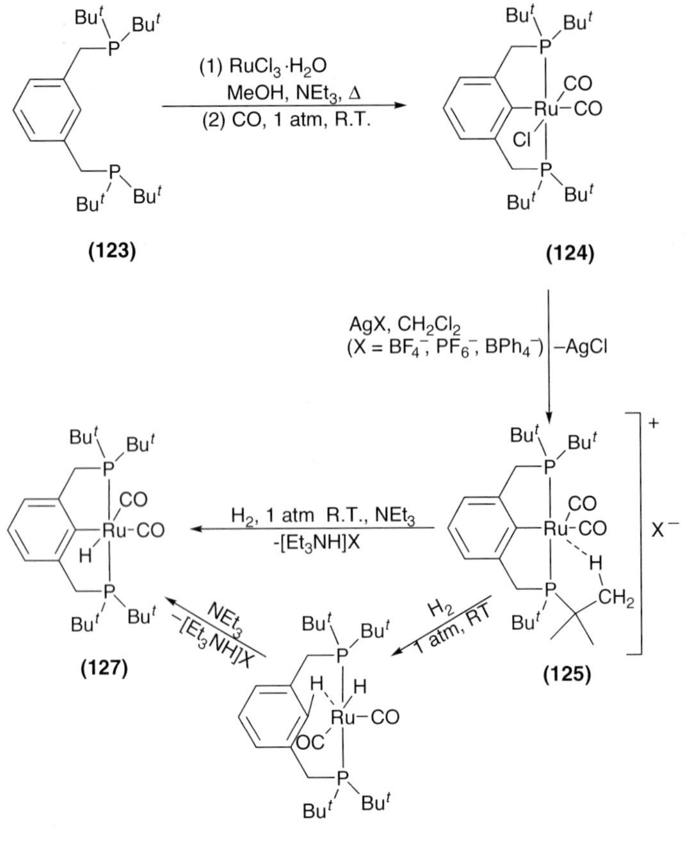

(1) $RuCl_3 \cdot H_2O$
MeOH, NEt_3, Δ
(2) CO, 1 atm, R.T.

(123)

(124)

AgX, CH_2Cl_2
($X = BF_4^-, PF_6^-, BPh_4^-$) $-AgCl$

H_2, 1 atm R.T., NEt_3
$-[Et_3NH]X$

H_2
1 atm, RT

NEt_3
$-[Et_3NH]X$

(127)

(125)

(126)

Scheme 2.56

(119)

(124)

(129)

(128)

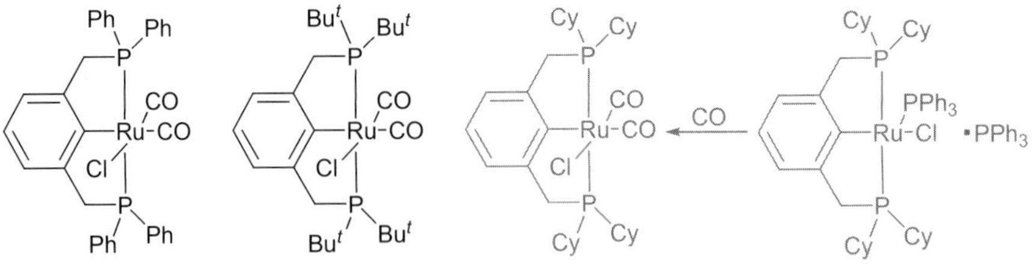

Scheme 2.57

the synthesis of the carbonyl pincer complex (**130**) from the reaction of the dinitrogen species (**129**) with carbon monoxide (Scheme 2.58). Complex (**130**) was employed successfully as catalyst precursor in the dehydrogenation of secondary alcohols to ketones and primary alcohols to esters.

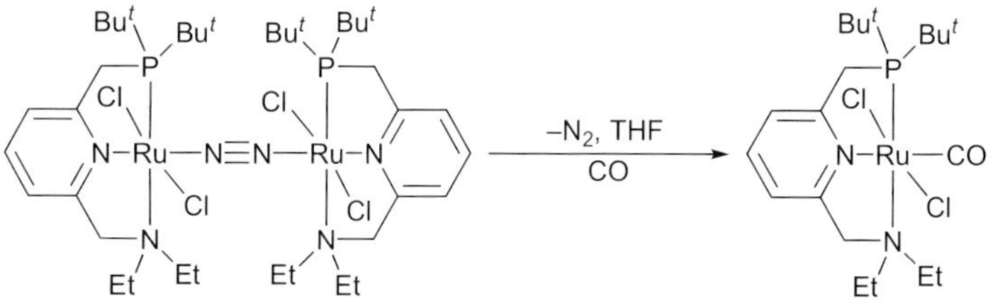

(129) (130)

Scheme 2.58

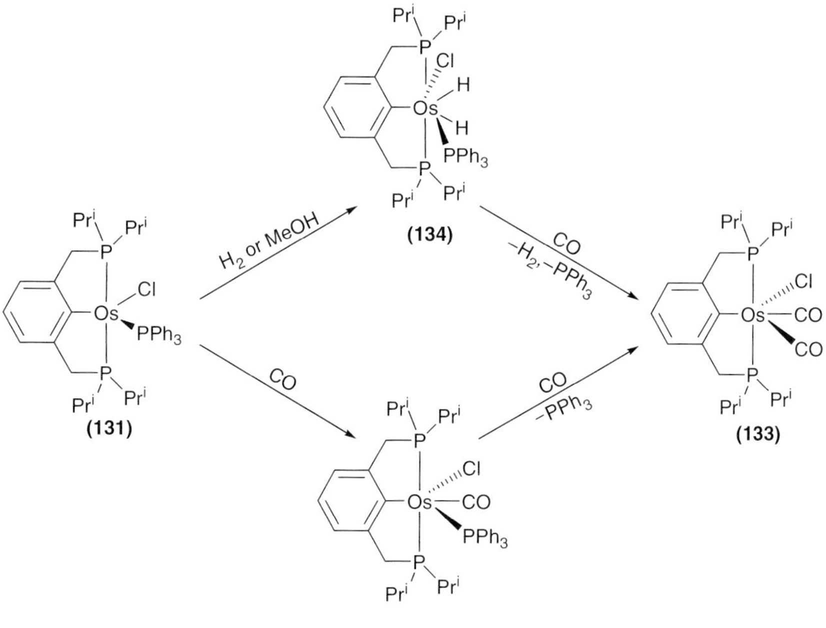

(131) (134) (132) (133)

Scheme 2.59

2.3.3
Osmium

The only study on the direct reaction of carbon monoxide with osmium pincer complexes was reported by Milstein and coworkers in 2001 [52]. In this case, the pincer (PCP)Os(II) complex (**131**) was reacted with CO in benzene at room temperature to afford quantitatively the carbon monoxide adduct (**132**). This species further reacted with an additional equivalent of CO to yield the bis-carbonyl (PCP)Os(II) pincer complex (**133**) with the elimination of PPh_3. Additionally, complex (**133**) can be directly generated by pressurizing the pincer compound (**131**) with 3 bar of CO or from the reaction of complex (**134**) with CO under mild conditions. The structure of complex (**132**) was unambiguously confirmed determined by single-crystal X-ray diffraction analysis.

2.4
Final Remarks

The present chapter demonstrates the great utility carbon monoxide has in the development of the chemistry of pincer compounds. The direct reaction of a number of these complexes with CO has provided derivatives that have allowed a better understanding of the electronics of different susbtituents in the pincer backbone, thus enabling the further rationalized design of new pincer ligands and complexes. At the same time, these studies permitted the isolation of important species and potential intermediates of a given process or transformations that have enabled different research groups around the world to optimize some of these important processes. In other cases, the carbonylated species attained are important in their own right, advancing the knowledge of the chemistry of pincer complexes and allowing envisioning the potential employment of these compounds in industrially relevant catalytic processes. Thus, it is expected that the development pincer compounds have witnessed in the past few years will continue, with carbon monoxide as common partner, for the study, understanding, and discovery of novel pincer species.

2.5
Acknowledgements

I gratefully acknowledge the support and enthusiasm of former and current group members and colleagues. The research from our group described in this chapter is supported by CONACYT (J41206-Q; F58692) and UNAM (IN114605; IN227008).

References

1 Elschenbroich, C. and Salzer, A. (2006) *Organometallics: A Concise Introduction*, 2nd edn, Wiley-VCH Verlag GmbH, Weinheim, Germany.

2 (a) Albrecht, M. and van Koten, G. (2001) *Angewandte Chemie-International Edition*, **40**, 3750.(b) van der Boom, M.E. and Milstein, D. (2003) *Chemical Reviews*, **103**, 1759. (c) Singleton, J.T. (2003) *Tetrahedron: Asymmetry*, **59**, 1837. (d) Szabo, K.J. (2006) *Synlett*, 811. (e) Morales-Morales D. and Jensen C.M. (eds) (2007) *The Chemistry of Pincer Compounds*, Elsevier, Amsterdam, The Netherlands.

3 Jensen, C.M. (1999) *Journal of the Chemical Society, Chemical Communications*, 2443.

4 Moulton, C.J. and Shaw, B.L. (1976) *Journal of the Chemical Society, Dalton Transactions*, 1020.

5 Giannoccaro, P., Vasapollo, G. and Sacco, A. (1980) *Journal of the Chemical Society, Chemical Communications*, 1136.

6 Sacco, A., Giannoccaro, P. and Vasapollo, G. (1984) *Inorganica Chimica Acta*, **83**, 1136.

7 Nobile, C.F., Vasapollo, G. and Sacco, A. (1985) *Journal of Organometallic Chemistry*, **296**, 435.

8 Sacco, A., Vasapollo, G., Nobile, C.F., Piergiovanni, A., Pellinghelli, M.A. and Lanfranchi, M. (1988) *Journal of Organometallic Chemistry*, **356**, 397.
9 Bennett, M.A., Jin, H. and Willis, A.C. (1993) *Journal of Organometallic Chemistry*, **451**, 249.
10 Steffey, B.D., Miedaner, A., Maciejewski-Farmer, M.L., Bernatis, P.R., Herring, A.M., Allured, V.S., Carperos, V. and DuBois, D.L. (1994) *Organometallics*, **13**, 4844.
11 Yun, J.G., Seul, J.M., Lee, K.D., Kim, S. and Park, S. (1996) *Bulletin of the Korean Chemical Society*, **17**, 311.
12 Ryu, S.Y., Yang, W., Kim, H.S. and Park, S. (1997) *Bulletin of the Korean Chemical Society*, **18**, 1183.
13 Cámpora, J., Palma, P., del Ŕo, D. and Álvarez, E. (2004) *Organometallics*, **23**, 1652.
14 Johansson, R. and Wendt, O.F. (2007) *Organometallics*, **26**, 2426.
15 Poverenov, E., Gandelman, M., Shimon, L.J. W., Rozenberg, H., Ben-David, Y. and Milstein, D. (2005) *Organometallics*, **24**, 1082.
16 Lin, G., Jones, N.D., Gossage, R.A., McDonald, R. and Cavell, R.G. (2003) *Angewandte Chemie-International Edition*, **42**, 4054.
17 Müller, G., Klinga, M., Leskelä, M. and Rieger, B. (2002) *Zeitschrift Fur Anorganische Und Allgemeine Chemie*, **628**, 2839.
18 Vasapollo, G., Giannoccaro, P., Nobile, C.F. and Sacco, A. (1981) *Inorganica Chimica Acta*, **48**, 125.
19 Nemeh, S., Jensen, C., Binamira-Soriaga, E. and Kaska, W.C. (1983) *Organometallics*, **2**, 1442.
20 Kaska, W.C., Nemeh, S., Shirazi, A. and Potuznik, S. (1988) *Organometallics*, **7**, 13.
21 See for instance (a) Peter, M.G. (1989) *Angewandte Chemie-International Edition*, **28**, 555. (b) Wan, P., Barker, B., Diao, L., Fischer, M., Shi, Y. and Yang, C. (1996) *Canadian Journal of Chemistry*, **74**, 465. (c) Sugumaran, M. (1987) *Bioorganic Chemistry*, **15**, 194.
22 Vigalok, A., Shimon, L.J.W. and Milstein, D. (1998) *Journal of the American Chemical Society*, **120**, 477.
23 Vigalok, A., Rybtchinski, B., Shimon, L.J.W., Ben-David, Y. and Milstein, D. (1999) *Organometallics*, **18**, 895.
24 Kossoy, E., Iron, M.A., Rybtchinski, B., Ben-David, Y., Shimon, L.J.W., Konstantinovski, L., Martin, J.M.L. and Milstein, D. (2005) *Chemistry – A European Journal*, **11**, 2319.
25 Salem, H., Ben-David, Y., Shimon, L.J.W. and Milstein, D. (2006) *Organometallics*, **25**, 2292.
26 French, C.M. and Milstein, D. (2006) *Journal of the American Chemical Society*, **128**, 12434.
27 French, C.M., Shimon, L.J.W. and Milstein, D. (2006) *Helvetica Chimica Acta*, **89**, 1730.
28 Morales-Morales D. and Jensen C.M. (eds) (2007) *The Chemistry of Pincer Compounds*, Elsevier, Amsterdam, The Netherlands, Chapter 9, p.151.
29 Miyazaki, F., Yamaguchi, K. and Shibasaki, M. (1999) *Tetrahedron Letters*, **40**, 7379.
30 Rubio, M., Suárez, A., Del Ŕo, D., Galindo, A., Álvarez, E. and Pizzano, A. (2007) *Journal of the Chemical Society, Dalton Transactions*, 407.
31 Morales-Morales, D., Redón, R., Wang, Z., Lee, D.W., Yung, C., Magnuson, K. and Jensen, C.M. (2001) *Canadian Journal of Chemistry*, **79**, 823.
32 Gu, X.-Q., Chen, W., Morales-Morales, D. and Jensen, C.M. (2002) *Journal of Molecular Catalysis A: Chemical*, **189**, 119.
33 Kanzelberger, M., Singh, B., Czerw, M., Krogh-Jespersen, K. and Golman, A.S. (2000) *Journal of the American Chemical Society*, **122**, 11017.
34 Mohammad, H.A.Y., Grimm, J.C., Eichele, K., Mack, H.-G., Speiser, B., Novak, F., Quintanilla, M.G., Kaska, W.C. and Mayer, H.A. (2002) *Organometallics*, **21**, 5775.
35 Lee, D.W., Jensen, C.M. and Morales-Morales, D. (2003) *Organometallics*, **22**, 4744.
36 Ben-Ari, E., Gandelman, M., Rozenberg, H., Shimon, L.J.W. and Milstein, D. (2003) *Journal of the American Chemical Society*, **125**, 4714.
37 Kanzelberger, M., Zhang, X., Emge, T.J., Goldman, A.S., Zhao, J., Incarvito, C. and Hartwig, J.F. (2003) *Journal of the American Chemical Society*, **125**, 13644.

38 Göttker-Schnetmann, I., White, P.S. and Brookhart, M. (2004) *Organometallics*, **23**, 1766.

39 See for instance Ono N. (ed.) (2001) *The Nitro Group in Organic Synthesis*, Wiley-VCH Verlag GmbH, New York, p.392.

40 Zhang, X., Emge, T.J., Ghosh, R., Krogh-Jespersen, K. and Goldman, A.S. (2006) *Organometallics*, **25**, 1303.

41 Sykes, A.C., White, P. and Brookhart, M. (2006) *Organometallics*, **25**, 1664.

42 Kloek, S.M., Heinekey, D.M. and Goldberg, K.I. (2006) *Organometallics*, **25**, 3007.

43 Ben-Ari, E., Cohen, R., Gandelman, M., Shimon, L.J.W., Martin, J.M.L. and Milstein, D. (2006) *Organometallics*, **25**, 3190.

44 Kuklin, S.A., Sheloumov, A.M., Dolgushin, F. M., Ezernitskaya, M.G., Peregudov, A.S., Petrovskii, P.V. and Koridze, A.A. (2006) *Organometallics*, **25**, 5466.

45 Trovitch, R.J., Lobkovsky, E. and Chirick, P. (2006) *Inorganic Chemistry*, **45**, 7252.

46 Benito-Garagorri, D., Wiedermann, J., Pollak, M., Mereiter, K. and Kirchner, K. (2007) *Organometallics*, **26**, 217.

47 Jia, G., Lee, H.M. and Williams, I.D. (1997) *Journal of Organometallic Chemistry*, **534**, 173.

48 Ashkenazi, N., Vigalok, A., Parthiban, S., Ben-David, Y., Shimon, L.J.W., Martin, J.M.L. and Milstein, D. (2000) *Journal of the American Chemical Society*, **122**, 8797.

49 Van der Boom, M.E., Iron, M.A., Atasoylu, O., Shimon, L.J.W., Rozenberg, H., Ben-David, Y., Konstantinovski, L., Martin, J.M. L. and Milstein, D. (2004) *Inorganica Chimica Acta*, **357**, 1854.

50 Amoroso, D., Jabri, A., Yap, G.P.A., Gusev, D.G., dos Santos, E.N. and Fogg, D.E. (2004) *Organometallics*, **23**, 4047.

51 Zhang, J., Gandelman, M., Shimon, L.J.W. and Milstein, D. (2007) *Journal of the Chemical Society, Dalton Transactions*, 107.

52 Gauvin, R.M., Rozenberg, H., Shimon, L.J. W. and Milstein, D. (2001) *Organometallics*, **20**, 1719.

3
Enantioselective Carbonylation Reactions

Carmen Claver, Cyril Godard, Aurora Ruiz, Oscar Pàmies, Montserrat Diéguez

3.1
Introduction

The main reactions in carbonylation are Rh-catalyzed hydroformylation and Pd-catalyzed hydroxy- and alkoxycarbonylation. In these reactions, olefins are converted into aldehydes, acids, or esters, using carbon monoxide and H_2, H_2O, or CH_3OH, respectively, in the presence of a metal complex catalyst, Rh or Pd. Although many similarities can be found for these reactions, the nature of the catalytic precursors based on Rh for hydroformylation or Pd for hydroxycarbonylation and alkoxycarbonylation, as well as the ligands required, determine important differences between both processes. Furthermore, with regard to enantioselective carbonylation, the "state of the art" is very different for both types of reactions. For these reasons, this chapter, which focuses mainly on the latest achievements and developments of the enantioselective carbonylation, includes two distinguished parts: rhodium-catalyzed asymmetric hydroformylation and palladium-catalyzed hydroxy- and alkoxycarbonylation. Each part focuses on the importance of the products obtained, mechanistic aspects, and catalytic systems that constitute modern carbonylation methods in enantioselective catalysts. When considering catalytic systems based on Rh or Pd, the type of ligand is the key for achieving adequate catalyst. Therefore, both parts of this chapter are classified according to the ligands that provide the best results in the activity and/or selectivity in carbonylation reaction.

3.2
Rhodium-Catalyzed Asymmetric Hydroformylation

3.2.1
Introduction

The hydroformylation reaction in which olefins are converted into aldehydes is the homogeneous transition metal catalyzed reaction used today for largest volume production. This reaction has been extensively studied, and nowadays a number of

Modern Carbonylation Methods. Edited by László Kollár

ISBN: 978-3-527-31896-4

efficient catalysts allow controlling the regioselectivity of the reaction of terminal and internal alkenes. Although earlier carbonylations were based on cobalt, nowadays, rhodium is the preferred catalyst because it requires lower pressure and affords higher selectivity, including chemoselectivity and enantioselectivity [1,2]. In recent years, extensive research aimed at producing only linear aldehydes has provided impressive results. The application of phosphines with a wide bite angle in rhodium-catalyzed hydroformylation of terminal alkenes allows practically total control of the regioselectivity [3]. Branched selective hydroformylation, although less studied, constitutes a useful tool for organic synthesis [4,5]. Furthermore, the asymmetric hydroformylation of olefins would be one of the most straightforward synthetic methods for the preparation of optically active aldehydes that are versatile intermediates for the synthesis of biologically active compounds as well as for other synthetic transformations [6,7]. Nonetheless, despite these advantages, hydroformylation has not been much used in the synthesis of fine chemicals, with the exception of applications in vitamins and flavors and fragrances [8]. This may be because of the difficulty in controlling the chemo-, regio-, and enantioselectivity simultaneously along the reaction. To date, much efforts in this field have been concentrated on the hydroformylation of vinyl arenes as a route to obtain enantiomerically enriched 2-aryl propionic acids, the profen class of nonstereoidal drugs. Interestingly, successful asymmetric hydroformylation has been obtained in the conversion of vinyl arenes [2]. The application of asymmetric hydroformylation to other substrates has been scarcely studied despite its potential for producing fine chemicals. In recent years, however, the interest of researchers in asymmetric hydroformylation has increased significantly, mainly in connection with the hydroformylation of vinyl acetate, unsaturated nitriles, or heterocyclic substrates because enantiopure diols and molecules of biological interest [2] can be easily obtained by these processes. In the early 1990s, several reports were published that described the state of the art in hydroformylation with rhodium and platinum systems [9–11]. After the discovery of the high ee using the rhodium/diphosphite and rhodium/phosphine-phosphite, with total conversion into aldehydes and high regioselectivities, rhodium systems became the catalysts of choice for asymmetric hydroformylation [1,2].

3.2.2
Catalytic Cycle and Mechanistic Highlights

The catalytic cycle for rhodium hydroformylation has been extensively studied mainly for the $RhH(CO)(PPh_3)_3$ catalysis. A general proposal [2,12] including the steps of the reaction is shown in Scheme 3.1.

For PPh_3 as the ligand, a common starting complex is $RhH(PPh_3)_3CO$, complex **3**, which under 1 bar of carbon monoxide forms complexes **2a** and **2b**, containing two phosphine ligands in equatorial positions or one in an apical position and the other in an equatorial position. Dissociation of either equatorial L or equatorial CO from **3** or **2** leads to the square–planar intermediates **4c** and **4t** (never observed) that have phosphines in *cis*- or *trans* configurations, respectively. Complexes **4** associate with

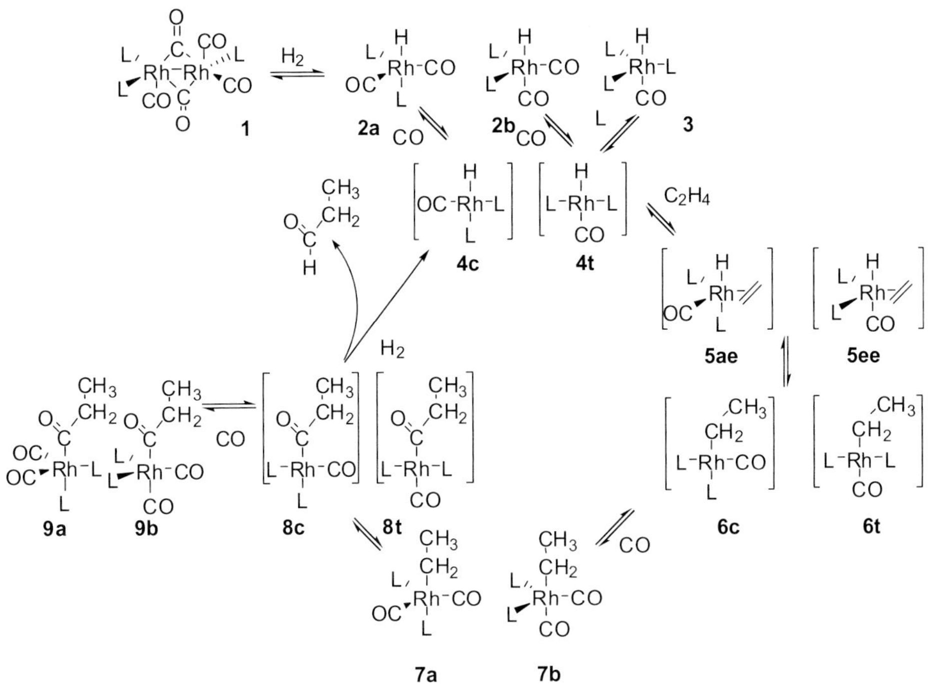

Scheme 3.1 General mechanism for rhodium hydroformylation of ethene.

ethene to give complexes **5**, again in two isomeric forms axial–equatorial and equatorial–equatorial, having a hydride in an apical position and an ethene coordinating in the equatorial plane. Complexes **5** undergo migratory insertion to give square–planar alkyl complexes **6c** and **6t**, which are in *cis-* or *trans* configuration, respectively. Complexes **6** can undergo β-hydride elimination, thus leading to isomerization when higher alkenes are used, or they can react with CO to form trigonal bipyramidal complexes **7**. Thus, under low pressure of CO more isomerization may be expected. At low temperatures (<70 °C) and a sufficiently high pressure of CO (>10 bar), the insertion reaction is usually irreversible and thus the regioselectivity of hydroformylation of 1-alkenes is determined at this point. Complexes **7** undergo the second migratory insertion in this scheme to form acyl complexes **8**. Complexes **8** can react either with CO to give saturated rhodium–acyl intermediates **9**, which have been observed spectroscopically, or with H_2 to give the aldehyde product and unsaturated intermediates **4**. At low hydrogen pressures and high rhodium concentrations, formation of dirhodium species such as **1** becomes significant. The nature of the phosphorus ligand as well as the conditions of the reaction has a profound effect on each individual step of the catalytic cycle, determining the activity and selectivity of the reaction. The careful choice of the ligand can allow the formation of the desired aldehyde with high regioselectivity. The kinetics of the hydroformylation reaction has also been reported to be different, depending on the nature of the catalyst [2]. The high-pressure NMR

(HP-NMR) and *in situ* high-pressure infrared (HPIR) characterizations of intermediates on asymmetric hydroformylation catalyzed by rhodium phosphine-phosphite, diphosphites and diphosphines allowed to elucidate mechanistic aspects of the reaction, especially the steps determining the reaction rate and the formation of the intermediates of the reaction [9,12].

3.2.3 Diphosphite Ligands

The first report on asymmetric hydroformylation using diphosphite ligands revealed no asymmetric induction [13]. In 1992, Takaya and coworkers published the results of the asymmetric hydroformylation of vinyl acetate (ee's up to 50%) using chiral diphosphites with binaphthol backbone [14]. In the same year, there was an important breakthrough when Babin and Whiteker at Union Carbide patented the asymmetric hydroformylation of various alkenes with ee's up to 90%, using bulky diphosphites **10a–c** derived from homochiral (2*R*,4*R*)-pentane-2,4-diol (Figure 3.1 and Table 3.1) [15]. Their results showed that (a) the presence of bulky substituents at the *ortho* positions of the biphenyl moieties is necessary for good regio- and enantioselectivities and (b) the presence of methoxy substituents at the *para* positions of the biphenyl moieties always produced better enantioselectivities than those observed for the corresponding *tert*-butyl-substituted analogues.

Inspired by the excellent early results obtained with the Union Carbide type ligands **10a–c**, other research groups have studied different modifications in these types of

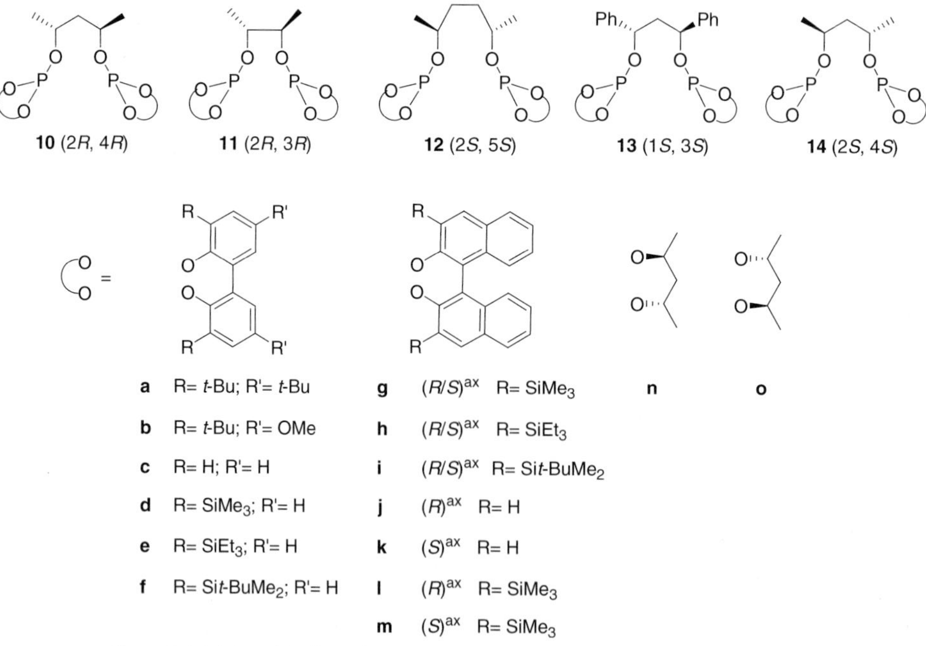

Figure 3.1 Diphosphite ligands **10–14**.

Table 3.1 Rh-catalyzed asymmetric hydroformylation of styrene using diphosphites **10a–c**.[a]

Ph⁀= —[H_2/CO; $[Rh(acac)(CO)_2]$ / **1**]→ Ph–CH(CH$_3$)–CHO (2-PP) + Ph–CH$_2$CH$_2$–CHO (3-PP)

Entry	Ligand	*T* (°C)	% 2-PP[b]	% ee[c]
1	**10a**	70	95	44
2	**10b**	70	93	61
3	**10c**	70	82	14
4	**10b**	25	98	90

[a] $[Rh(acac)(CO)_2]$ = 0.0135 mmol; ligand/Rh = 4; substrate/Rh = 1000; toluene = 15 ml; $P_{H_2O/CO}$ = 130 psi.
[b] Regioselectivity for 2-phenylpropanal.
[c] Enantiomeric excess.

ligand (Figure 3.1) [16,17]. In this context, they studied the influence of the bridge length, several phosphite moieties and backbone substituents, and the possibility of a cooperative effect between chiral centers on the performance of the catalysts.

The influence of the bridge length was studied with diphosphite ligands based on (2*R*,4*R*)-pentane-2,4-diol (ligands **10a** and **10b**), (2*R*,3*R*)-butane-2,3-diol (ligands **11a** and **11b**), and (2*S*,5*S*)-hexane-2,5-diol (ligands **12a** and **12b**). In general, ligands **10**, which have three carbon atoms in the bridge, provided higher enantioselectivities than ligands **11** and **12**, which have two and four carbon atoms in the bridge, respectively [17].

The effect of different phosphite moieties was studied with ligands **10d–o**. In general, sterically hindered phosphite moieties are necessary for high enantioselectivities [16,17]. Thus, ligands **10j**, **10k**, **10n**, and **10o** show low asymmetric induction (ee's up to 20%). Also, the results of using ligands **10d–i** indicated that varying the *ortho* substituents on the biphenyl and binaphthyl phosphite moieties has much effect on the asymmetric induction. The optimal steric bulk at the *ortho* positions therefore seems to be obtained with trimethylsilyl substitutents (i.e., ligand **10d** provided ee's up to 87% at 20 bar of syngas and 25 °C). This result is similar to the previously reported best result obtained with **10b**.

The influence of the backbone substituent was studied by comparing ligands **10a** and **10b** with ligands **13a** and **13b** (Figure 3.1). Surprisingly, the latter ligands, which have a more sterically hindered phenyl group, provided lower enantioselectivities than that of ligands **1** [17].

A possible cooperative effect between the different chiral centers was studied using ligands **10l–o** and **14l–o**. Initially, van Leeuwen and coworkers studied the cooperative effect between the chiral ligand bridge and the axially chiral binaphthyl phosphite moieties by comparing ligands **10l**, **10m**, **14l**, and **14m**. The hydroformylation results clearly indicate a cooperative effect that leads to a matched combination for ligand **10m** with (S^{ax}, 2*R*, 4*R*, S^{ax}) configurations (ee's up to 86%) (Table 3.2) [16]. Later, Bakos and coworkers studied ligands **10n**, **10o**, **14n**, and **14o** and found a similar

Table 3.2 Rh-catalyzed asymmetric hydroformylation of styrene using diphosphites **10l**, **10m**, **14l**, and **14m**.[a]

Entry	Ligand	TOF[b]	% 2-PP[c]	% ee[d]
1	**10l**	28	95	38 (*S*)
2	**10m**	17	88	69 (*S*)
3	**14l**	4	91	23 (*S*)
4	**14m**	45	94	40 (*R*)
5[e]	**10m**	11	92	86 (*S*)

[a] $[Rh(acac)(CO)_2] = 0.02$ mmol; ligand/Rh = 2.2; substrate/Rh = 1000; toluene = 20 ml; $P_{H_2O/CO} = 10$ bar; $T = 25\,°C$.
[b] TOF in mol styrene × mol $Rh^{-1} \times h^{-1}$ determined after 1 h reaction time.
[c] Regioselectivity for 2-phenylpropanal.
[d] Enantiomeric excess.
[e] $T = 15\,°C$.

cooperative effect between the chiral ligand bridge and the chiral phosphite moiety [17]. However, the matched combination afforded poorer results (ee's up to 17%) than those obtained with bulky biaryl phosphite ligands **10b**, **10d**, and **10m** (ee's up to 90%) because of the lower steric bulk of the Bakos' ligands (see above).

Interestingly, the hydroformylation results obtained with ligands **10b** and **10d**, which have conformationally flexible axially chiral biphenyl moieties, are similar to those obtained with ligand **10m**, which has conformationally rigid binaphthyl moieties. This indicates that diphosphite ligands containing the conformationally flexible axially chiral biphenyl moieties predominantly exist as single atropoisomer in the $[HRh(CO)_2(\text{diphosphite})]$ complexes when the right bulky substituents at *ortho* positions are present (see below) [16]. It is therefore not necessary to use expensive conformationally rigid binaphthyl moieties to reduce the degrees of freedom of the system.

To investigate whether a relationship exists between the solution structures of the hydridorhodium diphosphite species $[HRh(CO)_2(\text{diphosphite})]$ [18] and the catalytic performance, van Leeuwen and coworkers extensively studied the rhodium–diphosphite complexes formed under hydroformylation conditions by HP-NMR techniques. It is well known that these complexes have a trigonal bipyramidal (TBP) structure. Two isomeric structures of these complexes are possible, one containing the diphosphite coordinated in a bis-equatorial (ee) manner and one containing the diphosphite in an equatorial–axial (ea) manner (Figure 3.2) [1].

van Leeuwen's studies using diphosphite ligands **10** and **14** indicated that the stability and catalytic performance of the $[HRh(CO)_2(\text{diphosphite})]$ species depend strongly on the configuration of the 2,4-pentanediol ligand backbone and the chiral biaryl phosphite moieties. Thus, for example, ligands **10b**, **10d**, and **10m**, which form well-defined stable ee complexes, lead to good enantiomeric excesses, whereas enantioselectivities were low with ligands **10l** and **14m**, which form unidentified mixtures of complexes and ligand decomposition [16,19].

Another successful family of ligands was the sugar-based furanoside ligands **15–20** (Figure 3.3) in the Rh-catalyzed hydroformylation of vinyl arenes and dihydrofurans

Figure 3.2 ee and ea coordination modes of diphosphite ligands in the $[HRh(CO)_2(diphosphite)]$ complexes.

[20]. The modular construction of these ligands allowed fine tuning (a) the different configurations of the carbohydrate backbone and (b) the steric and electronic properties of the diphosphite substituents. These ligands showed both excellent enantioselectivities (up to 93%) and regioselectivities (up to 98.8%) under mild conditions (Tables 3.3 and 3.4).

The following were observed when using biphenyl-based ligands **15–20a–d** (Table 3.3) in Rh-catalyzed asymmetric hydrogenation of vinyl arenes:

(a) The presence of a methyl substituent at C-5 is necessary for high enantioselectivities and has a positive effect on rate (entries 3–12 versus 1 and 2).

(b) The level of enantioselectivity is influenced by a cooperative effect between stereocenters C-3 and C-5. Accordingly, ligands **18** and **19** provide better enantioselectivities than ligands **17** and **20** (Table 3.3; entries 6 and 8 versus 4 and 10).

(c) The absolute configuration of the product is governed by the configuration at the stereogenic center C-3. Accordingly, ligands **15, 18,** and **20**, with *S*-configuration at C-3, gave (*S*)-2-phenylpropanal (Table 3.3; entries 1, 5, 6, 9, 10, and 12), whereas

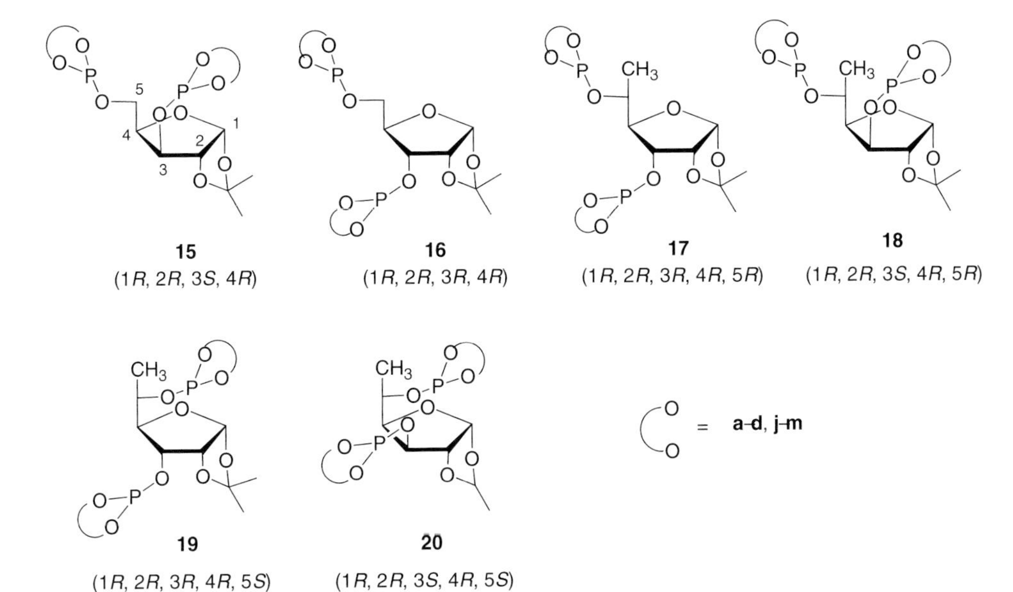

Figure 3.3 Furanoside diphosphite ligands **15–20**.

Table 3.3 Rh-catalyzed asymmetric hydroformylation of styrene using diphosphites **15–20a–d**.[a]

Entry	Ligand	TOF[b]	% 2-PP[c]	% ee
1	**15b**	5	97	60 (*S*)
2	**16b**	5	97	61 (*R*)
3	**17a**	14	97.1	46 (*R*)
4	**17b**	13	97.2	58 (*R*)
5	**18a**	19	98.4	74 (*S*)
6	**18b**	18	98.6	90 (*S*)
7	**19a**	16	98.7	76 (*R*)
8	**19b**	17	98.3	89 (*R*)
9	**20a**	15	97.4	52 (*S*)
10	**20b**	12	97.6	64 (*S*)
11	**17d**	10	98.1	62 (*R*)
12	**18d**	11	98.8	93 (*S*)

[a] $[Rh(acac)(CO)_2] = 0.0135$ mmol; ligand/Rh = 1.1; substrate/Rh = 1000; toluene = 15 ml; $P_{H_2O/CO} = 10$ bar; $T = 20\,°C$; $P_{CO}/P_{H_2} = 0.5$.
[b] TOF in mol styrene × mol $Rh^{-1} \times h^{-1}$ determined after 1 h reaction time.
[c] Regioselectivity for 2-phenylpropanal.

ligands **16**, **17**, and **19**, with *R*-configuration at C-3, gave (*R*)-2-phenylpropanal (Table 3.3; entries 2–4, 7, 8, and 11).

(d) As observed for the previously mentioned ligands **10a–d**, the substituents in the biaryl phosphite moieties are influenced. Thus, ligands **18b,d** and **19b,d**, with either methoxy substituents or trimethylsilyl groups, always produced the best enantioselectivities.

Furthermore, when using the binaphthyl-based ligands **15–20j–m** (Table 3.4), the results suggested that the absolute configuration of the product outcome is controlled by the configuration of the biaryl moieties. This suggests that the configuration of fluxional biphenyl moieties in ligands **17–20a–d** is controlled by the configuration of the stereogenic center C-3. The results also indicate a cooperative effect between the

Table 3.4 Rh-catalyzed asymmetric hydroformylation of styrene using diphosphites **15–20j–m**.[a]

Entry	Ligand	TOF[b]	% 2-PP[c]	% ee
1	**15j**	126	80	44 (*R*)
2	**15k**	85	83	37 (*S*)
3	**17j**	178	86	20 (*R*)
4	**17k**	158	84	5 (*S*)
5	**18j**	165	85	60 (*R*)
6	**18k**	153	85	25 (*S*)
7	**18m**	149	84	68 (*S*)

[a] $[Rh(acac)(CO)_2] = 0.0135$ mmol; ligand/Rh = 1.1; substrate/Rh = 1000; toluene = 15 ml; $P_{H_2/CO} = 10$ bar; $T = 40\,°C$; $P_{CO}/P_{H_2} = 0.5$.
[b] TOF in mol styrene × mol $Rh^{-1} \times h^{-1}$ determined after 1 h reaction time.
[c] Regioselectivity for 2-phenylpropanal.

chiral sugar backbone stereocenters (C-3 and C-5) and the axial chiral binaphthyl phosphite moieties. This cooperative effect, together with the previously observed cooperative effect between the backbone stereocenters C-3 and C-5, controls the enantioselectivity.

In summary, both *S*- and *R*-enantiomers of the product can be obtained with excellent regio- and enantioselectivity. These results are among the best ever reported for the asymmetric hydroformylation of vinyl arenes [1].

The characterization of rhodium complexes formed under hydroformylation conditions by NMR techniques and *in situ* IR spectroscopy showed that there is a relationship between the structure of $[HRh(CO)_2(P\text{–}P)]$ (P–P = **15**–**20**) species and their enantiodiscriminating performance. In general, enantioselectivities were highest with ligands with a strong ee coordination preference, while an equilibrium of species with ee and ea coordination modes considerably reduced the ee's [20].

Ligand **18a** was also successfully applied in the Rh-catalyzed asymmetric hydroformylation of 2,5-dihydrofuran and 2,3-dihydrofuran. Good enantioselectivities (up to 75% ee) and excellent regioselectivities (up to 99%) were achieved. Note that both enantiomers of tetrahydrofuran-3-carbaldehyde can be obtained using this ligand by simple substrate change from 2,5-dihydrofuran to 2,3-dihydrofuran [21].

Recent studies showed that the chiral phosphite kelliphite **21** (Figure 3.4) provided high enantioselectivities and excellent regioselectivity toward debranched isomer in vinyl acetate (ee's up to 88%) and allyl cyanide (ee's up to 78%) hydroformylation [22].

Through all these years, several authors have developed new diphosphite ligands with binaphthyl, spiro, pyranoside, and macrocyclic backbones (Figure 3.5) for asymmetric hydroformylation of vinyl arenes with low to moderate success (ee's from 36 to 76%) [23].

3.2.4
Phosphite-Phosphine Ligands

Asymmetric hydroformylation using phosphite-phosphine ligands was first reported by Takaya and coworkers [24]. With the aim of combining the effectiveness of the BINOL chemistry for asymmetric catalysis and that of the phosphite moiety for asymmetric hydroformylation, they developed the (*R*,*S*)-BINAPHOS ligand **30**, which turned out to be a very efficient ligand (Figure 3.6).

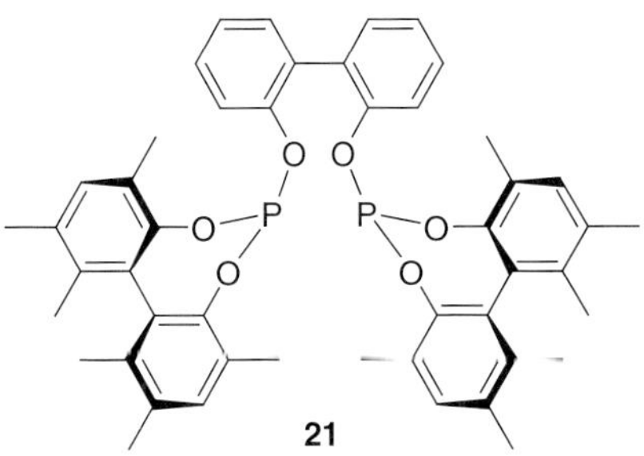

Figure 3.4 Diphosphite ligand **21**.

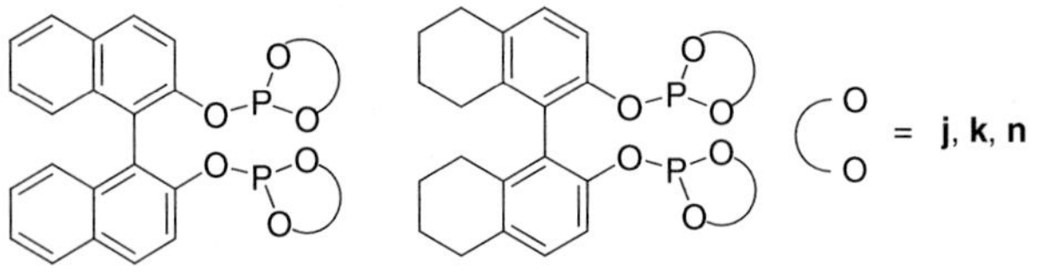

22 (*S*)^{ax} **23** (*R*)^{ax} **24** (*S*)^{ax} **25** (*R*)^{ax} (O–O = j, k, n) (ee's up to 37%)

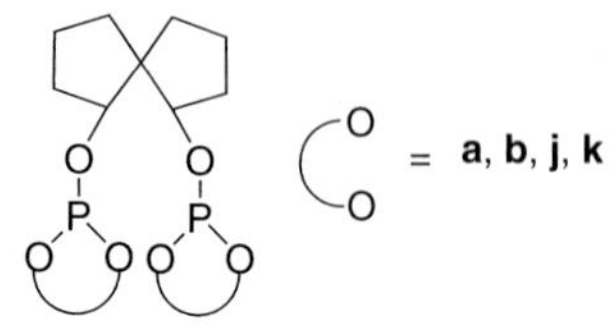

26 (1*S*, 5*S*, 6*R*) **27** (1*R*, 5*R*, 6*S*) (O–O = a, b, j, k) (ee's up to 70%)

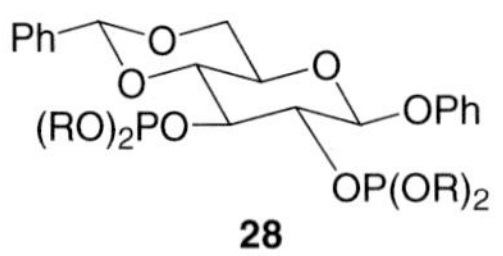

28 (ee's up to 36%)

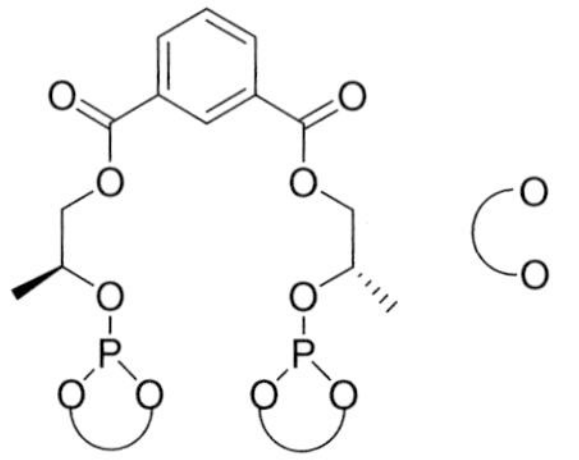

29 (O–O = j, k) (ee's up to 76%)

Figure 3.5 Miscellaneous diphosphite ligands **22–29**.

In the last few years, a wide range of structural variations has been reported. In this context, in 1997, Nozaki *et al.* used ligands **30–32** and found that enantioselectivity depends on the configuration of binaphthyl bridge, whereas the enantiomeric excess depends strongly on the configuration of both binaphthyl moieties (Figure 3.7) [25]. Enantioselectivity is therefore higher when the configurations of the two binaphthyl moieties are opposite (i.e., diastereoisomers *R,S* or *S,R*). Similar trends were observed with ligands **33** and **34**, which have a chiral biphenyl bridge.

Furthermore, it should be noted that the use of the diastereoisomers (*R,S*)- or (*S,R*)-BINAPHOS provide practically identical ee value, although inverted configuration (94%(S) and 95%(R)), as expected.

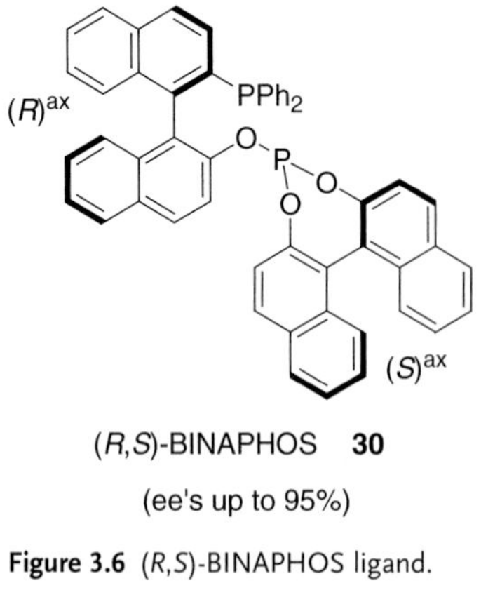

(*R*,*S*)-BINAPHOS **30** (ee's up to 95%)

Figure 3.6 (*R*,*S*)-BINAPHOS ligand.

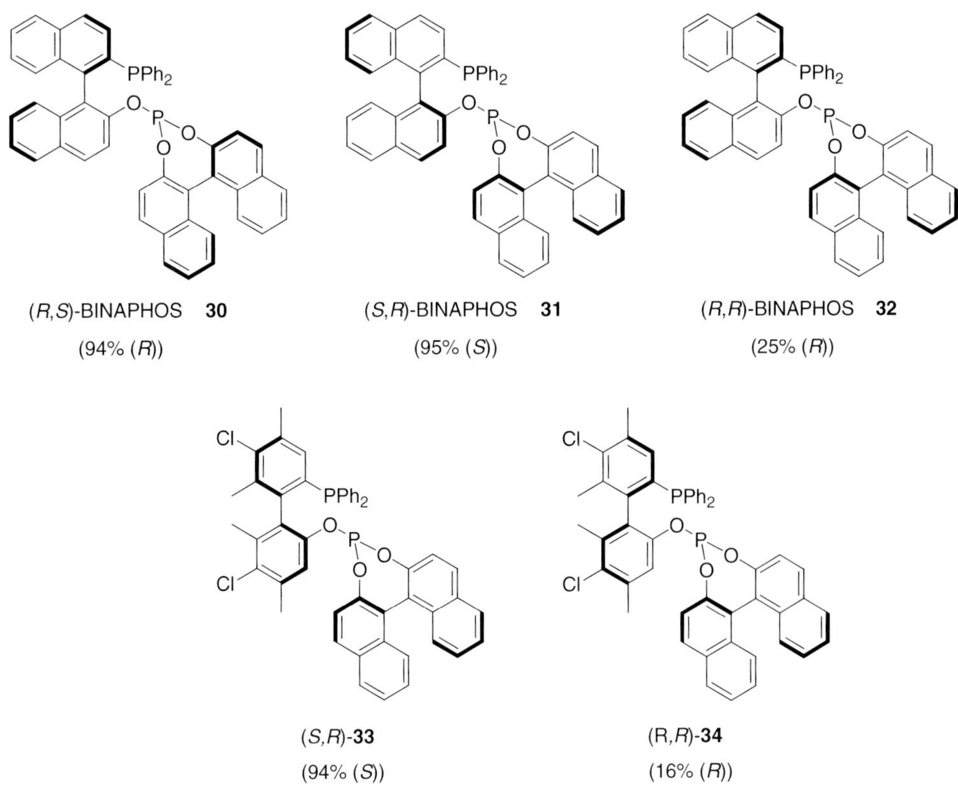

Figure 3.7 Rh-catalyzed asymmetric hydroformylation of styrene using ligands **31–34**. Enantioselectivities obtained at 100 bar of syngas and at 60 °C are shown in brackets.

To further understand the role of chirality at the bridge and axial chirality at the phosphite moiety in transferring the chiral information to the product outcome, ligands **35** and **36** were studied (Figure 3.8) [25]. Ligand **35**, which has an (*R*)-binaphthyl in the bridge, provides an ee of 83% (*R*). This value is close to that of (*R*,*S*)-BINAPHOS (94% (*R*) ee), suggesting that, in the formation of the Rh complex, the

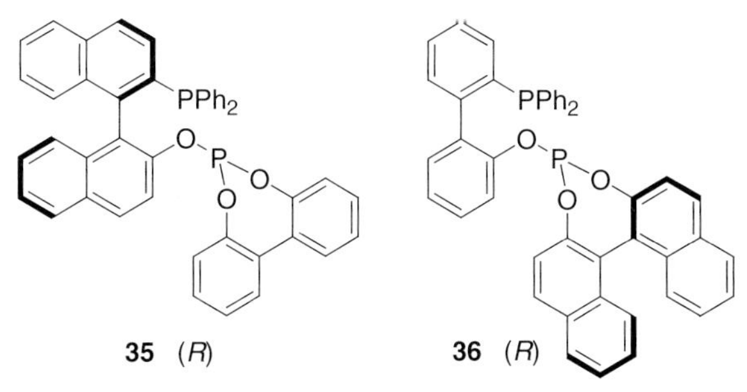

Figure 3.8 BINAPHOS-related ligands **35** and **36**.

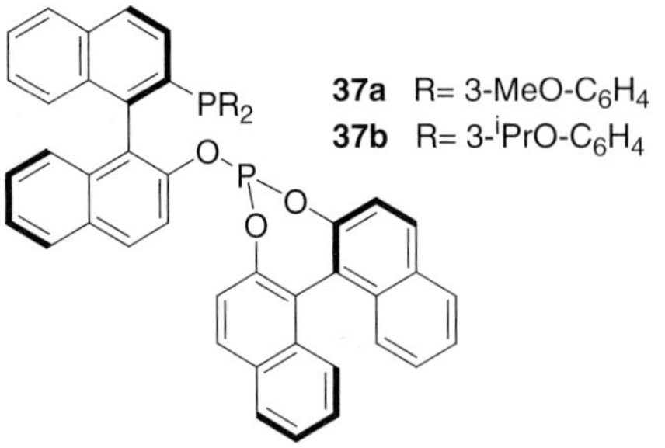

37a R= 3-MeO-C_6H_4
37b R= 3-^iPrO-C_6H_4

Figure 3.9 Phosphine-phosphite ligands **37**.

binaphthyl bridge controls the conformation of the biphenyl phosphite moiety. Likewise, ligand **36** provides an ee of 69% (*S*), suggesting that the binaphthyl phosphite moiety also controls the conformation of the biphenyl bridge upon coordination with rhodium. However, the control by the binaphthyl bridge is more efficient than that of the binaphthyl phosphite moiety.

The effect of several substituents in the phosphine moiety has been extensively studied by Nozaki and coworkers (Figure 3.9). Their results indicate that both regio- and enantioselectivity can be increased by suitable choice of the aryl phosphine group. The best combinations of regio- and enantioselectivity were therefore obtained with ligands **37a** and **37b** [26].

The characterization of the rhodium complexes formed under hydroformylation conditions by NMR techniques and *in situ* IR spectroscopy showed that there is a relationship between the structure of the [$HRh(CO)_2$(BINAPHOS)] species and their enantiodiscriminating performance. Thus, (*R,S*) and (*S,R*)-BINAPHOS ligands show high ea coordination preference with the phosphite moiety at the axial position. Meanwhile, the characterization of the (*R,R*)-BINAPHOS and (*S,S*)-BINAPHOS ligands suggests either a structural deviation of the monohydride complexes from an ideal TBP structure or an equilibrium between isomers [25,27].

Highly cross-linked polymer-supported BINAPHOS ligands were effective for the hydroformylation of styrene and other functionalized olefins (ee's up to 89%). Recovery and reuse of the catalyst was possible at low stirring conditions [28].

Perfluoroalkyl-substituted BINAPHOS ligand **37c** was also developed for asymmetric hydroformylation of vinyl arenes in $scCO_2$. With this ligand, high regio- and enantioselectivity (ee's up to 93.6%) were achieved without the need of hazardous organic solvents [29].

Thus, BINAPHOS is the most important ligand for asymmetric hydroformylation. Thus, this ligand provides higher enantioselectivities than diphosphine and diphosphite ligands for a wide variety of both functionalized and internal alkenes (Table 3.5).

Inspired by the excellent results using the BINAPHOS ligands, new phosphine-phosphite ligands with different backbones have been developed in recent years (Figure 3.10). Unfortunately, their Rh hydroformylation provided low to moderate enantioselectivities (ee's from 20 to 62%) [36].

Table 3.5 Several substrates efficiently hydroformylated using Rh–BINAPHOS system.

Substrate	Product	% ee[a]	Reference
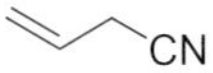 CN	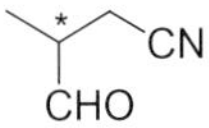 * CN CHO	66	[30]
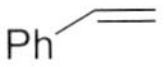 Ph	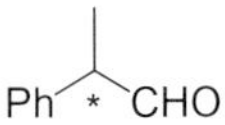 Ph * CHO	98.3	[26]
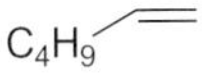 C_4H_9	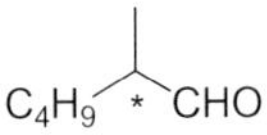 C_4H_9 * CHO	90	[26]
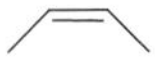	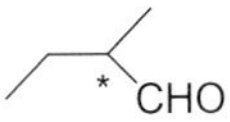 * CHO	89.9	[26]
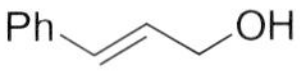 Ph OH	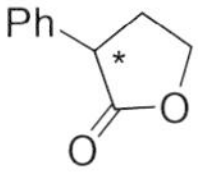 Ph * O O	88	[31]
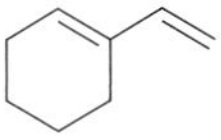	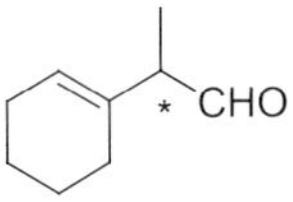 * CHO	97	[32]
 O	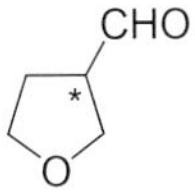 CHO * O	68	[33]
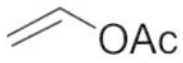 OAc	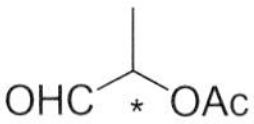 OHC * OAc	92	[25]
 S^tBu	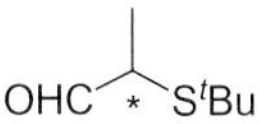 OHC * S^tBu	90	[34]
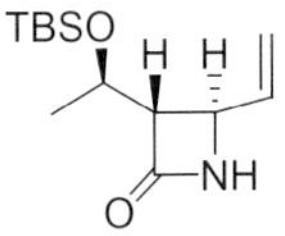 TBSO H H O NH	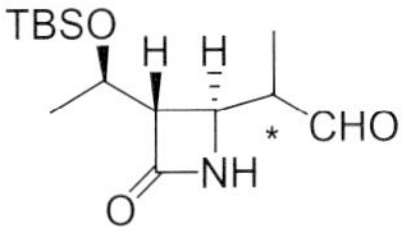 TBSO H H * CHO O NH	89[b]	[35]

[a]Enantiomeric excess.
[b]Diastereomeric excess.

3.2.5 Other Ligands

Other types of homodonor and heterodonor phosphorus ligands have also been developed for application in asymmetric hydroformylation.

In this context, several diphosphine ligands have been applied to this catalytic process. However, they do not achieve ee's as high as that obtained in the Rh system, with diphosphite or BINAPHOS. For example, in the Rh-catalyzed hydroformylation

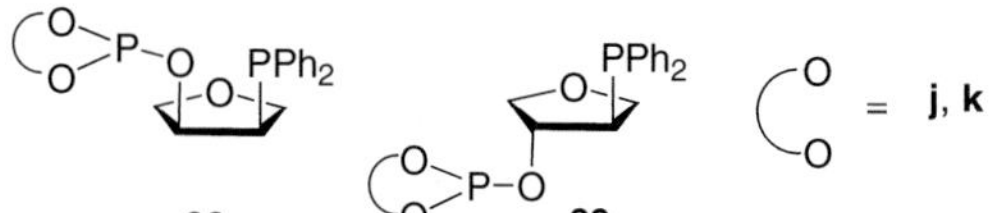

38 39 = j, k (ee's up to 44%)

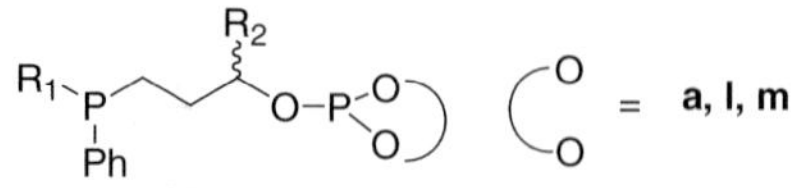

40 = a, l, m (ee's up to 62%)

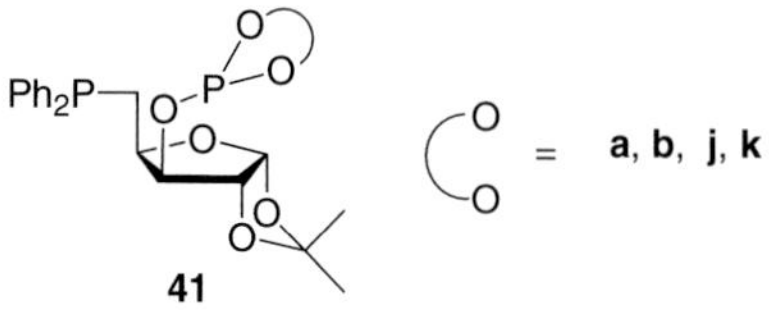

41 = a, b, j, k (ee's up to 49%)

42 = j, m (ee's up to 20%)

Figure 3.10 Miscellaneous phosphine-phosphite ligands.

of styrene, the Rh system containing ferrocenylethyldiphosphine derivative **43** (Figure 3.11) showed the highest value of ee's (60–76%), although the conversions were very low [37].

The bis-(diazaphospholidine) ESPHOS **44** provided high enantioselectivity (up to 89% ee) and regioselectivity (92–97% in the branched aldehyde) in the Rh-catalyzed hydroformylation of vinyl acetate (Figure 3.12) [38].

Recently, Zhang and coworkers developed a new phosphine-phosphoroamidite ligand based on BINAPHOS (Figure 3.13). This ligand provided excellent enantioselectivities (up to 99%) in the Rh-catalyzed hydroformylation of vinyl arenes and vinyl acetate [39].

Thus, several heterodonor ligands containing a phosphite moiety have been developed with little success (Figure 3.13) [40].

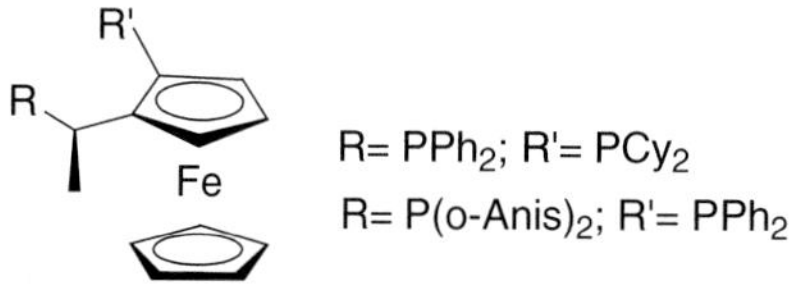

R= PPh_2; R'= PCy_2

R= $P(o\text{-}Anis)_2$; R'= PPh_2

43

Figure 3.11 Ferrocenylethyldiphosphine derivative **43**.

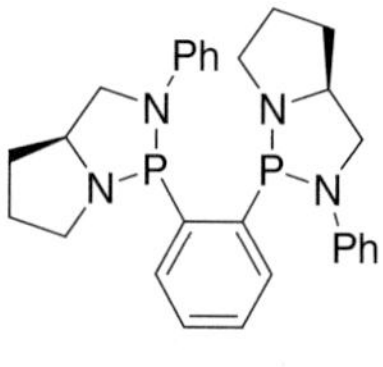

44 ESPHOS

Figure 3.12 Bis-(diazaphospholidine) ESPHOS ligand **44**.

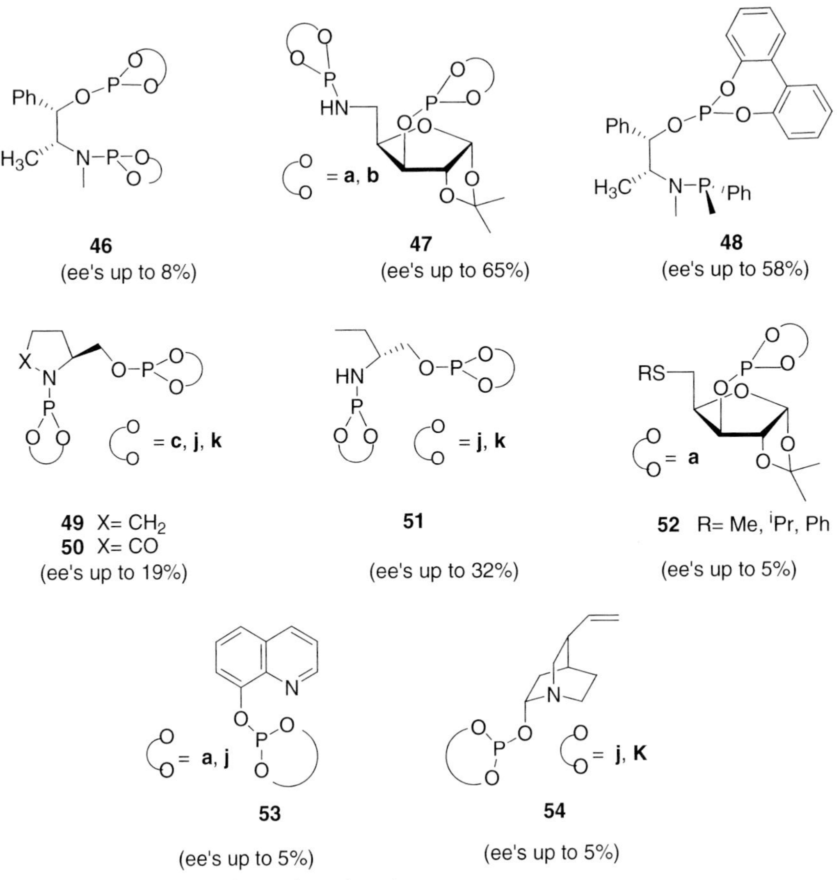

Figure 3.13 Miscellaneous heterodonor ligands **46–54**.

3.3 Pd-catalyzed Asymmetric Hydroxy- and Alkoxycarbonylation Reactions

Besides the term alkoxycarbonylation, hydroesterification and hydroalkoxycarbonylation have also been used in the literature to describe this reaction. Furthermore, more specific terms such as methoxycarbonylation can be found. In this chapter, the term alkoxycarbonylation will be used as a general term, and specific terms will be used to unambiguously define which reaction is meant.

3.3.1 Introduction

The asymmetric synthesis of carboxylic acids and their related esters is performed using olefins, carbon monoxide, and water or alcohols (represented as R^2OH in Scheme 3.2) in the presence of a chiral palladium catalyst. Considering that these chiral carboxylic acids are usually obtained from the oxidation of chiral aldehydes

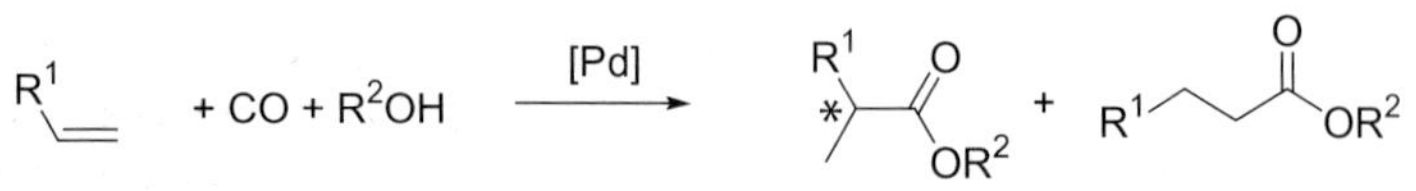

Scheme 3.2 General scheme for alkoxy- and hydroxycarbonylation of olefins.

synthesized by hydroformylation of vinyl arenes, the asymmetric hydroxycarbonylation and alkoxycarbonylation reactions attract much attention from both academic and industrial research groups.

However, these reactions are reportedly less successful than hydroformylation due to the difficulty in simultaneously obtaining both high regio- and enantioselectivities.

The alkoxycarbonylation of vinyl arenes is of particular relevance, as its products (2-aryl propionic acid and derivatives) are precursors for nonsteroidal anti-inflammatory drugs, particularly ibuprofen and naproxen [41].

3.3.2 Mechanism

For alkoxycarbonylation, two mechanisms have been suggested (Scheme 3.3) [42]. The catalytic cycle can start from either a hydridopalladium complex (Cycle A) or an alkoxycarbonyl–palladium species (Cycle B). In the hydride cycle, the first step is the insertion of alkene into the Pd–H bond to form an alkyl complex, followed by coordination and migratory insertion of CO to produce an acyl species. Alcoholysis of

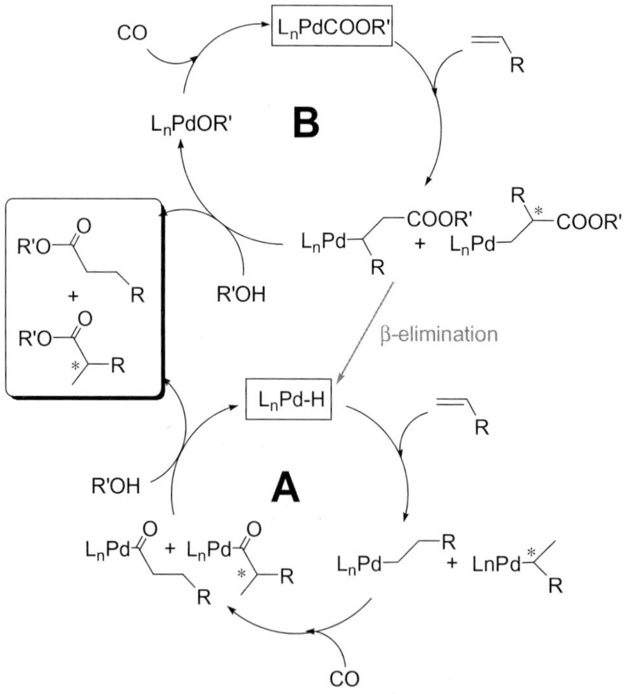

Scheme 3.3 Proposed mechanisms for the alkoxycarbonylation of olefins.

the Pd–acyl regenerates Pd–H complex and yields ester. In the alkoxycarbonyl cycle, alkene is inserted into the Pd–carbon bond of the alkoxycarbonyl–palladium complex, followed by alcoholysis to yield an alkoxy–palladium complex and an ester. Coordination and migratory insertion of CO then regenerate the initial alkoxycarbonyl–palladium complex. The production of the Pd–H species from complexes formed in Cycle B was also demonstrated to occur through the β-elimination of an unsaturated ester after alkene insertion. When the substrate is a vinyl arene, it should be taken into account that the branched alkyl intermediate can be stabilized through the formation of π-benzylic species with the two complexes in equilibrium [43].

The coexistence of these two cycles was suggested to be the cause of the regioselectivity of these reactions based on steric factors that would favor the linear insertion of styrene into a Pd-hydride bond rather than the branched insertion of styrene into a Pd-alkoxycarbonyl bond [44]. The regioselectivity of these reactions is of critical importance when it is to be performed in an asymmetric manner, as only the branched product contains a chiral center. The selective formation of the branched/linear product was shown to be closely related to the ligand properties and reaction conditions, especially in the presence of acids. The use of bidentate ligands generally leads to a greater amount of linear products, whereas catalytic systems bearing monodentate ligands usually favor the formation of branched products [45]. This difference in regioselectivity was suggested to be because of the ability of monodentate ligands to coordinate with the palladium center in a *cis*- or *trans* manner, although the alcoholysis step was later shown to require a *cis* coordination of phosphine ligands to be efficient [46]. The dissociation/association of phosphine ligands and counterions has also been suggested to influence regioselectivity [47]. In 1976, Sugi and Bando reported a study of ethoxycarbonylation of styrene using diphosphine palladium systems $PdCl_2(Ph_2P(CH_2)_nPPh_2)$ ($n = 1$–6 and 10) and showed that the length of the alkylic chain of the ligand dramatically influenced the regioselectivity [48]. When $n = 1, 6$, and 10, the branched ester was preferentially produced, as in the case of monophosphines. When $n = 2$ (dppe), no conversion was found, whereas the use of ligands with $n = 3, 4$, or 5 favored the production of the linear ester. Later, van Leeuwen and coworkers also demonstrated that the presence of electron-withdrawing substituents on diphosphine ligands can invert the regioselectivity of the methoxycarbonylation of styrene in favor of the branched ester [49]. Recent advances on the methanolysis step using DFT calculation methods have been reported [50].

Despite the growing number of mechanistic studies, the cause of regioselectivity of these reactions is still not established [43].

3.3.3 Bidentate Diphosphines

Asymmetric hydroxycarbonylation of olefins was first reported in 1973 using $PdCl_2$ and (−)-DIOP [51]. Later, catalysts containing bidentate diphosphine ligands had been frequently used for the asymmetric alkoxycarbonylation of alkenes, but these generally afforded low regioselectivity to the branched product [52]. In 1997, a $PdCl_2$–$CuCl_2$–chiral diphosphine **55** (DPPI) (Figure 3.14) was reported to achieve 98% ee

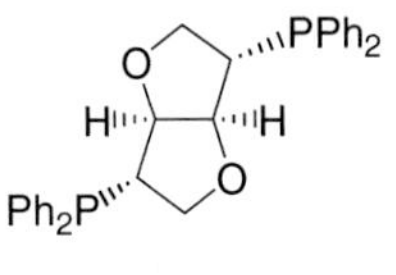

Figure 3.14 Chiral diphosphine **55** used by Zhou in the methoxycarbonylation of styrene.

and 99% regioselectivity to the branched ester, for methoxycarbonylation of styrene [53]. The reaction was performed at 80 °C under 50 atm of CO. However, for some reasons, no further development was described later on this catalytic system.

The use of $Pd(OAc)_2$–DPPI–*p*-TsOH catalytic system was also reported by the same authors for the methoxycarbonylation of norbornene, achieving 92% ee under 50 atm of CO at 120 °C [54]. The use of several chiral diphosphine was reported in the hydroxycarbonylation of styrene, but only moderate regioselectivity (up to about 30%) and ee's up to 11% were achieved [55].

The use of recoverable water-soluble diphosphine ligands in hydroxycarbonylation of vinyl arenes was reported to provide enantioselectivities up to 43% [56]. These chiral palladium-sulfonated diphosphine systems were shown to be active without the addition of acid. Heterogeneous catalytic systems formed by montmorillonite–diphenylphosphinepalladium(II) chloride in the presence of chiral mono- and bidentate phosphines were also reported in the methoxycarbonylation of styrene but afforded low enantioselectivities [57]. When the catalytic system containing the monodentate (*R*)-CH_3O–MOP ligand **56** (Figure 3.15) was used at 125 °C under 45 atm of CO and in the presence of concentrated HCl, total selectivity to the branched acid was achieved, but only 5% ee was obtained. Under the same conditions, the use of bidentate phosphine **57** (Figure 3.15) achieved total regioselectivity to the branched acid, together with 12% ee.

In 2003, Polo and coworkers reported the asymmetric methoxycarbonylation of acenaphthylene using BINAP as chiral ligand at 80 °C in the presence of *p*-TsOH and under 30 atm of CO [58]. Interestingly, using $PdCl_2(NCPh)_2$ as Pd precursor, ee's up to 45% with 12% conversion was achieved, whereas the use of $Pd(OAc)_2$ led to ee's up to only 34% but with 60% conversion under identical conditions. They described the existence of a degenerated substitution equilibrium between Pd^0L_n and the Pd^{II}-alkyl

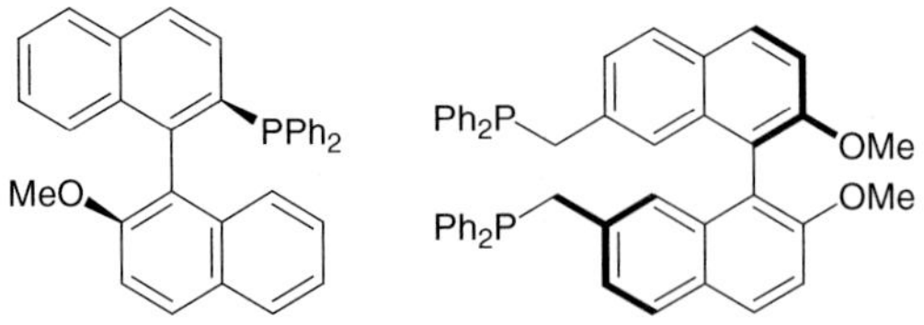

(*R*)-MeO-MOP
56

57

Figure 3.15 Ligands used by Nozaki in the heterogeneous system formed by montmorillonite–diphenylphosphinepalladium(II) systems.

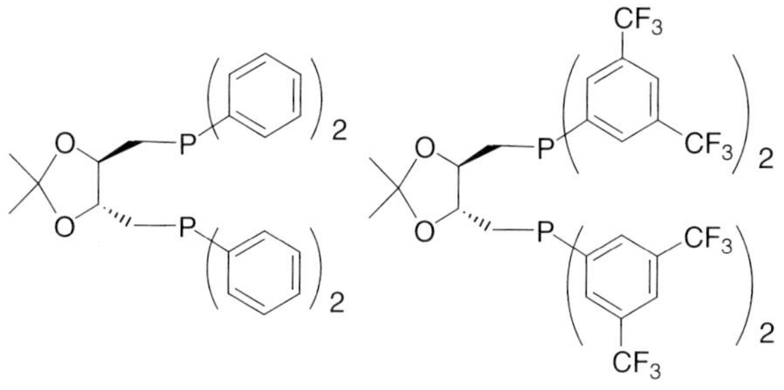

58
[49]
ee= 30%
l/b= 49/51

59
[49]
ee= 3%
l/b= 8/92

Figure 3.16 Ligands used by van Leeuwen and coworkers in the methoxycarbonylation of styrene.

species involving the inversion of the alkyl carbon, which produces a detrimental effect in the enantioselectivity of the reaction. In 2006, van Leeuwen and coworkers reported that the electronic properties of diphosphine ligands control the regioselectivity in the methoxycarbonylation of styrene under 30 atm of CO at 90 °C in the presence of HCl [49]. By varying the basicity of ligands (**58–59**), they observed that the formation of branched ester was favored when electron-poor diphosphines are used. However, in this case, the enantioselectivity of the reaction was found to decrease from 30 to 3%. (Figure 3.16).

3.3.4
Ferrocenyldiphosphines

In the methoxycarbonylation of styrene, the use of diphosphine ligands containing a ferrocenyl unit was reported to induce high enantioselectivity, although the regioselectivity to the branched product, as in general for diphosphine ligands, was usually low (Figure 3.17) [59]. In 1997, 86% ee was achieved by Inoue and coworkers together with a regioselectivity of 44% to the branched ester, using $Pd(OAc)_2$ as the palladium

60
[59a]
ee= 86%
l/b= 56/44

61
[59b]
ee= 64%
l/b= 60/40

62
[59c]
ee= 86%
l/b= 85/15

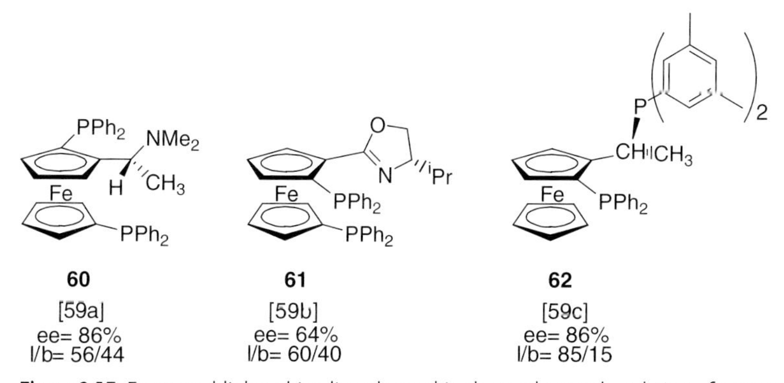

Figure 3.17 Ferrocenyldiphosphine ligands used in the methoxycarbonylation of styrene.

precursor in the presence of the chiral diphosphine (*S*,*R*)-BPPFA **60** (Figure 3.17) and *p*-TsOH under mild conditions (20 atm of CO at room temperature). However, in this case the branched product yield was 17%. When other ligands of the same family were used, lower ee's were obtained (up to 29%) and the regioselectivity was up to 38%. In 2003, Chan and coworkers reported the use of ferrocenylphosphine containing oxazoline moieties in the methoxycarbonylation of styrene and achieved 64% ee using the bidentate phosphine **61** (Figure 3.17) with $PdCl_2$ as the Pd source in the presence of *p*-TsOH at 50 °C under 170 atm of CO [59]. Although this enantiomeric excess was relatively high, it should be noted that only 40% regioselectivity was obtained and that the conversion was low (14%). More recently, Claver and coworkers reported the use of the families of ferrocenyldiphosphine from Solvias (*Josiphos*, *Mandyphos*, *Walphos*, and *Taniaphos*) in the same reaction [59]. High enantioselectivities (up to 86% using ligand **62**) and conversions (about 80%) were achieved, but the regioselectivities to branched ester were quite low (about 15%) in all cases.

3.3.5
Hemilabile P–N Ligands

To combine the properties of both mono- and bidentate phosphine ligands in terms of the results of regio- and enantioselectivity, respectively, the use of hemilabile ligands seems to be an obvious choice. The use of mixed bidentate pyridine phosphine ligands (Figure 3.18) was reported in 1996 by Chelucci *et al.* in the ethoxycarbonylation of styrene, yielding total selectivity to the branched ester with ee's up to 20% when the isolated precursor [$PdCl_2$ (**63**)] was used at 100 °C under 105 atm of CO for 10 days [60]. In view of the results, the authors concluded that under catalytic

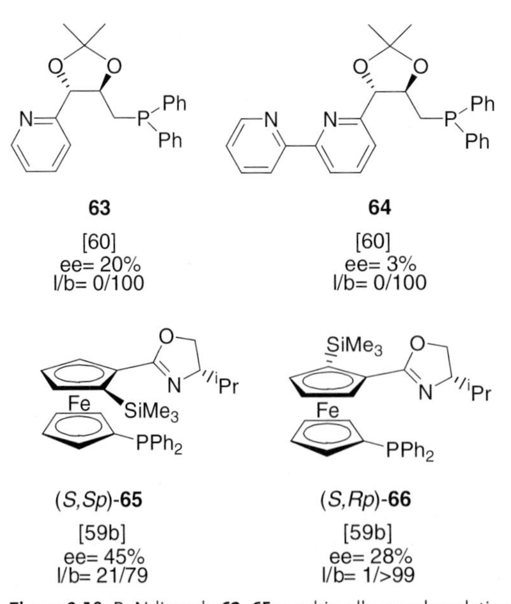

Figure 3.18 P–N ligands **62–65** used in alkoxycarbonylation reactions.

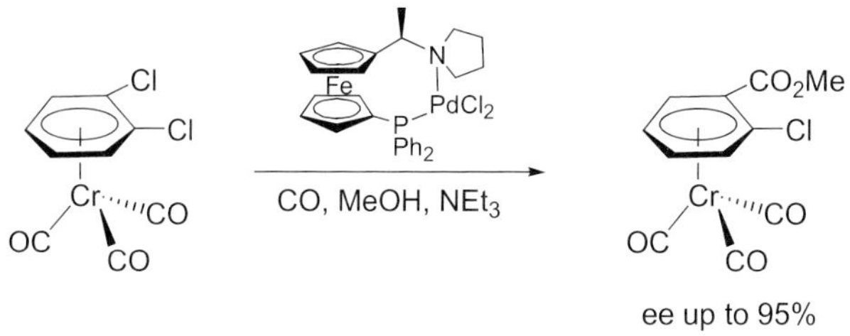

CO, MeOH, NEt_3

ee up to 95%

Scheme 3.4 Asymmetric methoxycarbonylation of 1,2-dichlorobenzene–$Cr(CO)_3$.

conditions, these ligands were coordinated in a monodentate manner. The use of the related ligand *dipydiphos* **64** yielded the same regioselectivity but a much lower enantioselectivity. Chan reported the use of P–N ferrocenylphosphineoxazoline (*S*, *Sp*)-**65** ligand (Figure 3.18) and achieved 45% ee together with a regioselectivity of 79% to the branched ester using a $PdCl_2$–$CuCl_2$–*p*-TsOH system [59]. Interestingly, when the diastereoisomer (*S*,*Rp*)-**66** ligand was used, the regioselectivity was found to increase (>99%) but both conversion and enantiomeric excess decreased considerably (Figure 3.18).

An innovative application of this type of ligands was reported by Schmaltz and Gotov [61]. They reported the use of a chiral ferrocenyl (*R*,*S*)-PPF-pyrrolidine system in the methoxycarbonylation of 1,2-dichlorobenzene-$Cr(CO)_3$ to introduce planar chirality in π-complexes (Scheme 3.4) and achieved up to 95% ee using the isolated precursor [$PdCl_2$(P–N)] under 1 atm of CO at 60 °C in the presence of NEt_3.

3.3.6
Monodentate Ligands

In view of the results obtained with diphosphine ligands in terms of regioselectivity in the alkoxycarbonylation of vinyl arenes, monodentate ligands have attracted much attention in recent years. In 1982, Cometti and Chiusoli reported the use of neomenthyldiphenylphosphine **67** (NMDPP) as the chiral ligand in the asymmetric methoxycarbonylation of styrene with 52% ee using $Pd(dba)_2$ as precursor in the presence of trifluoroacetic acid at 50 °C under atmospheric pressure of CO [62]. In 1990, Alper reported high enantioselectivity (91% ee) and total regioselectivity to the branched acid (64% of isolated yield) using the $PdCl_2$–$CuCl_2$–HCl–**68** (BNPPA) system (Figure 3.19) for the hydroxycarbonylation of 2-vinyl-6-methoxynaphtalene under 1 atm of a mixture of CO and O_2 at room temperature [63]. When *p*-isobutylstyrene was used as the substrate, the same catalytic system afforded 84% ee under the same conditions. Using the same ligand, the methoxycarbonylation of styrene was recently reported by Yang and Jiang, but only 38% ee was achieved under identical conditions [64].

In 1997, Nozaki *et al.* used (*S*,*S*)-phospholane **69** ligand (Figure 3.19) [57], but although this system afforded high regioselectivity, the enantioselectivity was low (2%). Later, the same authors reported on the application of palladium complexes with binaphthol-derived phosphines (Figure 3.19) in the methoxycarbonylation of 2-vinyl-6-methoxynaphtalene under 30 atm of CO at 40 °C, achieving 53% of ee using

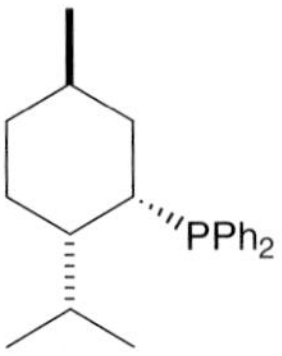

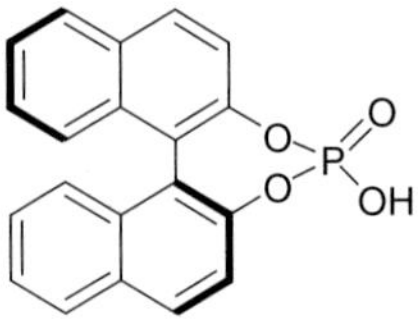

67 [62] ee= 52 % b/l = 94/6

68 [63] ee= 91%

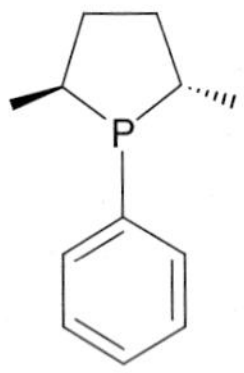

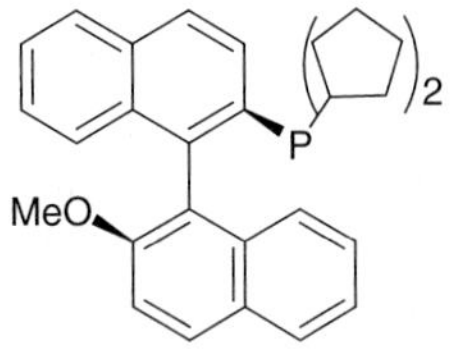

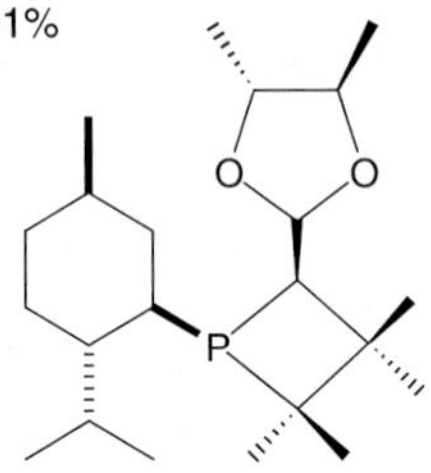

69 [25] ee = 2.4 % b/l = 98/2

70 [65] ee= 53 % b/l = 100/–

71 [66] ee= 29% b/l= 97/3

Figure 3.19 Monodentate phosphines used in asymmetric methoxycarbonylation of vinyl arenes.

70 for the branched ester, (*S*)-naproxen methyl ester, as the only reaction product [65]. In 2005, Claver and coworkers showed that systems containing phosphetane ligands could also yield high regioselectivity to the branched ester and ee's up to 29% in the methoxycarbonylation of styrene, using PCl_2 (**71**) as the precursor at 70 °C under 35 atm of CO (Figure 3.19) [66].

3.3.7
Asymmetric Bis-Alkoxycarbonylation of Alkenes

The enantioselective synthesis of optically active butanedioic acid derivatives is of interest as they are important intermediates of pharmaceuticals [67] and building blocks for rennin inhibitors [68]. These products are formed by the bis-alkoxycarbonylation of alkenes (Scheme 3.5), a reaction that was first reported by Heck in the early 1970s [42]. However, the first asymmetric version of this reaction was described more than 20 years later [69].

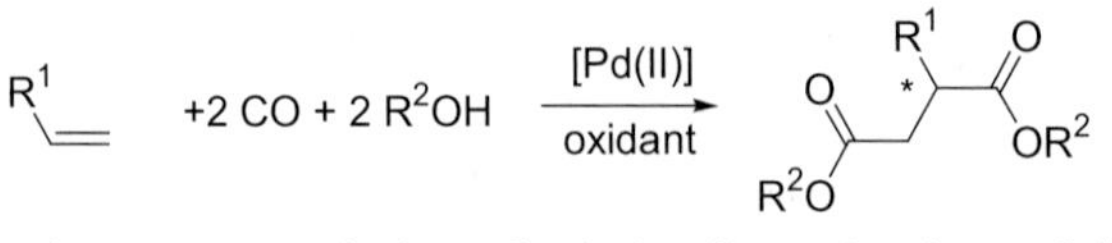

$$R^1\text{-CH=CH}_2 + 2\,CO + 2\,R^2OH \xrightarrow[\text{oxidant}]{[Pd(II)]} R^2O_2C\text{-}CH_2\text{-}C^*H(R^1)\text{-}CO_2R^2$$

Scheme 3.5 General scheme for the bis-alkoxycarbonylation of alkenes.

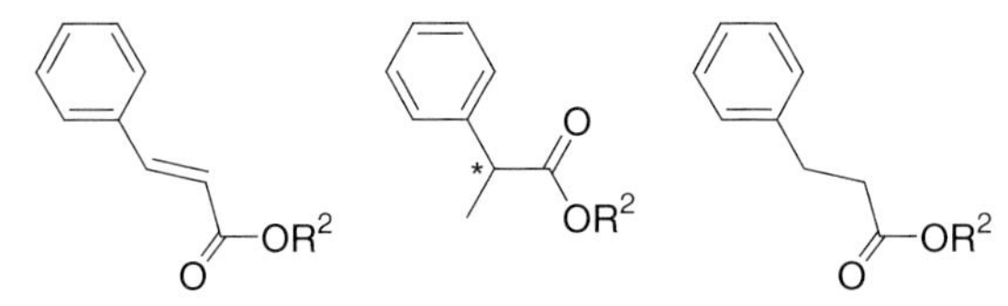

Figure 3.20 By-products generally formed during the bis-alkoxycarbonylation of styrene.

In this reaction, styrene is usually the substrate, and palladium systems containing bidentate phosphine ligands are used as catalysts in the presence of an oxidant such as benzoquinone in stoichiometric amount. The role of the oxidant is to reoxidize Pd(0) species formed during the process. In his initial report, Consiglio used a Pd system containing the (*R,R*)-DIOP ligand but only low enantioselectivity was achieved [69]. The use of atropisomeric diphosphine ligands yielded high enantiomeric excess but low chemoselectivity. The selective formation of bis-alkoxycarbonylation product is often a problem in this reaction, and by-products the methyl cinnamate, the methyl-2-phenylpropionate, and the methyl-3-phenylpropionate usually formed during this process (Figure 3.20).

The former product is derived from β-elimination reaction, whereas the two latter products are obtained by mono-alkoxycarbonylation of the substrate. Some oligomeric species are also often formed.

The use of chiral phosphine sulfides [70] and bisoxazoline [71] was later reported, but rather low ee's were obtained for succinate. In 1999, Consiglio and coworkers reported the screening of a series of ligands (Figure 3.21) in the enantioselective bismethoxycarbonylation of alkenes [72]. When styrene was used as a substrate, the catalytic systems containing ligands of the types BIPHEP and CH_3PHOS–BIPHEP yielded high enantioselectivities (92 and 81% respectively) with chemoselectivities of

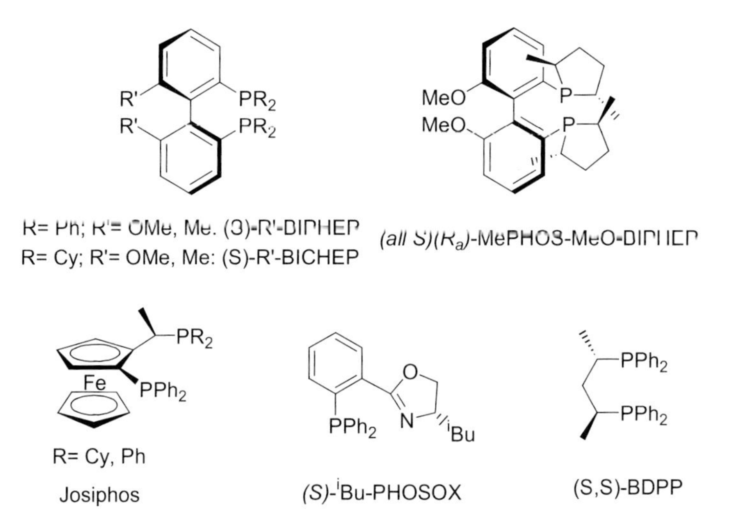

Figure 3.21 Ligands used by Consiglio in enantioselective bis-alkoxycarbonylation of alkenes.

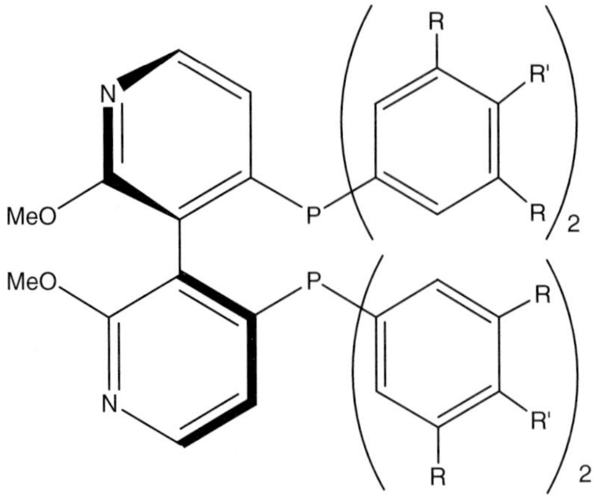

R= R′= H
R= Me, R'=H
R=H, R'= Me

Figure 3.22 Chiral dipyridylphosphine ligands used by Chan and coworkers in the bis-methoxycarbonylation of styrene.

about 50% at 50 °C under 350 atm of CO and in the presence of benzoquinone (ratio BQ/styrene = 1). Using other ligands, the conversion was very low (<1%). When propene was the substrate, the highest enantioselectivity (60%) was achieved using the (*S*)-CH_3O–BICHEP ligand, although the chemoselectivity and conversion were poor (13 and 23%, respectively). When 4-methyl-1-pentene was used as the substrate, chemoselectivity up to 79% was achieved, but the enantioselectivity was found to be rather low (up to 14%).

More recently, Chan and coworkers reported the use of chiral dipyridylphosphines in the enantioselective bis-methoxycarbonylation of styrene (Figure 3.22) and reached up to 84% ee and 79% chemoselectivity to dimethyl-2-phenylsuccinate at 50 °C under 152 atm of CO in the presence of benzoquinone (ratio BQ/styrene 0.5) [73].

Despite these promising results, only a few articles have been reported so far on this asymmetric carbonylation reaction.

3.4 Conclusion

Although much progress has been made over the last years in controlling the regioselectivity of the alkoxycarbonylation reaction, stereoselectivity, which is crucial in terms of pharmaceutical applications, is still to be largely improved. However, because of this industrial interest, a number of patents have already been published [74].

References

1 van Leeuwen, P.W.N.M. and Claver, C. (eds) (2000) *Rhodium Catalyzed Hydroformylation*, Kluwer Academic Press, Dordrecht.

2 Claver, C., Dieguez, M., Pamies, O. and Castillon, S., (2006) *Topics in Organometallic Chemistry, Catalytic Carbonylation Reactions, Asymmetric Hydroformylation*, vol. 18, Springer GmbH, Berlin, p. 35.

3 van Leeuwen, P.W.N.M., Casey, C.P. and Whiteker, G.T. (2000) *Rhodium Catalyzed Hydroformylation* (eds P.W.N.M. van Leeuwen and C. Claver), Kluwer Academic Press, Dordrecht, Chapter 4, p. 63.

4 Breit, B. and Seiche, W. (2001) *Synthesis*, 1.

5 Castillón, S. and Férnandez, E. (2000) *Rhodium Catalyzed Hydroformylation* (eds P.W.N.M. van Leeuwen and C. Claver), Kluwer Academic Press, Dordrecht, Chapter 6, p. 145.

6 Botteghi, C., Marchetti, M. and Paganelli, S. (1998) Transition metals for organic synthesis, in *Building Blocks and Fine Chemicals*, vol. 1 (eds M. Beller and C. Bolm), Wiley-VCH Verlag GmbH, Weinheim, Germany, p. 25.

7 Nozaki, K. and Ojima, I. (2000) *Catalytic Asymmetric Synthesis*, 2nd edn (ed. I. Ojima), Wiley-VCH Verlag GmbH, New York, Chapter 7.

8 Chapuis, C. and Jacoby, D. (2001) *Applied Catalysis*, **229**, 93.

9 Agbossou, F.F., Carpentier, J.F. and Mortreux, A. (1995) *Chemical Reviews*, **95**, 2485.

10 Gladiali, S., Bayón, J.C. and Claver, C. (1995) *Tetrahedron: Asymmetry*, **7**, 1453.

11 Beller, M., Cornils, B., Frohming, C.D. and Kohlpaintner, C.W. (1995) *Journal of Molecular Catalysis*, **104**, 17.

12 For a complete discussion including the detailed aspects of the different steps in this catalytic cycle see van Leeuwen, P.W.N.M. (2004) *Homogeneous Catalysis: Understanding the Art*, Kluwer Academic Publishers, Dordrecht, Chapter 8, p. 141.

13 Wink, J.D., Kwok, T.J. and Yee, A. (1990) *Inorganic Chemistry*, **29**, 5007.

14 Sakai, N., Nozaki, K., Mashima, K. and Takaya, H. (1992) *Tetrahedron: Asymmetry*, **3**, 581.

15 Babin, J.E. and Whiteker, G.T. (1993) WO 93/03839 (to Union Carbide Chemicals & Plastics Technology Corporation), *Chemical Abstracts*, 1993, **119**, P159872h.

16 Buisman, G.J.H., van deer Veen, L.A., Klootwijk, A., de Lange, W.G.J., Kamer, P.C.J., van Leeuwen, P.W.N.M. and Vogt, D. (1997) *Organometallics*, **16**, 2929.

17 (a) Buisman, G.J.H., Vos, E.J., Kamer, P.C.J. and van Leeuwen, P.W.N.M. (1995) *Journal of the Chemical Society, Dalton Transactions*, 409. (b) Cserépi-Szûcs, S., Tóth, I., Párkányi, L. and Bakos, J. (1998) *Tetrahedron: Asymmetry*, **9**, 3135.

18 $HRh(CO)_2$(diphosphite) species are known to be in the resting state in the hydroformylation reaction. See for instance Ref. [1].

19 Buisman, G.J.H., van der Veen, L.A., Kamer, P.C.J. and van Leeuwen, P.W.N.M. (1997) *Organometallics*, **16**, 5681.

20 (a) Diéguez, M., Pàmies, O., Ruiz, A., Castillón, S. and Claver, C. (2000) *Journal of the Chemical Society, Chemical Communications*, 1607. (b) Diéguez, M., Pàmies, O., Ruiz, A., Castillón, S. and Claver, C. (2001) *Chemistry – A European Journal*, **7**, 3086. (c) Diéguez, M., Pàmies, O., Ruiz, A. and Claver, C. (2002) *New Journal of Chemistry*, **26**, 827. (d) Pàmies, O., Net, G., Ruiz, A. and Claver, C. (2000) *Tetrahedron: Asymmetry*, **11**, 1097. (e) Diéguez, M., Ruiz, A. and Claver, C. (2003) *Journal of the Chemical Society, Dalton Transactions*, 2957. (f) Buisman, G.J.H., Martin, M.E., Vos, E.J., Klootwijk, A., Kamer, P.C.J. and van Leeuwen, P.W.N.M. (1995) *Tetrahedron: Asymmetry*, **8**, 719.

21 Diéguez, M., Pàmies, O. and Claver, C. (2005) *Journal of the Chemical Society, Chemical Communications*, 1221.

22 Cobley, C.J., Gardner, K., Klosin, J., Praquin, C., Hill, C., Whiteker, G.T., Zanotti-Gerosa, A., Petersen, J.L. and Abboid, K.A. (2004) *The Journal of Organic Chemistry*, **69**, 4031.

23 (a) Cserépi-Szûcs, S., Huttner, G., Zsolnai, L. and Bakos, J. (1999) *Journal of Organometallic Chemistry*, **586**, 70. (b) Cserépi-Szûcs, S., Huttner, G., Zsolnai, L., Szölösy, A., Hegebüs, C. and Bakos, J. (1999) *Inorganica Chimica Acta*, **296**, 222. (c) Vegehetto, V., Scrivanti, A. and Matteoli, U. (2001) *Catalysis Communications*, **2**, 139. (d) Jiang, Y., Xue, S., Li, Z., Deng, J., Mi, A. and Chan, A.S.C. (1998) *Tetrahedron: Asymmetry*, **9**, 3185. (e) Jiang, Y., Xue, S., Yu, K., Li, Z., Deng, J., Mi, A. and Chan, A.S.C. (1999) *Journal of Organometallic Chemistry*, **586**, 159. (f) Kadyrov, R., Heller, D. and Selke, R. (1999) *Tetrahedron: Asymmetry*, **9**, 329. (g) Freixa, Z. and Bayón, J.C. (2001) *Journal of the Chemical Society, Dalton Transactions*, 2067.

24 Sakai, N., Mano, S., Nozaki, K. and Takaya, H. (1993) *Journal of the American Chemical Society*, **115**, 7033.

25 Nozaki, K., Sakai, N., Nanno, T., Higashijima, T., Mano, S., Horiuchi, T. and Takaya, H. (1997) *Journal of the American Chemical Society*, **119**, 4413.

26 Nozaki, K., Matsuo, T., Shibahara, F. and Hiyama, T. (2001) *Advanced Synthesis & Catalysis*, **343**, 61.

27 Nozaki, K., Matsuo, T., Shibayara, F. and Hiyama, T. (2003) *Organometallics*, **22**, 594.

28 (a) Nozaki, K., Itoi, Y., Shibayara, F., Shirakawa, E., Ohta, T., Takaya, H. and Hiyama, T. (1998) *Journal of the American Chemical Society*, **120**, 4051. (b) Nozaki, K., Shibahara, F., Itoi, Y., Shirakawa, E., Ohta, T., Takaya, H. and Hiyama, T. (1999) *Bulletin of the Chemical Society of Japan*, **72**, 1911.

29 (a) Franciò, G. and Leitner, W. (1999) *Journal of the Chemical Society, Chemical Communications*, 1663. (b) Franciò, G., Wittmann, K. and Leitner, W. (2001) *Journal of Organometallic Chemistry*, **621**, 130.

30 Lambers-Verstappen, M.M.H. and de Vries, J.G. (2003) *Advanced Synthesis & Catalysis*, **345**, 478.

31 Nozaki, K., Li, W., Horiuchi, T. and Takaya, H. (1997) *Tetrahedron Letters*, **38**, 4611.

32 (a) Horiuchi, T., Ohta, T., Nozaki, K. and Takaya, H. (1996) *Journal of the Chemical Society, Chemical Communications*, 155. (b) Horiuchi, T., Ohta, T., Shirakawa, E., Nozaki, K. and Takaya, H. (1997) *Tetrahedron: Asymmetry*, **53**, 7795.

33 Horiuchi, T., Ohta, T., Shirakawa, E., Nozaki, K. and Takaya, H. (1997) *The Journal of Organic Chemistry*, **62**, 4285.

34 Nanno, T., Sakai, N., Nozaki, K. and Takaya, H. (1995) *Tetrahedron: Asymmetry*, **6**, 2583.

35 Nozaki, K., Li, W., Horiuchi, T. and Takaya, H. (1996) *The Journal of Organic Chemistry*, **61**, 7658.

36 (a) Deeremberg, S., Kamer, P.C.J. and van Leeuwen, P.W.N.M. (2000) *Organometallics*, **19**, 2065. (b) Pàmies, O., Net, G., Ruiz, A. and Claver, C. (2001) *Tetrahedron: Asymmetry*, **12**, 3441.(c) Arena, C.G., Faraone, F., Graiff, C. and Tiripicchio, A. (2002) *European Journal of Inorganic Chemistry*, 711.

37 Rampf, F.A. and Herrmann, W.A. (2000) *Journal of Organometallic Chemistry*, **601**, 138.

38 (a) Breeden, S., Cole-Hamilton, D.J., Foster, D.F., Schwarz, G.J. and Wills, M. (2000) *Angewandte Chemie-International Edition*, **39**, 4106. (b) Clarkson, G.J., Ansel, J.R., Cole-Hamilton, D.J., Pogorzelec, P.J., Whittell, J. and Mills, M. (2004) *Tetrahedron: Asymmetry*, **15**, 1787.

39 Yan, Y. and Zhang, X. (2006) *Journal of the American Chemical Society*, **128**, 7198.

40 (a) Lot, O., Suisse, I., Mortreux, A. and Agbossou, F. (2000) *Journal of Molecular Catalysis A: Chemical*, **164**, 125. (b) Diéguez, M., Ruiz, A. and Claver, C. (2001) *Tetrahedron: Asymmetry*, **12**, 2827. (c) Naili, S., Suisse, I., Mortreux, A., Agbossou-Niedercorn, F. and Nowogrocki, G. (2001) *Journal of Organometallic Chemistry*, **628**, 114. (d) Cessarotti, E., Araneo, S., Rimoldi, I. and Tassi, S. (2003) *Journal of Molecular*

Catalysis A: Chemical, **204–205**, 211. (e) Ewalds, R., Eggeling, E.B., Hewat, A.C., Kamer, P.C.J., van Leeuwen, P.W.N.M. and Voght, D. (2000) *Chemistry – A European Journal*, **6**, 1496. (f) Pàmies, O., Diéguez, M., Net, G., Ruiz, A. and Claver, C. (2000) *Organometallics*, **19**, 1488. (g) Saluzzo, C., Breuzard, J., Pellet-Rostaing, S., Vallet, M., Guyader, F.L. and Lemaire, M. (2002) *Journal of Organometallic Chemistry*, **643–644**, 98. (h) Franció, G., Drommi, D., Graiff, C., Faraone, F. and Tiripicchio, A. (2002) *Inorganica Chimica Acta*, **338**, 59.

41 Beller, M., Seayad, J., Tillack, A. and Jiao, H. (2004) *Angewandte Chemie-International Edition*, **43**, 3368.

42 (a) Knifton, J. (1976) *The Journal of Organic Chemistry*, **41**, 793. (b) Cavinato, G. and Toniolo, L. (1990) *Journal of Organometallic Chemistry*, **398**, 187. (c) Milstein, D. (1988) *Accounts of Chemical Research*, **21**, 428. (d) Kanawa, M., Nakamura, S., Watanabe, E. and Urata, H. (1997) *Journal of Organometallic Chemistry*, **542**, 185.

43 del Río, I., Claver, C. and van Leeuwen, P.W.N.M. (2001) *European Journal of Inorganic Chemistry*, **2719**, and references therein.

44 (a) Ali, B.E. and Alper, H. (1993) *Journal of Molecular Catalysis*, **80**, 377. (b) Fuchikami, T., Ohishi, K. and Ojiva, I. (1983) *The Journal of Organic Chemistry*, **48**, 3803. (c) Pisano, C., Mezzetti, A. and Consiglio, G. (1992) *Organometallics*, **11**, 20.

45 (a) Consiglio, G. and Marchetti, M. (1976) *Chimia*, **30**, 26. (b) Consiglio, G., Nefkens, S. C.A., Pisano, C. and Wenzinger, F. (1991) *Helvetica Chimica Acta*, **74**, 323.

46 van Leeuwen, P.W.N.M., Zuideveld, M.A., Swennenhuis, B.H.G., Freixa, Z., Kamer, P. C.J., Goubitz, K., Fraanje, J., Lutz, M. and Spek, A.L. (2003) *Journal of the American Chemical Society*, **125**, 5523.

47 del Río, I., Ruiz, N. and Claver, C. (2000) *Inorganic Chemistry Communications*, **3**, 166.

48 Sugi, Y. and Bando, K. (1976) *Chemistry Letters*, 727.

49 Guiu, E., Caporali, M., Muñoz, B., Müller, C., Lutz, M., Spek, A.L., Claver, C. and van Leeuwen, P.W.N.M. (2006) *Organometallics*, **25**, 3102.

50 (a) Zuidema, E., Bo, C. and van Leeuwen, P. W.N.M. (2007) *Journal of the American Chemical Society*, **129** (13), 3989. (b) Donald, S.M.A., Macgregor, S.A., Settels, V.D., Cole-Hamilton, J. and Eastham, G.R. (2007) *Journal of the Chemical Society, Chemical Communications*, **6**, 562.

51 Botteghi, C., Consiglio, G. and Pino, P. (1973) *Chimia*, **27**, 477.

52 Nozaki, K. and Ojima, I. (2000) *Catalytic Asymmetric Synthesis* (ed. I. Ojima), 2nd edn, John Wiley & Sons, Ltd, New York, p. 448.

53 Zhou, H., Hou, J., Cheng, J., Lu, S., Fu, H. and Wang, H. (1997) *Journal of Organometallic Chemistry*, **543**, 227.

54 Zhou, H., Hou, J., Chen, J., Lu, S., Fu, H. and Wang, H. (1998) *Chemistry Letters*, **19** (3), 247.

55 del Río, I., Ruiz, N., Claver, C., van der Veen, L.A. and van Leeuwen, P.W.N.M. (2000) *Journal of Molecular Catalysis A: Chemical*, **161**, 39.

56 Miquel-Serrano, M.D., Aghmiz, A., Diéguez, M., Masdeu-Bultó, A.M., Claver, C. and Sinou, D. (1999) *Tetrahedron: Asymmetry*, **10**, 4463.

57 Nozaki, K., Kantam, M.L., Horiuchi, T. and Takaya, H. (1997) *Journal of Molecular Catalysis A: Chemical*, **118**, 247.

58 Gironès, J., Duran, J., Polo, A. and Real, J. (2003) *Journal of the Chemical Society, Chemical Communications*, 1776.

59 (a) Oi, S., Nomura, M., Aiko, T. and Inoue, Y. (1997) *Journal of Molecular Catalysis A: Chemical*, **115**, 289. (b) Wang, L., Kwok, W. H., Chan, A.S.C., Tu, T., Hou, X. and Dai, L. (2003) *Tetrahedron: Asymmetry*, **14**, 2291. (c) Godard, C., Ruiz, A. and Claver, C. (2006) *Helvetica Chimica Acta*, **89**, 1610.

60 Chelucci, G., Cabras, M.A., Botteghi, C., Basoli, C. and Marchetti, M. (1996) *Tetrahedron: Asymmetry*, **7**, 885.

61 Gotov, N. and Schmaltz, H.-G. (2001) *Organic Letters*, **3**, 1753.

62 Cometti, G. and Chiusoli, G.P. (1982) *Journal of Organometallic Chemistry*, **236**, C31.

63 Alper, H. and Hamel, N. (1990) *Journal of the American Chemical Society*, **112**, 2803; Alper, H. (1991) WO 9103452.

64 Yang, K. and Jiang, X. (2005) *Chemical Journal on Internet*, **7**, 14.

65 Kawashima, Y., Okano, K., Nozaki, K. and Hiyama, T. (2004) *Bulletin of the Chemical Society of Japan*, **77**, 347.

66 Muñoz, B., Marinetti, A., Ruiz, A., Castillon, S. and Claver, C. (2005) *Inorganic Chemistry Communications*, **8**, 1113.

67 Kleeman, J. and Engel, A. (1982) *Pharmazeutische Wirkstoffe*, Thieme Verlag, Stuttgart.

68 Yoshikawa, K., Inoguchi, K. and Achiwa, K. (1990) *Heterocycles*, **31**, 1413. (b) Ito, Y., Kamijo, T., Harada, H., Matsuda, F. and Terashima, S. (1990) *Tetrahedron Letters*, 2731. (c) Inoguchi, K., Morimoto, T. and Achiwa, K. (1989) *Journal of Organometallic Chemistry*, **370**, C9. (d) Jendralla, H. (1991) *Tetrahedron Letters*, **32**, 3671. (e) Kammermeier, B., Beck, G., Holla, W., Jacobi, D., Napierski, B. and Jendralla, H. (1996) *Chemistry – A European Journal*, **2**, 307.

69 Nefkens, S.C.A., Sperrle, M. and Consiglio, G. (1993) *Angewandte Chemie-International Edition in English*, **2**, 1719.

70 Hayashi, M., Takezaki, H., Hashimoto, Y., Takaoki, K. and Saigo, K. (1998) *Tetrahedron Letters*, **39**, 7529.

71 Takeuchi, S., Ukaji, Y. and Inomata, K. (2001) *Bulletin of the Chemical Society of Japan*, **74**, 955.

72 Sperrle, M. and Consiglio, G. (1999) *Journal of Molecular Catalysis A: Chemical*, **143**, 263.

73 Wang, L., Kwok, W., Wu, J., Guo, R., Au-Yeung, T.T.-L., Zhou, Z., Chan, A.S.C. and Chan, K.-S. (2003) *Journal of Molecular Catalysis A: Chemical*, **196**, 171.

74 Drent, E. (1984) EP 0106379; Stephan, M.M. and Mohar, B. (2006) WO 2006136695; Klosin, J., Whiteker, G. and Cobley, C. (2006) WO 2006116344; Holz, J., Boermer, A., Almena, J.J., Kadyrov, R., Monsees, A. and Riermeier, T. (2006) WO 2006100169; J., Boermer, A., Zayas, O., Almena Perea, J.J., Kadyrov, R., Monsees, A. and Riermeier, T. (2006) WO 2006100165; Zhang, X. and Liu, D. (2005) WO 2005058917; Zhang, X. and Tang, W. (2007) WO 2005117907; Zhang, X. (2004) US 2004072680; Zhang, X. and Tang, W. (2003) WO 2003042135; Zhang, X. (2002) WO 2002040491; Zhang, X. (2001) WO 2001034612; Zhang, X. (2001) WO 2001021625; Zhang, X. (1997) WO 9713763; Alper, H. (1989) EP 305089; Botteghi, C., Consiglio, G. and Pino, P. (1978) CH 598184; Consiglio, G. (1977) CH 588438.

4
Microwave-Promoted Carbonylations

Johan Wannberg, Mats Larhed

4.1
Introduction

The increased requirements to improve the drug discovery process have sparked a move away from traditional organic synthesis toward small-scale, high-speed synthesis [1,2]. Thus, to allow fine-tuning of the molecular composition of a lead scaffold with the ultimate goal of obtaining improved biological properties, very large sets of modified analogues are required. As a direct consequence, an efficient and rapid compound production has become a prerequisite. Organic transformations must be rapidly executed and products quickly isolated and purified. Such demands have driven the development of novel technologies, which have begun to accelerate compound generation. Controlled high-density microwave heating is one method that enhances the productivity of the preparative chemists, developed primarily to reduce the reaction times and not to directly aid product isolation. With modern microwave synthesizers, reaction times can often be reduced from hours to minutes or seconds and chemistry previously considered impractical or unattainable can now be accessed [3–5]. The possibility to speed up the processing time is of special importance with otherwise slow transition metal catalyzed reactions [6].

Procedures used by modern high-throughput chemists are mainly limited to the manipulation of solid and liquid reagents/reactants unless the most advanced equipments are utilized. The microwave heating technology should, however, also be useful for gaseous reactions. The identification of solid or liquid reagents releasing gases upon heating or irradiation, or the development of dedicated equipment for smooth and safe gas handling, is therefore most important for the future of microwave chemistry.

Although there are many carbonylative synthetic transformation methods available, they require the use and handling of toxic carbon monoxide gas [7,8]. This reduces the utility of carbon monoxide in general, and particularly in small-scale high-speed microwave applications using septum-sealed single-use reaction vials. A convenient way of avoiding many of these problems would be to exploit a solid or liquid reagent with the ability to release carbon monoxide *in situ* during a reaction [9].

Modern Carbonylation Methods. Edited by László Kollár

ISBN: 978-3-527-31896-4

When *in situ* methods are used to generate CO, chemical waste may be produced and the atom economy is therefore not optimal. This is, of course, a major problem on an industrial scale but for small-scale applications in research laboratories, the increased safety and ease-of-handling are strongly advantageous. However, it is obvious that the concept of *in situ* liberation would be even more appealing if inexpensive organic materials, preferably the solvent itself, could serve as the source of carbon monoxide. Or alternatively, safe, swift, and straightforward procedures for microwave processing of CO-pressurized reaction vessels are developed. Thus, the authors find it highly promising that the number of reports of successful carbonylative microwave applications is presently increasing at a steady pace. Besides providing a general description of working reaction examples, we would like to start this chapter by giving a brief discourse of basic microwave heating theory.

4.2 Microwave Heating in Organic Chemistry

Electromagnetic radiation with a frequency of 0.3–300 GHz ($\lambda = 1$–0.001 m) is called microwave radiation. The microwave part of the electromagnetic spectrum lies between the more energetic infrared radiation and the less energetic radio waves (Figure 4.1). Microwaves are used in radar, satellite communication, land-based communication links spanning moderate distances (cell phones) and other applications, not forgetting microwave ovens.

Microwaves are able to heat a chemical reaction mixture (or food) by two general mechanisms: dipolar polarization and ionic conductance [10–12]. All matters that contain dipoles and/or charged species can absorb microwave energy and convert it into heat. This is because dipoles and ions constantly try to align themselves to the electric component of the oscillating electromagnetic field, resulting in rotation of molecules and oscillation of ions. The electromagnetic energy absorbed in this process is hence first converted into kinetic energy, which is then lost as heat through molecular friction.

To achieve efficient heating, it is important that the frequency of the applied radiation is within certain limits. If it is too low, the dipoles will realign too quickly with the electric field and completely follow the field fluctuations, resulting in poor

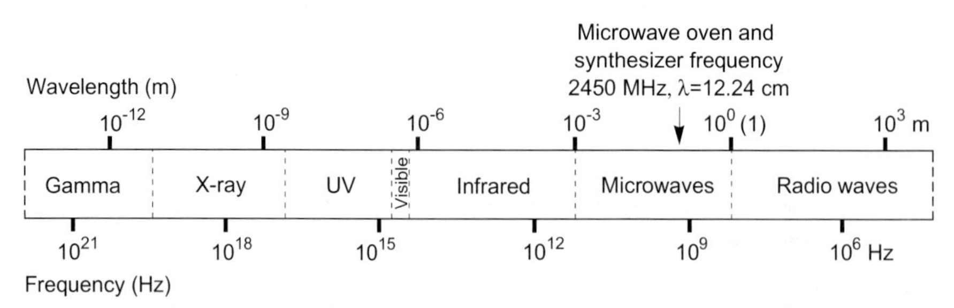

Figure 4.1 The electromagnetic spectrum including the microwave frequency used for standard microwave heating.

heating. If it is too high the dipoles do not have time to realign themselves at all to the alternating field, which means no motion is created and therefore no heat. The frequency used by domestic microwave ovens and synthesizers (2.45 GHz, $\lambda = 12.2$ cm) is located between these extremes where dipoles have time to partly realign with the oscillating electric field but cannot follow the field fluctuations, resulting in an effective heat generation [5,10–12].

The generally accepted advantages of microwave-assisted organic synthesis compared to conventional heating techniques are speed, convenience, and energy efficiency. If at least one component of the reaction mixture can interact with microwaves, a very high heating rate can be obtained. A consequence of the direct bulk heating generated by microwave irradiation is an energy-efficient and uniform heating of the whole reaction system. In contrast, during conventional heating the energy must first be transferred from the heat source to the wall of the reaction vessel and then to the reaction medium. This leads to so-called wall effects, for example, the catalyst deactivation on the hot vessel wall [12].

Solvents irradiated by microwaves can generally be heated well above their boiling points at atmospheric pressure using open vessels. This is because the heating being faster than the convection to, and loss of heat from, the surface of the solvent/reaction mixture. Thus, a higher temperature is reached before bubbles nucleates. This superheating alone (up to 20 °C above the boiling point) can increase reaction rates considerably [3–5,12]. One should remember that according to the rate law, a 10 °C increase in reaction temperature will roughly double the reaction rate. If closed vessel conditions are used, and this is the most common approach in modern microwave chemistry, the "pressure cooker" effect can lead to even more dramatic rate enhancements.

4.3 Microwave-Promoted Carbonylations

The remaining part of this chapter reviews published small-scale methods (commonly 0.5–5 ml reaction volumes) for carbonylative reactions driven by controlled microwave energy using standard single-mode reactors, an area which is strongly dominated by palladium-catalyzed examples using aryl halides (or pseudohalides) as starting materials. The content is subdivided according to the employed method to provide the essential CO moiety, starting with metal carbonyl and formamide sources of carbon monoxide, and followed by applications using CO-prepressurized reaction containers.

4.3.1 Microwave-Promoted Carbonylations Using $Mo(CO)_6$ as a Source of Carbon Monoxide

The first example of a microwave-promoted carbonylation was published by Kaiser *et al.* [13]. In this investigation, the palladium-catalyzed carbonylation of aryl halides

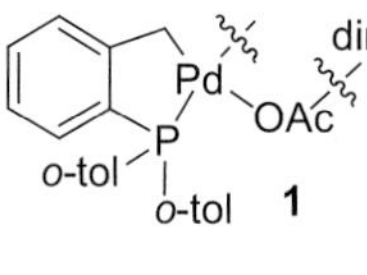

Figure 4.2 Herrmann's palladacycle (**1**).

was used to quickly generate benzamides while using $Mo(CO)_6$ as a CO-releasing solid. The use of $Mo(CO)_6$ as a solid CO-releasing agent in microwave heated carbonylations was inspired by the CO release observed while using $Mo(CO)_6$ as a precatalyst in microwave-promoted enantioselective molybdenum-catalyzed allylic alkylations [14]. In the initial carbonylation report, *n*-butylamine, piperidine, and water were used as nucleophiles. *para*-Substituted aryl iodides with Pd/C or *para*-substituted aryl bromides with Herrmann's palladacycle (**1**) (*trans*-di(μ-acetato)-bis[*o*-(di-*o*-tolylphosphino)benzyl]dipalladium(II), Figure 4.2) [15] and *rac*-BINAP (2,2′-bis(diphenylphosphino)-1,1′-binaphthyl) were used as arylpalladium precursors. Diglyme served as the solvent and aqueous K_2CO_3 as the base.

Microwave irradiation at 150 °C for 15 min produced butyl- and piperidyl benzamides in 65–83% yield (Scheme 4.1). In one example where the amine was omitted and ethylene glycol was used as cosolvent, the corresponding benzoic acid was produced in 87% yield.

The following year the first microwave-promoted carbonylative synthesis of benzoic esters from aryl halides and alcohols was reported [16]. The paper on alkoxycarbonylation presented a method where aryl iodides were converted into the corresponding benzoic esters using $Mo(CO)_6$ and Pd/C with DIEA (diisopropylethylamine) as base and DMAP (4-dimetylaminopyridine) as nucleophilic catalyst in a 1 : 1 mixture of the alcohol (*n*-butanol, *t*-butanol, benzylalcohol, or 2-TMS-ethanol) and 1,4-dioxane. Microwave heating at 150 °C for 15 min furnished ester products in 33–89% isolated yields. Using **1** as precatalyst, the alkoxycarbonylation of aryl bromides was also possible, although an increase in temperature of 180–190 °C was required and yields ranged between 64 and 72% [16] (Scheme 4.2).

Also in 2003, a more thorough investigation of microwave-promoted carbonylations using $Mo(CO)_6$ was presented [17]. The initial protocol worked reasonably well with highly nucleophilic amines but was unproductive for weaker nucleophiles such as anilines. With model substrates 2-iodotoluene and aniline, the aminocarbonylation to yield 2-methylbenzanilide was investigated using different solvents, bases, palladium catalysts, and CO sources (metal carbonyls). Etherous solvents (diglyme,

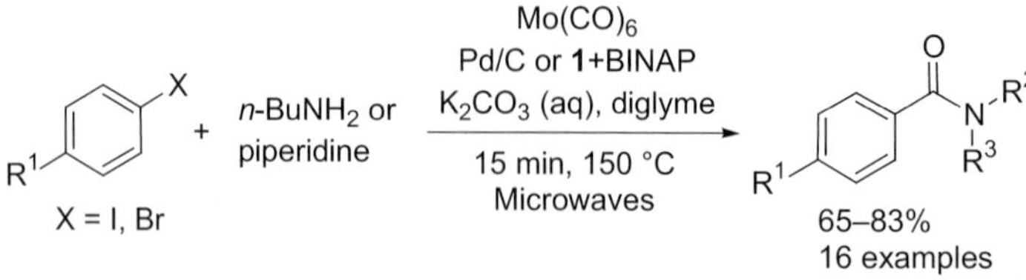

Scheme 4.1

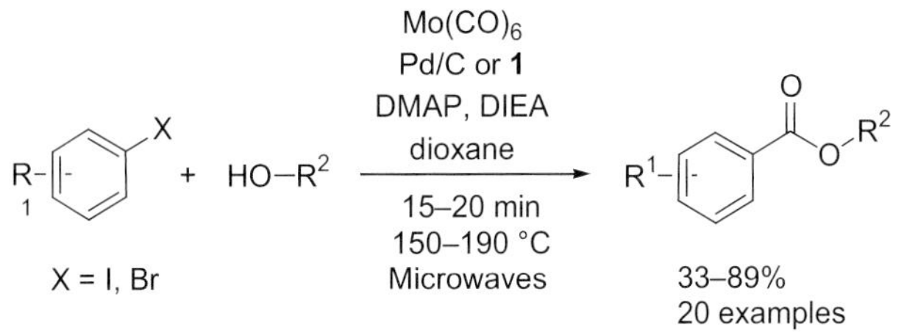

$Mo(CO)_6$
Pd/C or **1**
DMAP, DIEA
dioxane
15–20 min
150–190 °C
Microwaves

R1-Ar-X + HO–R^2 → R^1-Ar-C(O)O-R^2

X = I, Br

33–89%
20 examples

Scheme 4.2

DME, 1,4-dioxane, and THF) were found superior to DMF, acetonitrile, and toluene and, accordingly, THF was chosen as a suitable solvent.

The choice of base had dramatic effects on the reactivity. Inorganic carbonate bases and tertiary aliphatic amines were not useful in aminocarbonylations with aniline as a nucleophile, at least not at temperatures up to 150 °C. When using the nucleophilic catalysts imidazole or DMAP as bases, full conversion of 2-iodotoluene could be achieved at high temperatures (150 °C). Another effect noted from the pressure profiles recorded by the microwave synthesizer was that when adding imidazole, pyridine, or DMAP to the reaction mixture, the release of gaseous carbon monoxide was substantially faster than from pure thermal decomposition. This chemical liberation of CO occurred below 150 °C although full conversion of the aryl halide at these temperatures was not observed. The strong amidine base DBU (1,8-diazabicyclo[5.4.0]undec-7-ene) has been used in conventional carbonylations with aniline [18]. When DBU was employed as a base with $Mo(CO)_6$, an extremely rapid release of CO was observed. Moreover, complete consumption of 2-iodotoluene was achieved at much lower temperatures. Evidently DBU functions as both an effective carbon monoxide releasing promoter and an efficient base in the aminocarbonylation reactions. The initially used palladium acetate was found to be equally good or better than other phosphine free and phosphine containing palladium (pre) catalysts examined. With optimized conditions, a reaction of 2-iodotoluene, aniline (3 equiv.), $Pd(OAc)_2$ (10 mol%), DBU (3 equiv.), and $Mo(CO)_6$ (1 equiv.) in THF at 100 °C for 15 min afforded complete consumption of aryl iodide and an 84% isolated yield of 2-methyl-*N*-phenylbenzamide.

The selected conditions were investigated with other potential solid sources of carbon monoxide. Pressure curves from these reactions indicated gas release in all cases. All of the Group 6 metal carbonyls ($Cr(CO)_6$, $Mo(CO)_6$, and $W(CO)_6$) turned out to be effective sources of carbon monoxide, although $Mo(CO)_6$ gave the highest yield. Using iron carbonyls was unproductive and reactions with $Co_2(CO)_8$ produced some products, but aryl halide starting material was also consumed in a side reaction forming 2,2′-dimethyl-benzophenone (see Section 4.3.2).

The optimized $Mo(CO)_6$-based aminocarbonylation protocol was applied in the synthesis of a variety of benzamides. Both electron-rich and electron-poor aryl iodides proved to be efficient coupling partners. In the set of amines examined in this reaction, the previously investigated piperidine produced improved yields. Furthermore, aniline and benzylamine also reacted easily. The sterically hindered

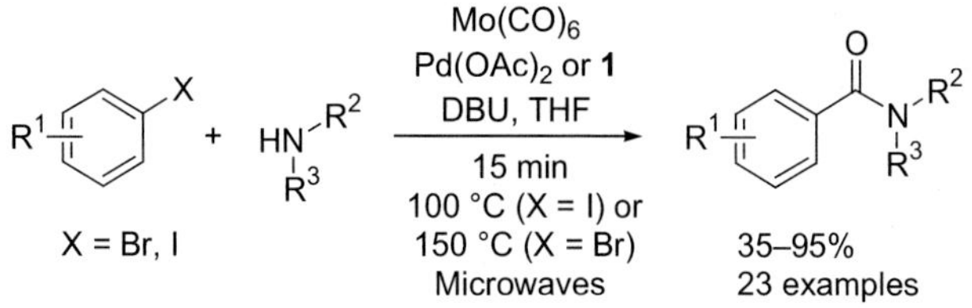

Scheme 4.3

t-butylamine and the heat sensitive 2-aminothiazole afforded lower yields of products (Scheme 4.3).

Aryl bromides were also examined as coupling partners. In these aminocarbonylations, Herrmann's palladacycle **1** was found to be an effective precatalyst at 150 °C. These conditions were applied with good results on a selection of aryl bromides and amines. The results followed the same trend as for the corresponding aryl iodides with yields from 35 to 92%, although a useful yield could not be obtained with 2-aminothiazole [17].

The positive results achieved in developing aminocarbonylation protocols prompted the evaluation of other nucleophiles in similar microwave-promoted carbonylations. Surprisingly, few reports concerning the direct carbonylation of aryl or vinyl halides with hydrazides to the corresponding *N,N′*-diacylhydrazines are found in the literature and an effort was made to modify the previously reported aminocarbonylation protocol [17] for the carbonylative synthesis of *N,N′*-diacylhydrazines [19].

By simply exchanging the nucleophiles to various hydrazides and raising the temperature to 110 °C, full conversion of a number of aryl iodides was achieved in 15 min of microwave heating and the corresponding *N,N′*-diacylhydrazine products were isolated in moderate to good yields (30–71%). The moderate yields obtained, despite complete consumption of aryl iodide, were explained by formation of primary benzamide and benzonitrile side products in various amounts, probably because of the decomposition of the diacylhydrazine products. Attempts to minimize side product formation, by shortening reaction times, generally improved the isolated yields (36–78%) (Scheme 4.4) [19].

The use of aryl bromides as substrates was also explored. Unfortunately, the high temperature protocol (150 °C) using **1** as the precatalyst produced a large number of side products. When **1** was combined with the phosphonium salt $[(t\text{-Bu})_3PH]BF_4$ (**2**, a stable source of tri-*t*-butylphosphine) [20], a very active catalytic system was generated, affording complete conversion of aryl bromides within 5 min at 130 °C.

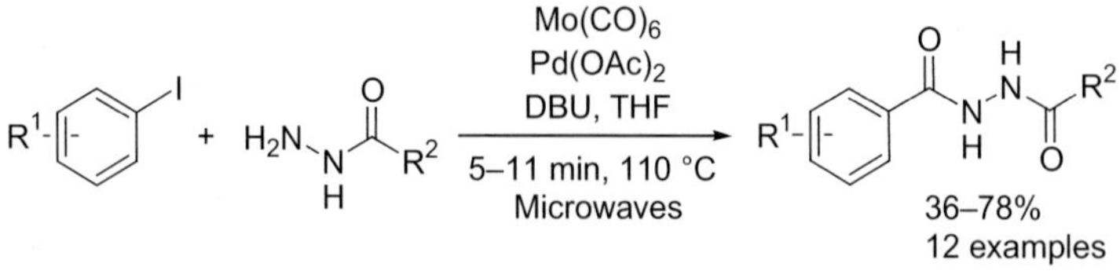

Scheme 4.4

Scheme 4.5

The protocol was examined by the reaction of three different aryl bromides with benzhydrazide producing isolated yields of 43–57% [18].

$Mo(CO)_6$ has also been used in other types of microwave-promoted carbonylative chemistry. In 2005, Larhed and coworkers presented a carbonylative cyclization protocol whereby *ortho*-bromo- and *ortho*-chlorostyrenes were converted into indan-1-ones [21]. Optimized reaction conditions for *o*-Br-styrenes involved $Pd(OAc)_2$ and **2** as catalytic system with $Mo(CO)_6$ as CO source, *n*-Bu_4NCl as additive and pyridine as base in 1,4-dioxane. Microwave-heating at 150 °C for 30 min provided indanones in 59–82% yield (Scheme 4.5) [21].

The good results obtained with *o*-Br-styrenes prompted an attempt to expand the scope of the reaction to include *o*-Cl-styrenes as well. It was found that switching the palladium source to **1** and elevating the reaction temperature to 170 °C resulted in full conversion of the chlorides, although the yields were relatively low (25–51%, five examples). The modest yields could not solely be explained by dechlorination. At this high temperature, polymerization of starting materials was thought to have occurred.

Expanding this microwave-promoted carbonylative cyclization strategy further, a set of *ortho*-bromoaryl enamides was prepared by regioselective internal Heck arylation of enamides with *ortho*-bromoaryl triflates. These were then cyclized to the corresponding 3-acylaminoindan-1-ones (47–88% yield) using a slightly higher temperature (160 °C) with other conditions remaining the same as for the simple *o*-bromostyrenes (Scheme 4.6) [21].

Acylsulfonamides have important applications as carboxylic acid bioisosteres in medicinal chemistry. Thus, it was of great interest to examine sulfonamides as nucleophiles in microwave-heated carbonylations. Starting from the conditions reported for the hydrazidocarbonylations, it was revealed that by simply changing the solvent from THF to 1,4-dioxane, aryl iodides and primary sulfonamides were

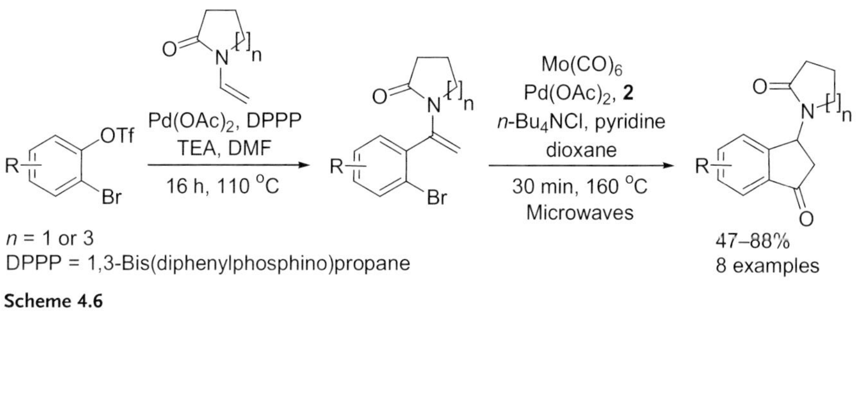

Scheme 4.6

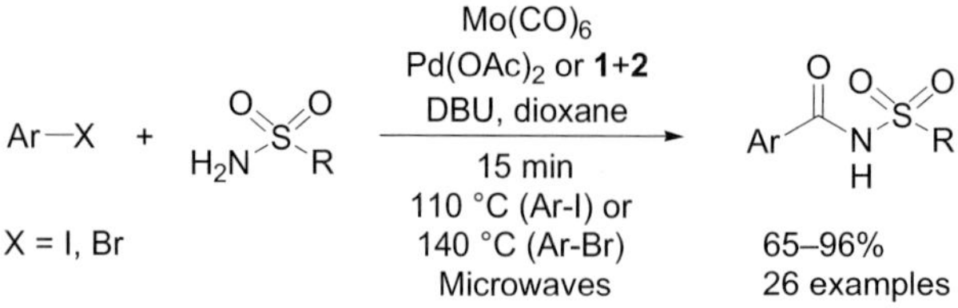

Ar—X + H_2NSO_2R → (Mo(CO)$_6$, Pd(OAc)$_2$ or **1+2**, DBU, dioxane; 15 min, 110 °C (Ar-I) or 140 °C (Ar-Br), Microwaves) → ArC(O)NHSO$_2$R

X = I, Br

65–96%
26 examples

Scheme 4.7

conveniently converted into acylsulfonamides in 65–88% isolated yields (Scheme 4.7) [22]. A secondary sulfonamide resulted in a yield of 47% of the corresponding acylsulfonamide.

Aryl bromides could also be applied as starting materials when using the combination of **1** and **2** as the catalytic system. The carbonylation of Ar–Br with sulfonamides at 140 °C actually provided superior yields (79–96%) compared to the corresponding Ar–I at 110 °C (Scheme 4.7).

The presented protocol was applied to the synthesis of a building block in the synthetic route to an HCV NS3 protease inhibitor (Scheme 4.8). Gratifyingly, no racemization of the norvaline α-carbon occurred in the carbonylation step as only one diastereomer of the inhibitor was detected [22].

In a 2005 paper published by Prof Kappe's group describing palladium-catalyzed functionalizations of 4-arylquinolin-2(1*H*)-ones, one example of a microwave-promoted aminocarbonylation using Mo(CO)$_6$ appeared. 6-Bromo-4-phenylquinolin-2(1*H*)-one was converted into the corresponding *N*-benzylamide in 61% yield (Scheme 4.9) [23].

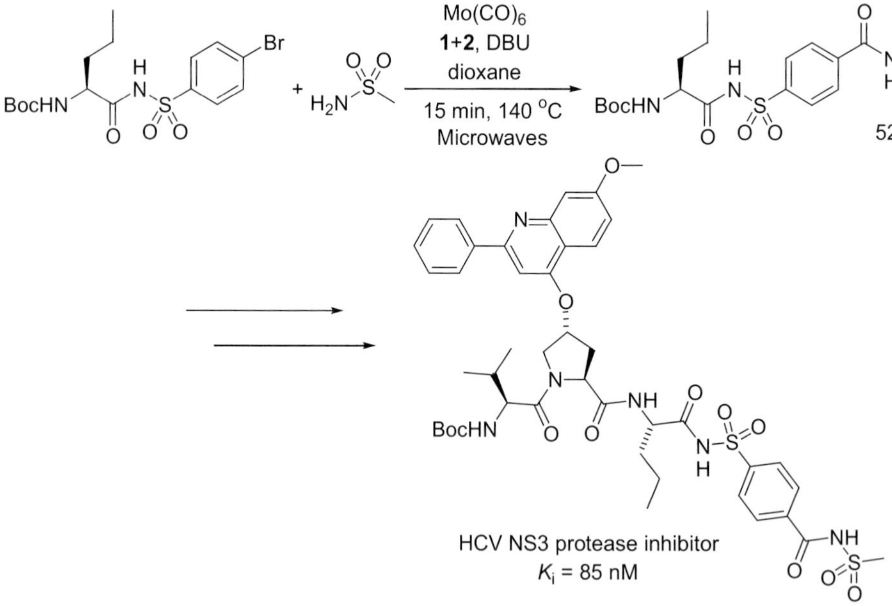

Mo(CO)$_6$
1+2, DBU
dioxane
15 min, 140 °C
Microwaves

BocHN H N S O O O Br + H_2N S O O

52%

HCV NS3 protease inhibitor
K_i = 85 nM

Scheme 4.8

Mo(CO)$_6$, **1+2**
DBU, MeCN
25 min, 170 °C
Microwaves

61%

Scheme 4.9

Compounds containing the dihydropyrimidone (DHPM) scaffold are easily accessed through the acid-catalyzed three-component cyclocondensation known as the Biginelli reaction [24]. In their 2005 collaborative study, the groups of Kappe and Larhed described microwave-promoted, metal-catalyzed functionalizations of 4-aryl-dihydropyrimidones [25]. Three 4-phenyl-3,4-dihydropyrimidin-2(1*H*)-ones (*ortho-*, *meta-*, and *para*-bromo substituted) were used as starting materials for different palladium-catalyzed reactions, including microwave-promoted carbonylations using $Mo(CO)_6$ as the carbon monoxide source. The effective combination of **1** and **2** also gave useful yields of benzamides (56–87%) and moderate yields of benzoic esters (42–52%) from the *meta-* and *para*-bromo substituted starting materials using 3 equiv. of amine or 5 equiv. of alcohol after 15 min of microwave heating at 130–140 °C (Scheme 4.10). By using methanol as solvent, full conversion to the methyl esters could be observed within 15 min at temperature as low as 110 °C (yields 71 and 77%). One example of a hydrazidocarbonylation generated a modest 35% isolated yield. The *ortho*-bromo starting material proved difficult, generating complex product mixtures and low yields of carbonylation products (two examples, 21 and 24% yield) (Scheme 4.10) [25].

The aminocarbonylation strategy was next utilized in a 2005 study for the rapid synthesis of two novel series of HIV-1 protease inhibitors [26]. C_2-symmetric, 1,2-dihydroxyethylene-based core structures containing aryl halide-substituted P1/P1′ side chains were transformed in very small scale to 21 different (bis)-benzamide analogues that were tested for their affinity to the HIV-1 protease. Interesting compounds were synthesized again on a larger scale for retesting and characterization using protected variants of the starting materials, producing isolated yields

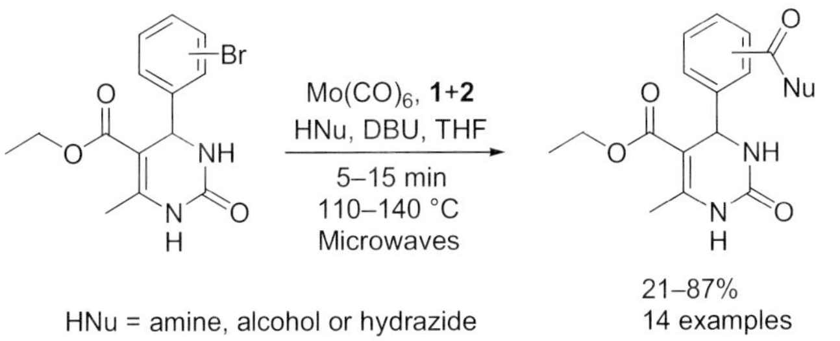

Mo(CO)$_6$, **1+2**
HNu, DBU, THF
5–15 min
110–140 °C
Microwaves

HNu = amine, alcohol or hydrazide

21–87%
14 examples

Scheme 4.10

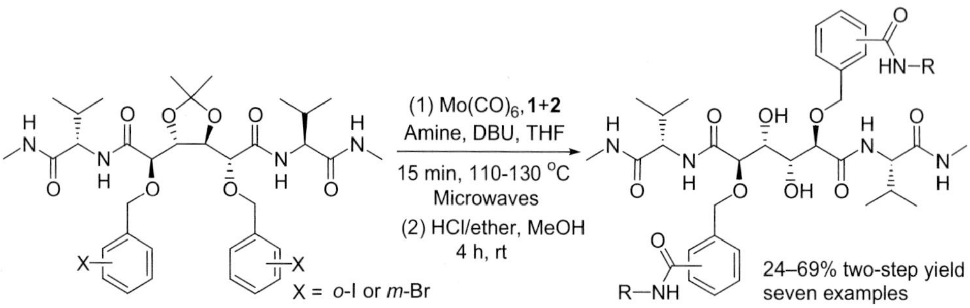

Scheme 4.11

between 24 and 69% of these sophisticated, bis-amidated HIV-1 protease inhibitors (Scheme 4.11). The potency of P1/P1′ *ortho*-substituted anilide derivatives indicated that larger groups than previously expected were tolerated in the space spanning the S1–S3 and S1′–S3′ pockets of the enzyme [26].

In 2005, Cao and Xiao presented a palladium-catalyzed carbonylative cyclization reaction in which $Mo(CO)_6$, iodophenols, and alkynes were converted into cromen-2-one derivatives in the presence of DIEA and DMAP by microwave heating at 160 °C for 30 min in dioxane (63–78% isolated yields). When terminal alkynes were used, cromen-2-ones were formed exclusively, but disubstituted alkynes produced 3–4 : 1 mixtures of the cromen-2-one and cromen-4-ones (Scheme 4.12) [27].

Wu and Larhed reported in 2005 that microwave-promoted aminocarbonylations could be carried out efficiently in pure water [28]. It was found that in the presence of an amine, the aminocarbonylation of an aryl bromide dominated over competing hydroxycarbonylation. By examining the impact of different molar ratios of 1-naphthyl bromide and *n*-butylamine on the yield of corresponding amide and acid products, it was found that when a 1 : 2 ratio of Ar–Br/amine was used, high yields of amides were obtained within 10 min of microwave heating at 170 °C using **1** as palladium source and Na_2CO_3 as base. The scope of these conditions was first examined using a variety of different aryl bromides with *n*-butylamine and piperidine as nucleophiles. Subsequently, a wide range of amines was explored with 4-bromotoluene as reactant. The conclusion from these experiments was that electron-poor, neutral, and electron-rich, as well as sterically hindered aryl bromides all worked well with both *n*-butylamine and secondary piperidine. Sterically hindered amines also worked well and even with *t*-butylamine, the aminocarbonylation competed favorably with hydroxycarbonylation, providing the *t*-butylamide in 60% yield (Scheme 4.13) [28].

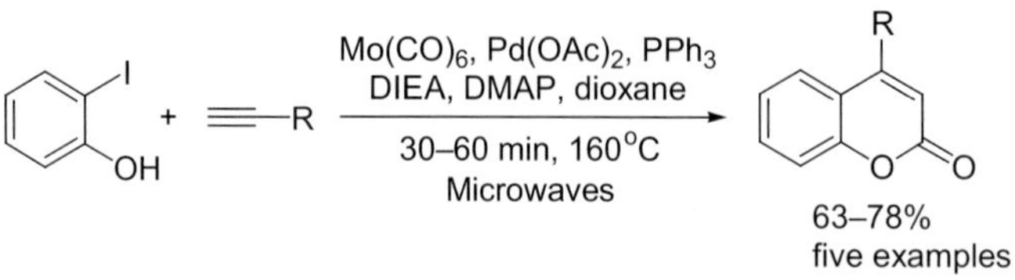

Scheme 4.12

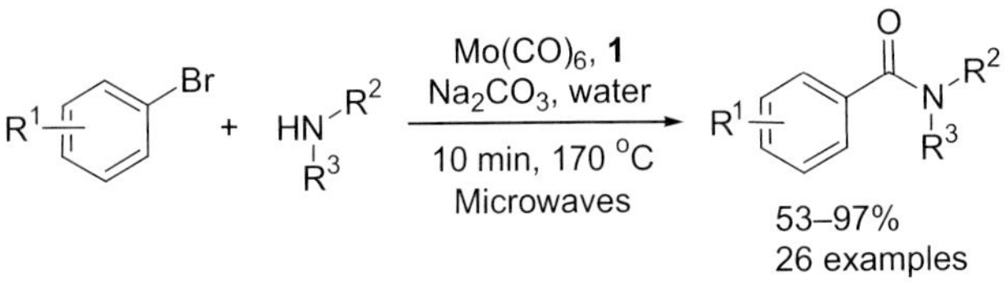

Scheme 4.13

Thereafter, a full paper describing aminocarbonylations in water was published [29]. In addition to presenting the results achieved with aryl bromides, aminocarbonylation protocols for aryl iodides and aryl chlorides were also disclosed. More than 90 successful aminocarbonylations in water were presented (Scheme 4.14), including a medicinal chemistry application where this amidation procedure was used to synthesize a potent HIV-1 protease inhibitor from the aryl bromide (Scheme 4.15) [29].

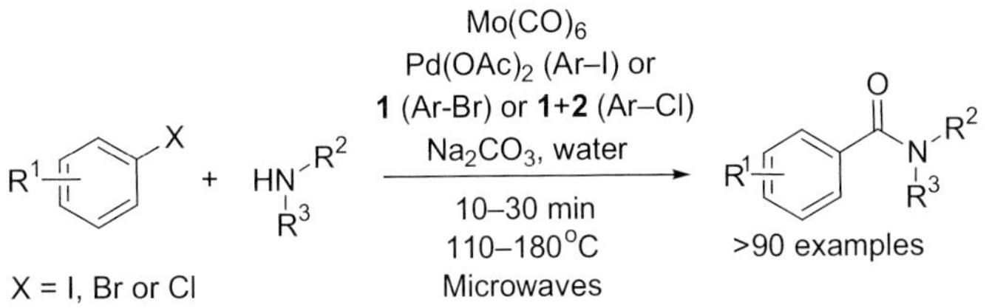

Scheme 4.14

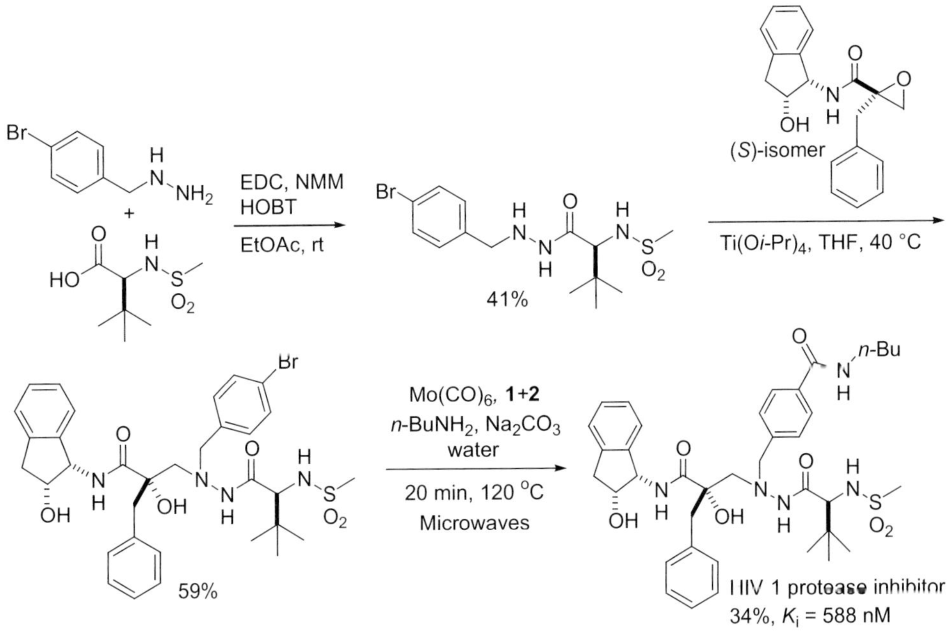

Scheme 4.15

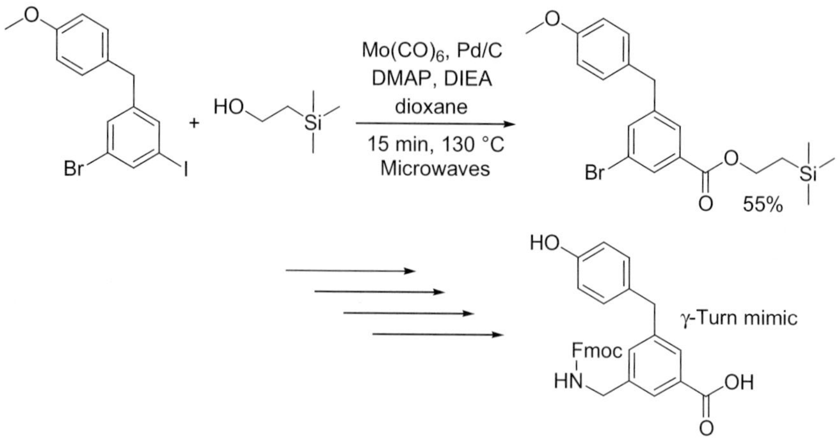

Scheme 4.16

In 2005, the same group applied the alkoxycarbonylation protocol developed by Georgsson *et al.* [16] to synthesize a small aromatic scaffold designed to function as a peptidomimetic when incorporated into the angiotensin II sequence [30]. Bromo-3-iodo-5-(4-methoxybenzyl)benzene was carbonylated using 2-trimethylsilylethanol with $Mo(CO)_6$ as the CO source to the corresponding ester in 55% yield. After a palladium-catalyzed cyanation followed by nitrile reduction, amine protection, and a combined ester–ether deprotection, the peptidomimetic building block was ready to be used in peptide couplings (Scheme 4.16). The resulting pseudopeptide exhibited nanomolar affinity to the AT2 receptor and it was concluded that the 1,3,5-trisubstituted-benzene scaffold might serve as a useful γ-turn mimic [30].

In an early 2006 paper, microwave-promoted aminocarbonylations of aryl chlorides were explored [31]. Again, the combination of **1** and **2** was shown to produce a highly active catalytic system and at a temperature of 170 °C, a diverse set of aryl chlorides and amines could be converted into the corresponding benzamides within 25 min of microwave irradiation in 51–91% yield. Satisfyingly, very challenging aryl chlorides (e.g., 4-chloroanisole and 2-chloro-1,3-xylene) could be employed with good results (Scheme 4.17). Also one example of a benzoic ester synthesis was presented. 4-Chlorobenzotrifluoride was converted into the corresponding *n*-butylbenzoate in 67% yield after 60 min of microwave heating at 170 °C [31].

In 2006, Silvani and coworkers described the use of $Mo(CO)_6$ in the development of straightforward hydroxycarbonylation of aryl and vinyl triflates to the corresponding carboxylic acids [32]. Applying the method reported previously to an aryl triflate

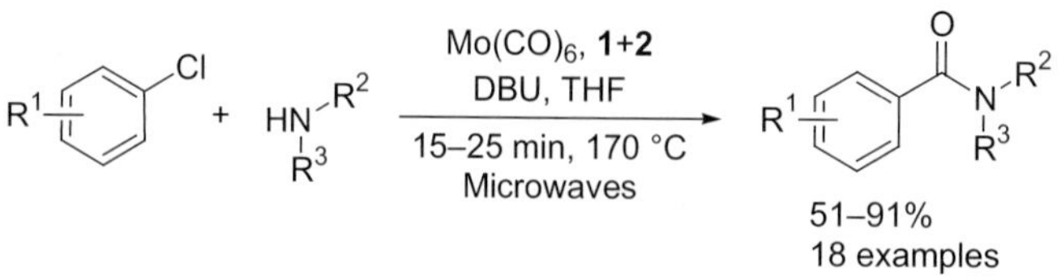

Scheme 4.17

$$\underset{\text{R = aryl, vinyl}}{\text{R}\!\!\diagup^{\text{OTf}}} \xrightarrow[\text{20 min, 150 °C, Microwaves}]{\text{Mo(CO)}_6\text{, Pd(OAc)}_2\text{, DPPF, H}_2\text{O, pyridine}} \text{RCOOH}$$

26–97%
21 examples

Scheme 4.18

produced significant amounts of the corresponding phenol (hydrolyzed triflate) as side product. After experimenting with different solvents and bases, it was disclosed that water, as a combined solvent/nucleophile with pyridine as base and $Pd(OAc)_2$/DPPF as catalytic system, effectively converted aryl and vinyl triflates into 21 different carboxylic acids within 20 min of microwave heating at 150 °C in 15–97% yield (Scheme 4.18, DPPF = 1,1′-bis(diphenylphosphino)ferrocene). Interestingly, with these conditions 4-bromophenyl triflate was selectively converted into 4-bromobenzoic acids with no side products resulting from activation of the bromine [32].

In a 2006 paper outlining the rapid microwave synthesis of novel cyclic sulfamide HIV-1 protease inhibitors, the aminocarbonylation protocol presented in ref. 17 was applied in the decoration of mono- and bis-aryl bromide containing cyclic sulfamide starting materials [33]. Aniline and benzylamine were used as nucleophiles for the production of two symmetric bis-functionalized and two unsymmetrical monosubstituted compounds in good yields (59–80%) (Scheme 4.19). Unfortunately, of these anilide or benzylamide containing compounds only one displayed weak HIV-1 protease inhibition [33].

The first $Mo(CO)_6$-promoted, microwave-assisted protocols for aminocarbonylation of aryl bromides and iodides to primary benzamides were presented in 2006 [34]. Hydroxylamine hydrochloride was used as a solid ammonia equivalent as it was found that hydroxylamine was smoothly reduced *in situ* by $Mo(CO)_6$, generating free ammonia that could undergo the carbonylation. The conditions found suitable for this reaction included an aryl halide, $Mo(CO)_6$, $NH_2OH{\cdot}HCl$, **1**, **2**, DBU, and DIEA in dioxane followed by 20 min of microwave heating at 110 °C for Ar–I and 150 °C for Ar–Br. Eight primary benzamides were synthesized from aryl bromides in 70–81% yield and the same amides from aryl iodides in 76–84% yield (Scheme 4.20). The method was also used in the synthesis of weak HIV-1 protease inhibitors (Scheme 4.21) [34].

A palladium-catalyzed, $Mo(CO)_6$ and microwave-mediated aminocarbonylation was used by Gupton *et al.* in 2006 to synthesize Rigidin and Rigidin E. An iodo-substituted pyrrole was converted to the corresponding *N*-methylamides and *N*-dimethoxybenzylamide in 65 and 80% yield, respectively (Scheme 4.22) [35].

Early in 2007, Letavic and Ly published a letter describing aminocarbonylations of heteroaromatic bromides using $Mo(CO)_6$ as the CO source, thereby further demonstrating the usefulness of this type of reaction for the rapid generation of a series of druglike structures. Heteroaromatic bromides used in these examples were different pyridines and pyrimidines, 3-bromoisoquinoline, 4-bromoindole, 2-bromothiazole, and 5-bromo-1-methyl-1*H*-imidazole. Sixteen heteroaromatic

Scheme 4.19

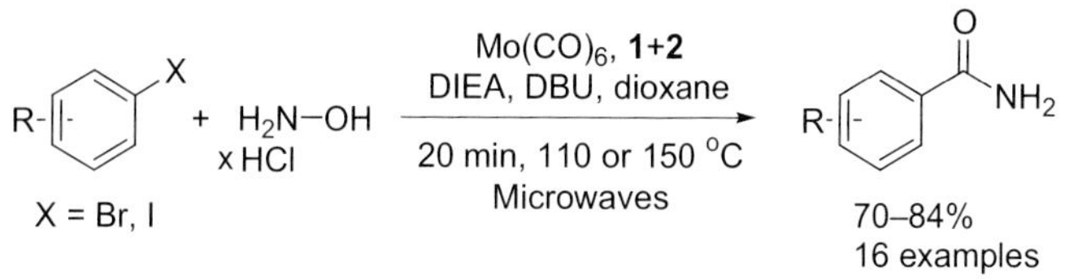

R–Ar–X + $H_2N{-}OH$ x HCl → $Mo(CO)_6$, **1+2**, DIEA, DBU, dioxane; 20 min, 110 or 150 °C, Microwaves → R–Ar–$CONH_2$

X = Br, I

70–84%
16 examples

Scheme 4.20

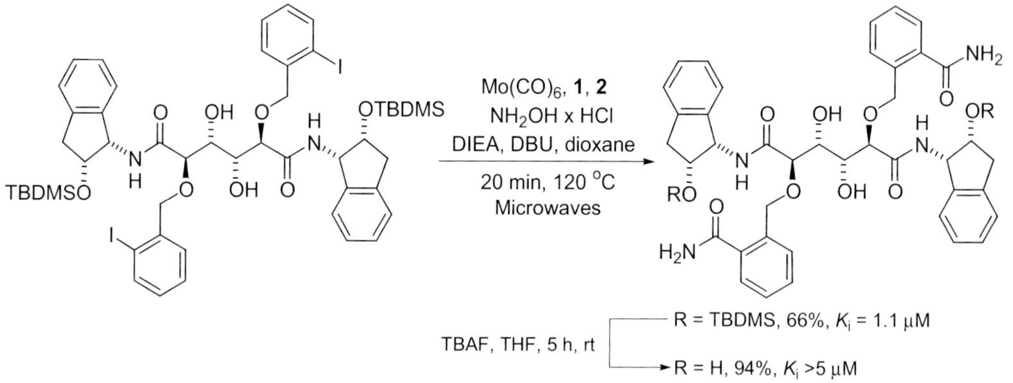

$Mo(CO)_6$, **1**, **2**
NH_2OH x HCl
DIEA, DBU, dioxane
20 min, 120 °C
Microwaves

TBAF, THF, 5 h, rt

R = TBDMS, 66%, K_i = 1.1 μM

R = H, 94%, K_i >5 μM

Scheme 4.21

amides were synthesized in 34–97% yield after 6 min of microwave heating at 125 °C using the well-established **1** + **2** combination as catalytic system (Scheme 4.23) [36].

The important role aryl triflates play as arylpalladium precursors in small-scale organic-, medicinal-, and high-throughput chemistry encouraged the Larhed group to investigate also this class of starting materials [37]. Although well established as useful substrates for aminocarbonylations in organic solvents, aryl triflates are prone to hydrolysis in aqueous, alkaline media at elevated temperatures. It is noteworthy that aryl triflates have been reported to furnish benzoic acids in water under carbonylative conditions [32]. In Scheme 4.24, a facile and gas-free protocol for the aminocarbonylation of aryl triflates in water is disclosed, to a large extent avoiding the

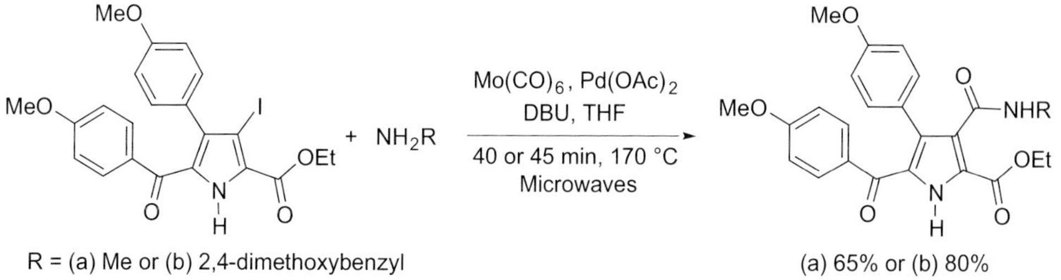

+ NH_2R

$Mo(CO)_6$, $Pd(OAc)_2$
DBU, THF
40 or 45 min, 170 °C
Microwaves

R = (a) Me or (b) 2,4-dimethoxybenzyl

(a) 65% or (b) 80%

Scheme 4.22

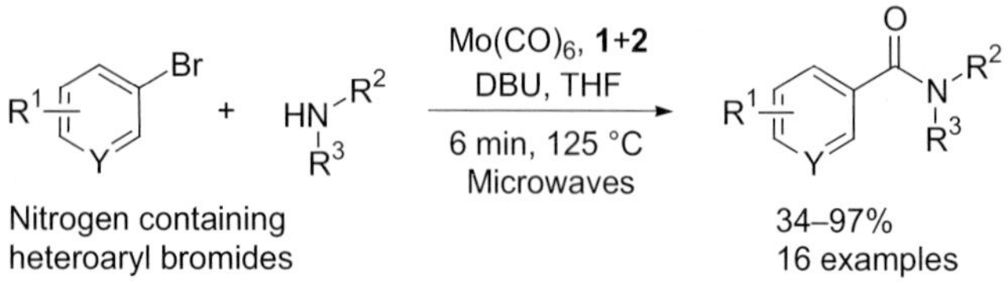

Mo(CO)6, **1+2**
DBU, THF
6 min, 125 °C
Microwaves

Nitrogen containing heteroaryl bromides

34–97%
16 examples

Scheme 4.23

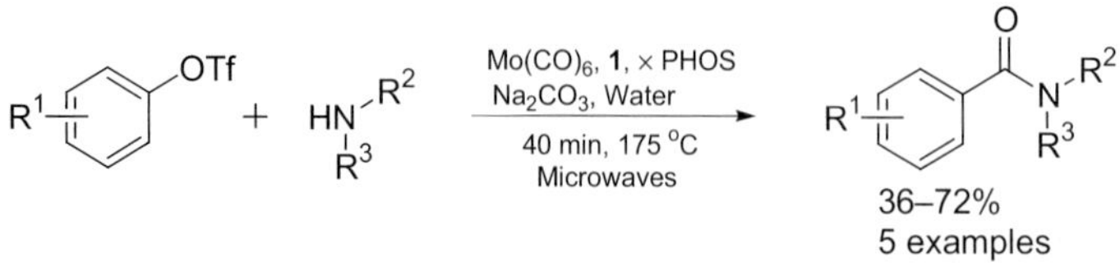

Mo(CO)6, **1**, × PHOS
Na2CO3, Water
40 min, 175 °C
Microwaves

36–72%
5 examples

Scheme 4.24

formation of benzoic acid side products. The reactions were carried out under high-density microwave irradiation using $Mo(CO)_6$ as a convenient *in situ* CO-liberator, promoting the transformations in a mere 40 min. A number of structurally diverse aryl triflates and nucleophiles were incorporated into the studies and target benzamides were isolated in useful yields and purities (Scheme 4.24) [37].

4.3.2 Microwave-Promoted Carbonylations Using $Co_2(CO)_8$ as a Reaction Mediator

In 2003, an extremely fast protocol for the cobalt carbonyl mediated formation of symmetric diaryl ketones from aryl halides was disclosed by Larhed and coworkers [38]. Microwave irradiation of aryl iodides and $Co_2(CO)_8$ in acetonitrile for 10 s or less was enough to produce high yields (57–97%) of symmetric diarylmethanones (Scheme 4.25). Please note that the $Co_2(CO)_8$-mediated chemistry was performed without an additional transition metal catalyst.

After short heating, the temperature generally peaked at 130 °C and the pressure never exceeded 12 bar. A general condition for aryl bromides was not found but two examples were presented, one producing 43% yield after 30 s at maximum 130 °C and the other 64% after 30 min heating at maximum 120 °C [38].

In 2005, Larhed's group presented extremely fast cobalt carbonyl mediated syntheses of ureas from primary amines [39]. A protocol was developed where a primary amine, $Co_2(CO)_8$ (0.66 equiv.) and triethylamine (2 equiv.) in acetonitrile

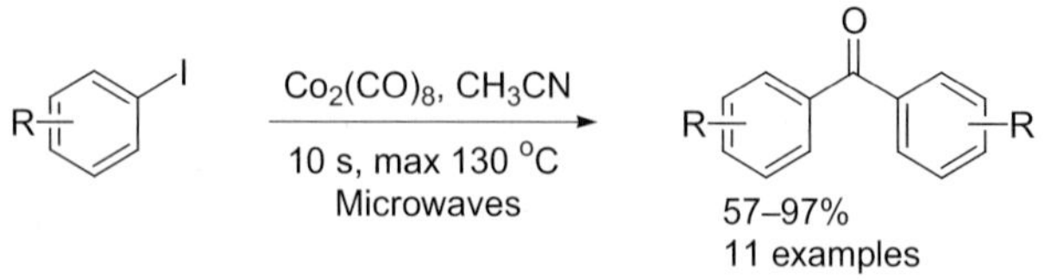

$Co_2(CO)_8$, CH_3CN
10 s, max 130 °C
Microwaves

57–97%
11 examples

Scheme 4.25

$Co_2(CO)_8$, CH_3CN, NEt_3; 10 s to 40 min, 110–150°C, Microwaves; 10–86%, 10 examples

Scheme 4.26

were subjected to microwave heating. In many cases, 10 s of irradiation was enough to produce the symmetric urea in high yields whereas other amines required a reaction time up to 40 min (Scheme 4.26).

The authors proposed that an isocyanate was initially formed as an intermediate, which was then captured by a remaining free amine to form the urea product. This proposed isocyanate intermediate was exploited in attempts to synthesize unsymmetrical ureas. There, an excess of a secondary amine was intended to capture the isocyanate, forming the unsymmetrical- and minimizing-symmetrical urea formation. Indeed, four unsymmetrical ureas were isolated in 41–55% yield when 10 s of microwave irradiation with 5 equiv. of a secondary amine was used [39].

4.3.3
Microwave-Promoted Carbonylations Using the Solvent as a Source of Carbon Monoxide

An alternative strategy for performing microwave carbonylations without directly using carbon monoxide is to use formic acid derivatives as the source of CO. In fact, common solvents such as DMF or formamide are known to thermally decompose in the presence of a strong base to carbon monoxide and the corresponding amine [40]. The carbon monoxide released in this manner has been used in palladium-catalyzed carbonylation of aryl halides. The nucleophile involved can be either amine derived from the CO source or externally added.

In 2002, Alterman and coworkers exploited this concept and developed a microwave-promoted carbamoylation protocol using DMF as the *in situ* carbon monoxide liberator in septum-sealed reaction vials [40]. It was discovered that aryl dimethyl amides were accessible from the corresponding bromides in the presence of KO*t*-Bu and the nucleophilic catalyst, imidazole (Scheme 4.27). Moreover, tertiary benzamides other than dimethylamides were prepared by adding 3 equiv. of an external primary or secondary amine. This *in situ* carbon monoxide generation methodology

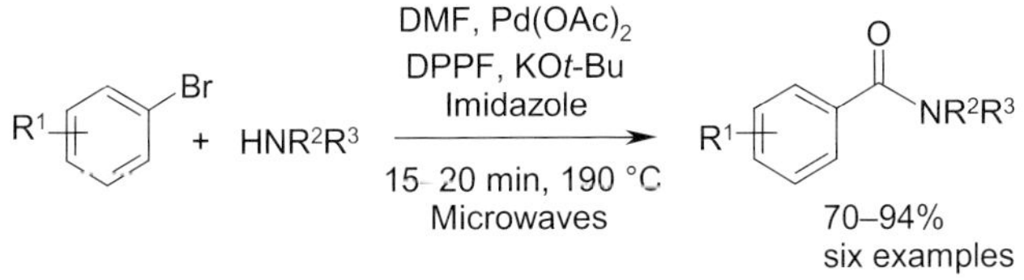

+ HNR^2R^3; DMF, $Pd(OAc)_2$, DPPF, KO*t*-Bu, Imidazole; 15–20 min, 190 °C, Microwaves; NR^2R^3; 70–94%, six examples

Scheme 4.27

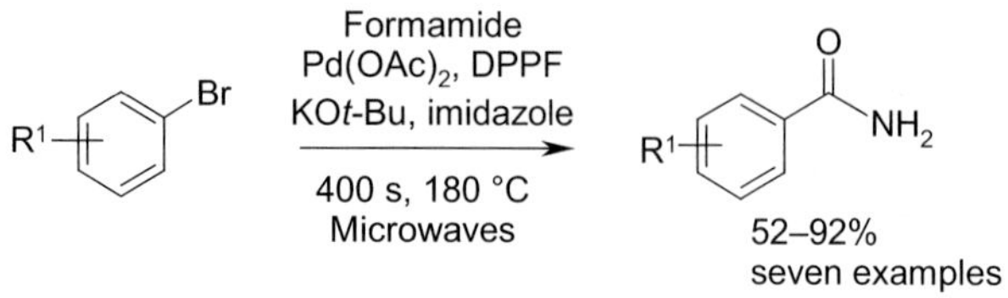

Formamide
$Pd(OAc)_2$, DPPF
KO*t*-Bu, imidazole
400 s, 180 °C
Microwaves

R^1 Br → R^1 $CONH_2$

52–92%
seven examples

Scheme 4.28

works efficiently with bromobenzene and more electron-rich aryl bromides, but aminocarbonylation of electron-poor aromatic systems does not occur [40].

As previously discussed, the palladium-catalyzed aminocarbonylation of aryl halides in the presence of carbon monoxide gas and amines constitutes a versatile methodology for the selective and direct synthesis of secondary and tertiary benzamide derivatives. Unfortunately, the synthesis of primary amides is recognized as considerably more difficult, attributed, in part, to the lower nucleophilicity of ammonia in combination with the less convenient handling of gaseous ammonia. As with DMF, the solvent formamide has been shown to be an excellent source of carbon monoxide that also serves as an effective ammonia synthon [41]. Scheme 4.28 illustrates the reported results in which, again, KO*t*-Bu and imidazole were used as essential activating agents.

4.3.4
Microwave-Promoted Carbonylations Using Reaction Vessels Prepressurized with Carbon Monoxide

The last alternative strategy to conduct microwave-assisted carbonylations in sealed vessels involves the use of carbon monoxide prepressurized reaction tubes. The main advantage of using gaseous CO is that neither does it generate any metal waste nor does it require the use of special activators to chemically release CO from an organic source. The drawbacks are the demand for specialized equipment such as an external pressure controlling system equipped with a gas-loading interface, a heavy-walled reaction vessel, and an exit tube for venting the vial at the end of the reaction. The use of this type of purpose-built instrumentation has recently been reported both by Leadbeater and Taddei and coworkers [42–44].

Alkoxycarbonylations of aryl iodides were performed using alcohol as the solvent and nucleophile in sealed tubes preloaded with 10 bar CO [43]. Reactions delivered high yields of ester products after 20 min of microwave heating at 125 °C employing only 0.1 mmol of old-fashioned palladium acetate as the catalyst (Scheme 4.29). The authors found it possible to scale up the reaction to 4 mmol without a drop in yield provided the heating time was extended to 30 min. Given the fact that the microwave synthesizer could accommodate eight reaction vessels in parallel, it would be possible to heat 32 mmol of starting material in one run [43].

Despite its great importance in industrial-scale processes, the organic chemistry community has not fully exploited the hydroformylation reaction in designing chemical routes. The explanation lies in the demanding experimental conditions

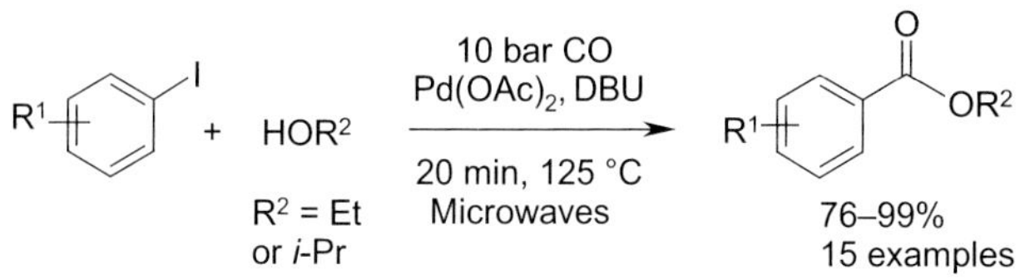

R^1-C$_6$H$_4$-I + HOR^2 → (10 bar CO, Pd(OAc)$_2$, DBU; 20 min, 125 °C, Microwaves) → R^1-C$_6$H$_4$-C(O)OR^2

R^2 = Et or *i*-Pr

76–99%
15 examples

Scheme 4.29

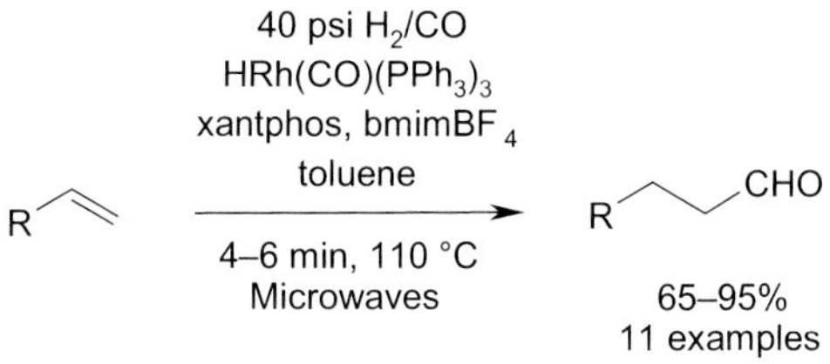

R-CH=CH$_2$ → (40 psi H_2/CO, HRh(CO)(PPh$_3$)$_3$, xantphos, bmimBF$_4$, toluene; 4–6 min, 110 °C, Microwaves) → R-CH$_2$CH$_2$-CHO

65–95%
11 examples

Scheme 4.30

generally applied, using gaseous CO and H_2 at high pressure and temperature. By adapting a standard multimode microwave reactor for gas transfer, hydroformylation reactions were carried out using syngas (CO/H_2 1 : 1) at an initial pressure of only 40 psi [44]. Linear aldehydes were isolated in high yields from terminal alkenes after only 4 min of microwave irradiation at 110 °C using commercially available HRh(CO)(PPh$_3$)$_3$/xantphos as the catalytic system. Addition of the thermostable ionic liquid [bmim][BF$_4$] was necessary to heat the reaction mixture owing to the low tan δ of the toluene solvent (Scheme 4.30) [44].

4.4 Conclusion

From its humble beginnings in mid-1980s [45,46], microwave-accelerated organic synthesis has evolved rapidly in the past 20 years. From being a curiosity, the technology has now spread all over the world. The popularity has grown in parallel with the improvements in performance, control, reproducibility, and safety of the dedicated microwave instrumentation developed in recent years. Competition between the manufacturers has also lead to a reduction of reactor prices. If this trend continues, microwave synthesizers may gradually become the standard tool for heating organic reactions, both in industry and academia.

Microwave-assisted carbonylation reactions have developed rapidly over the past 5 years – from the first reports in 2002, to around 30 publications in the beginning of 2007. Indeed, it is today possible to perform most types of metal-catalyzed carbonylative reactions in sealed vessels with microwave heating and either a suitable CO source or by using appropriate gas-loading accessories and prepressurized heavy-walled reaction vessels. The high-speed chemist can now take advantage of unique carbonylative transformations using only a fraction of previously required handling and process time, an important feat as many carbonylation reactions are known to be both time-consuming and requiring careful optimization. Finally, we believe the combination of microwave heating and CO chemistry will important not only in the field of drug discovery but also for many other types of small-scale organic synthesis.

A Note of Caution

For safety reasons, the heating of sealed vessels, regardless of the method applied, should always be performed carefully only by authorized and skilled personnel using appropriate equipment.

References

1 Larhed, M. and Hallberg, A. (2001) Microwave-assisted high-speed chemistry. A new technique in drug discovery. *Drug Discovery Today*, **6**, 406–416.

2 Kappe, C.O. and Dallinger, D. (2006) The impact of microwave synthesis on drug discovery. *Nature Reviews Drug Discovery*, **5**, 51–63.

3 Lidström, P., Tierney, J., Wathey, B. and Westman, J. (2001) Microwave assisted organic synthesis – a review. *Tetrahedron*, **57**, 9225–9283.

4 Kappe, C.O. (2004) Controlled microwave heating in modern organic synthesis. *Angewandte Chemie-International Edition*, **43**, 6250–6284.

5 Kappe, C.O. and Stadler, A. (2005) *Microwaves in Organic and Medicinal Chemistry*, Wiley-VCH Verlag GmbH, Weinheim, Germany.

6 Olofsson, K., Nilsson, P. and Larhed, M. (2006) Microwave-assisted transition metal-catalyzed coupling reactions, in *Microwaves in Organic Synthesis*, 2nd edn (ed. A. Loupy), Wiley-VCH Verlag GmbH, pp. 685–725.

7 Beller, M., Cornils, B., Frohning, C.D. and Kohlpaintner, C.W. (1995) Progress in hydroformylation and carbonylation. *Journal of Molecular Catalysis A: Chemical*, **1**, 17–85.

8 Skoda-Foldes, R. and Kollar, L. (2002) Synthetic applications of palladium catalysed carbonylation of organic halides. *Current Organic Chemistry*, **6**, 1097–1119.

9 Morimoto, T. and Kakiuchi, K. (2004) Evolution of carbonylation catalysis: no need for carbon monoxide. *Angewandte Chemie-International Edition*, **43**, 5580–5588.

10 Mingos, D.M.P. and Baghurst, D.R. (1991) Applications of microwave dielectric heating effects to synthetic problems in chemistry. *Chemical Society Reviews*, **20**, 1–47.

11 Gabriel, C., Gabriel, S., Grant, E.H., Halstead, B.S.J. and Mingos, D.M.P. (1998) Dielectric parameters relevant to microwave dielectric heating. *Chemical Society Reviews*, **27**, 213–224.

12 Strauss, C.R. and Trainor, R.W. (1995) Invited review – developments in microwave-assisted organic chemistry. *Australian Journal of Chemistry*, **48**, 1665–1692.

13 Kaiser, N.-F.K., Hallberg, A. and Larhed, M. (2002) *In situ* generation of carbon monoxide from solid molybdenum hexacarbonyl. A convenient and fast route to palladium-catalyzed carbonylation reactions. *Journal of Combinatorial Chemistry*, **4**, 109–111.

14 Kaiser, N.F.K., Bremberg, U., Larhed, M., Moberg, C. and Hallberg, A. (2000) Fast, convenient and efficient molybdenum-catalyzed asymmetric allylic alkylation under noninert conditions: an example of microwave-promoted fast chemistry.

Angewandte Chemie-International Edition, **39**, 3596–3598.

15 Herrmann, W.A., Brossmer, C., Reisinger, C.P., Riermeier, T.H., Öfele, K. and Beller, M. (1997) Palladacycles: efficient new catalysts for the Heck vinylation of aryl halides. *Chemistry – A European Journal*, **3**, 1357–1364.

16 Georgsson, J., Hallberg, A. and Larhed, M. (2003) Rapid palladium-catalyzed synthesis of esters from aryl halides utilizing Mo $(CO)_6$ as a solid carbon monoxide source. *Journal of Combinatorial Chemistry*, **5**, 350–352.

17 Wannberg, J. and Larhed, M. (2003) Increasing rates and scope of reactions: sluggish amines in microwave-heated aminocarbonylation reactions under air. *The Journal of Organic Chemistry*, **68**, 5750–5753.

18 Perry, R.J. and Wilson, B.D. (1993) Palladium-mediated carbonylation and coupling reactions of iodobenzene and aniline – model reactions for the preparation of aromatic polyamides. *Macromolecules*, **26**, 1503–1508.

19 Herrero, M.A., Wannberg, J. and Larhed, M. (2004) Direct microwave synthesis of *N,N′*-diacylhydrazines and Boc-protected hydrazides by *in situ* carbonylations under air. *Synlett*, 2335–2338.

20 Netherton, M.R. and Fu, G.C. (2001) Air-stable trialkylphosphonium salts: simple, practical, and versatile replacements for air-sensitive trialkylphosphines. Applications in stoichiometric and catalytic processes. *Organic Letters*, **3**, 4295–4298.

21 Wu, X., Nilsson, P. and Larhed, M. (2005) Microwave-enhanced carbonylative generation of indanones and 3-acylaminoindanones. *The Journal of Organic Chemistry*, **70**, 346–349.

22 Wu, X., Rönn, R., Gossas, T. and Larhed, M. (2005) Easy-to-execute carbonylations: microwave synthesis of acyl sulfonamides using $Mo(CO)_6$ as a solid carbon monoxide source. *The Journal of Organic Chemistry*, **70**, 3094–3098.

23 Glasnov, T.N., Stadlbauer, W. and Kappe, C. O. (2005) Microwave-assisted multistep synthesis of functionalized 4-arylquinolin-2 (1*H*)-ones using palladium-catalyzed cross-coupling chemistry. *The Journal of Organic Chemistry*, **70**, 3864–3870.

24 Kappe, C.O. (2000) Recent advances in the Biginelli dihydropyrimidine synthesis. New tricks from an old dog. *Accounts of Chemical Research*, **33**, 879–888.

25 Wannberg, J., Dallinger, D., Kappe, C.O. and Larhed, M. (2005) Microwave-enhanced, and metal-catalyzed functionalizations of the 4-aryl-dihydropyrimidone template. *Journal of Combinatorial Chemistry*, **7**, 574–583.

26 Wannberg, J., Kaiser, N.-F.K., Vrang, L., Samuelsson, B., Larhed, M. and Hallberg, A. (2005) High-speed synthesis of potent C2-symmetric HIV-1 protease inhibitors by *in situ* aminocarbonylations. *Journal of Combinatorial Chemistry*, **7**, 611–617.

27 Cao, H. and Xiao, W.-J. (2005) Microwave-accelerated, palladium-catalyzed carbonylative cyclization reactions of 2-iodophenol with alkynes. Rapid and efficient synthesis of chromen-2-one derivatives. *Canadian Journal of Chemistry*, **83**, 826–831.

28 Wu, X. and Larhed, M. (2005) Microwave-enhanced aminocarbonylations in water. *Organic Letters*, **7**, 3327–3329.

29 Wu, X., Ekegren, J.K. and Larhed, M. (2006) Microwave-promoted aminocarbonylation of aryl iodides, aryl bromides and aryl chlorides in water. *Organometallics*, **25**, 1434–1439.

30 Georgsson, J., Sköld, C., Plouffe, B., Lindeberg, G., Botros, M., Larhed, M., Nyberg, F., Gallo-Payet, N., Gogoll, A., Karlén, A. and Hallberg, A. (2005) Angiotensin II pseudopeptides containing 1,3,5-trisubstituted benzene scaffolds with high AT2 receptor affinity. *Journal of Medicinal Chemistry*, **48**, 6620–6631.

31 Lagerlund, O. and Larhed, M. (2006) Microwave-promoted aminocarbonylations of aryl chlorides using $Mo(CO)_6$ as a solid carbon monoxide source. *Journal of Combinatorial Chemistry*, **8**, 4–6.

32 Lesma, G., Sacchetti, A. and Silvani, A. (2006) Palladium-catalyzed hydroxycarbonylation of aryl and vinyl triflates by *in situ* generated carbon monoxide under microwave irradiation. *Synthesis*, 594–596.

33 Gold, H., Ax, A., Vrang, L., Samuelsson, B., Karlén, A., Hallberg, A. and Larhed, M. (2006) Fast and selective synthesis of novel cyclic sulfamide HIV-1 protease inhibitors under controlled microwave heating. *Tetrahedron*, **62**, 4671–4675.

34 Wu, X., Wannberg, J. and Larhed, M. (2006) Hydroxylamine as an ammonia equivalent in microwave-enhanced aminocarbonylations. *Tetrahedron*, **62**, 4665–4670.

35 Gupton, J., Banner, E., Scharf, A., Norwood, B., Kanters, R., Dominey, R., Hempel, J., Kharlamova, A., Bluhn-Chertudi, I., Hickenboth, C., Little, B., Coppock, M., Krumpe, K., Burnham, B., Holt, H., Du, K., Keertikar, K., Diebes, A., Ghassemi, S. and Sikorski, J. (2006) The application of vinylogous iminium salt derivatives to an efficient synthesis of the pyrrole containing alkaloids rigidin and rigidin E. *Tetrahedron*, **62**, 8243–8255.

36 Letavic, M.A. and Ly, K.S. (2007) Microwave assisted, palladium catalyzed aminocarbonylations of heteroaromatic bromides using solid $Mo(CO)_6$ as the carbon monoxide source. *Tetrahedron Letters*, **48**, 2339–2343.

37 Odell, L., Savmarker, S. and Larhed, M. Facile aminocarbonylations of aryl triflates; in perparation.

38 Enquist, P.A., Nilsson, P. and Larhed, M. (2003) Ultrafast chemistry: cobalt carbonyl-mediated synthesis of diaryl ketones under microwave irradiation. *Organic Letters*, **5**, 4875–4878.

39 Enquist, P.A., Edin, J., Nilsson, P. and Larhed, M. (2005) Super fast cobalt carbonyl-mediated synthesis of urea compounds. *Tetrahedron Letters*, **46**, 3335–3339.

40 Wan, Y.Q., Alterman, M., Larhed, M. and Hallberg, A. (2002) Dimethylformamide as a carbon monoxide source in fast palladium-catalyzed aminocarbonylations of aryl bromides. *The Journal of Organic Chemistry*, **67**, 6232–6235.

41 Wan, Y.Q., Alterman, M., Larhed, M. and Hallberg, A. (2003) Formamide as a combined ammonia synthon and carbon monoxide source in fast palladium-catalyzed aminocarbonylations of aryl halides. *Journal of Combinatorial Chemistry*, **5**, 82–84.

42 Kormos, C.M. and Leadbeater, N.E. (2006) Microwave-promoted hydroxycarbonylation in water using gaseous carbon monoxide and pre-pressurized reaction vessels. *Synlett*, 1663–1666.

43 Kormos, C.M. and Leadbeater, N.E. (2007) Alkoxycarbonylation of aryl iodides using gaseous carbon monoxide and pre-pressurized reaction vessels in conjunction with microwave heating. *Organic and Biomolecular Chemistry*, **5**, 65–68.

44 Petricci, E., Mann, A., Rota, A., Schoenfelder, A. and Taddei, M. (2006) Microwaves make hydroformylation a rapid and easy process. *Organic Letters*, **8**, 3725–3727.

45 Gedye, R., Smith, F., Westaway, K., Ali, H., Baldisera, L., Laberge, L. and Rousell, J. (1986) The use of microwave ovens for rapid organic synthesis. *Tetrahedron Letters*, **27**, 279–282.

46 Giguere, R.J., Bray, T.L., Duncan, S.M. and Majetich, G. (1986) Application of commercial microwave ovens to organic synthesis. *Tetrahedron Letters*, **27**, 4945–4948.

5
Recent Advances in Two-Phase Carbonylation

Detlef Selent

5.1
Introduction

Two-phase, or biphasic, catalysis has gained increasing attention over the past two decades. One driving force undoubtedly was the successful implementation of large-scale applications in industry. Rhodium-catalyzed aqueous biphasic propylene hydroformylation, performed by the Ruhrchemie/Rhône-Polenc process, represents the most important carbonylation reaction today, with a total outcome of more than 9.2 million tons of products per year [1].

The highly interesting concept of two-phase catalysis combines two main features. First, the catalyst is solubilized in a liquid phase, thus to be dispersed homogeneously on a molecular level as a prerequisite for high activity and selectivity. Second, for easier separation of catalyst and product, an additional phase is presented or formed during the reaction. This latter phase may consist of the product or mixtures of substrate, product and a cosolvent. Separation of the catalyst and its product will be possible, if sufficiently differing partition coefficients of catalyst and product exist. As an additional precondition for effective catalysis, substrate diffusion to the catalyst must not be severely hindered. Some solubility of substrate in the catalyst containing phase and appropriate methodology, bringing the catalyst and substrate together at an extended phase boundary will be advantageous. Improved and environmentally friendlier technology is still under development. Newly developed methodology may ask for an extension of the classical description that defines two-phase catalysis as a purely liquid/liquid case. In 2005, the progress in the field was an important part of textbooks on multiphase catalysis [2]. A set of excellent reviews on multiphase catalysis was published recently, including latest developments in two-phase catalysis with a wider scope of chemistry involved [3].

This chapter will focus on biphasic carbonylation reactions that are divided by reaction types. Mainly results from 2005 and younger literature are covered.

Modern Carbonylation Methods. Edited by László Kollár

ISBN: 978-3-527-31896-4

5.2
Carbonylation Reactions

5.2.1
Hydroformylation

For organic/aqueous biphasic catalysis, hydroformylation of higher, very sparingly water-soluble olefins is still a challenge. One of the several approaches under investigation is micellar catalysis. A pronounced effect on biphasic hydroformylation of 1-dodecene have dicationic surfactants [4]. Critical micelle concentrations (cmc) of $C_{16}H_{33}(CH_3)_2N^+{-}X{-}N^+(CH_3)_2C_{16}H_{33}$ $(Br^-)_2$ type gemini dicationics with X = $-(CH_2)_{2,4,6}-$ and *p*-xylyl (G(xyl), see Scheme 5.1) are nearly an order of magnitude lower than that of a usual cationic such as cetyltrimethylammonium bromide (CTAB), indicating that the spacer between the nitrogen atoms, and also the number of long-chain alkyl groups attached to nitrogen, does strongly influence the structure of the micelles formed. This gave input to the catalytic activity of 1-dodecene hydroformylation with a Rh/TPPTS catalyst formed from $[RhCl(CO)(TPPTS)_2]$/16 TPPTS at 100 °C, 20 bar syngas and $[Rh] = 10^{-3}\,mol\,l^{-1}$. Much above cmc, turnover frequencies increased with enhancing the surfactant concentration within the range of $1–6 \times 10^{-3}\,mol\,l^{-1}$. Dicationics gave an improvement in the reaction compared to common CTAB. Maximum turnover frequencies (TOFs) of ~1000 h^{-1} (CTAB: 800 h^{-1}) were measured. However, highly surface active $(C_{16}H_{33})_2N^+(CH_3)_2Br^-$ gave an even higher rate of 1200 h^{-1}, as shown in the Figure 5.1. Recycling experiments verified a decay of conversion from 76 to 65%, including four catalyst recycles. Leaching determined was 0.18% in total rhodium to the organic phase per cycle at a P/Rh ratio of 18. With more TPPTS added up to a ratio P/Rh = 54 leaching to the organic phase was prevented almost completely. As an alternative, it is possible to extract rhodium from the organic phase with concentrated aqueous TPPTS solution. However, results suggest that the rhodium is lost mechanically or the catalyst is deactivated irreversibly during the extraction routine.

An enhancement of the reaction rates was also achieved in micellar catalysis of hydroformylation of terminal C_8 to C_{18} olefins with the Rh/TPPTS catalyst employed in the aqueous/organic phase system without additional organic solvent in the presence of surfactants of the type $R^1R^2N(CH_3)_2Br$ (e.g., $R^1 = C_{22}H_{45}$, $R^2 = C_nH_{2n+1}$, $n = 2, 4, 8, 12, 16$) [5]. Surprisingly, high turnover frequencies of 7865 h^{-1}, measured for 1-dodecene in the biphasic medium, were comparable with that of the homogeneous reaction. Essential for high activities were two long alkyl chains present in the

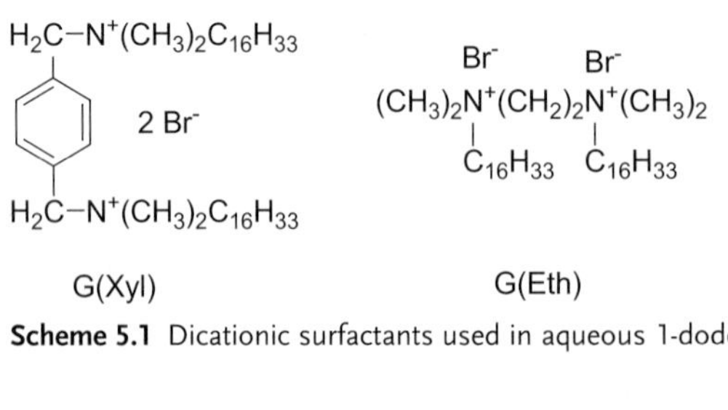

Scheme 5.1 Dicationic surfactants used in aqueous 1-dodecene hydroformylation [4].

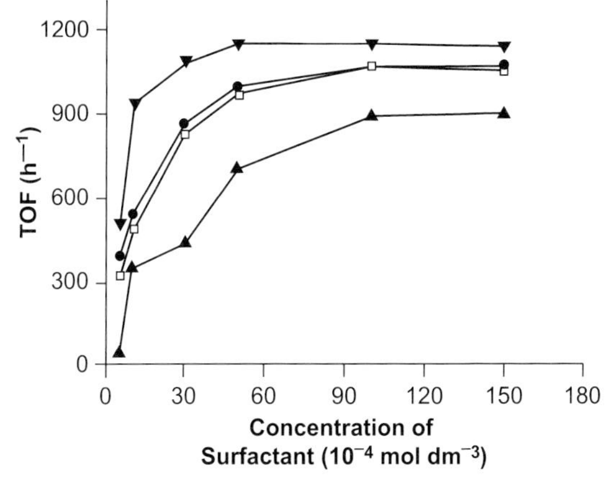

Figure 5.1 Influence of spacer in dicationics and of a second long alkyl chain in a classical monocationic on TOF of aqueous 1-dodecene hydroformylation [4]: (▲) CTAB, (•) G(Xyl), (□) G(Eth), (▼) $(C_{16}H_{33})_2N^+(CH_3)_2Br^-$.

ammonium cation; furthermore, a trend is observed suggesting the synergistic, substrate–detergent interaction that leads to the acceleration of the reaction. Interestingly, alkylimidazolium salts, added to a similar reaction mixture and catalyst, act as surfactants and accelerate the aqueous biphasic hydroformylation of 1-hexene, 1-octene, and 1-decene [6]. Especially, OctMimBr applied above the critical micelle concentration of 2.16×10^{-2} mol l^{-1} provided a good combination of rate enhancement, phase separation after reaction, and low rhodium leaching of less than 0.5 ppm. With 1-octene, at 100 °C, 20 bar CO/H_2 and $[Rh]_{H_2O}$ 1.25×10^{-3} mM, an initial TOF of 1105 h^{-1} was measured. Aldehyde yields up to 90% resulted with the formation of 5.7% isomerized octenes within 3 h.

A concept focused on a micelle or a hydrophobic domain formed from a soluble phosphine-modified polymer, which represents a microreactor containing the catalyst fixed in the hydrophobic core, was published [7]. In this concept, advantage is taken from enhanced reaction rates as an effect of high local hydrophobic substrate concentration, providing better catalyst/substrate contact. An effort was made to synthesize and characterize block and random copolymers of oxazolidine derivatives. Nonyl substituents introduced allowed for the adjustment of hydrophobicity of the polymer. A comparison of phosphane-modified and nonmodified copolymers revealed suppressed olefin isomerization during aqueous biphasic rhodium-catalyzed hydroformylation of 1-octene with conversion-independent regioselectivity, $n/i = 3$, for the intramicellar catalyst only. Micellar systems derived from the random copolymer showed superior activity over the block copolymer, as TOFs of 3700 h^{-1} were observed at 100 °C, 50 bar, $[Rh] = 2 \times 10^{-4}$ mol l^{-1}, $[polymer] = 8 \times 10^{-4}$ mol l^{-1} (P/Rh = 12.4–17.6). This was attributed to the higher surface activity of the random polymer. However, easier phase separation was achieved for the block copolymer (see Scheme 5.2) derived catalyst, which therefore was used in recycling experiments.

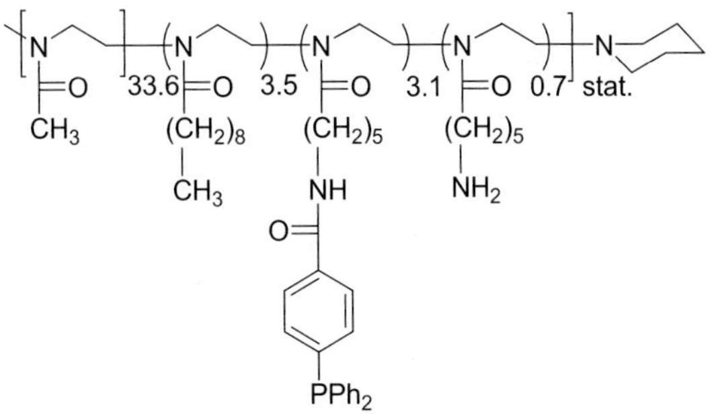

Scheme 5.2 Phosphine-modified block copolymer used for rhodium-catalyzed aqueous hydroformylation of 1-octene [7].

Within four cycles, the best catalyst almost completely lost its activity. The authors speculated about oxidation during recycling and decomposition of the hydrido rhodium complex in the presence of water.

Stabilizing the catalyst against acids was needed for the Rh/BINAS-catalyzed aqueous hydroformylation of internal olefins (see Table 5.1) [8]. Reaction rates were low (averaged TOF = 62 h^{-1} for 2-pentene), but very high regioselectivities of 99% toward the terminal aldehydes were obtained for the hydroformylation of 2-pentene and 2-octene, respectively, under optimized reaction conditions. Controlling pH was found to be essential to increase both the selectivity and the aldehyde yield. Best results were obtained in solutions buffered at pH 8–9, or with additional triethanolamine or TMEDA employed to trap formic acid suggested to be formed in a side reaction.

The water-soluble pyrazolato complex $[Rh(\mu\text{-}Pz)(CO)(TPPTS)]_2$ was used as precursor for olefin hydroformylation in an aqueous heptane solvent system [9]. Without additional ligand, olefin isomerization dominated hydroformylation at 7 bar CO/H_2 (1 : 1) giving large amounts of 2-hexene. Isomerization was suppressed most effectively at 49.8 bar, leading to aldehyde chemoselectivities >90% and a 2.4–3.5 *n*/*iso* ratio. It was demonstrated that aerobic recycling of the aqueous catalyst phase by

Table 5.1 Aqueous Rh/BINAS-catalyzed hydroformylation of different internal olefins[a] [8].

Substrate	*T* (°C)	p_0 [CO]	p_0 [H_2]	p_R[b] (bar)	*t* (h)	Aldehyde (%)	*n*/*i*	Alcohol (%)
2-Butene[c]	125	2	10	19	24	68	98 : 2	4
2-Pentene	125	2	10	18	24	74	99 : 1	3
Octenes[d,e]	125	2	10	16	24	10	97 : 3	0.1
Octenes[d,e]	125	2	10	16	72	33	98 : 2	2
2-Octene[d]	140	2	10	16	24	47	99 : 1	3

[a] Reaction conditions: aqueous phase, buffer (pH 8), 73 mmol olefin, olefin : BINAS : Rh = 2000 : 5 : 1.
[b] Pressure at reaction temperature.
[c] 144 mmol 2-butene (*E*/*Z*-mixture), 2-butene : BINAS : Rh = 4000 : 5 : 1.
[d] Aqueous phase: 19 ml H_2O + 1 ml triethanolamine + 20 ml PEG 300.
[e] Octene mixture (1-oct. : 2-oct. : 3-oct. : 4-oct. = 4 : 46 : 34 : 13).

Table 5.2 Comparison of recycling experiments performed under air and inert gas using $[Rh(\mu\text{-}Pz)(CO)(TPPTS)]_2$ as a precatalyst for aqueous 1-hexene hydroformylation [9].

	Aerobic			Anaerobic				
Routine cycle	**0**	**1**	**2**	**0**	**1**	**2**	**3**	**4**
Conversion (%) of 1-hexene	100	10	3	100	99	99	99	98
n/i ratio	2.1	2.6	2.8	3.2	2.6	2.5	2.5	2.4

simple decantation does alter the outcome of the reaction drastically. Within two recycles, the conversion of 1-hexene obtained with the 0.6 mmolar catalyst solution dropped from 100 to 3%. In contrast, full conversion was achieved for cumulated batches with the same catalyst when new substrate was added for several times. Obviously, air contact of recycled catalyst solution had a negative influence (Table 5.2).

One possible way to overcome the known limitations of mass transfer is to enhance the surface area between the catalyst-containing aqueous phase and organic substrate by microemulsions (MEs) which spontaneously form by mixing water, a nonpolar organic liquid, and an appropriate detergent in a determined ratio. On the macroscopic level, MEs appear to be homogeneous; however, microscopically, these ternary mixtures are structured in hydrophilic, water-rich, and hydrophobic domains separated by an amphiphilic membrane. Structure and fraction of the domains depend on the surfactant concentration. As the solubility of usual amphiphiles in water decreases with increasing temperature, the phase behavior of MEs strongly depends on temperature. It is possible to form three-phase systems at low amphiphile concentration; when temperature is raised and a new ME is formed that coexists with one oil-rich and one water-rich phase known as nonpolar and polar phases, respectively. It is possible to solubilize these phases in the microemulsion at a higher detergent concentration. Hydroformylation of 1-octene was performed with an Rh/TPPTS catalyst dissolved in water, P : Rh = 10, 100 ppm Rh, with 60 wt% of olefin acting as reactant and the organic phase, and several nonionic decanol ethoxylates of different hydrophilic–hydrophobic balance including $CH_3(CH_2)_9O(CH_2CH_2O)_7H$ [10]. High averaged rates of 4314 h^{-1} (2 h reaction time) were observed for this detergent applied in concentrations around 1 wt% (see Figure 5.2). Detergents with $x = 5$, 11 tended to foam. Rates comparable to 1-octene were also obtained with higher olefins such as 1-dodecene and 1-tetradecene, characterized by low water solubility. This leads to the conclusion, that reaction does not take place in the aqueous phase. Furthermore, the decrease in the reaction rate observed with enhancing TPPTS/Rh ratio could be shown to be because of blocking the rhodium center instead of an salt effect. A detailed comparison for the same reaction performed by other researchers applying various media, and with varied TPPTS/Rh ratio, showed an initial rate of the microemulsion reaction superior to the biphasic one [11].

Apparently no olefin containing microemulsions were formed in the presence of randomly methylated cyclodextrins (Rame-CDs, Scheme 5.3). The olefin is solubilized to the aqueous phase for hydroformylation with a rhodium/water soluble

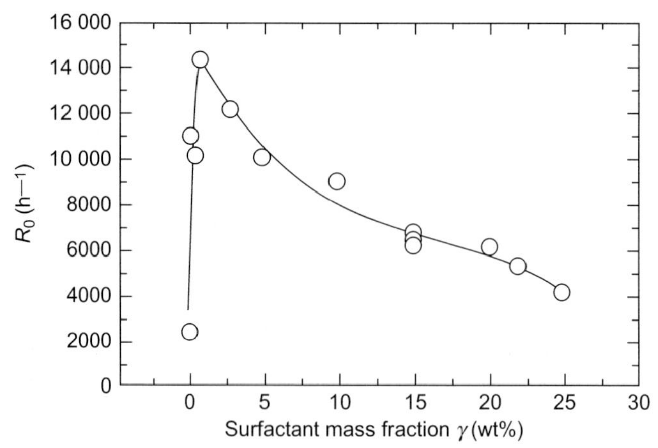

Figure 5.2 Influence of concentration of $CH_3(CH_2)_9O(CH_2CH_2O)_7H$ on the initial rate of nonanal formation from hydroformylation of 1-octene in a microemulsion from water, olefin, and nonionic [10]. ($p = 70$ bar, $T = 100\,°C$, 100 ppm wt rhodium, TPPTS : Rh = 10).

phosphane catalyst. The tris(*m*-carboxyphenyl) phosphane trilithium salt (*m*-TPPTC) used showed higher surface activity that might be responsible for its higher efficiency compared to TPPTS in the hydroformylation reaction of 1-octene, 1-decene, and 1-dodecene [12]. When a ratio of substrate/Rh of 500 was applied, olefin conversions were high after 3 h at 40 bar, 80 °C, with the methylated β-CD providing better results for 1-decene and 1-dodecene, respectively.

2D T-ROESY ^{1}H NMR spectroscopy performed for β-CD/*m*-TPPTC revealed strong correlation peaks between the *ortho*- and *meta*-protons of phosphane with the inner protons H-3 and H-5 of CD, indicating a host–guest interaction between the CD cavity and phosphane. Physicochemical measurements indicate that CDs prefer adsorption at the aqueous/air interface, with their cavity perpendicularly oriented to the interface [13]. Both effects support substrate diffusion of the aqueous-supported

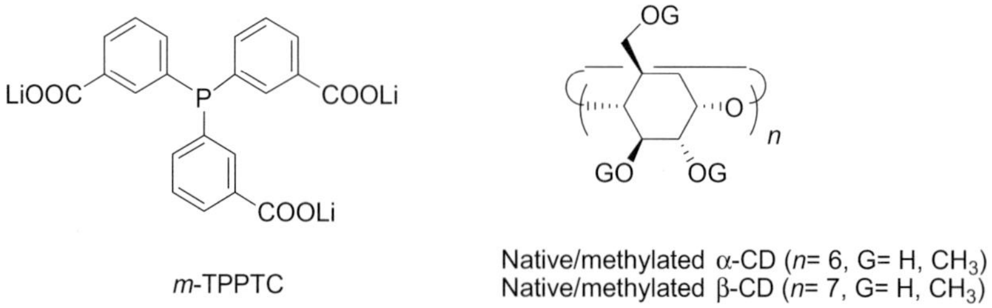

LiOOC P COOLi COOLi

OG GO OG O *n*

m-TPPTC

Native/methylated α-CD (n= 6, G= H, CH_3)
Native/methylated β-CD (n= 7, G= H, CH_3)

Scheme 5.3 The hydrosoluble ligand *m*-TPPTC and types of cyclodextrins used in aqueous hydroformylation of long chain olefins [12].

catalyst. Reaction rates of hydroformylation as well as Tsuji–Trost reactions are linearly correlated to the surface excess of the CD applied. Separation of products is achieved by decantation.

Strongly coordinating, bidentate water-soluble phosphines such as sulfonated xantphos show a less pronounced interaction with Rame-β-CD [14]. Therefore, the formation of coordinatively unsaturated and less regioselective catalytic species owing to phosphine loss is unlikely. Moreover, the additional steric hindrance of the CD cavity is discussed to interfere with catalysis giving enhanced regioselectivity by compelling the substrate to react preferably with the terminal carbon. A systematic study was carried out on the behavior of more flexible bidentate α,ω-alkylene bis (diarylphosphines), $(m\text{-}NaO_3SC_6H_4)_2P{-}(CH_2)_n{-}P(m\text{-}C_6H_4{-}SO_3Na)_2$ ($n = 2$, DPPETS; 3, DPPPTS; 4, DPPBTS) in the presence of cyclodextrins [15]. Interaction was studied by NMR titrations and two-dimensional T-ROESY ^{1}H NMR spectroscopy. All ligands form 1 : 1 inclusion complexes with β-CD. Association constants of about $K = 1600\ M^{-1}$ were determined, thus verifying a much stronger interaction compared to sulfoxantphos, $K = 80\ M^{-1}$. Magnetic coupling similar to the TPPTC ligand was observed between the inner $H_{3,5}$ protons of the CD and aromatic protons of the ligand, strongly suggesting a ligand intrusion from the secondary face of β-CD (complex A, Scheme 5.4). Additionally, interaction toward the alkyl chain of the diphosphine was also observed. Most surprisingly, T-ROESY spectra observed for DPPBTS/β-CD gave proof for the formation of a structurally different 1 : 1 complex occupying the primary face of the CD (complex B, Scheme 5.4). The effects on catalysis of aqueous biphasic 1-decene hydroformylation are comparable for the three different ligands tested. Rame-α- as well as Rame-β-CD enhance the regioselectivity as well as reaction rate in aldehydes. Especially, when Rame-β-CD was present, reduced *n*-aldehyde regioselectivity was observed. This suggests a CD-promoted monodissociation of bidentate ligand. Indeed, as shown by NMR, the free diphenylphosphino group could be trapped by the CD cavity.

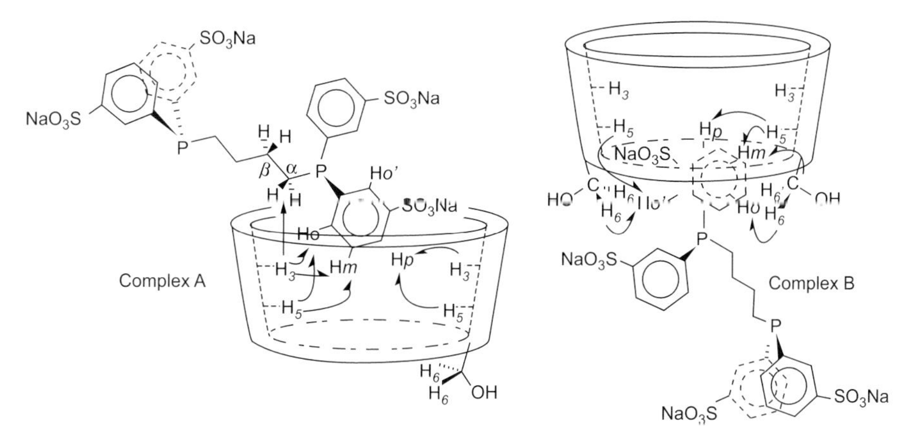

Scheme 5.4 Structures proposed for β-CD/DPPBTS inclusion complexes. The main interactions observed in T-ROESY ^{1}H NMR spectra are indicated [15].

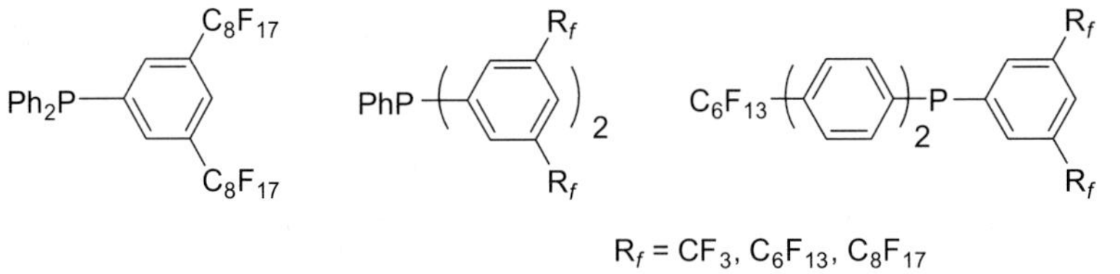

$R_f = CF_3, C_6F_{13}, C_8F_{17}$

Scheme 5.5 Ligands prepared for the fluorous biphasic hydroformylation of 1-octene [16].

For fluorous biphasic catalysis, a set of triarylphosphines $Ph_xP[(3,5\text{-}R_F)_2C_6H_4]_{3-x}$ and $(4\text{-}C_6F_{13})C_6H_4P[(3,5\text{-}R_F)_2C_6H_4]_2$ ($x = 1,2$; $R_F = CF_3$, C_6F_{13}, C_8F_{17}), the first bearing two perfluorinated alkyl groups on at least one of the aryl rings have been used as ligands (see Scheme 5.5) [16]. Newly prepared $(4\text{-}C_6F_{13})C_6H_4P[(3,5\text{-}C_6F_{13})_2C_6H_4]_2$ 1 : 99 has the highest reported partition coefficient of 1 : 99 between toluene and perfluorodimethylcyclohexane ever reported. Short trifluoromethyl groups are not sufficiently fluorophilic to promise high potential in fluorous biphasic catalysis for corresponding ligands. This is illustrated by the partition coefficient of 48/52 determined for $(4\text{-}C_6F_{13})C_6H_4P[(3,5\text{-}CF_3)_2C_6H_4]_2$ in the same solvent combination. The long perfluorinated chains (ponytails) are also responsible for the trend from pure *cis*-coordination to *trans*-coordination of the fluorosubstituted phosphine ligand in $[PtCl_2L_2]$ type complexes as a result of steric congestion. Though coupling constants J_{PPt} increase with the number of fluorinated groups attached, there is no significant difference measured when short and long perfluoroalkyl groups are substituted by each other. These phosphines were used in fluorous biphasic rhodium-catalyzed hydroformylation of 1-octene at 20 bar CO/H_2 (1 : 1) at 70 °C in perfluorodimethylcyclohexane. From gas uptake at constant pressure, first-order rate constants between 1.9 and 3.7 s^{-1} were measured, with the highest activity, but also the highest leaching to the organic product, showing the catalyst derived from the trifluoromethyl substituted ligands. Minimum rhodium loss of 7292 ppm wt was achieved with $P(C_6H_4\text{-}4\text{-}C_6F_{13})_3$ at $P/Rh = 10$.

In addition, a multifold trifluoromethylsubstitution at triphenylphosphine benefits the Rh-catalyzed 1-octene hydroformylation. The sixfold trifluoromethyl substituted triarylphosphine $P[C_6H_3\text{-}3,5\text{-}(CF_3)_2]_3$ was compared with similar but less substituted ligands in different reaction media [17]. It showed the highest value for the turnover frequency = 9820 mol h^{-1} in supercritical carbon dioxide. When employing threefold trifluoromethyl substituted $P(C_6H_4\text{-}3\text{-}CF_3)_3$, an n : *iso* ratio of 4.6–4.8 was measured for both $scCO_2$ and hexane. These values are significantly higher than those obtained with triphenylphosphine (n : *iso* = 3.1–3.3). In terms of reaction rate and regioselectivity, comparable results in supercritical carbon dioxide and hexane were obtained whereas toluene provides a more different reaction medium.

The extension of classical aqueous biphasic propylene hydroformylation using additional supercritical carbon dioxide or supercritical substrate itself may provide a better catalytic performance when homogeneity of the reaction mixture at operating temperature is attained. This was investigated with $[Rh(acac)(CO)_2]$/20 TPPTS

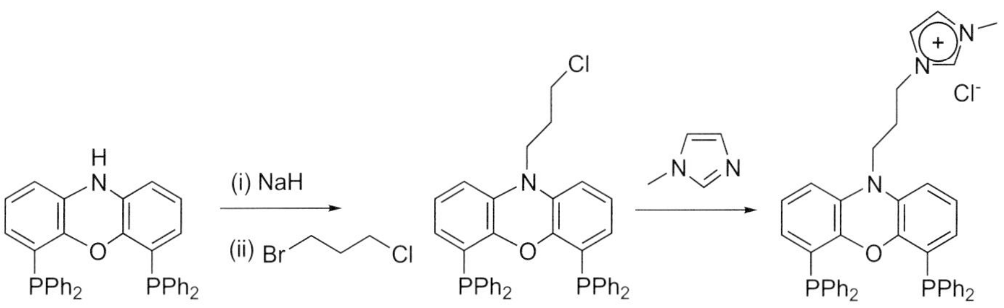

Scheme 5.6 Synthesis of xantphos-type ligand for hydroformylations in $scCO_2$ [20].

applied as a catalyst in small H_2O/ethanol (1 : 1) volumes and the reaction carried out batchwise at 55 °C/40 bar syngas ($scCO_2$) or 117 °C/40 bar syngas (scC_3H_6) [18]. The catalytic activity and aldehyde regioselectivity were found superior to the classical system. The n/i ratio for the $scCO_2/H_2O$ and scC_3H_6/H_2O reactions were described to reach values of 4.3 and 8.4, respectively, but do not meet the ratio of 21 formerly reported for the propylene hydroformylation via the RCH/RP process [19]. However, the use of the supercritical fluid technology is seen to be especially promising for hydroformylation of higher olefins.

Continuous hydroformylation of 1-octene using supercritical carbon dioxide was performed with imidazolium-salt-modified diphosphine ligands of the xantphos type (Scheme 5.6), the latter inducing high n/iso ratios about 40 [20]. As solvent, the steady-state mixture of octane and nonanal can also be used. Supercritical CO_2 is used to feed the system with olefin substrate and to extract the product. Under optimum conditions, the rhodium catalyst is sufficiently insoluble in $scCO_2$. A fragile balance of factors has to be held to successfully perform a continuous reaction: overall pressure and temperature, substrate feed, syngas flow, and the structure of the ligand. A detailed study using the ionic ligands [Rmim][$Ph_2P(3\text{-}C_6H_4SO_3)$] ($R$ = propyl, pentyl, octyl) bearing the phosphine in the anionic part revealed that the pentyl derivative provides an optimum of both, lipophilicity and polarity (Scheme 5.7) [21]. Thus, sufficient catalyst solubility in the product phase was achieved, without significant rhodium leaching by $scCO_2$. In the IL system, 517 turnovers per hour were measured. The challenge of this methodology is unmasked especially in the IL-free system. Enhanced 1-octene flow causes subsequent enrichment of the olefin, precipitation of catalyst and therefore a decay of activity.

To solve the problem of product/catalyst separation in homogeneous catalysis, "thermophoric" solvent mixtures providing a homogeneous solution at reaction

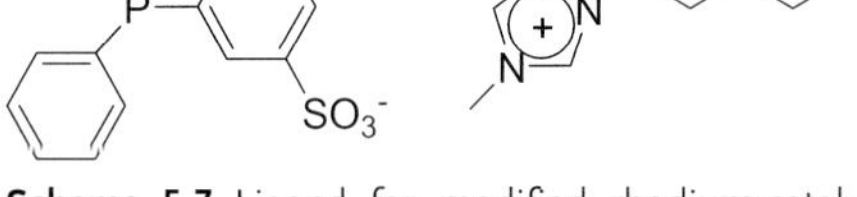

Scheme 5.7 Ligand for modified rhodium-catalyzed 1-octene hydroformylation in nonanal as the starting solvent with a continuous flow of $scCO_2$ [21].

temperature but exhibiting miscibility gaps at the desired workup temperature, together with appropriate partition coefficients for the components to separate, can also be applied. Usually, a high-boiling polar solvent is used as the catalyst "carrier" that forms a homogeneous mixture with the substrate/product containing apolar solvent at the reaction temperature. As a further precondition, polar products must not prevent phase separation upon cooling. For example, for isomerizing hydroformylation of *trans*-4-octene in propylene carbonate (PC)/dodecane mixtures benefit was taken from the reduced solubility of the sterically demanding diphosphite BIPHEPHOS/Rh catalyst in the hydrocarbon solvent [22]. Unfortunately, the substrate and aldehyde both influence the miscibility gap inversely to the desired way (see Figure 5.3).

At high reaction temperatures of 125 °C, the gap is increased in the presence of 4-octene, whereas at low separation temperatures of 25 °C, the gap decreases in the presence of more polar nonanal. However, an optimized solvent system could be identified by using a mediator solvent. Thirty grams of a propylene carbonate/isododecane/*N*-methylpyrrolidone-2 (50/10/40; wt%) mixture allowed complete conversion of 2.24 g *trans*-4-octene (*trans*-4-octene : Rh = 194) at 125 °C/10 bar initial syngas within 4 h. Product separation was accompanied by a minor rhodium and phosphorous ligand leaching, respectively, in the range of 0.1 and 0.5% per cycle. A similar approach used the ternary-solvent system PEG 4000/toluene/*n*-heptane for hydroformylation of *p*-isobutyl styrene applying a rhodium/phosphite catalyst bearing oligoethylene glycol ether chains [23]. At 50 bar/130 °C, a TOF of 388 h^{-1} was

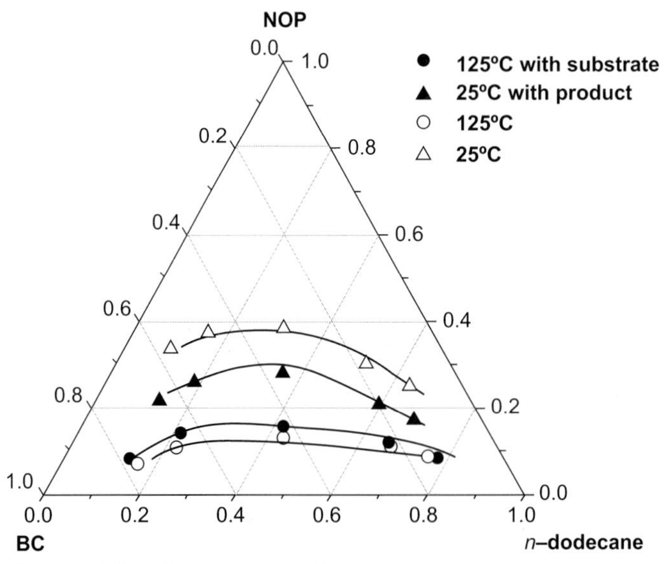

Figure 5.3 Effect of *trans*-4-octene and nonanal on the thermophoric solvent system butylene carbonate (BC), *N*-octyl pyrrolidone (NOP), and dodecane. Addition of 15 wt% 4-octene at 125 °C (reaction temperature) and of 17 wt% of nonanal at 25 °C (separation temperature) [22].

measured, to be compared with 160 h^{-1} found for 1-dodecene hydroformylation with $Rh/P[O(CH_2CH_2O)_8CH_3]_3$ in the PEG 400/*n*-heptane/1,4-dioxane solvent mixture. The most striking feature of the latter reaction was the recyclability and robustness of the catalyst. Twenty-three cycles were possible without significant changes in conversion and selectivity, suggesting that the catalyst loss and alteration is a minor problem [24]. Other solvent mixtures such as DMSO/*n*-heptane have also been tested. With a catalyst formed from $[HRh(CO)(PPh_3)_3]/12\ P(OPh)_3$, aldehyde yields in 1-octene hydroformylation remained constant above 90% over eight cycles representing averaged TOFs of 660 h^{-1}. However, three equivalents of $P(OPh)_3$ per rhodium were added after each run to maintain the performance of the catalyst [25].

5.2.2 Hydroaminomethylation

The unmodified rhodium-catalyzed hydroaminomethylation of 1-octene and morpholine was investigated in ternary thermophoric solvent systems [26]. Therefore, hexane or dodecane, together with propylene carbonate and a mediator solvent, provided complete miscibility at the desired reaction temperature of 125 °C. *N*- and iso-*N*-nonylmorpholine products adopted the nonpolar phase after cooling and could be separated (Scheme 5.8). Applying a 1-octene/morpholine ratio of 1.5, 96% amine selectivity was the highest achieved in the PC/dodecane/1,4-dioxane (1 : 0.55 : 1.3, volume ratio) polar/nonpolar/mediator solvent system. Rhodium loss was measured by ICP–OES to be less than 1.5% and was correlated to the mediator solvent polarity. The less polar the mediator was the more of it dissolved in the product containing nonpolar phase and the more rhodium was lost.

Better retention of rhodium was observed for the similar reaction between 1-dodecene and morpholine performed biphasically at 30 bar CO/H_2, 130 °C, in the presence of an ionic liquid such as $[bmim](p\text{-}CH_3C_6H_4SO_3]$ and a modified Rh/BISBIS catalytic system [27]. The product phase was separated from the catalyst ionic liquid phase by centrifugation and showed 0.03% of rhodium leaching only. Importantly, the authors pointed toward problems that may occur during recycling. Thus, partial oxidation of the bidentate phosphine ligand took place. This had influence only in the linear/branched ratio of the amine product whereas results on conversion, amine selectivity, and undesired parallel reactions such as olefin isomerization and olefin hydrogenation remained unchanged. The authors carried out the same reaction under comparable conditions in the aqueous/organic system also [28]. Thus, comparison of the 1-dodecene reaction with morpholine revealed an amine ratio of $n/i = 63.8$ (aqueous, BISBIS : Rh = 5) and 41.6 (IL, BISBIS : Rh = 2.5), therefore gave similar regioselectivity and overall performance.

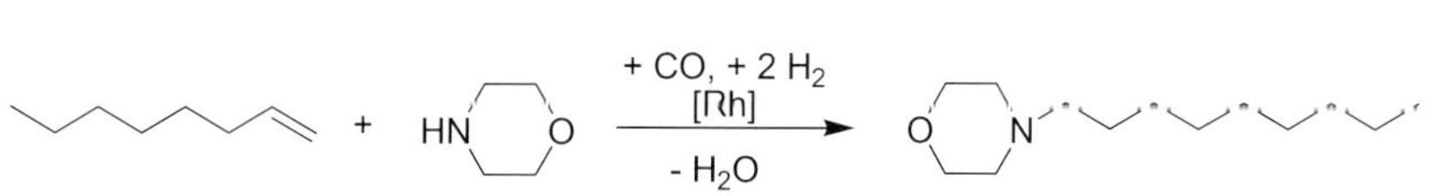

+ HN O — + CO, + 2 H_2 / [Rh] / - H_2O → O N

Scheme 5.8 Hydroaminomethylation of 1-octene with morpholine [26].

5.2.3 Hydroesterification (hydroalkoxycarbonylation) and Related Reactions

Several researchers have applied an indirect biphasic methodology for reactions in ionic liquids, starting with a homogeneous reaction followed by stepwise extraction of product. This approach was used for the hydroesterification reaction of styrene derivatives with methanol/CO catalyzed by $[Cl_2Pd(PPh_3)_2]/2\ PPh_3$ [29]. After reaction at 50–90 °C and 10.3 bar CO pressure, the reaction mixture was extracted with hexanes to obtain *n*- and isophenylpropionic acid methylester products and then the catalyst-containing IL phase was recycled after addition of extra phosphine (Scheme 5.9).

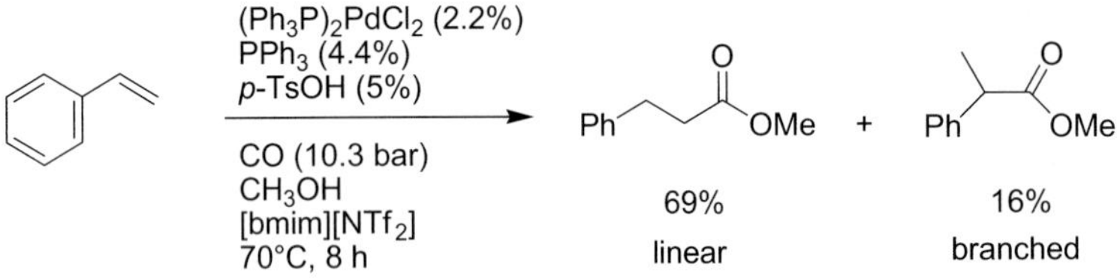

$(Ph_3P)_2PdCl_2$ (2.2%)
PPh_3 (4.4%)
p-TsOH (5%)
CO (10.3 bar)
CH_3OH
$[bmim][NTf_2]$
70°C, 8 h

Ph, O, OMe + Ph, O, OMe

69% linear

16% branched

Scheme 5.9 Hydroesterification of styrene [29].

The best results were obtained in a mixture of methanol with 1-ethyl-3-methylimidazolium ethylsulfate. At 90 °C and 200 psi CO pressure, 70% ester with an l : b ratio of 6.8 was obtained. The highest yield was found in pure methanol, whereas regioselectivity increased significantly with added IL. The cumulated recycling experiments (see Table 5.3) revealed yields and regioselectivity with inverse tendency. This was attributed to a mechanical loss of catalyst containing ionic liquid phase and the subsequent enrichment of the reaction mixture by PPh_3. Similar reactions were performed with broader scope of modifying phosphine ligands and alcohol substrate, respectively. Thus, the reaction of styrene was tested in $[bmim]PF_6$ and $[bmim]BF_4$ with $PdCl_2(PPh_3)_2$ as a precatalyst [30]. Reactions were conducted at 100 °C, $p_{CO} = 100$ bar in the presence of diphosphines $Ph_2(CH_2)_nPPh_2$ ($n = 2$, dppe; 3, dppp; 4, dppb) after which products were extracted with toluene. For dppb, 100% conversion and 91% *n*-selectivity were obtained with ethanol in the hexafluorophosphate

Table 5.3 Recycling of IL catalyst solution used in the hydroesterification of styrene[a] [29].

Run	Yield[b]	l : b[c]
1	85	4.3
2	83	5.0
3	64	5.6
4	50	7.2
5	60	6.8
6	30	9.8

[a] 4 ml $[C_4mim][NTf_2]$, 4 ml methanol, 70 °C, 13.7 bar CO, 0.25 ml styrene.
[b] Determined by GC.
[c] Linear : branched ratio determined by GC.

ionic liquid used. In tetrafluoroborate, in contrast, only 4% conversion was determined with no predominant regioselectivity (*n*/*iso* = 1 : 1). The optimum conditions to be applied for high conversion and selectivity were found to differ strongly, depending on the alcohol substrate, ionic liquid, and phosphine used. A possible reason for the relatively low rates observed was the gas diffusion control because of the viscosity of the ionic liquid, especially when reactions are carried out at lower temperature to diminish solvent decomposition and subsequent catalyst deactivation (Table 5.4).

A more detailed investigation of the influence of the structure of the ionic liquid [Rmim]X on the styrene hydroethoxycarbonylation catalyzed by $PdCl_2(PPh_3)_2$ under conditions identical to that described above also showed the nature of substituents at the imidazolium cation to be significant. One hundred percent iso-selectivity was obtained for R = acetonyl, benzyl. This is interpreted in terms of an enhanced ability of the IL toward metal coordination when electron-withdrawing substituents are present in the solvent [31]. The resulting monophosphine complex gives high iso-selectivity, as it is known to give, from catalysis in organic solvents also. With bidentate bis(diphenylphosphino) ferrocene, dppf, present as additional ligand in equimolar quantities, again [acetonyl-mim][PF_6] produces the most pronounced effect, now giving fully inverse regioselectivity with 100% of linear product. Styrene polymerization was observed for all batches to a higher extent compared to reactions performed in organic solvents. Attempts to recycle the catalyst showed strong leaching of triphenyl phosphine. A decrease in activity was observed even with phosphine added prior to the next cycle, comparable to the same effect described in reference [29].

An inverse aqueous/organic biphasic approach for separation of catalyst and product is available with amphiphilic phosphines bearing amino groups. *N*-Bis (*N'*,*N'*-diethyl-2-aminoethyl)-4-aminomethylphenyl diphenylphosphine (N3P) was used in aqueous acidic media to modify and solubilize the palladium catalyst formed *in situ* from [$PdCl_2(PhCN)_2$] for hydrocarboxylation of 2-pentenoic acid to adipic acid [32]. Reaction carried out at 100 °C, p_{CO} = 50 bar showed best

Table 5.4 Anion and ligand influence on hydroethoxycarbonylation of styrene in ionic liquids[a] [30].

Alcohol	Solvent [bmim]X	Ligand added[b]	Conversion[c] (%)	R_{br} (%)[c]
EtOH	X = BF_4	—	59	81
EtOH	X = BF_4	dppb	4	50
EtOH	X = PF_6	—	3	17
EtOH	X = PF_6	dppb	100	9
BzOH	X = BF_4	—	93	100
BzOH	X = BF_4	dppb	8	100
BzOH	X = PF_6	—	22	45
BzOH	X = PF_6	dppb	29	52

[a] $PdCl_2(PPh_3)_2$ precatalyst, Pd : styrene = 1 : 50, T = 100 °C, p_{CO} = 100 bar, reaction time = 24 h.
[b] One equivalent per Pd.
[c] GC.

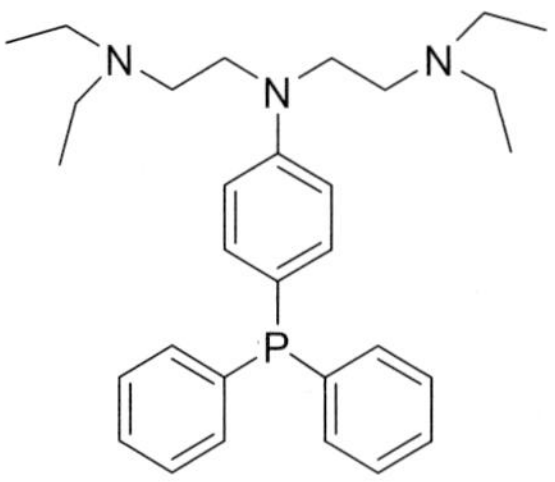

Scheme 5.10 Ligand N3P used for the aqueous hydrocarboxylation of 2-pentenoic acid to adipic acid in the presence of $[PdCl_2(PhCN)_2]$ [32].

performances in the presence of noncoordinating sulfonic acid anions, for example, in an aqueous solution of methanesulfonic acid adjusted to pH 1.8. The separation procedure required addition of base and pH ~12. Extracting deprotonated ligand together with catalyst to an organic phase left the product dianion in the aqueous phase for workup. For the reuse of catalyst, the organic phase was extracted with aqueous acid again. Reaction of 1-octene and styrene was also successful; however, the workup followed usual routine with the catalyst remaining in the water phase. High-pressure NMR spectroscopy (performed in a sapphire NMR tube) of the acidified catalyst solution at 50 bar carbon monoxide gave evidence for the formation of $[Pd(0)(N3P)_3]$ and $[Pd(0)(N3P)_4]$ complexes that reacted smoothly with 3-butene-1-ol to produce acyl and hydroxybutyl palladium(II) species, respectively (Scheme 5.10).

The first hydrocarboxylation of 1-octene with $[PdCl_2(PhCN)_2]$ performed in microemulsions of water with $scCO_2$ showed improved selectivity to the desired acids compared to aqueous/organic biphasic catalysis [33]. This was possible with tris (*p*-trifluoromethylphenyl) phosphine as well as when fluorinated surfactant was applied. At 90 °C and $p_{CO} = 30$ bar turnover frequencies $\sim 5\ h^{-1}$ were observed. Chemoselectivity to the desired nonanoic acid reached 90% with the *n*/*iso* ratio varying from 60 : 40 to 89 : 11. Isolation of products required extraction with diethyl ether.

5.2.4 Amidocarbonylation and Cyclocarbonylation

The photo-induced carbonylations of amines in the presence of aryl and alkyl iodides in ionic liquids gave amides as well as α-keto amides smoothly with palladium carbene complexes as catalysts under irradiation with a xenon arc lamp [34]. Though no instant biphasic procedure was used, extraction of products was easy with cyclohexane as a nonpolar phase. Solvents were degassed from air and saturated with argon before they were used. Interestingly, results showing the solubility of $[bmim]PF_6$ and $[bmim]NTf_2$ in various organic solvents can explain catalyst losses in recycling routines, especially if diethyl ether is used as extractant (see Figure 5.4).

2-Allylphenols were cyclized in the presence of syngas to five- to seven-membered ring lactones in ionic liquid media with a catalyst derived from $Pd_2(dba)_2$ and

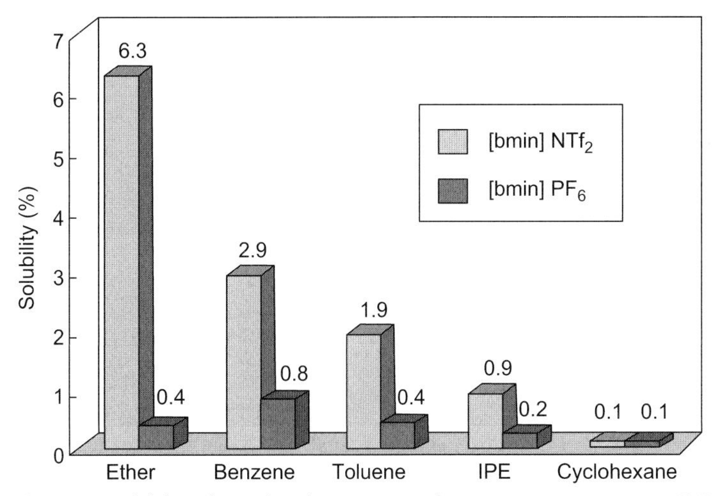

Figure 5.4 Solubility of ionic liquids in organic solvents at room temperature [34].

1,4-bis-(diphenylphopsphino)butane (dppb) [35]. At 90 °C full conversion was achieved for the most of the structurally diverse allylphenols within 20 h. Whereas the seven-membered ring lactone was the desired reaction product, five- and six-membered products were also formed because of an initial double bond isomerization catalyzed by the mutually formed Pd(0) *N*-heterocyclic carbene catalyst. With the same catalyst, 2-vinylphenols, 2-aminostyrenes, and 2-allylanilines could be cyclized under comparable reaction conditions. After extraction of organic products with toluene, the catalyst-containing phase was reused at least six times without any significant loss of performance (Scheme 5.11).

Palladium-catalyzed amidocarbonylation of aldehyde amide mixtures was performed in ionic liquids to obtain *N*-acyl-α-amino acids [36]. It was found that strongly acidic ionic liquids such as [Rmim]HSO_4 (R = $-(CH_2)_3SO_3H$) act as powerful cocatalysts giving higher yields compared to traditionally added sulfuric acid. A subsequent biphasic approach was applied to the workup of catalytic batches and the catalyst-containing IL phase reused without further addition of palladium, phosphine, or cocatalyst.

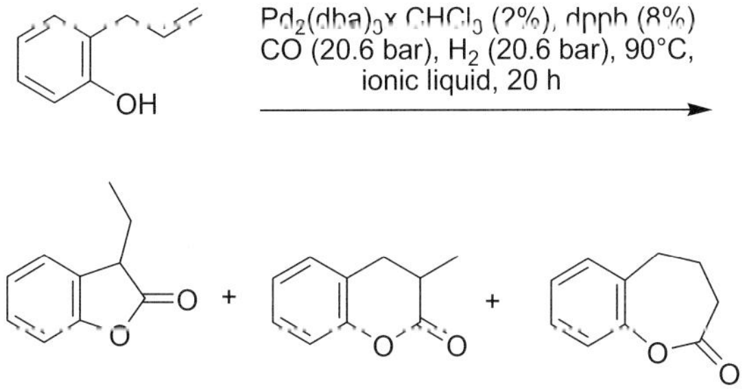

OH

$Pd_2(dba)_3 \times CHCl_3$ (2%), dppb (8%)
CO (20.6 bar), H_2 (20.6 bar), 90°C,
ionic liquid, 20 h

O + O O + O O

Scheme 5.11 Palladium-catalyzed cyclocarbonylation of allylphenol [35].

5.3 Methodology and Stability of Catalysts

The setup and workup of biphasic batches are often challenging at the laboratory level. As described in several contributions, the desired complete separation of the catalyst and products in many cases is not achieved by simple decantation. Some of the main problems identified in connection with separation and recycling are the need of an additional solvent for extraction, formation of emulsions, leaching or other mechanical loss of the catalyst, altering of the organometallic catalyst by depressurization from the reaction pressure and by contact with air, and insufficient inertness of the solvent. One other important point regarding reactions involving gases is sufficient gas diffusion to the place where reaction occurs. Additionally, for gas mixtures applied, unequal gas solubility in the distinct liquid phases will afford the ratio of partial pressures different from that in the gas feed. Data on static solubility for gases can be found in the literature even for ionic liquids. Solubility for carbon monoxide at 22 °C/1.0 bar is 1.47 mM in [bmim]PF_6, compared to 7.30 mM in toluene [37]; further data are available [38,39]. Convenient solutions for mixing in the laboratory imply gas-entry stirrers that act as rotating gas feeding frits, as well as new devices introduced by microreactor technology. A comparison between magnetically stirred batch procedure and methodology using a microflow system was recently reported for Pd-catalyzed carbonylative Sonogashira coupling of aryl iodides and phenylacetylene in ionic liquid (Scheme 5.12) [40].

The microflow system (see Figure 5.5) uses micromixers and was proved to generate a series of small alternating plugs of CO, IL, and the substrate mixture present within the residence time unit at room temperature. By heating to reaction temperature of 120 °C, substrates dissolved in the catalyst-containing IL phase. The result achieved illustrates very impressively the impact of physics on the chemistry observed. Whereas batch reactions showed low selectivity owing to CO starvation that led to extensive aryl iodide acetylene coupling, the microflow system allowed for 100% selectivity to the desired carbonylated product. Obviously only the microflow system was capable of generating a sufficiently high specific gas/liquid interfacial area for diffusion and providing the amount of CO required in time.

New concepts or reaction media may also result in serious effects toward catalysis. Thus, the use of ionic liquids on the basis of C^2-unsubstituted imidazolium salts as solvents for the palladium-catalyzed methoxycarbonylation of iodobenzene leads to protolysis of the Pd-phenyl intermediate and the undesired formation of benzene [41]. Chlorobenzene is also formed in the presence of imidazolium chlorides. This explains why phosphites added did not show modifying properties, because they are

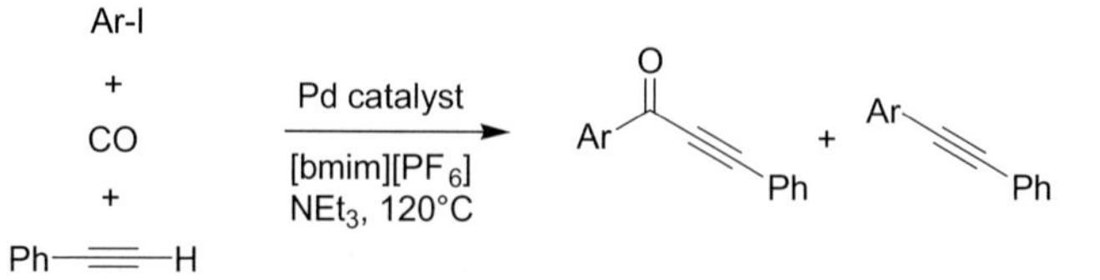

Ar-I + CO + Ph—≡—H $\xrightarrow[\text{[bmim][PF}_6\text{], NEt}_3\text{, 120°C}]{\text{Pd catalyst}}$ Ar–C(=O)–C≡C–Ph + Ar–C≡C–Ph

Scheme 5.12 Pd-catalyzed carbonylative coupling of aryliodides with phenyl acetylene [40].

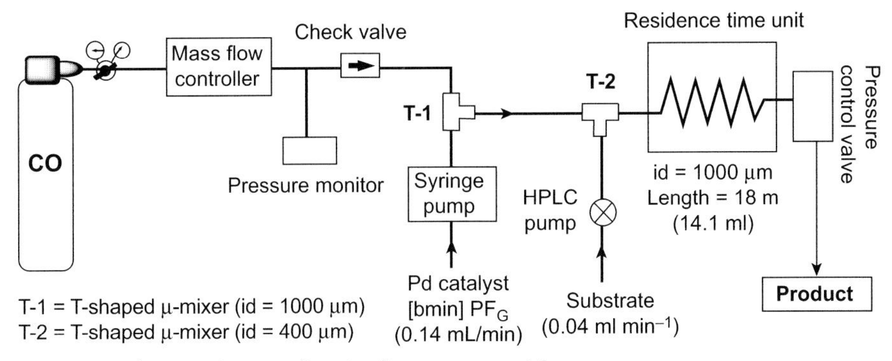

Figure 5.5 Schematic diagram of a microflow system used for carbonylative coupling of aryl iodides with phenyl acetylene [40].

trapped with the subsequent formation of phosphonium salts. If C^2-methylsubstituted imidazolium salts are used, no inhibiting effect was observed. Best precursors for the catalyst were bis(*N*-alkylpyridinium) tetrahalogeno palladates(II).

Interestingly, compressed carbon dioxide can cause melting point depressions of 100 K of simple organic salts, thus expanding the availability range of ionic liquids for use in biphasic catalysis. For the simple salt $[NBu_4][BF_4]$, a liquid phase below 75 °C was achieved [42].

5.4 Innovative Concepts for Catalyst Separation in Biphasic Homogeneous Catalysis

For several reactions studied, classical liquid/liquid two-phase catalysis will not provide the expected results with respect to catalyst performance, product separation, and catalyst recycling. Therefore, improvement of the known methodology and development of new concepts are still under way to make catalysis more effective and to find more versatile catalysts. Thus, a catalyst cartridge system for catalyst recycling based on a biphasic procedure including a $scCO_2$ switch that allows for a repeated reaction, separation sequence under mild conditions in organometallic catalysis was reported [43]. Rhodium catalysts formed from $[Rh(acac)(CO)_2]$ and CH_3O-$PEG_{750}$$-$$PPh_2$ were applied in hydroformylation of structurally diverse olefins. Reaction was carried out in the pure substrate at 70 °C, $p(CO + H_2) = 50$ bar, substrate/Rh = 1000. Complete conversion was achieved after 2 h of reaction time. Addition of CO_2 at 50 °C up to a density of 0.5 g ml^{-1} led to precipitation of the catalyst. The product was separated within the CO_2 phase, as the catalyst-containing residue was extracted with CO_2 also by this methodology ensuring that initial catalytic performance was preserved. The effectiveness of the experimental design was demonstrated not only by using one and the same catalyst in sequential batches for one reaction type, hydroformylation, with the substrate changed, but also by changing chemistry to the one needing a similar catalyst. Thus, the recyclable cartridge catalyst could be used for hydroformylation, hydrogenation, and hydroboration simply with

$scCO_2$ product extraction steps in-between and without the necessity of a certain order of the distinct catalytic reactions.

Other new developments from the substrates point of view are purely biphasic or even homogeneous, but compare with heterogeneous catalysis with respect to handling. Dendrimeric polyphosphines with a Fe_3O_4 inorganic magnetic core are reported to show solubility in organic solvents. They are formed by coprecipitation from Fe(II) and Fe(III) salt solutions in basic conditions to form magnetite nanoparticles of 8–12 nm diameter, coated with silica. Subsequent organic modification, starting with the reaction of surface hydroxyl groups with triethoxy-(3-aminopropyl)silane as the first step, has been performed to build up a dendritic structure with peripheric diphenylphosphine moieties. By combining with [RhCl(cod)]$_2$, the hydroformylation of olefins becomes possible with such kind of "homogenized" heterogeneous catalyst, which was separated easily with an external magnetic field applied [44]. Another bridge between biphasic homogeneous and heterogeneous catalysis is provided by the supported ionic liquid phase (SILP) catalysis. Advantage is taken from better handling, while the enhanced volume fraction of the diffusion layer allows for a less hindered mass transport. For example, a rhodium-sulfoxantphos complex was dissolved in [bmim][n-$C_8H_{17}OSO_3$] and this solution was highly dispersed on silica to be used as a catalyst in the hydroformylation of 1-butene [45]. Such a catalyst can be used in gas-phase reactions and also with condensed substrates. Very recent development demonstrates that the combination of $scCO_2$ technology with SILP catalysis gives a SILP process with supercritical fluid transport (SILP–SCF) [46]. Supercritical carbon dioxide strongly improves the solubility and the diffusion behavior of apolar substrates in ionic liquids. With [Octmim][Tf_2] as supported solvent of choice, continuous flow hydroformylation was achieved with the substrate 1-octene mobilized by $scCO_2$. Rates observed reached 800 h^{-1} with the catalyst remaining stable for at least 40 h.

References

1 Bohnen, H.W. and Cornils, B. (2002) *Advances in Catalysis*, **47**, 1–64.

2 Cornils, B., Herrmann, W.A., Horváth, I.T., Leitner, W., Mecking, S., Olivier-Bourbigou, H. and Vogt, D. (eds)(2005) *Multiphase Homogeneous Catalysis*, Wiley-VCH Verlag GmbH & Co KGaA, Weinheim, Germany.

3 *Advanced Synthesis and Catalysis* (2006) **348** (12 and 13 special issues): *Multiphase Catalysis, Green Solvents and Immobilization*, 1317–1771. See also Bektesevic, S., Kleman, A.M., Marteel-Parrish, A.E. and Abraham, M.A. (2006) *Journal of Supercritical Fluids*, **38**, 232–241.

4 Li, M., Fu, H., Yang, M. Zheng, H., He, Y., Chen, H. and Li, X. (2005) *Journal of Molecular Catalysis A: Chemical*, **235**, 130–136.

5 Fu, H., Li, M., Chen, H. and Li, X. (2006) *Journal of Molecular Catalysis A: Chemical*, **259**, 156–160.

6 Desset, S.L., Cole-Hamilton, D.J. and Foster, D.F. (2007) *Journal of the Chemical Society, Chemical Communications*, 1933–1935.

7 Bortenschlager, M., Schöllhorn, N., Wittmann, A. and Weberskirch, R. (2007) *Chemistry – A European Journal*, **13**, 520–528.

8 Klein, H., Jackstell, R. and Beller, M. (2005) *Journal of the Chemical Society, Chemical Communications*, **17**, 2283–2285.

9 Baricelli, P.J., López-Linares, F., Bruss, A., Santos, R., Lujano, E. and Sánchez-Delgado, R.A. (2005) *Journal of Molecular Catalysis A: Chemical*, **239**, 130–137.

10 Miyagawa, Ch.C., Kupka, J., and Schumpe, A. (2005) *Journal of Molecular Catalysis A: Chemical*, **234**, 9–17.

11 Ünveren, H.H.Y. and Schomäcker, R. (2006) *Catalysis Letters*, **110**, 195–201.

12 Tilloy, S., Genin, E., Hapiot, F., Landy, D., Fourmentin, S., Genêt, J.-P., Michelet, V. and Monflier, E. (2006) *Advanced Synthesis and Catalysis*, **348**, 1547–1552.

13 Leclercq, L., Bricout, H., Tilloy, S. and Monflier, E. (2007) *Journal of Colloid and Interface Science*, **307**, 481–487.

14 Leclercq, L., Hapiot, F., Tilloy, S., Ramkisoensing, K., Reek, J.N.H., van Leeuwen, P.W.N.M. and Monflier, E. (2005) *Organometallics*, **24**, 2070–2075.

15 Tilloy, S., Crowyn, G., Monflier, E., van Leeuwen, P.W.N.M. and Reek, J.N.H. (2006) *New Journal of Chemistry*, **30**, 377–383.

16 Adams, D.J., Bennett, J.A., Cole-Hamilton, D.J., Hope, E.G., Hopewell, J., Kight, J., Pogorzelec, P. and Stuart, A.M. (2005) *Journal of the Chemical Society, Dalton Transactions*, 3862–3867.

17 Koeken, A.C.J., van Vliet, M.C.A., van den Broeke, L.J.P., Deelman, B.-J. and Keurentjes, J.T.F. (2006) *Advanced Synthesis and Catalysis*, **348**, 1553–1559.

18 Jingchang, Z., Hongbin, W., Hongtao, L. and Weiliang, C. (2006) *Journal of Molecular Catalysis A: Chemical*, **260**, 95–99.

19 Herrmann, W.A., Albanese, G.P., Manetsberger, R.B., Lappe, P. and Bahrmann, H. (1995) *Angewandte Chemie – International Edition in English*, **34**, 811–813.

20 Webb, P.B., Kunene, Th.E. and Cole-Hamilton, D. (2005) *Green Chemistry*, **7**, 373–379.

21 Webb, P.B. and Cole-Hamilton, D.J. (2004) *Journal of the Chemical Society, Chemical Communications*, 612–613.

22 Behr, A., Henze, G., Obst, D. and Turkowski, B. (2005) *Green Chemistry*, **7**, 645–649.

23 Feng, C., Wang, Y., Jiang, J., Yang, Y., Yu, F. and Jin, Z. (2006) *Journal of Molecular Catalysis A: Chemical*, **248**, 159–162.

24 Yang, Y., Jiang, J., Wang, Y., Liu Ch. and Jin, Z. (2007) *Journal of Molecular Catalysis A: Chemical*, **261**, 288–292.

25 Tijani, J. and ElAli, B. (2006) *Applied Catalysis A*, **303**, 158–165.

26 Behr, A. and Roll, R. (2005) *Journal of Molecular Catalysis A: Chemical*, **239**, 180–184.

27 Wang, Y.Y., Luo, M.M., Lin, Q., Chen, H. and Li, X.J. (2006) *Green Chemistry*, **8**, 545–548.

28 Wang, Y., Chen, J., Luo, M., Chen, H. and Li, X. (2006) *Catalysis Communications*, **7**, 979–981.

29 Klingshirn, M.A., Rogers, R.D. and Shaughnessy, K.H. (2005) *Journal of Organometallic Chemistry*, **690**, 3620–3626.

30 Rangits, G. and Kollár, L. (2005) *Journal of Molecular Catalysis A: Chemical*, **242**, 156–160.

31 Rangits, G. and Kollár, L. (2006) *Journal of Molecular Catalysis A: Chemical*, **246**, 59–64.

32 Karlsson, M., Ionescu, A. and Andersson, C. (2006) *Journal of Molecular Catalysis A: Chemical*, **259**, 231–237.

33 Tortosa-Estorach, C., Ruiz, N. and Masdeu-Bultó, A.M. (2006) *Journal of the Chemical Society, Chemical Communications*, 2789–2791.

34 Fukuyama, T., Inouye, T. and Ryu, I. (2007) *Journal of Organometallic Chemistry*, **692**, 685–690.

35 Ye, F. and Alper, H. (2006) *Advanced Synthesis and Catalysis*, **348**, 1855–1861.

36 Zhu, B. and Jiang, X. (2006) *Synlett*, **17**, 2795–2798.

37 Ohlin, C.A., Dyson, P.J. and Laurenczy, G. (2004) *Journal of the Chemical Society, Chemical Communications*, 1070–1071.

38 Jacquemin, J., Husson, P., Majer, V. and Gomes, M.M.F.C. (2006) *Fluid Phase Equilibria*, **240**, 87–95.

39 Still, C., Salmi, T., Maeki-Arvela, P., Eraenen, E., Murzin, D.Y. and Lehtonen, J.

(2006) *Chemical Engineering Science*, **61**, 3698–3704.

40 Rahman, T., Fukuyama, T., Kamata, N., Sato, M. and Ryu, I. (2006) *Journal of the Chemical Society, Chemical Communications*, 2236–2238.

41 Zawartka, W., Trzeciak, A.M., Ziółkowski, J. J., Lis, T., Ciunik, Z. and Pernak, J. (2006) *Advanced Synthesis and Catalysis*, **348**, 1689–1698.

42 Scurto, A.M. and Leitner, W. (2006) *Journal of the Chemical Society, Chemical Communications*, 3681–3683.

43 Solinas, M., Jiang, J., Stelzer, O. and Leitner, W. (2005) *Angewandte Chemie – International Edition*, **44**, 2291–2295.

44 Abu-Reziq, R., Alper, H., Wang, D. and Post, M.L. (2006) *Journal of the American Chemical Society*, **128**, 5279–5282.

45 Haumann, M., Dentler, K., Joni, J., Riisager, A. and Wasserscheid, P. (2007) *Advanced Synthesis and Catalysis*, **349**, 425–431.

46 Hintermair, U., Zhao, G., Santini, C.C., Muldoon, M.J. and Cole-Hamilton, D.J. (2007) *Journal of the Chemical Society, Chemical Communications*, 1462–1464.

6
Catalytic Carbonylations in Ionic Liquids

Crestina S. Consorti, Jairton Dupont

6.1
Introduction

Multiphase organometallic catalysis, in particular, liquid–liquid biphasic catalysis involving two immiscible phases, may offer the possibility of circumventing the problems associated with the homogeneous process such as product separation, catalyst recycling, and the use of organic solvents. The concept of this system implies that the catalyst is soluble in only one phase, whereas the substrates/products remain in the other phase. The reaction can take place in one (or both) of the phases or at the interface. In most cases, the catalyst phase can be reused and the products/substrates are simply removed from the reaction mixture by decantation.

Nowadays, there are several alternatives under investigation as fluid for multiphase catalysis, including the resurgence of water, perfluorinated hydrocarbons, and supercritical fluids, in particular CO_2. Indeed, the advent of water-soluble organometallic complexes, especially those based on sulfonated phosphorus-containing ligands, has enabled various biphasic catalytic reactions to be conducted on an industrial scale, in particular, for the hydroformylation of olefins.

However, the use of water as catalyst immobilizing phase has its limitations. (i) It is a highly polar and coordinating protic solvent. (ii) It is reactive toward many organometallic complexes and substrates. (iii) From an environmental perspective, trace amounts of organic compounds in water are notoriously difficult to remove. (iv) The synthesis of specially designed water-soluble ligands/organometallic complexes is essential for its use.

In the 1990, Chauvin and coworkers have introduced ionic liquids (ILs) – especially those derived from the combination of quaternary ammonium salts and weakly coordinating anions – as immobilizing agents for various "classical" transition metal catalyst precursors in reactions [1]. In particular, these liquids provide more adequate and favorable environment for carbonylation reactions as compared to those performed in classical organic solvents or water. The vast majority of these compounds a) are effectively nonvolatile (most of them exhibit negligible vapor pressure);

Modern Carbonylation Methods. Edited by László Kollár

ISBN: 978-3-527-31896-4

b) are nonflammable and liquid over a wide range of temperatures; c) are more viscous and more dense than most classical organic solvents; d) have higher thermal, electrochemical, and chemical stabilities compared to those of classical organic solvents and dissolve a very broad range of organic, inorganic, biological, and organometallic compounds and polymeric materials. Moreover, their miscibility with these substances can be finely tuned by changing the nature of the cation and/or anion; the hydrophobicity can also be modulated by the judicious choice of the cation and/or anion or by changing the temperature of the process. Finally, they are easily prepared from commercially available reagents through classical synthetic procedures and several of these liquids are now commercially available.

Among the various advantages for using ionic liquids as immobilizing agents for catalytic carbonylations, the following are worth mentioning:

(i) catalyst stabilization and recycle;
(ii) product separation (e.g., distillation, decantation, extraction with supercritical carbon dioxide ($scCO_2$), and continuous-flow processes);
(iii) controlling selectivity through ionic reaction pathways;
(iv) multiphase carbonylations with long-chain alkenes that usually are not suitable for processing in aqueous-phase organometallic catalysis;
(v) modulating solubility with most of the organic substrates;
(vi) carbonylations via C–H bond activation.

Multiphase carbonylation catalysis performed in ionic liquids (ILs) can lead to various phase systems in which the catalyst should reside in the IL. Before the reaction starts, and in the absence of carbon monoxide or other gaseous reactants, two systems can usually be formed: a single-phase system, in which the substrates are soluble in the ionic liquid, or a biphasic system, in which one or all of the substrates reside preferentially in an organic phase. With the addition of a gaseous reactant, two-phase and three-phase systems can be formed. At the end of the reaction, three systems can be formed: a single-phase system; a two-phase system in which the residual substrates are soluble in the ionic catalytic solution and the products reside preferentially in the organic phase; and a three-phase system, formed, for example, by an ionic catalytic solution, an organic phase containing the desired product, and a third phase containing the by-products. All these systems are obtained in carbonylation reactions, and they will be discussed in detail further in the text.

6.2
Brief History

Carbonylation reactions catalyzed by transition metals were one of the first applications of ionic liquids as liquid supports for multiphase processes. It was showed earlier in 1972 that the hydroformylation of ethylene and alkoxycarbonylation of 1-hexene could be performed by $PtCl_2$ dissolved in molten $[Et_4N][SnCl_3]$, at temperatures between 60 and 100 °C [2]. In 1981, it was reported that ruthenium compounds immobilized in ionic liquids based on the tetrabutylphosphonium cation are able to catalyze the hydrogenation of carbon monoxide to ethylene glycol at 220 °C [3]. This was shortly followed by the investigation of Fischer–Tropsch catalysis promoted by $Ir_4(CO)_{12}$ in molten $AlCl_3/NaCl$ at 175 °C [4]. At the end of 1980s, Knifton extended his work on the use of Ru dissolved in molten salts for the hydroformylation of alkenes [5,6]. The catalyst precursor was prepared by the dispersion of RuO_2 hydrate, $Ru(acac)_2$, and $Ru_3(CO)_{12}$ cluster in low-melting phosphonium salts such as tetrabutylphosphonium bromide (melting point 100–103 °C). Hydroformylation of terminal or internal olefins with this catalytic system under moderate syngas pressures is facile, and the major products are linear alcohols. Aldehydes may also be predominant under certain reaction conditions, and internal olefins can be converted into oxo products. Spectroscopic investigation indicated that the major species in solution is the $[H(Ru_3(CO)_{11}]^-$ polynuclear anion.

The modification of the catalytic system by the combination of bidentate ligands such as 2,2′-bipyridine and 1,2-bis(diphenylphosphino)ethane allows the regioselective hydroformylation of internal olefins to linear alcohols and aldehydes with selectivities greater than 99%. The species involved in this modified catalytic system is probably the anion cluster $[HRu_3(CO)_9(L\text{–}L)]^-$ (L–L = bidentate ligand) that is known to catalyze this reaction under one-phase conditions with selectivities in linear products up to 95%. The anionic ruthenium catalytically active species in the tetrabutylphosphonium ionic liquid are high stable, probably because of the formation of ion pairs of the type $[P(Bu)_4]^+[Ru]^-$. This property allows easy separation of the products (even employing thermal separation techniques) and reuse of the ionic catalytic solution. The mechanism suggested for this process is similar to that proposed for homogeneous conditions.

Although these highly promising results were published more than 20 years ago, this study was not followed up till the advent of low melting point, less water- and air-sensitive 1-*n*-butyl-3-methylimidazolium ionic liquids associated with hexafluorophosphate (PF_6^-), tetrafluoroborate (BF_4^-), and bis(trifluoromethanesulfonyl)imidate (NTf_2^-) anions (Scheme 6.1).

Doubtless, the hydroformylation reaction was the most investigated transition metal catalyzed carbonylation reactions in ionic liquids. This is mainly because of the industrial importance of this reaction that is in operation in various plants, which in some of them operates in aqueous biphasic regimes. The intense academic and industrial interest is mainly because of the limitations of the current aqueous-phase process to short-chain ($<C_6$) alkenes and the very low solubility of heavier olefins in water for an effective and reasonable reaction rate to occur.

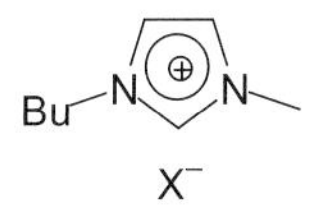

X = PF_6^-, BF_4^-, NTf_2^-

Scheme 6.1 1-*n*-Butyl-3-methylimidazolium ILs.

Other carbonylation reactions – especially those based on palladium catalyst precursors – such alkoxyaminocarbonylation of aryl halides and alkenes have been investigated in ionic liquids.

6.3 Hydroformylation

6.3.1 Classical Rh and Pt Phosphine Catalyst Precursors

Classical transition metal catalyst precursors are, in most of the cases, "soluble" in imidazolium ILs and are not removed from the ionic solution by a major part of organic compounds. Indeed, various catalytic processes can be "directly" transposed in ionic liquids, such as those based on homogeneous transition metal catalyst precursors and colloids, with more advantages than with those performed in organic solvents or water [7]. This is one of the main advantages of ILs in organometallic catalysis, that is, it allows direct transposition of well-known homogeneous processes for liquid–liquid biphasic conditions without the use of the specially designed ligands/complexes that are necessary for the catalytic processes of aqueous, perfluorinated, or supercritical fluids. However, in the case of carbonylation reactions, the classical metal catalysts are removed from the ionic phase by the products formed (aldehydes, alcohols, esters, and amides), and ionically modified ligands such as phosphines have be used to minimize the catalyst leaching [8].

For example, hexanals are poorly soluble in the ionic liquids; a biphasic hydroformylation of 1-pentene catalyzed by $[Rh(CO)_2(acac)]$ associated with PPh_3 dissolved in $BMI{\cdot}PF_6$ yields hexanals with a 99% yield (l/b (*n*-hexanal/*i*-hexanal) = 3, Table 6.1, entry 1) and a TOF of 333 h^{-1} under relatively mild reaction conditions [9]. However, in this particular case, a small part of the rhodium catalyst is extracted from the organic phase. The extraction of the catalyst could almost be completely suppressed by the use of polar ligands such as sodium salts of monosulfonated (tppms) and trisulfonated (tppts) triphenylphosphine ligands (Table 6.1, entry 2). In all hydroformylation reactions performed with Rh complexes associated with classical ligands, a significant amount of catalyst leaching was detected. Therefore, the use of modified phosphine ligands with polar groups such as sulfonates or quaternary ammonium salts associated with the rhodium precursors is essential for avoiding metal leaching in the hydroformylation reactions.

The formation of platinum–diphosphine complexes, presumably involved in the hydroformylation reaction, was investigated in $BMI{\cdot}PF_6$ ionic liquid by NMR. The $PtCl(SnCl_3)(bdpp)$ and $Pt(SnCl_3)_2(bdpp)$ complexes (where bdpp = 2,4-bis(diphenylphosphino)pentane) were identified in the insertion reaction of tin(II)chloride into the Pt–Cl bond of $PtCl_2(bdpp)$ complex. The addition of PPh_3 to $PtCl_2(bdpp)$ resulted in the formation of $[Pt(bdpp)(PPh_3)Cl]^+$ complex cation as a minor component, which turned to be a major component in the presence of tin(II)chloride while it formed trichlorostannate counterion. Both types of reactions show high similarity

Table 6.1 Examples of hydroformylation reactions catalyzed by metal complexes [M] associated with classical neutral N- and P-containing ligands in ionic ligands.

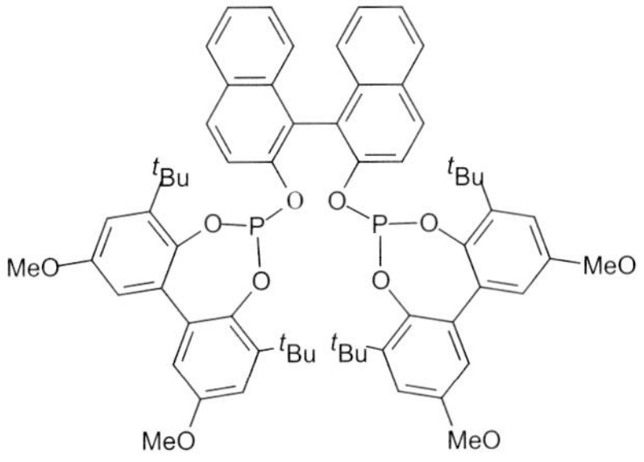

1

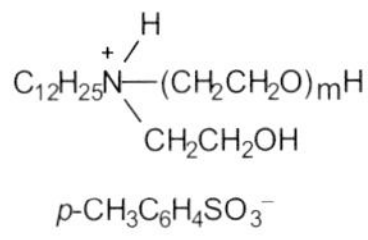

2

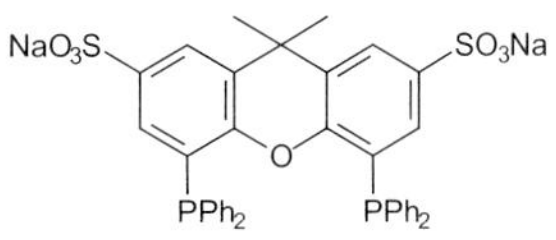

3

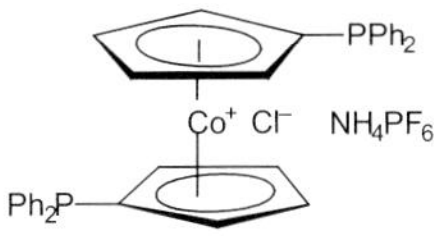

4

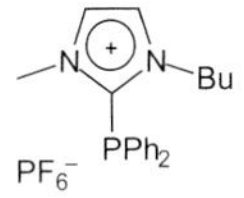

5

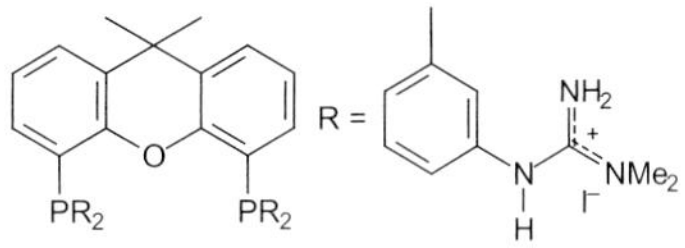

6

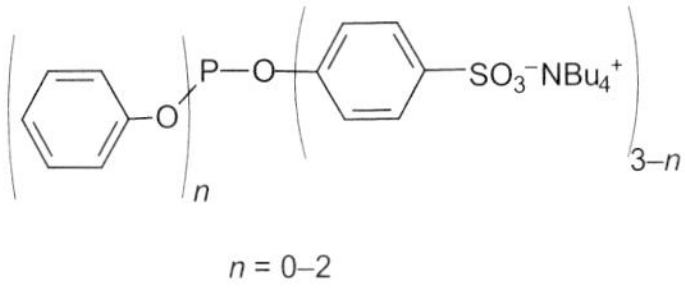

7

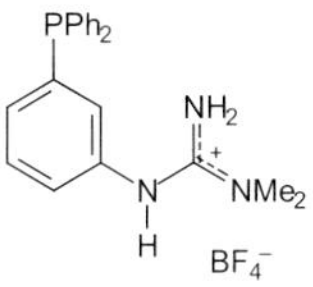

8

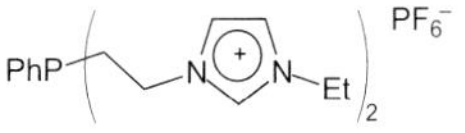

9

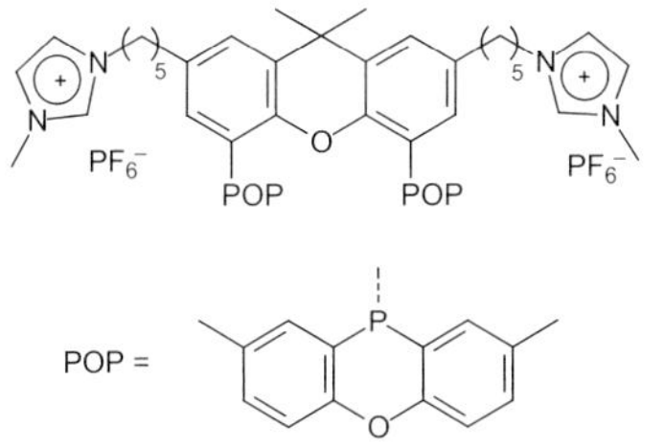

10

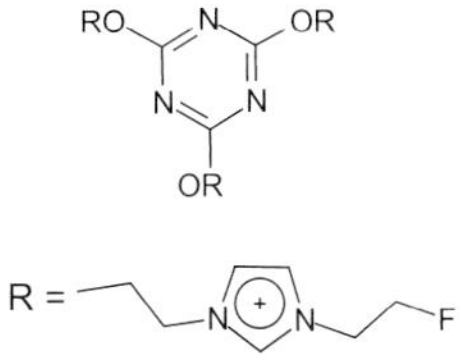

11

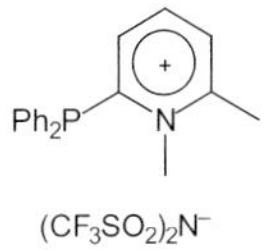

12

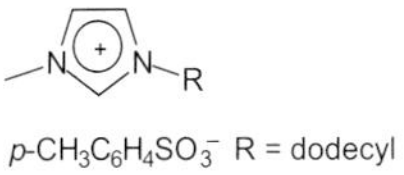

13

Entry	IL	Catalyst precursor	Ligand	Alkene	*P* (bar)	*T* (°C)	Conversion (%)	l/b ratio	Reference
1	BMI·PF_6	$Rh(CO)_2(acac)$	PPh_3	1-Pentene	20	80	99	3.0	[9]
2	BMI·PF_6	$Rh(CO)_2(acac)$	tppms	1-Pentene	20	80	16	3.8	[9]
3	$P(^{n}Bu)_4Ts$	$Rh_2(OAc)_4$	PPh_3	1-Hexene	40	120	100	2.0	[11]
4	BMI·PF_6	$Rh(CO)_2(acac)$	PPh_3	Methyl-3-pentenoate	10	110	99	0.1	[12]

(continued)

Table 6.1 (*Continued*)

Entry	IL	Catalyst precursor	Ligand	Alkene	*P* (bar)	*T* (°C)	Conversion (%)	l/b ratio	Reference
5	$BMI{\cdot}PF_6$	$Rh(CO)_2(acac)$	**1**	Methyl-3-pentenoate	10	110	98	1.0	[12]
6	$BMI{\cdot}PF_6$	$Rh(CO)_2(acac)$	**3**	1-Octene	50	80	90	4.8	[13]
7	**2**	$RhCl_3$	PPh_3	1-Tetradecene	50	105	95	1.0	[14]
8	**2**	$RhCl_3$	tppts	1-Tetradecene	50	105	94	0.4	[14]
9	$BMI{\cdot}PF_6$	$Rh(CO)_2(acac)$	**4**	1-Octene	10	100	n.r.[a]	16.2	[15]
10	$BMI{\cdot}PF_6$	$Rh(CO)_2(acac)$	**5**	1-Octene	30	100	n.r.	1.1	[16]
11	$BMI{\cdot}PF_6$	$Rh(CO)_2(acac)$	**6**	1-Octene	30	100	38	21.3	[17]
12	$BMI{\cdot}PF_6$	$Rh(CO)_2(acac)$	**7**	1-Hexene	20	80	96	12.6	[18]
13	$BMI{\cdot}BF_4$	$Rh(CO)_2(acac)$	**8**	1-Hexene	20	80	77	1.7	[18]
14	$BMI{\cdot}PF_6$	$Rh(CO)_2(acac)$	**9**	1-Octene	30	100	n.r.	2.8	[19]
14	$BMI{\cdot}PF_6$	$Rh(CO)_2(acac)$	**10**	1-Octene	17	100	98	44.0	[20,21]
15	$BMI{\cdot}BF_4$	$Rh(CO)_2(acac)$	tppti[b]	1-Hexene	20	100	n.r.	2.2	[22]
16	$BMI{\cdot}PF_6$	$Rh(CO)_2(acac)$	tppti	1-Hexene	41	100	n.r.	2.6	[22]
17	$BMI{\cdot}BF_4$	$RhCl_3$	tppts	1-Butene				9.1	[23]
18	**11**	$Rh(CO)_2(acac)$	**12**	1-Octene	69	80	60	18.0	[24]
19	**13**	$RhH(CO)(tppts)_3$	tppts	1-Decene	30	100	97	2.5	[25]
20	4-MBPCl/ $SnCl_2$	$PtCl_2(PPh_3)_2$	PPh_3	Methyl-3-pentenoate	50	120	6.3	1.3	[26]
21	CH_2Cl_2	$PtCl_2(PPh_3)_2$	PPh_3	Methyl-3-pentenoate	50	120	1.5	0.8	[26]
22	4-MBPCl/ $SnCl_2$	$PtCl_2(PPh_3)_2$	PPh_3	1-Octene	90	120	20	24	[26]

[a] Not reported.
[b] Tri(*m*-sulfonyl)triphenylphosphine 1,2-dimethyl-3-butyl-imidazolium salt.

with those obtained in conventional organic solvents. The partial decomposition of the hexafluorophosphate counterion in the presence of the smallest amount of water (and tin(II)chloride) was also studied [10].

6.3.2
Ionic Liquids, Catalyst Recycle, Selectivity, and Product Separation

The most used ionic liquids in the hydroformylation are those based on the 1-*n*-butyl-3-methylimidazolium cation, in particular, associated with hexafluorophosphate anion ($BMI{\cdot}PF_6$, see Table 6.1). However, the activity and selectivity in the hydroformylation of 1-hexene catalyzed by rhodium–TPPTS complexes in $BMI{\cdot}BF_4$ were much higher than those reported in other ionic liquids. Under optimum conditions, the TOF of 1-hexene and selectivity for aldehyde were 1508 h^{-1} and 92%, respectively. The high activity of the catalyst was ascribed to the much higher solubility of hydrogen [27] and rhodium–TPPTS complexes in $BMI{\cdot}BF_4$ than in $BMI{\cdot}PF_6$ [28]. Other ionic liquids have also been used, such as those containing polyether chains attached to ammonium salts **2** (entries 7 and 8, Table 6.1) [14], phosphonium salts (entry 3, Table 6.1) [11], tris[oxoethyl(trimethyl)ammonium]triazine derivatives

11 (entry 18, Table 6.1) [24], 1-alkyl-3-methylimidazolium tosylates **13** (entry 19, Table 6.1) [25], and organostannate melts (entries 20–22, Table 6.1) [26,29].

The ionic phosphine ligands used are usually the same as employed in aqueous-phase carbonylation reactions, such as sulfonated triphenyl phosphines [22,23,25,28], phosphates [18], or xantphos [13]. Moreover, especially designed ligands have also been used, such as those containing cobaltocenium backbone [15], imidazolium [20,21] or guanidinium [17] cation attached to the xantphos skeleton or diphenylphosphine moiety [16,19], and diphenylphosphino-*N*-methylpyridinium bis(trifluoromethanesulfonyl)amide [24]. The Rh leaching in these cases is lower than 0.1%, and the recovered ionic catalytic solution can be reused at least four times without any significant changes in activity and selectivity.

The observed reaction rates for the hydroformylation in ionic liquids are usually quite superior to those performed in aqueous-phase regimes, and the selectivities are similar to those obtained in homogeneous conditions.

The catalytic activity and selectivity of the Rh-catalyzed hydroformylation of alkenes in ionic liquids are apparently dependent on the solubility of the alkenes, carbon monoxide, hydrogen, and the primary products in the ionic liquid catalytic phase. For example, an increase in hydrogen partial pressure (H_2/CO ratio) has no effect on the reactions performed in imidazolium ionic liquids in contrast to some hydroformylation reactions performed in conventional organic solvents that can be sensitive to such changes in hydrogen partial pressure. This behavior was assigned to the limited solubility of hydrogen in the ionic liquid phase, which is much lower than that of carbon monoxide. The changes in regioselectivity have been associated with the formation of the ee and ea $RhH(CO)_2$(diphosphine) catalytic species that are the same as observed in conventional organic solvents. The syngas pressure has a direct influence on the ee–ea equilibrium; when the pressure is increased, the concentration of the ea complex also increases, thereby reducing the selectivity of *n*-aldehyde. The increase in the reaction temperature causes an augmentation in the selectivity of *n*-aldehyde. It was demonstrated that the rate of hydroformylation of 5-hexen-2-one does not correlate with the CO solubility, as expected from the determined relative solubility of CO compared to H_2 [30]. Interestingly, the solubility of carbon monoxide in various ionic liquids and in some organic solvents has been determined using high-pressure ^{13}C NMR spectroscopy.

The regioselectivity, that is, the ratio between linear (l) and branched (b) products (aldehydes and/or alcohols) of the hydroformylations catalyzed by metal complexes – mainly those based on Rh(I) catalyst precursors – is usually quite similar to that obtained in the reactions catalyzed by the same catalyst precursors in classical organic solvents or aqueous biphasic regimes. However, the chemoselectivity of the hydroformylation in ionic liquids is in most of the cases different from that obtained in organic solvents or water. In ionic liquids, hydrogenated alkenes are usually not observed, but alkene isomerization is significant and it is greater in the hydroformylation of heavier alkenes that are more soluble in the ionic liquids than in lighter alkenes. The isomerization process increases with the reduction of the syngas pressure, and the best l/b selectivity without compromising on the conversion is usually between 10 and 20 bar (H_2/CO, 1/1).

The products – mainly aldehydes and unreacted or isomerized alkene – are simply removed by decantation and filtration, extraction with the aid of an organic solvent immiscible with the ionic liquid or using supercritical carbon dioxide, and distillation.

The distillation process is particularly suitable in the cases where the products are miscible with the ionic liquid, as in the case of the Rh-catalyzed hydroformylation of methyl-3-pentenoate in which the reaction mixture is monophasic. The use of an ionic liquid as a solvent results in the almost complete retention of the regioselectivity, which is influenced by the ligand, and in significant enhancement of the lifetime and overall productivity of the catalyst. The catalyst recycling and product isolation were achieved by a distillation process under reaction conditions. In these cases, the immobilized catalyst is stabilized by the ionic liquid under the thermal stress of the distillation. The catalyst can be reused several times without additional regeneration process and without loss in activity and selectivity [12].

A more efficient protocol has been developed on the basis of continuous-flow homogeneous hydroformylation of alkenes using a phase-separable system. In these systems, rhodium catalysts modified with either phosphite or sulfonated phosphines are dissolved in an ionic liquid phase, and the substrates, gases, and products are transported in and out of the reactor by a supercritical CO_2 phase vector [31].

For example, $BMI{\cdot}Ph_2PC_6H_4SO_3$ associated with $[Rh_2(OAc)_4]$ has been used as a catalyst precursor for the hydroformylation of 1-nonene in the $scCO_2$ $BMI{\cdot}PF_6$ biphasic system with flushing of the products from the reactor with $scCO_2$. The catalytic activity remains high for 12 runs (turnover number = 160–320 h^{-1}), the l/b ratio falls slowly (from 3.7 to 2.5), the isomerization increases, and, after the ninth run, Rh leaching becomes significant. Rhodium complexes modified by simple trialkyl phosphines can also be used to carry out homogeneous hydroformylation in $scCO_2$ [32]. The catalyst derived from PEt_3 is more active and slightly more selective for the linear products in $scCO_2$ than in toluene under the same reaction conditions (100 °C, 40 bar of CO/H_2 (1 : 1)). Under subcritical conditions, alcohols from hydrogenation of the first formed aldehydes are the main products, whereas above a total pressure of 200 bar, where the solution remains supercritical (monophasic) throughout the reaction, aldehydes are obtained with 97% selectivity.

In these systems, the rate is apparently largely influenced by the solubility of the alkene in the ionic liquid that is not very high in $BMI{\cdot}PF_6$. Therefore, the partitioning of an alkene toward the gaseous phase in the $BMI{\cdot}PF_6/scCO_2$ system results in rates in the continuous-flow process that are determined largely by mass-transport limitations. Indeed, the use of ionic liquids in which alkenes are more soluble, such as those containing longer alkyl chains attached to the imidazolium cation and associated with the bis(trifluoromethanesulfonyl)amide (NTf_2) anion, increases the reaction rate. Indeed, dramatic increases in rate are observed in the hydroformylation of 1-dodecene in the continuous-flow system (the conversion can be higher than 80%) using 1-*n*-decyl-1-methylimidazolium bis(trifluoromethanesulfonyl)amide ($DMI{\cdot}NTf_2$) ionic liquid [33].

Interestingly, large melting point depressions were observed for simple ammonium and phosphonium salts in the presence of compressed CO_2, bringing them well

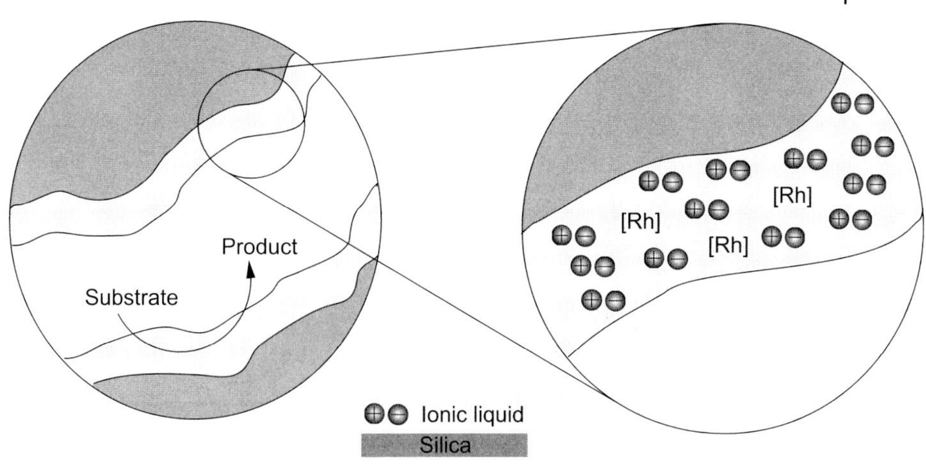

Figure 6.1 Schematic representation of SILP catalyst (adapted from Ref. [41]).

within the range of typical ionic liquids. These materials were used for rhodium-catalyzed hydroformylation of 2-vinyl-naphthalene using a CO_2-induced molten sample of $NBu_4{\cdot}BF_4$ as a carrier phase at temperatures in the range of 55–75 °C (100 °C below the normal salt melting point) [34].

Another highly efficient protocol for the hydroformylation consists in the combination of an ionic liquid with a solid support material (Figure 6.1). This process denominated supported ionic liquid phase (SILP) catalysis is a concept that combines the advantages of ionic liquids with those of heterogeneous support materials and allows the use of fixed-bed reactors for continuous reactions.

These materials are prepared by the covalent attachment of ionic liquids to the support surface or by simple deposition of the ionic liquid phases containing catalytically active species on the surface of the support (usually silica-based or polymeric materials including membranes). In various cases, the procedure involves the simple dissolution of a sulfonated phosphine-modified rhodium catalyst into a supported ionic liquid, while the alkene constitutes the organic phase. This method reduces the amount of ionic liquid and allows for a facile and efficient separation of products from catalyst. In comparison to traditional biphasic systems, higher catalytic activity and lower metal leaching can be obtained by appropriately tuning the experimental conditions [35–41].

SILP catalysts composed of monophosphines (PPh_3 and TPPTS) dissolved in BMI·X (X = PF_6 or BF_4) on a 1-*n*-butyl-3-[3-(triethoxysilanyl)propyl]imidazolium-modified silica gel support have been prepared and used in rhodium-catalyzed 1-hexene hydroformylation [39]. The SILP catalysts were found to have higher activity than analogous biphasic systems; however, a significant amount of the metal catalyst leached into the product phase at high conversions (rhodium loss of up to 2.1 mol%), because of the depletion of the ionic liquid phase from the support. Importantly, even at lower conversion, pronounced catalyst deactivation was

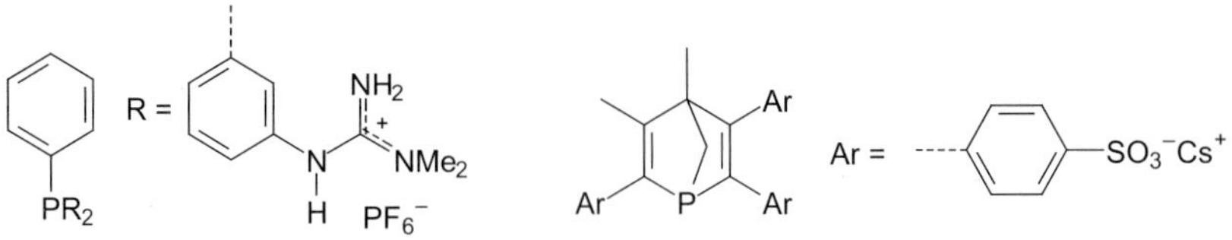

Scheme 6.2 Bis(*m*-phenylguanidinium) phenylphosphine and NORBOS ligands.

observed during recycling, which was independent of the presilylation of the support. This deactivation shortened the lifetime and limited the applicability of the catalysts significantly.

Analogous drawbacks were observed in SILP catalyst systems prepared by impregnation of silica gel with [Rh(CO)$_2$(acac)] and sulfoxantphos **3** dissolved in BMI·PF$_6$ or BMI·*n*-C$_8$H$_{17}$OSO$_3$. This Rh/silica SILP catalysts were active for the hydroformylation of propene, exhibiting TOFs of up to 37 h^{-1} and with a selectivity of up to 96% for the linear product, the conditions being typically 3–4 h under 10 bar at 100 °C. IR studies indicated an *in situ* formation of the catalytically active [HRh(**3**)(CO)$_2$] complex [42]. Similar catalyst complex preformation was also reported during the liquid biphasic 1-octene/BMI·PF$_6$ hydroformylation [17], suggesting that neither the ionic liquid nor the heterogenization itself affected the prereaction. Similarly, SILP catalysts were made by immobilizing Rh–monophosphine complexes of bis (*m*-phenylguanidinium) phenylphosphine **14** and NORBOS **15** ligands (Scheme 6.2) in BMI·PF$_6$, on a silica gel. These catalysts were used in the continuous gas- and liquid-phase hydroformylation of propene and 1-octene, exhibiting TOFs up to 88 h^{-1} for SILP Rh-**2** catalysts, with a moderate l/b ratio of 2.8 [43].

However, SILP catalyst system in which the catalyst remains active, highly selective, and stable over extended periods in a continuous gas-phase propene hydroformylation can be accessed by the immobilization of Rh/sulfoxantphos **3** dissolved in BMI·*n*-C$_8$H$_{17}$OSO$_3$ with a partly dehydroxylated silica support [44]. The effect of the support is directly related to the irreversible reaction of the ligand with the acidic silanol surface groups, before and during catalysis. Therefore, not only the ionic liquid solvent but also a relatively large excess of phosphine ligand are the prerequisites for active, highly selective, and long-term stable SILP catalysts for compensating for some detrimental surface reactions. Moreover, decreasing the number of surface silanol groups on the support material by thermal treatment is a crucial parameter in the preparation of stable hydroformylation systems that are selective and active for at least 60 h onstream. Spectroscopic investigations indicates that the catalysis in the supported ionic liquid layer is homogeneous and that the complex formation in the SILP Rh-**3**/IL/SiO$_2$ catalysts is very similar to that previously reported for Rh-**1** and analogous rhodium–xanthene-based systems in organic and ionic liquid solvents [45]. The determined activation energy of 63.3 ± 2.1 kJ mol^{-1} is in good agreement with the known results from biphasic hydroformylation processes [46].

Carriers other than silica have also been used for the preparation of SILP catalyst for the alkenes hydroformylation. Simple immobilized water-soluble TPPTS–Rh complex dissolved in 1,1,3,3-tetramethylguanidinium lactate ionic liquid on MCM-41 mesoporous silicas exhibited high performance and stability for the hydroformylation of 1-hexene, and the catalyst system could be reused many times without reducing the activity and selectivity. In these cases, considerable amounts of IL and Rh species were essentially located in the inner channel of MCM-41. The SILP process performed using SiO_2 as the carrier presented much lower 1-hexene conversion as compared to the SILPC using MCM-41 under the same reaction conditions [47,48].

6.3.3 Pt–Sn and Ru Catalyst Precursors

The regioselective hydroformylation of functionalized and nonfunctionalized olefins can also be performed by platinum compounds [26] in chlorostannate ionic liquids as solvents for homogeneous catalysis (entries 20–22, Table 6.1). Dissolved in chlorostannate ionic liquids, the Pt catalyst shows enhanced stability and selectivity in the hydroformylation of methyl-3-pentenoate compared to the identical reaction in conventional organic solvents. The moderate Lewis acidity of these ionic liquids allows the activation of the Pt catalyst combined with tolerance of the functional groups in the substrate. In the case of 1-octene hydroformylation, a biphasic reaction system could be performed using the chlorostannate ionic liquid.

The proposed active species *cis*-$[Pt(PPh_3)_2Cl(SnCl_3)]$ and *cis*-$[Pt(PPh_3)_2(SnCl_3)_2]$ formed from the hydroformylation catalyst precursor *cis*-$[Pt(PPh_3)_2Cl_2]$ in the presence of $SnCl_2$ have been identified [29].

More interestingly, $Ru_3(CO)_{12}$ dissolved in 1,3-dialkylimidazolium-based ionic liquids, in particular those associated with chloride anion [49], effectively catalyzed the hydroformylation of various kinds of alkenes with carbon dioxide to give the corresponding alcohols (Table 6.2). Compared to the conventional reaction, this reaction proceeded in the biphasic system, where the chemoselectivity in the

Table 6.2 $Ru_3(CO)_{12}$-catalyzed hydroformylation of 1-hexene with carbon dioxide (4 bar)/hydrogen (4 bar).

Entry	Ionic liquid	*T* (°C)	Conversion (%)	Heptanal yield (%)	Hexane yield (%)	Reference
1	BMI·Cl	140	97	84	11	[49]
2	$BMI\cdot BF_4$	140	96	63	26	[49]
3	$BMI\cdot PF_6$	140	95	3	86	[49]
4	$BMI\cdot Cl/BMI\cdot BF_4$	160	95	71	9	[51]
5	$BMI\cdot Cl/BMI\cdot PF_6$	160	93	50	7	[51]
6	$BMI\cdot Cl/BMI\cdot NTf_2$	160	94	82	9	[51]

hydroformylation was improved owing to the occurrence of hydrogenation of aldehydes in preference to undesired hydrogenation of 1-hexene [50]. Mixture of BMI·Cl with BMI·PF_6, BMI·BF_4, and BMI·NTf_2 can also be used in these processes [51] (Table 6.2). The selectivity can be modulated by the reaction temperature. At lower temperatures, the yield of heptanol decreased because of a decrease in the reaction rate, whereas at higher temperatures, the hydrogenation of 1-hexene was enhanced, thereby decreasing the yield of heptanol. Even at lower temperatures, it is difficult to obtain heptanal selectively, because it is readily hydrogenated to form heptanol in the ionic liquid media.

6.4 Aryl Halides and Alcohols

Palladium-catalyzed carbonylation of aryl halides with nucleophiles such as alcohols, amines, and water can be performed in ionic liquid media. Several systems have been designed so that the ionic phase can be isolated and recycled. Once carbonylation substrates/products form homogeneous mixtures with ionic liquids, the experimental protocols for catalyst/ionic liquid mixture recycling involve separation of the product by either distillation or extraction procedures using organic solvents or supercritical CO_2.

The methoxycarbonylation of bromobenzene yielding methyl benzoate promoted by $Pd(OAc)_2$ can be performed in ionic liquid media (BMI·BF_4 and BMI·PF_6, entry 1, Table 6.3) with better yields when compared with the classical homogeneous system using methanol as a solvent. A 20-fold excess PPh_3 relative to palladium was found to be necessary to prevent catalyst decomposition to metallic palladium; the ionic

Table 6.3 Carbonylation of aryl halides promoted by palladium complexes.

Entry	IL	Catalyst precursor	R–X	NuH	$P_{(CO)}$ (bar)	T (°C)	Yield (%)	Reference
1	BMI·BF_4	$Pd(OAc)_2/PPh_3$	PhBr	CH_3OH	30	150	82	[52]
2	BMI·PF_6	$Pd(OAc)_2/PPh_3$	PhI	H_2O	30	120	83	[53]
3	NBu_4·Cl	$PdCl_2[COD]$	PhI	CH_3OH	5	80	93	[54]
4	NBu_4·Br	**16**	$CH_3COPhBr$	BuOH	8	130	76	[57]
5	BMI·BF_4	**16**	$CH_3COPhBr$	BuOH	8	130	3	[57]
6	BMI·Cl	**16**	$CH_3COPhBr$	BuOH	8	130	16	[57]
7	BMI·PF_6	Colloidal	PhI	CH_3OH	5	90	83	[55]
8	BMI·BF_4	$Pd(OAc)_2$/DPPBA[a]	RI[b]	Methyl glycinate	1	100	100[c]	[58]
9	BMI·PF_6	$Pd(OAc)_2/PPh_3$	RI[b]	Morpholine	1	100	100	[59]
10	BMI·PF_6	$Pd(PPh_3)_2Cl_2$	Bromostyrene	H_2O	20	100	68	[61]

[a] 4-(Diphenylphosphino)benzoic acid.
[b] 17-Iodo-androst-16-ene.
[c] Conversion.

liquid/catalyst mixture can be recycled up to six times upon isolation of the reaction products by distillation or ether extraction, although a diminishment in catalytic activity is observed by the second recycle. [52] A similar catalytic system can be used for the hydroxycarbonylations of aryl halides and benzyl chloride [53]. Thus, the hydroxycarbonylation of iodobenzene proceed in BMI·PF_6 ionic liquid with 83% yield (entry 2, Table 6.3). Immobilized in imidazolium-based ionic liquids, the Pd-catalyst shows enhanced catalytic activity toward benzoic acid compared to the identical reaction in conventional organic solvents. Upon extraction of the reaction product with water, the catalytic system remains active and can be recycled up to four times.

The beneficial effect of the addition of tetraalkylammonium salts was observed in the methoxycarbonylation of iodobenzene with $PdCl_2(COD)$ and $PdCl_2(P(OPh)_3)_2$ [54]. In a reaction catalyzed by $PdCl_2(COD)$ at 40 °C and 1 atm of CO in methanol, the ester yield increased from 14 to 52% when Bu_4N·Cl was added, with no formation of palladium black. Almost quantitative yield of ester was obtained in molten Bu_4N·Cl at 80 °C and 5 atm of CO (entry 3, Table 6.3).

A systematic study of the influence of different ionic liquids on the activity of the methoxycarbonylation of iodobenzene promoted by Pd colloids reveals the structure dependence on the catalytic activity: tetraalkyl ammonium and pyridinium salts are much more effective additives than imidazolium-based ILs to stabilize palladium colloids. The yield of methoxycarbonylation reaction catalyzed by Pd colloid decreases in the following order: NBu_4·Br > NBu_4·I > NBu_4·Cl > BMpy·PF_6 > NEt_4·Br > BMpy·Cl > BMpy·BF_4 > NEt_4·CI > BMI·PF_6 > BMI·BF_4 > BMI·Cl [55].

A more detailed investigation reveals a strong inhibition effect caused by imidazolium halides in the methoxycarbonylation of iodobenzene promoted by palladium (II) complexes [56]. No catalytic activity was observed when BMI·Cl and BMI·Br were used as the reaction media. However, the inhibiting effect is not observed when *N*-butyl-4-methylpiridinium chloride, BMI·BF_4, BMI·PF_6, or H2-protected imidazolium halides (1,2-dimethyl-3-butylimidazolium bromide) were used, suggesting that the combination of halide ions and the acidic imidazolium H2 is responsible for the lack of catalytic activity observed. Moreover, benzene formation was detected during the methoxycarbonylation of iodobenzene, which lead the authors to propose the deactivation mechanism shown in Figure 6.2.

The inhibition of methoxycarbonylation in BMI·Cl media may not be general because the Pd–carbene catalyst **16** catalyzes the butoxycarbonylation of 4-Br-acetophenone in 16% yield (entries 4–6, Table 6.3), whereas 3% yield in the carbonylation product is obtained in BMI·BF_4 media. Among the ILs, tetrabutylammonium bromide (NBu_4·Br) was found to give better results (entry 4, Table 6.3) [57] (Scheme 6.3).

Carbonylation of 17-iodo-5α-androst-16-ene in the presence of methyl glycinate or morpholine using $Pd(OAc)_2$/phosphine can be performed in BMI·BF_4 and BMI·PF_6 media (entries 8–9, Table 6.3), and the IL–catalyst mixture can be reused up to five times before losses in catalytic activities were observed [58,59]. The methodology could be extended for the aminocarbonylation of other steroids with 17-iodo-5a-androst-16-ene functionality. Other interesting examples of carbonylation reactions

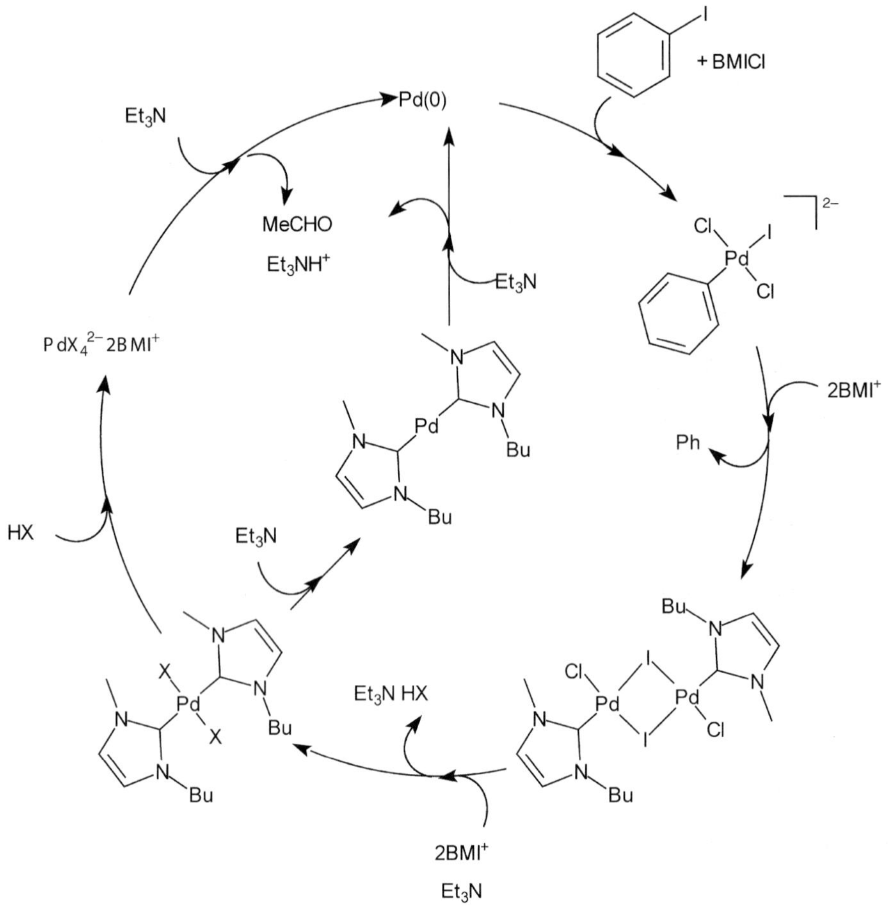

Figure 6.2 Proposed deactivation of phosphane-free catalyst precursor in BMI·Cl media.

in IL media include the carbonylation of iodobenzenes or iodoalkenes with prolinates in BMI·PF_6 media [60] and the hydroxycarbonylation of bromostyrene promoted by $PdCl_2(PPh_3)_2$, which yields cinnamic acid with excellent *E*/*Z* stereoselectivity (up to 99 : 1) in the ionic liquid BMI·PF_6 (entry 10, Table 6.3) [61].

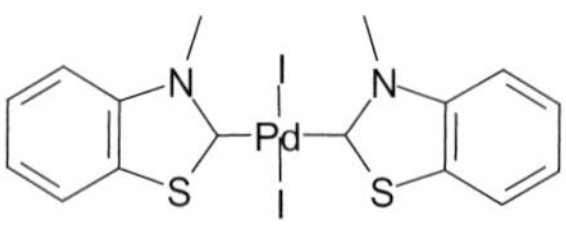

16

Scheme 6.3 Carbene complex **16**.

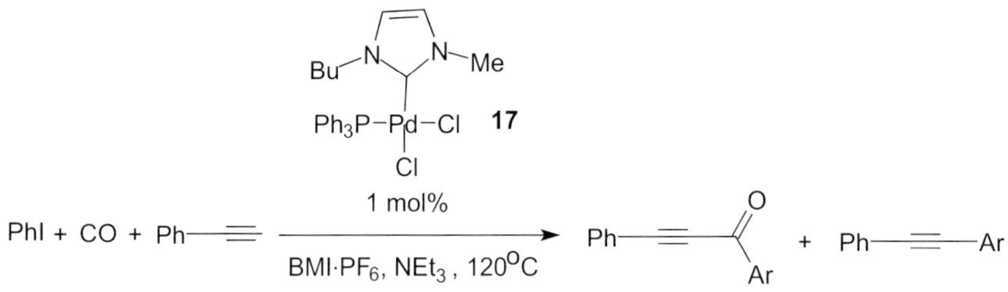

PhI + CO + Ph—≡ → (17, 1 mol%; $BMI\cdot PF_6$, NEt_3, 120 °C) Ph—≡—C(O)Ar (A) + Ph—≡—Ar (B)

17: Ph_3P–Pd(Cl)(Cl)–(1-Bu-3-Me-imidazol-2-ylidene)

Entry	CO pressure	System	Product (yield)	
1	5 atm	Microflow	**A** (83%)	**B** (—)
2	5 atm	Batch	**A** (25%)	**B** (60%)

Scheme 6.4 Pd-catalyzed carbonylative three-component coupling of iodobenzene, phenylacetylene, and CO.

The three-component coupling reaction of aryl iodides with terminal alkynes and CO promoted by Pd complexes was also tested in IL media [62,63]. Good yields of α,β-acetylenic ketones are obtained in $BMI\cdot PF_6$ at high CO pressures, whereas Sonogashira product is dominant using 5 atm of CO. This selectivity drop is probably controlled by the low CO diffusion into the IL at low CO pressures, the effect of which is avoided if the reaction is performed in a low-pressure microflow system (Scheme 6.4) [63]. No Sonogashira coupling product is observed in this system even at CO pressures as low as 5 bar.

The linear-to-branched regioselectivity on the hydroxycarbonylation of 1-phenylethanol promoted by $Pd(OAc)_2$–TsOH system (Scheme 6.5) is significantly affected by the nature of the IL media. In both BMI·Br and $NBu_4\cdot Br$, the acid product is formed in good yields (86 and 81%, respectively). Interestingly, the selectivity to branched acid was higher in $NBu_4\cdot Br$ medium (1 : 2 = 1.1) than in BMI·Br (1 : 2 = 0.4) [64]. A similar catalytic system was employed for the hydroesterification of *t*-butyl alcohol with ethanol [65]. Because of the immiscibility of IL–ethyl *t*-valerate, a biphasic system is obtained, such that and the reaction products can easily be separated from the reaction mixture.

Acidic ILs promote the carbonylation of tertiary alcohols with CO (Koch carbonylation, Scheme 6.6) [66]. The reaction is performed in a biphasic mode, using decane as a second phase, therefore allowing the separation of reaction products. The pivalic acid yield reaches 23% at 150 °C and 80 bar CO.

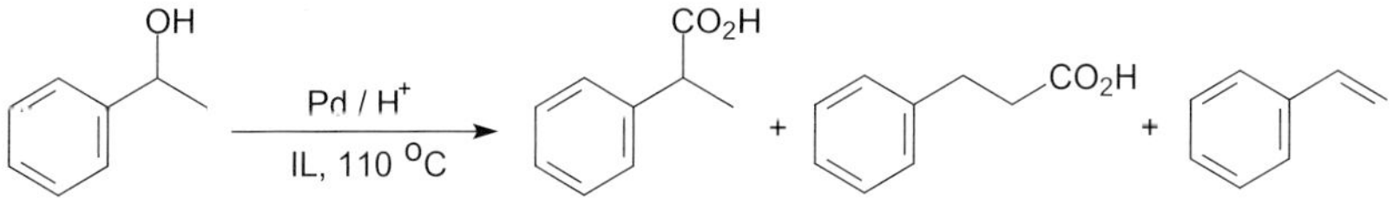

PhCH(OH)CH₃ → (Pd / H^+; IL, 110 °C) PhCH(CH₃)CO_2H + $PhCH_2CH_2CO_2H$ + PhCH=CH₂

Scheme 6.5 Hydroxycarbonylation of 1-phenylethanol.

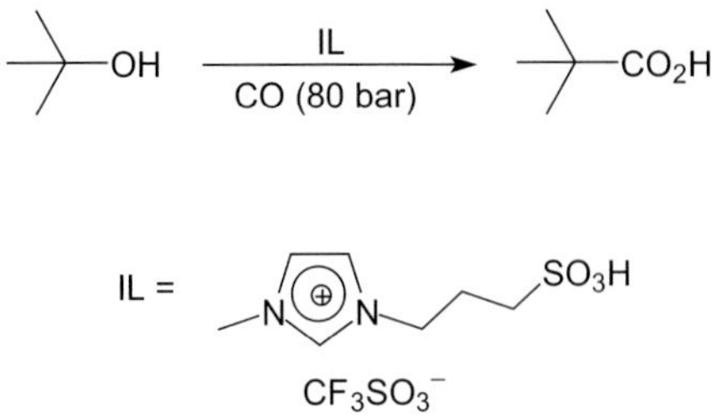

$$t\text{-BuOH} \xrightarrow[\text{CO (80 bar)}]{\text{IL}} t\text{-BuCO}_2\text{H}$$

IL = 1-methyl-3-(4-sulfobutyl)imidazolium $CF_3SO_3^-$

Scheme 6.6 Koch carbonylation of *t*-BuOH in acidic IL medium.

The carbonylation of formaldehyde and sequential esterification with methanol in BMI·PF_6 IL promoted by *p*-toluenesulfonic acid/CF_3SO_3Ag produces methyl glycolate with a significant improvement in the catalytic activity when compared with solid acid catalysts in organic solvents [67]. This effect is rationalized in terms of the stabilization effect of the IL on the acylium ion intermediate, $HOCH_2CO^+$.

The troublesome process of product separation and catalyst recycling in carbonylation reactions using ionic liquids can be considerably simplified by using a solid ionic phase [68,69] or by introducing of an inert solid support [70]. The continuous liquid-phase carbonylation of methanol has been performed using the rhodium carbonyl iodide complex $[Rh(CO)_2I_2]^-$ immobilized on a methylpyridinium cation resin [68,69]. The catalytic activity remains constant for the 2000-h operation with virtually no Rh leaching. IL-impregnated silica was used as a solid support for the Monsanto-type catalyst system $[Rh(CO)_2I_2]$–BMI·I [70].

6.5 Carbonylation of Amines

The most promising approaches for the preparation of technologically important ureas and carbamates by nonphosgene routes are the catalytic oxidative carbonylation of amines in the presence of an alcohol or the reductive carbonylation of nitro compounds [71].

BMI·BF_4 IL provides a useful medium for Pd(phen)Cl_2-catalyzed oxidative carbonylation of aniline using a CO/O_2 mixture (Scheme 6.7) [72]. The *N*-phenylamide product can be recovered from the reaction mixture by precipitation with water and the catalyst–IL system reused. The poor efficiency of this system in IL-free media is likely to be related to the low solubility of the palladium phenanthroline complex; at 175 °C, 45 bar CO, and 5 bar O_2, 99% aniline conversion is attained in BMI·BF_4.

The oxidative carbonylation of anilines can be performed in milder reaction conditions by using selenium compounds [73]. The most critical disadvantage of

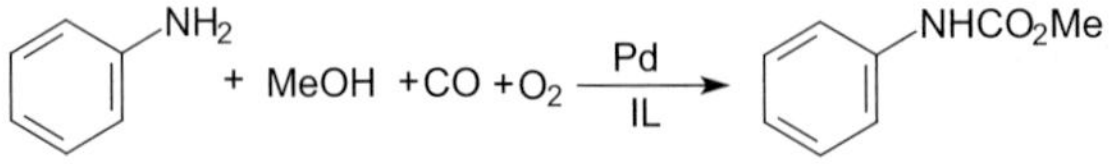

$$PhNH_2 + MeOH + CO + O_2 \xrightarrow[\text{IL}]{\text{Pd}} PhNHCO_2Me$$

Scheme 6.7 Oxidative carbonylation of aniline in BMI·BF_4 IL.

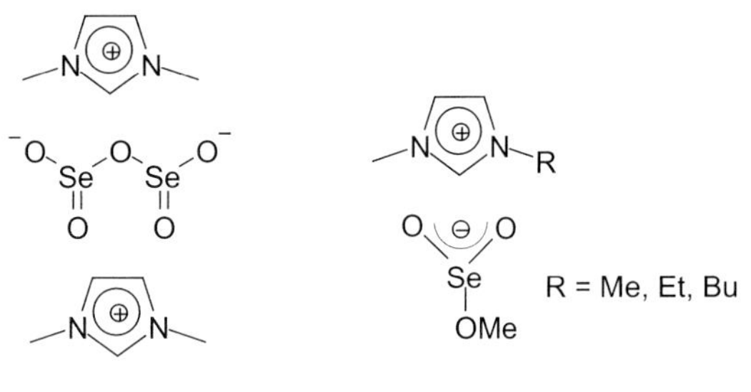

R = Me, Et, Bu

18 **19**

Scheme 6.8 Task-specific ILs.

this system – the toxic volatile selenium by-products – is considerably reduced by using task-specific ionic liquids. The selenium-containing ILs **18** and **19** show high activity for the oxidative carbonylation of aniline, even at temperatures as low as 40 °C [74,75]. Acid-functionalized ionic liquids (Scheme 6.8) improve both catalytic activity and selectivity on the reductive carbonylation of nitrobenzene promoted by Pd(phen) Cl_2 (Scheme 6.9) [76]. The elemental sulfur-catalyzed carbonylation of nitroaromatics with carbon monoxide in BMI·BF_4 ionic liquid results in much higher yields than those obtained in common organic solvents [77].

An improvement in the SILP catalysis was introduced by synthesizing highly dispersed ionic liquid catalysts through physical confinement or encapsulation of ionic liquids (with or without metal complex) in a silica gel matrix through the sol–gel process [78]. Several ILs (i.e., EMI·BF_4, BMI·BF_4, DMI·BF_4, and BMI·PF_6) or IL mixtures with $Pd(PPh_3)_2Cl_2$ and $Rh(PPh_3)_3Cl$ can be encapsulated in a silica gel matrix although EMI and BMI cations can be completely removed from the support by solvent washing. High catalytic activities are obtained for the carbonylation of aniline and nitrobenzene yielding diphenyl urea or carbonylation of aniline with CO/O_2 [78]. A similar catalytic system catalyzes the carbonylation of amines and nitrobenzene without using explosive CO/O_2 mixture, yielding the correspondent ureas [79].

Alternative sources for carbonyl groups were also investigated. Dimethyl carbonate, for example, efficiently replaces the dangerous CO/O_2 mixture in the carbonylation of aliphatic amines to yield alkylcarbamates using ILs as reaction media and catalyst [80]. The solid carbamate can be recovered by simple filtration from the biphasic mixture of dimethyl carbonate/IL after reaction. The IL phase can be

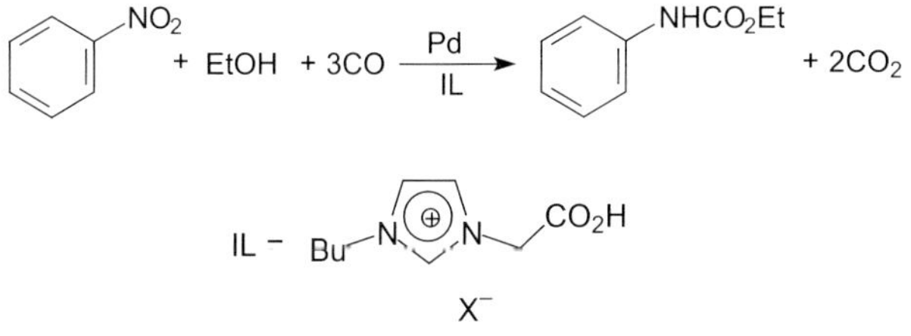

$$PhNO_2 + EtOH + 3CO \xrightarrow[IL]{Pd} PhNHCO_2Et + 2CO_2$$

IL = Bu–Im⁺–$CH_2CH_2CO_2H$ X^-

Scheme 6.9 Reductive carbonylation of nitrobenzene to ethylphenylcarbamate.

recovered and reused without losing catalytic activity. The synthesis of dimethyl carbonate can be accomplished with a recyclable catalytic system based on the oxidative carbonylation of methanol with $PdCl_2$ in BMI·PF_6 IL [81].

Ionic liquids have also been found to be efficient reaction media for the activation of CO_2 [82–84]. Diurea derivatives are easily obtained from amines and carbon dioxide with a CsOH/ionic liquid catalyst system [83]. *N*-Substituted 4-methylene-2-oxazolidinones are also efficiently synthesized in IL media by reaction of carbon dioxide with propargylic alcohol and different amines [82]. Other examples include the cycloaddition of CO_2 to epoxides to produce cyclic carbonates [85] and synthesis of cyclic urethanes [86,87].

6.6
Carbonylation of C=C and C≡C bonds (Hydroesterification and Aminocarbonylation, Pauson–Khand, and Copolymerization)

Much attention has been focused on the palladium-catalyzed Reppe-type carbonylation (Scheme 6.10) [88–92]. A strong influence of the imidazolium-based IL structure on the branched-to-linear selectivity was observed in the hydroalkoxycarbonylation of styrene with alcohols (ethanol, *n*- and *i*-propyl alcohol, benzyl alcohol, *n*-octyl alcohol, etc.). While reactions performed in BMI·PF_6 IL result mainly in the formation of the linear esters, a completely different picture is observed in BMI·BF_4 IL: the carbonylation of styrene with *n*-octyl alcohol gives 100% linear and 83% branched selectivities in BMI·PF_6 and BMI·BF_4, respectively [89]. The branched selectivity is also favored in ILs derived from the methylimidazolium cation with 3-benzyl or 1-acetonyl groups [91].

BMI·PF_6 or BMI·NTf_2 ILs were successfully employed for the palladium-catalyzed cyclocarbonylation of 2-allylphenols or anilines (Scheme 6.11) [93]. High yields of lactones or lactams are obtained with good selectivities. Moreover, the recovered IL catalytic solution can be reused without any significant changes in activity and selectivity.

The carbonylation of terminal 3-alkyn-1-ols and 1-alkyn-4-ols by $Pd(OAc)_2$ associated with 2-(diphenylphosphino)pyridine (2-$PyPPh_2$) in BMI·BF_4 quantitatively and

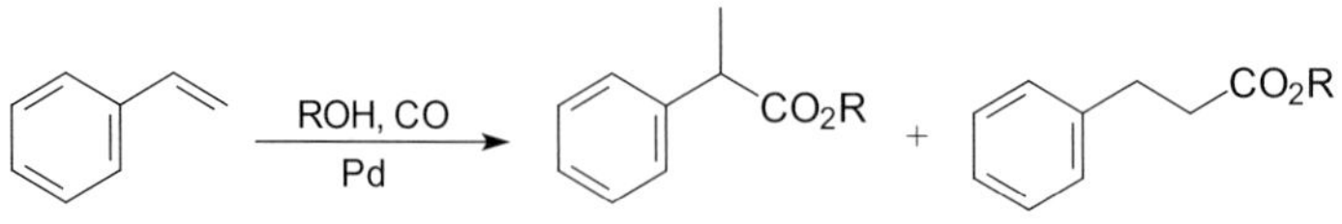

Scheme 6.10 Pd-catalyzed hydroalkoxycarbonylation of styrene.

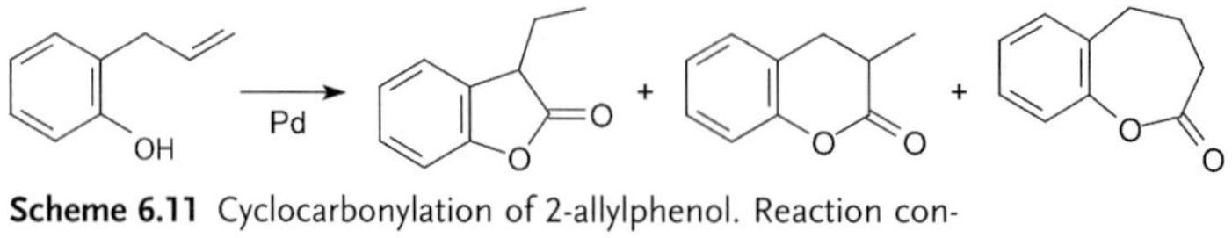

Scheme 6.11 Cyclocarbonylation of 2-allylphenol. Reaction conditions: 2 mol% Pd, 20 bar CO, and 20 bar H_2.

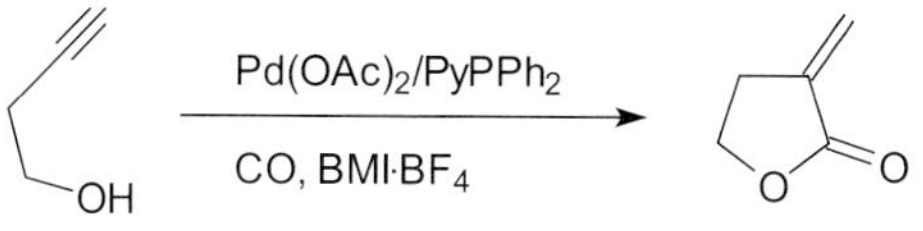

Pd(OAc)2/PyPPh2

CO, BMI·BF4

Scheme 6.12 Carbonylation of homopropargyl alcohol.

selectively affords exo-α-methylene γ- and δ-lactones (Scheme 6.12). The IL–catalyst mixture shows low recyclability after product distillation, since the catalytic activity drops to 37% yield in the third run [94].

The palladium-catalyzed aminocarbonylation of alkynes proceeds efficiently in $BMI{\cdot}NTf_2$ IL [95]. The IL acts as the reaction medium and additive because improved product yields and regioselectivities are obtained compared to conventional solvents, without the need of any acid additive (Scheme 6.13). 2-Substituted acrylamides are obtained using $Pd(OAc)_2$/dppp as the catalyst in $BMI{\cdot}NTf_2$, whereas only traces of the carbonylation product were detected in THF, DMF, or $BMI{\cdot}BF_4$ IL.

ILs provide a useful medium for the palladium-catalyzed copolymerization of styrene and carbon monoxide [96]. The catalytic system exhibits excellent polyketone productivity (up to 24 kg of polymer g^{-1} Pd) with a low palladium loading (styrene/Pd = 100 000) at 40 bar CO. Besides the enhanced productivity over conventional solvents, IL catalytic solutions show no drop in yields and polydispersity after at least four successive recycles.

The cyclization of an enyne with carbon monoxide (Pauson–Khand reaction, Scheme 6.14) promoted by $Co_2(CO)_8$ in ILs gave cyclopentenones in moderate to

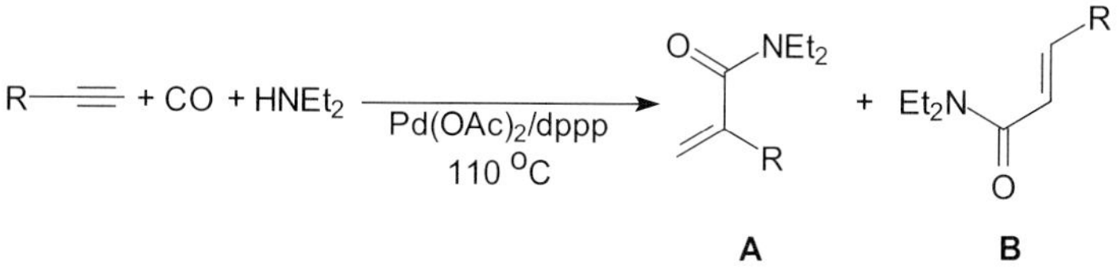

R—≡ + CO + HNEt2 → (Pd(OAc)2/dppp, 110 °C) A + B

Solvent	Yield (%)	A : B
$BMI{\cdot}NTf_2$	66	99 : 1
$BMI{\cdot}BF_4$	Trace	
THF	0	
DMF	0	

Scheme 6.13 Palladium-catalyzed aminocarbonylation of alkynes.

+ CO → (10 mol% $Co_2(CO)_8$, IL)

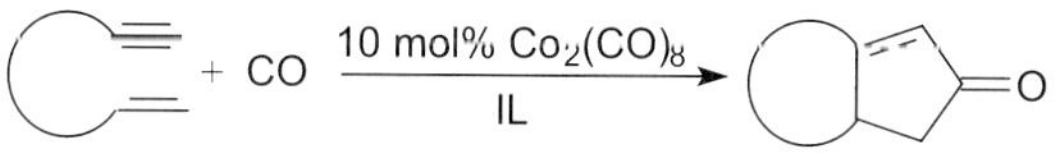

Scheme 6.14 Catalytic Pauson–Khand annelation in ILs.

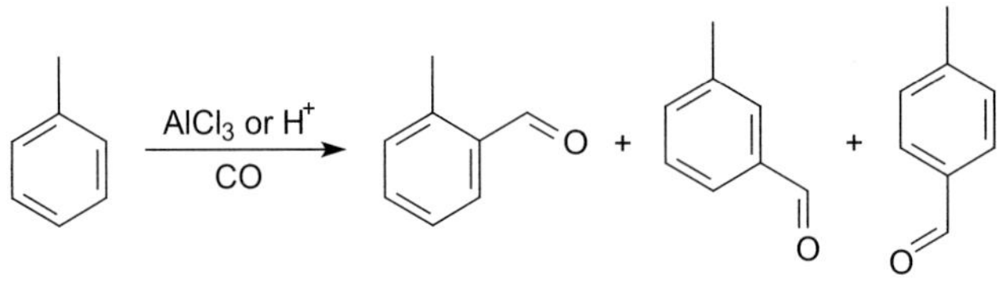

Scheme 6.15 Carbonylation of toluene.

good yields [97]. The low-affinity $Co_2(CO)_8$ for the IL phase reflected in severe cobalt leaching, and, in turn, no recycling of the catalyst mixture was performed.

6.7
Via C—H Bond Activation

Chloroaluminate ionic liquids were the first reported compounds of this class, and they are still among the best investigated imidazolium ionic liquids [98]. These ILs, specifically the compositions called acidic ([cation]Cl/$AlCl_3$ ratio <1), have been reported as catalysts for alkylation of benzene with dodecene [99], Friedel–Crafts sulfonylation [100], Friedel–Crafts alkylations and acylations [101], and oligomerizaton of olefins [102].

Ionic liquids containing chloroaluminate ($AlCl_4$, Al_2Cl_7) anions are strong Lewis acids, and if protons are present they behave as superacids. ILs formed by the combination of either EMI·Cl and BMI·Cl with $AlCl_3$ and HCl can function as both a catalyst and a solvent for acid-catalyzed toluene carbonylation at room temperature (CO pressures from 10 to 76 atm) [103,104]. Kinetic data showed that increasing HCl partial pressure on chloroaluminate IL or $AlCl_3$ concentration increased the initial reactivity of the toluene carbonylation reaction [104]. The regioselectivity to the three isomers (*p*-tolualdehyde, *o*-tolualdehyde, and *m*-tolualdehyde) showed no significant dependence with the $AlCl_3$ molar fraction or HCl pressure (90/7/3% at 100% conversion) (Scheme 6.15). The toluene conversions reached 97% in 4 h (76 atm CO).

A similar selectivity in the carbonylation of toluene, although with lower yields, was obtained with NTf_2 pyridinium salts in combination with $AlCl_3$ [105]. A maximum molar ratio of 2 mol $AlCl_3$ per mole of chloride IL is possible, until 5 mol $AlCl_3$ per NTf_2 IL can be solubilized, yielding a much more acidic medium.

Acidic ILs promote the benzaldehyde formation by direct carbonylation of benzene [106]. Up to 91% yield with 96% benzaldehyde selectivity was obtained with BMI·Br/$AlCl_3$ IL. Under similar reaction conditions, using $AlCl_3$ alone as Lewis acid catalyst, benzaldehyde is formed in much lower yields (Table 6.4).

6.8
Stoichiometric Reactions and Mechanism

ILs have also been employed as an effective medium for the reduction–carbonylation metal complexes such as of Cp_2TiCl_2. The room temperature melt was composed of

Table 6.4 Effect of IL in the carbonylation of benzene.

$$\text{C}_6\text{H}_6 \xrightarrow[\text{CO (40 bar)}]{\text{AlCl}_3\text{, IL}} \text{C}_6\text{H}_5\text{CHO}$$

Entry	Catalyst	Conversion (%)	Yield (%)	Selectivity
1	BMI·Br	Trace	Trace	—
2	BMI·Br/$AlCl_3$ (1 : 1)	3	2	83
3	BMI·Br/$AlCl_3$ (1 : 3)	95	91	96
4	$AlCl_3$	45	43	96

aluminum chloride and 1-ethyl-3-methylimidazolium chloride (EMI·Cl) with an $AlCl_3$: EMI·Cl molar ratio greater than 1, termed as an acidic melt. Thus, the addition of sodium metal to an acidic EMI·$AlCl_4$ produces an active aluminum reagent that effectively reduces Cp_2TiCl_2 to the Ti(III) complex $Cp_2Ti(AlCl_4)_2^-$. Exposure of this TI(III) complex to carbon monoxide gas produced a Ti(III) carbonyl compound $Cp_2Ti(CO)_2^+$ that can be subsequently reduced to $Cp_2Ti(CO)_2$ [107].

Au nanoparticles with controllable sizes were successfully prepared through carbonylation and reduction of hydrochloroauric acid with CO and H_2O in BMI·BF_4 ionic liquid. Different types of Au nanoparticles with average diameters of 1.7–12.8 nm were obtained with different preparation conditions [108].

6.9 Conclusions and Perspectives

The use of ionic liquids as immobilizing agents for catalytic carbonylations has various advantages over those performed in classical organic solvents, water, or perfluorinated fluids: catalyst stabilization and recycling; product separation (e.g., distillation, decantation, extraction with supercritical carbon dioxide, and continuous-flow processes); selectivity controlled through ionic reaction pathways; multiphase carbonylations with long chains alkenes that usually are not suitable for process in aqueous-phase organometallic catalysis; modulate solubility with most of the organic substrates; and carbonylations via C—H bond activation. Moreover, the advent of SILP catalysts allows the development of highly efficient protocols for the catalytic carbonylations. Therefore, the combination of an ionic liquid with a solid support material allows combining the advantages of ionic liquids with those of heterogeneous support materials and allows the use of fixed-bed reactors for continuous reactions.

Acknowledgments

We thank CAPES, CNPq, and Humboldt Foundation for partial financial support to CSC and JD.

References

1 Chauvin, Y., Gilbert, B. and Guibard, I. (1990) *Journal of the Chemical Society, Chemical Communications*, 1715–1716.
2 Parshall, G.W. (1972) *Journal of the American Chemical Society*, **94**, 8716–8719.
3 Knifton, J.F. (1981) *Journal of the American Chemical Society*, **103**, 3959–3961.
4 Collman, J.P., Brauman, J.I., Tustin, G. and Wann, G.S. (1983) *Journal of the American Chemical Society*, **105**, 3913–3922.
5 Knifton, J.F. (1987) *Journal of Molecular Catalysis*, **43**, 65–78.
6 Knifton, J.F. (1988) *Journal of Molecular Catalysis*, **47**, 99–116.
7 Dupont, J., de Souza, R.F. and Suarez, P.A.Z. (2002) *Chemical Reviews*, **102**, 3667–3691.
8 Wasserscheid, P. and Waffenschmidt, H. (2002) *ACS Symposium Series*, **818**, 373–386.
9 Chauvin, Y., Mussmann, L. and Olivier, H. (1996) *Angewandte Chemie-International Edition in English*, **34**, 2698–2700.
10 Rangits, G., Petocz, G., Berente, Z. and Kollar, L. (2003) *Inorganica Chimica Acta*, **353**, 301–305.
11 Karodia, N., Guise, S., Newlands, C. and Andersen, J.A. (1998) *Journal of the Chemical Society, Chemical Communications*, 2341–2342.
12 Keim, W., Vogt, D., Waffenschmidt, H. and Wasserscheid, P. (1999) *Journal of Catalysis*, **186**, 481–484.
13 Dupont, J., Silva, S.M. and de Souza, R.F. (2001) *Catalysis Letters*, **77**, 131–133.
14 Kong, F.Z., Jiang, J.Y. and Jin, Z.L. (2004) *Catalysis Letters*, **96**, 63–65.
15 Brasse, C.C., Englert, U., Salzer, A., Waffenschmidt, H. and Wasserscheid, P. (2000) *Organometallics*, **19**, 3818–3823.
16 Brauer, D.J., Kottsieper, K.W., Liek, C., Stelzer, O., Waffenschmidt, H. and Wasserscheid, P. (2001) *Journal of Organometallic Chemistry*, **630**, 177–184.
17 Wasserscheid, P., Waffenschmidt, H., Machnitzki, P., Kottsieper, K.W. and Stelzer, O. (2001) *Journal of the Chemical Society, Chemical Communications*, 451–452.
18 Favre, F., Olivier-Bourbigou, H., Commereuc, D. and Saussine, L. (2001) *Journal of the Chemical Society, Chemical Communications*, 1360–1361.
19 Kottsieper, K.W., Stelzer, O. and Wasserscheid, P. (2001) *Journal of Molecular Catalysis A: Chemical*, **175**, 285–288.
20 Bronger, R.P.J., Silva, S.M., Kamer, P.C.J. and van Leeuwen, P.W.N.M. (2002) *Journal of the Chemical Society, Chemical Communications*, **8**, 3044–3045.
21 Bronger, R.P.J., Silva, S.M., Kamer, P.C.J. and van Leeuwen, P.W.N.M. (2004) *Journal of the Chemical Society, Dalton Transactions*, 1590–1596.
22 Mehnert, C.P., Cook, R.A., Dispenziere, N.C. and Mozeleski, E.J. (2004) *Polyhedron*, **23**, 2679–2688.
23 Gong, Y.H., Xue, H.R., Xie, Z.K., Jin, Z.S., Yang, F. and He, M.Y. (2004) *Chinese Journal of Organic Chemistry*, **24**, 1108–1110.
24 Omotowa, B.A. and Shreeve, J.M. (2004) *Organometallics*, **23**, 783–791.
25 Lin, Q., Fu, H.Y., Xue, F., Yuan, M.L., Chen, H. and Li, X.J. (2006) *Acta Physico-Chimica Sinica*, **22**, 465–469.
26 Wasserscheid, P. and Waffenschmidt, H. (2000) *Journal of Molecular Catalysis A: Chemical*, **164**, 61–67.
27 Berger, A., de Souza, R.F., Delgado, M.R. and Dupont, J. (2001) *Tetrahedron: Asymmetry*, **12**, 1825–1828.
28 Zheng, H.J., Li, M., Chen, H., Li, R.X. and Li, X.J. (2005) *Chinese Journal of Catalysis*, **26**, 4–6.
29 Illner, P., Zahl, A., Puchta, R., van Eikema Hommes, N., Wasserscheid, P. and van Eldik, R. (2005) *Journal of Organometallic Chemistry*, **690**, 3567–3576.
30 Ohlin, C.A., Dyson, P.J. and Laurenczy, G. (2004) *Journal of the Chemical Society, Chemical Communications*, **10**, 1070–1071.

31 Sellin, M.F., Webb, P.B. and Cole-Hamilton, D.J. (2001) *Journal of the Chemical Society, Chemical Communications*, 781–782.

32 Sellin, M.F., Bach, I., Webster, J.M., Montilla, F., Rosa, V., Aviles, T., Poliakoff, M. and Cole-Hamilton, D.J. (2002) *Journal of the Chemical Society, Dalton Transactions*, 4569–4576.

33 Webb, P.B., Sellin, M.F., Kunene, T.E., Williamson, S., Slawin, A.M.Z. and Cole-Hamilton, D.J. (2003) *Journal of the American Chemical Society*, **125**, 15577–15588.

34 Scurto, A.M. and Leitner, W. (2006) *Journal of the Chemical Society, Chemical Communications*, 3681–3683.

35 Bianchini, C. and Giambastiani, G. (2003) *Chemtracts*, **16**, 301–309.

36 Mehnert, C.P. (2004) *Chemistry – A European Journal*, **11**, 50–56.

37 Riisager, A., Fehrmann, R., Haumann, M. and Wasserscheid, P. (2006) *European Journal of Inorganic Chemistry*, 695–706.

38 Feng, C.L., Wang, Y.H. and Jin, Z.L. (2005) *Progress in Chemistry*, **17**, 209–216.

39 Mehnert, C.P., Cook, R.A., Dispenziere, N.C. and Afeworki, M. (2002) *Journal of the American Chemical Society*, **124**, 12932–12933.

40 Webb, P.B., Kunene, T.E. and Cole-Hamilton, D.J. (2005) *Green Chemistry*, **7**, 373–379.

41 Riisager, A., Fehrmann, R., Haumann, M. and Wasserscheid, P. (2006) *Topics in Catalysis*, **40**, 91–102.

42 Riisager, A., Wasserscheid, P., van Hal, R. and Fehrmann, R. (2003) *Journal of Catalysis*, **219**, 452–455.

43 Riisager, A., Eriksen, K.M., Wasserscheid, P. and Fehrmann, R. (2003) *Catalysis Letters*, **90**, 149–153.

44 Riisager, A., Fehrmann, R., Flicker, S., van Hal, R., Haumann, M. and Wasserscheid, P. (2005) *Angewandte Chemie-International Edition*, **44**, 815–819.

45 Silva, S.M., Bronger, R.P.J., Freixa, Z., Dupont, J. and van Leeuwen, P. (2003) *New Journal of Chemistry*, **27**, 1294–1296.

46 Riisager, A., Fehrmann, R., Haumann, M., Gorle, B.S.K. and Wasserscheid, P. (2005) *Industrial & Engineering Chemistry Research*, **44**, 9853–9859.

47 Yang, Y., Lin, H.Q., Deng, C.X., She, J.R. and Yuan, Y.Z. (2005) *Chemistry Letters*, **34**, 220–221.

48 Yang, Y., Deng, C.X. and Yuan, Y.Z. (2005) *Journal of Catalysis*, **232**, 108–116.

49 Tominaga, K. and Sasaki, Y. (2004) *Chemistry Letters*, **33**, 14–15.

50 Tominaga, K.I. and Sasaki, Y. (2004) *Studies in Surface Science and Catalysis*, **153**, 227–232.

51 Tominaga, K. (2006) *Catalysis Today*, **115**, 70–72.

52 Mizushima, E., Hayashi, T. and Tanaka, M. (2001) *Green Chemistry*, **3**, 76–79.

53 Mizushima, E., Hayashi, T. and Tanaka, M. (2004) *Topics in Catalysis*, **29**, 163–166.

54 Trzeciak, A.M., Wojtkow, W. and Ziolkowski, J.J. (2003) *Inorganic Chemistry Communications*, **6**, 823–826.

55 Wojtkow, W., Trzeciak, A.M., Choukroun, R. and Pellegatta, J.L. (2004) *Journal of Molecular Catalysis A: Chemical*, **224**, 81–86.

56 Zawartka, W., Trzeciak, A.M., Ziolkowski, J.J., Lis, T., Ciunik, Z. and Pernak, J. (2006) *Advanced Synthesis & Catalysis*, **348**, 1689–1698.

57 Calo, V., Giannoccaro, P., Nacci, A. and Monopoli, A. (2002) *Journal of Organometallic Chemistry*, **645**, 152–157.

58 Muller, E., Peczely, G., Skoda-Foldes, R., Takacs, E., Kokotos, G., Bellis, E. and Kollar, L. (2005) *Tetrahedron: Asymmetry*, **61**, 797–802.

59 Skoda-Foldes, R., Takacs, E., Horvath, J., Tuba, Z. and Kollar, L. (2003) *Green Chemistry*, **5**, 643–645.

60 Takacs, E., Skoda-Foldes, R., Acs, P., Muller, E., Kokotos, G. and Kollar, L. (2006) *Letters in Organic Chemistry*, **3**, 62–67.

61 Zhao, X.D., Alper, H. and Yu, Z.K. (2006) *Journal of Organic Chemistry*, **71**, 3988–3990.

62 Fukuyama, T., Yamaura, R. and Ryu, I. (2005) *Canadian Journal of Chemistry*, **83**, 711–715.

63 Rahman, M.T., Fukuyama, T., Kamata, N., Sato, M. and Ryu, I. (2006) *Journal of the Chemical Society, Chemical Communications*, 2236–2238.

64 Lapidus, A., Eliseev, O., Bondarenko, T. and Stepin, N. (2006) *Journal of Molecular Catalysis A: Chemical*, **252**, 245–251.

65 Qiao, K. and Deng, Y.Q. (2002) *New Journal of Chemistry*, **26**, 667–670.

66 Qiao, K. and Yokoyama, C. (2006) *Catalysis Communications*, **7**, 450–453.

67 Li, T., Souma, Y. and Xu, Q. (2006) *Catalysis Today*, **111**, 288–291.

68 Yoneda, N., Minami, T., Shiroto, Y., Hamato, K. and Hosono, Y. (2003) *Journal of the Japan Petroleum Institute*, **46**, 229–239.

69 Yoneda, N. and Hosono, Y. (2004) *Journal of Chemical Engineering of Japan*, **37**, 536–545.

70 Riisager, A., Jorgensen, B., Wasserscheid, P. and Fehrmann, R. (2006) *Journal of the Chemical Society, Chemical Communications*, 994–996.

71 Shi, F., Gu, Y.L., Zhang, Q.H. and Deng, Y.Q. (2004) *Catalysis Surveys from Asia*, **8**, 179–186.

72 Shi, F., Peng, J.J. and Deng, Y.Q. (2003) *Journal of Catalysis*, **219**, 372–375.

73 Mei, J.T., Yang, Y., Xue, Y. and Lu, S.W. (2003) *Journal of Molecular Catalysis A: Chemical*, **191**, 135–139.

74 Kim, H.S., Kim, Y.J., Lee, H., Park, K.Y., Lee, C. and Chin, C.S. (2002) *Angewandte Chemie-International Edition*, **41**, 4300–4303.

75 Kim, H.S., Kim, Y.J., Bae, J.Y., Kim, S.J., Lah, M.S. and Chin, C.S. (2003) *Organometallics*, **22**, 2498–2504.

76 Shi, F., He, Y.D., Li, D.M., Ma, Y.B., Zhang, Q.H. and Deng, Y.Q. (2006) *Journal of Molecular Catalysis A: Chemical*, **244**, 64–67.

77 Wang, X.F., Li, P., Yuan, X.H. and Lu, S.W. (2006) *Journal of Molecular Catalysis A: Chemical*, **255**, 25–27.

78 Shi, F., Zhang, Q., Li, D. and Deng, Y. (2005) *Chemistry – A European Journal*, **11**, 5279–5288.

79 Shi, F., Zhang, Q., Gu, Y. and Deng, Y. (2005) *Advanced Synthesis & Catalysis*, **347**, 225–230.

80 Sima, T., Guo, S., Shi, F. and Deng, Y.Q. (2002) *Tetrahedron Letters*, **43**, 8145–8147.

81 Jiang, T., Han, B., Zhao, G., Chang, Y., Gao, L., Zhang, J. and Yang, G. (2003) *Journal of Chemical Research – Part S*, 549–551.

82 Zhang, Q.H., Shi, F., Gu, Y.L., Yang, J. and Deng, Y.Q. (2005) *Tetrahedron Letters*, **46**, 5907–5911.

83 Shi, F., Deng, Y., SiMa, T., Peng, J., Gu, Y. and Qiao, B. (2003) *Angewandte Chemie-International Edition*, **42**, 3257–3260.

84 Fujita, S., Kanamaru, H., Senboku, H. and Arai, M. (2006) *International Journal of Molecular Sciences*, **7**, 438–450.

85 Sun, J.M., Fujita, S. and Arai, M. (2005) *Journal of Organometallic Chemistry*, **690**, 3490–3497.

86 Kawanami, H., Matsumoto, H. and Ikushima, Y. (2005) *Chemistry Letters*, **34**, 60–61.

87 Gu, Y.L., Zhang, Q.H., Duan, Z.Y., Zhang, J., Zhang, S.G. and Deng, Y.Q. (2005) *Journal of Organic Chemistry*, **70**, 7376–7380.

88 Zim, D., de Souza, R.F., Dupont, J. and Monteiro, A.L. (1998) *Tetrahedron Letters*, **39**, 7071–7074.

89 Rangits, G. and Kollar, L. (2005) *Journal of Molecular Catalysis A: Chemical*, **242**, 156–160.

90 Balázs, A., Benedek, C. and Törös, S. (2006) *Journal of Molecular Catalysis A: Chemical*, **244**, 105–109.

91 Rangits, G. and Kollar, L. (2006) *Journal of Molecular Catalysis A: Chemical*, **246**, 59–64.

92 Wang, Y.Y., Luo, M.M., Lin, Q., Chen, H. and Li, X.J. (2006) *Green Chemistry*, **8**, 545–548.

93 Ye, F.G. and Alper, H. (2006) *Advanced Synthesis & Catalysis*, **348**, 1855–1861.

94 Consorti, C.S., Ebeling, G. and Dupont, J. (2002) *Tetrahedron Letters*, **43**, 753–755.

95 Li, Y., Alper, H. and Yu, Z.K. (2006) *Organic Letters*, **8**, 5199–5201.

96 Hardacre, C., Holbrey, J.D., Katdare, S.P. and Seddon, K.R. (2002) *Green Chemistry*, **4**, 143–146.

97 Mastrorilli, P., Nobile, C.F., Paolillo, R. and Suranna, G.P. (2004) *Journal of Molecular Catalysis A: Chemical*, **214**, 103–106.

98 Welton, T. (1999) *Chemical Reviews*, **99**, 2071–2083.

99 Qiao, K. and Deng, Y.Q. (2001) *Journal of Molecular Catalysis A: Chemical*, **171**, 81–84.

100 Nara, S.J., Harjani, J.R. and Salunkhe, M.M. (2001) *Journal of Organic Chemistry*, **66**, 8616–8620.

101 Boon, J.A., Levisky, J.A., Pflug, J.L. and Wilkes, J.S. (1986) *Journal of Organic Chemistry*, **51**, 480–483.

102 Stenzel, O., Brull, R., Wahner, U.M., Sanderson, R.D. and Raubenheimer, H.G. (2003) *Journal of Molecular Catalysis A: Chemical*, **192**, 217–222.

103 Angueira, E.J. and White, M.G. (2005) *Journal of Molecular Catalysis A: Chemical*, **238**, 163–174.

104 Angueira, E.J. and White, M.G. (2005) *Journal of Molecular Catalysis A: Chemical*, **227**, 51–58.

105 Brausch, N., Metlen, A. and Wasserscheid, P. (2004) *Journal of the Chemical Society, Chemical Communications*, 1552–1553.

106 Zhao, W.J. and Jiang, X.Z. (2006) *Catalysis Letters*, **107**, 123–125.

107 Carlin, R.T. and Fuller, J. (1997) *Inorganica Chimica Acta*, **255**, 189–192.

108 Guo, S., Shi, F., Gu, Y.L., Yang, J. and Deng, Y.Q. (2005) *Chemistry Letters*, **34**, 830–831.

7
Carbonylation of Alkenes and Dienes

Tamás Kégl

The first example for carbonylation reaction of olefins, discovered by Otto Roelen and patented in 1938, was the addition of carbon monoxide and dihydrogen to an olefin double bond in the presence of transition metal catalysts (Equation 7.1).

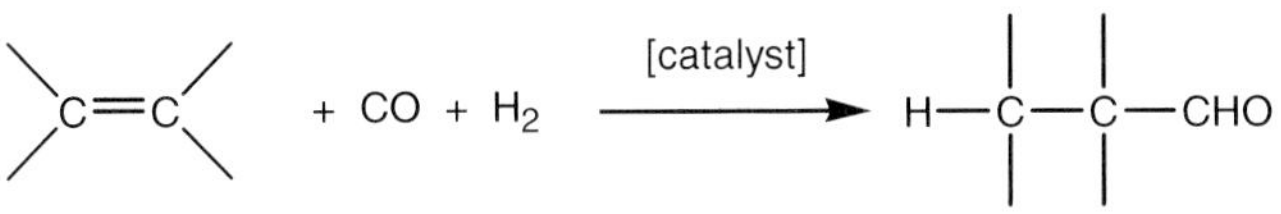

$$>C=C< \; + CO + H_2 \xrightarrow{[catalyst]} H-\overset{|}{\underset{|}{C}}-\overset{|}{\underset{|}{C}}-CHO \tag{7.1}$$

Replacing hydrogen by water yields carboxylic acids and the modified reaction is named hydrocarboxylation (Equation 7.2). Another example of synthesis taking place with very similar mechanism is hydroesterification (also termed hydroalkoxycarbonylation) that uses alcohol as hydrogen source, yielding esters as products (Equation 7.3).

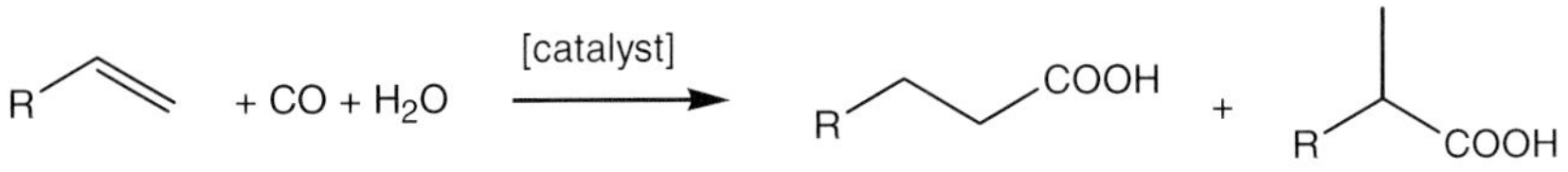

$$RCH=CH_2 + CO + H_2O \xrightarrow{[catalyst]} RCH_2CH_2COOH + RCH(CH_3)COOH \tag{7.2}$$

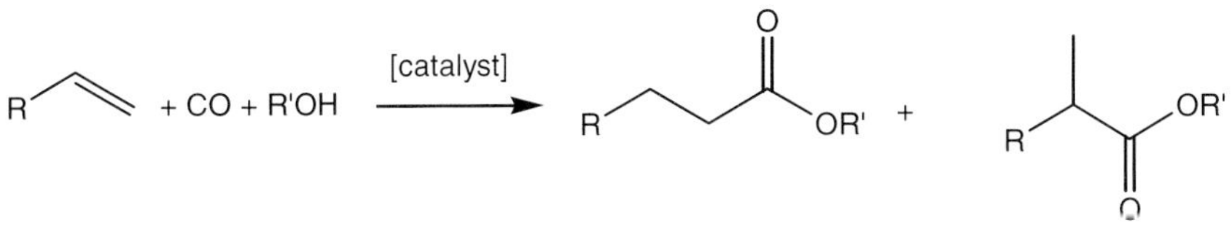

$$RCH=CH_2 + CO + R'OH \xrightarrow{[catalyst]} RCH_2CH_2C(O)OR' + RCH(CH_3)C(O)OR' \tag{7.3}$$

In this chapter, some examples of the reactions depicted in Equations 7.1–7.3 and some illustrations of one-pot reactions mainly based on hydroformylation will be reviewed. The scientific literature is covered from 2002 till April 2007.

Modern Carbonylation Methods. Edited by László Kollár

ISBN: 978-3-527-31896-4

7.1 Hydroformylation of Alkenes and Dienes

In the beginning, only cobalt catalysts were utilized for hydroformylation under relatively vigorous conditions. The moderate selectivity of the generally more desired linear isomers, the substantial formation of by-products, and the low stability of the catalysts forced more appropriate catalytic systems to be developed. Using donor ligands such as phosphanes resulted in an increase in the linear selectivity. Rhodium-containing catalysts, however, permitted the use of much milder conditions in combination with suitable ligands. Other transition metals such as ruthenium, palladium, platinum, and iridium also tended to show activity for hydroformylation, albeit a vast majority of works on hydroformylation still concentrated on rhodium-containing systems.

7.1.1 Cobalt Catalysts

The initial rate of $Co_2(CO)_8$-catalyzed cyclohexene hydroformylation, triethyl orthoformate carbonylation, and $CoH(CO)_4$ formation from $Co_2(CO)_8$ and H_2 is reduced by the addition of dinitrogen, argon, or xenon. It is assumed that the additional gas competes with one or more reactants for a coordinatively unsaturated site responsible for their activation, thus affecting the reaction rate [1].

The carbonylation reaction of propylene oxide in the presence of various $[\text{Lewis acid}]^+[Co(CO)_4]^-$ salts was investigated using *in situ* attenuated total reflection infrared (ATR-IR) spectroscopy. β-Alkoxy-acyl-cobalttetracarbonyl species were found to be key intermediates from which two reaction routes start depending on the applied Lewis acid. Labile Lewis acid–alkoxy combinations primarily favor the production of lactone products [2].

The reaction of olefins (1-octene, 3,3-dimethylbutene, cyclohexene) introduced into a preequilibrated $Co_2(CO)_8 + H_2$ system was investigated by high-pressure infrared spectroscopy. Based on the observed induction period for the formation of the corresponding aldehyde, the decrease in $HCo(CO)_4$ concentration and increase in $Co_2(CO)_8$ concentration during the induction period, the active catalytic species of the type $H_xCo_y(CO)_z$ was proposed [3]. Experiments with isotope mixtures of H_2/D_2 in the gas phase during the various steps of the reaction showed that the ratio of H/D isotopes in the hydrocarbon portion of the aldehyde product correlates with the $HCo(CO)_4/DCo(CO)_4$ ratio in solution, whereas the RC(=O)H/RC(=O)D product ratio correlates with the H_2/D_2 in the gas phase. It was concluded that the dominant pathway for the hydrogenolysis step in this type of hydroformylation is the direct reaction of hydrogen or deuterium with the acyl complex intermediate [4].

The effect of argon (280 bar) on the rate of aldehyde formation in 1-hexene hydroformylation catalyzed by $Co_2(CO)_8$ at 50 °C, 70 bar $P(CO)$ and 85 bar $P(H_2)$ in toluene was investigated by high-pressure FT-IR spectroscopy. The initial rate of the reaction was found to be reduced by the presence of argon. A similar effect was observed in the $RhH(CO)(PPh_3)_3$-catalyzed cyclohexene hydroformylation reaction [5].

The phosphane-modified cobalt-catalyzed hydroformylation of 1-dodecene was investigated using **1** and **2** as the modifier. The composition of the cobalt species in the reaction mixture was deduced from high-pressure IR and NMR spectra [6].

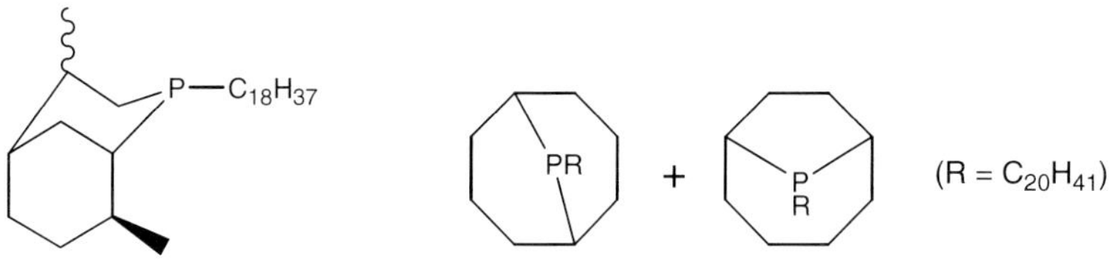

1 (P—$C_{18}H_{37}$)

2 (PR + PR; $R = C_{20}H_{41}$)

The modified cobalt-catalyzed 1-octene hydroformylation using phosphanes derived from limonene was investigated at 170 °C and 85 bar $H_2:CO = 2:1$. The composition of cobalt complexes in the reaction mixtures was deduced from IR and NMR spectra and correlated to the products of the hydroformylation reaction [7].

Each of the previously proposed steps in cobalt-catalyzed hydroformylation was confirmed by the detection of intermediates using *para*-hydrogen-induced polarization [8]. Thermodynamic parameters relevant to the phosphane-modified cobalt-catalyzed hydroformylation reaction were reported. Thus, equilibrium constants for the hydrogenation of $Co_2(CO)_6L_2$ to yield $HCo(CO)_3L$ (L = *tert*-phosphanes) were determined using *in situ* ^{1}H and ^{31}P NMR spectroscopy between 75 and 175 °C for various solvents and phosphane ligands. Based on the analysis of ^{31}P NMR linewidth, lower limits were established for the catalytically relevant Co—Co and Co—H bond energies in the case of $L = {}^nBu_3P$(Co–Co $\geq$ 23 kcal/mol and Co–H $\geq$ 60 kcal/mol) relative to the previously reported values for the case of L = CO (Co—Co = 19 ± 2 kcal mol^{-1} and Co—H 59 ± 1 kcal mol^{-1}) [9].

Flash photolysis with time-resolved infrared (TRIR) spectroscopy was used to elucidate the photochemical reactivity of the hydroformylation catalyst precursor $Co_2(CO)_6(PCH_3Ph_2)_2$. Depending on reaction conditions, the net products of photolysis varied significantly [10].

7.1.2 Rhodium Catalysts

The effects of ammonium chromate, ammonium dichromate, ammonium molybdate, and ammonium tungstate as additives on the catalytic performance of rhodium catalysts for the hydroformylation of C_8-olefins and 1-dodecene were investigated. Modification of the rhodium catalyst with ammonium salts resulted in moderate increase in aldehyde yields and decrease in rhodium losses in the distillation process for the separation of products from the catalyst [11].

Hydroformylation of **3** at 80 °C and 80 bar $CO:H_2/1:1$ pressure using 5 mol% $RhCl(PPh_3)_3$ as the catalyst precursor was found to give an isomeric mixture of **4** and

5 quantitatively (Equation 7.4), which can be converted into the corresponding alcohols [12].

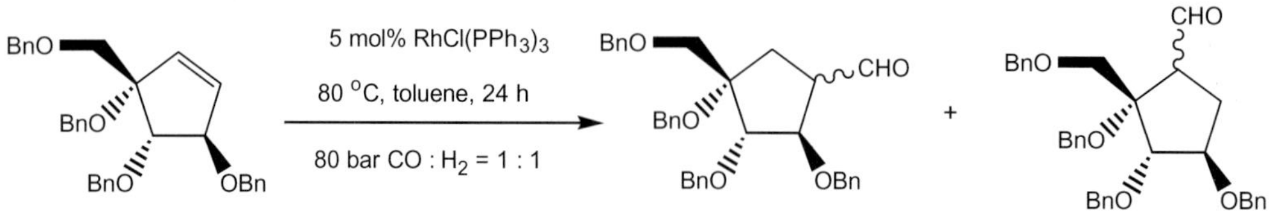

(7.4)

The unmodified rhodium-catalyzed hydroformylation of 2-phenylsulfonylbicyclo [2.2.1] alkenes was investigated. It was found that the steric properties of the sulfonyl substituent, more than the electronic ones, influence the regioselectivity of the process [13]. Highly regioselective hydroformylation of unsaturated esters was achieved when a reactive, ligand-modified rhodium catalyst was employed near ambient temperatures (15–50 °C) and pressures over 30 bar. The use of 1,3,5,7-tetramethyl-2,4,8-trioxa-6-phosphaadamantane was found to show distinct advantages over other commonly applied phosphanes in terms of reaction rate and regioselectivity and chemoselectivity. Hydroformylation of 1,1-di- and 1,1,2-trisubstituted unsaturated esters using this catalytic system yields quaternary aldehydes [14].

The naturally occurring cinchona alkaloids, cinchonidine, quinine, and quinidine, were hydroformylated selectively to the corresponding terminal aldehyde derivatives with 87, 71, and 85% isolated yields, respectively, using $Rh(CO)_2(acac)$/tetraphosphite/**5** catalyst system at 90 °C and 20 bar $CO:H_2 = 1:1$ in toluene [15].

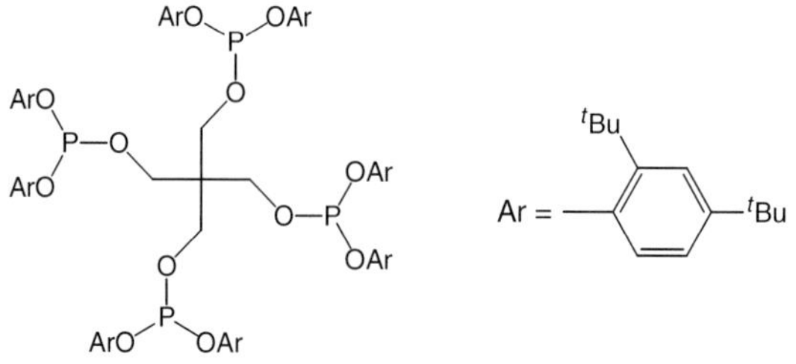

The hydroformylation of vegetable oils (soybean, high oleic safflower, safflower, and linseed) using $Rh(CO)_2(acac)$ as the catalyst precursor in the presence of PPh_3 or $(PhO)_3P$ was studied. The ligand $(PhO)_3P$ resulted in a lesser reactivity compared to TPP in contrast to the rates of bulky phosphite ligands reported in the literature [16].

Diphenylphosphine functionalized polyhedral oligomeric silsesquioxane dendrimers were used as ligands for the rhodium-catalyzed hydroformylation of 1-octene. High regioselectivity to the linear nonanal (linear : branched ratio = 14 : 1) was observed with only one structure within the first- and second-generation dendrimer having a spacer of five atoms between the phosphorus atoms and a carbon–silicon linkage. Lower selectivities were obtained using dendrimers having spacers of three or seven atoms between the phosphorus atoms and a carbon–silicon linkage [17]. Dendritic ligands with triphenylphosphanes at the periphery were applied to rhodium-catalyzed hydroformylation of 1-octene and styrene at 80 °C and 20 bar $CO:H_2$ 1 : 1 [18]. The rhodium precursor $[Rh(COD)Cl]_2$ complexed on phosphonated dendronized magnetic nanoparticles was found to be highly branched selective hydroformylation catalyst for various vinyl arenes. At 50 °C and 69 bar $CO:H_2 = 1:1$ up to 100 : 1 = branched : linear ratios were achieved [19].

Polyether phosphites with over 19 ethylene glycol units were used as ligands in the rhodium-catalyzed nonaqueous hydroformylation of 1-decene. The catalysts were recovered by precipitation from the reaction mixture after reaction on cooling to room temperature or lower. The precipitated catalysts could be reused up to six times without any decrease in activity. *In situ* formed polyetherphosphite/$Ru_3(CO)_{12}$ catalyst was found to be active in hydroformylation of 1-decene in *n*-heptane solution at 130 °C and 50 bar [20].

The influence of the properties of ligands **6**, **7**, **8**, and **9** on the catalyst performance in the rhodium-catalyzed hydroformylation and deuterioformylation of 1-octene and 1-hexene, respectively, has been investigated. The pyrrolyl substituents were found to result in a very active catalyst for the hydroformylation of 1-octene. Especially the bidentate pyrrolyl-containing ligand **9** forms a catalyst that shows high activity together with a high regioselectivity for the linear aldehyde and moderate amount of 2-octenes. In the deuterioformylation of 1-hexene using the ligand **9**, it was found that the hydride migration is a reversible step under the conditions studied (80 °C, [Rh] = 0.2 mM, $P(CO) = P(D_2) = 10$ bar, [1-hexene] = 0.81 M, in benzene) for both linear and branched alkylrhodium species [21].

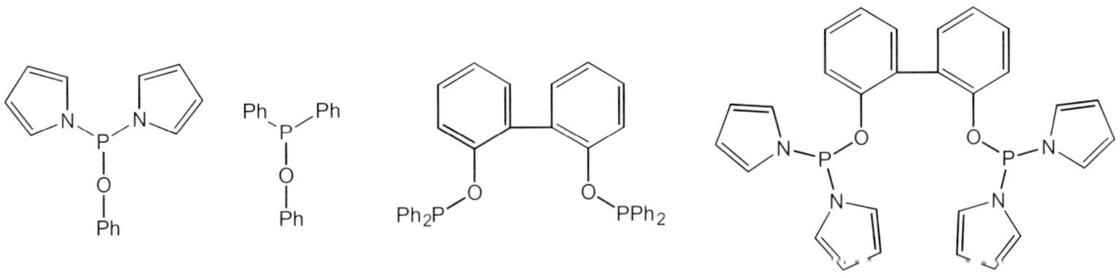

6 **7** **8** **9**

A new dirhodium(I) bisimidazolium–carbene complex (**10**) was used as a catalyst precursor in hydroformylation of vinyl arenes, 1-octene, and 2,5-dihydrofuran giving high selectivities for the branched aldehyde isomer (up to 100% at 25 °C and 30 bar pressure) when vinyl arenes were used as substrates. The high-pressure NMR spectroscopy provided evidence that the dinuclear unit containing the carbene ligand is maintained under catalytic conditions [22].

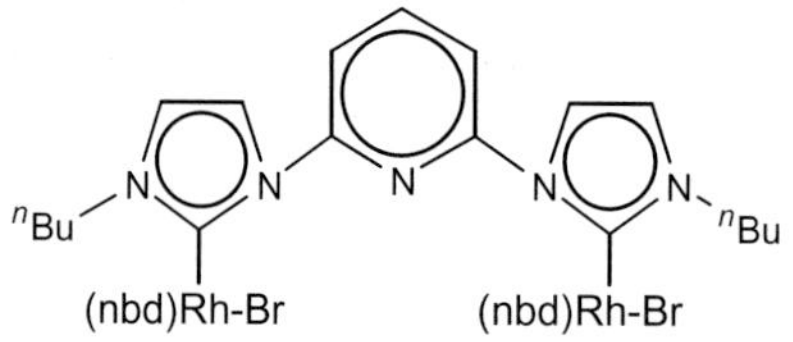

10

Four different rhodium–carbene complexes were tested as catalyst precursors in the hydroformylation reaction of 1-octene. Using 0.02 mol% [RhBr(1,3-dimesityl-3,4,5,6-tetrahydropyrimidin-2-ylidene)(COD) in toluene at 100 °C and 50 bar CO : H_2 = 1 : 1, TOFs up to 1500 h^{-1} were observed [23].

Diphosphane ligands **11** and **12** with bisphenol backbones were tested in rhodium-catalyzed hydroformylation of 1-octene. Especially, with ligand **6**, high activities (TOF = 3485 $mol_{1-octene}\,mol_{Rh}^{-1}\,h^{-1}$ at 80 °C and 20 bar, ligand : Rh = 6 : 1, 1-octene : Rh = 4000 : 1) were obtained [24].

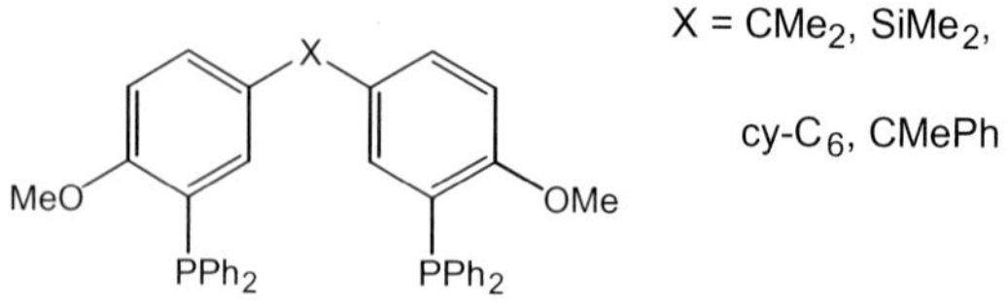

X = CMe_2, $SiMe_2$, cy-C_6, CMePh

11

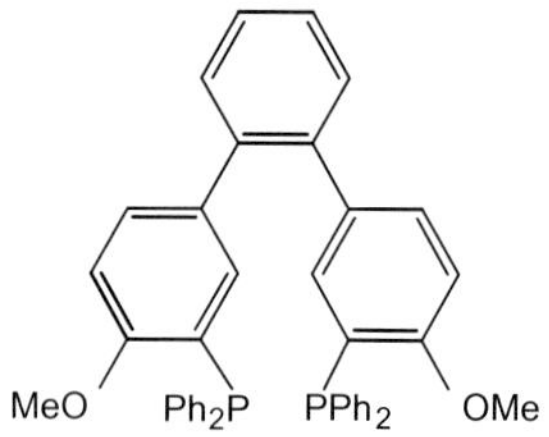

12

Rhodium complexes of bidentate phosphorus ligands assembled on a dimeric zinc (II) porphyrin template from functionalized monomeric ligands were tested in the hydroformylation reaction of 1-octene and styrene. Enhanced selectivities were observed compared to the nontemplate analogue [25]. A new supramolecular approach was used to obtain chelating bidentate phosphite ligands for regioselective rhodium-catalyzed hydroformylation. The self-organizing multicomponent assembly consists of two (zinc(II) porphyrin) phosphite and three 1,4-diazabicyclo[2.2.2]octane, which coordinates rigidly to a Rh(acac) part as a chelate. The rhodium catalyst based on such an assembly was studied in the rhodium-catalyzed hydroformylation of 1-octene. Very high activity and a linear-to-branched ratio of 15.1 was found at 80 °C [26]. The direct introduction of formyl groups into β-vinyl-metalloporphyrins via a PPh_3-modified rhodium-catalyzed hydroformylation reaction was described. The regioselectivity of the reaction was found to remarkably depend on the metal center of the porphyrin, yielding 100% of the branched aldehyde with zinc(II) complexes and 75% with the nickel(II) complex [27].

High chemoselectivities toward hydroformylation and high regioselectivities toward the branched aldehyde isomer were obtained in the hydroformylation of styrene at 40 °C and 50 bar CO : H_2 = 1 : 1 in the presence of an *in situ* rhodium catalyst containing sterically hindered 1-arylphosphole ligands [28].

A new concept for the construction of bidentate ligands employing self-assembly of two monodentate ligands through hydrogen bonding was used to obtain highly active and highly regioselective catalysts for the *n*-selective hydroformylation of terminal alkenes. Thus, a catalytic system composed of Rh : 6-diphenylphosphanyl-2-pyridone : 1-octene 1 : 20 : 7000, [1-octene] = 1.4 M was found to give at 65 °C and 10 bar CO : H_2 = 1 : 1 in toluene 56% conversion in 4 h and a linear : branched aldehyde ratio of 97 : 3 [29].

A thermoregulated phase-separable catalyst formed *in situ* from P[p-C_6H_4O($CH_2CH_2O)_{10}H]_3$ and $RhCl_3 \cdot 3H_2O$ was used in the hydroformylation of diisobutylene. Under the optimum conditions, 93.1% conversion of diisobutylene and 82.5% yield of aldehydes were observed. The catalyst could be efficiently recycled up to three times without any loss of activity [30].

The addition of 30% water by volume to acetone was found to create a simple polar-phase solvent system that produces 30–115% rate enhancements for the hydroformylation of 1-hexene with a variety of monometallic rhodium phosphane catalysts, and a 265% rate increase for a dirhodium tetraphosphane catalyst possessing bridging and terminal carbonyl groups along with improved chemoselectivity to aldehyde products [31].

Dinuclear aryloxide- and carboxylate-bridged rhodium complexes were tested as catalysts in the hydroformylation of styrene and 1-octene. A regioselectivity toward the branched aldehyde up to 97 and 45%, respectively, was found [32].

The hydroformylation of the water-soluble substrates, 4-penten-1-ol and 3-buten-1-ol in aqueous solution using HRh(CO)$(TPPTS)_3$ as the catalyst was investigated. Activation parameters and reaction selectivity for the hydroformylation of 4-penten-1-ol were found to depend on the ionic strength of the solution. As sodium sulfate was added, the activation energy increased. The linear–branched selectivity was strongly influenced by the ionic strength and temperature. The reaction could be directed to yield a product distribution of modest linearity (75%) or an exceptionally high ratio of the branched product (98%) as a cyclic 2-hydroxy-3-methyltetrahydropyran [33].

High regioselectivity and stereoselectivity were achieved in rhodium-catalyzed *ortho*-diphenylphosphanylbenzoate (*o*-DPPB)-directed hydroformylation reactions of various 1,3-disubstituted and monosubstituted allylic *o*-DPPB esters [34] (Equation 7.5).

O(*o*-DPPB), Bn, Me → 1.8 mol% $Rh(CO)_2$(acac), 3 mol% $P(OPh)_3$, 40 bar CO:H_2 = 1:1; 30 °C; 44 h → (*o*-DPPB)O, O, Bn, Me + (*o*-DPPB)O, Me, Bn, O

90 : 10

(*anti*:*syn* = 98:2)

(7.5)

Rhodium catalyst modified with tetrakis-(2,4-di-*t*-butylphenyl)-4,4′-biphenylphosphonite was found to exhibit an excellent catalytic activity as well as stability for the hydroformylation of mixed olefins, especially butane dimer [35].

A new class of phosphabarrelene (**13**)-rhodium catalysts was described that allow the hydroformylation of internal alkenes with very high activity and proceed free of alkene isomerization [36].

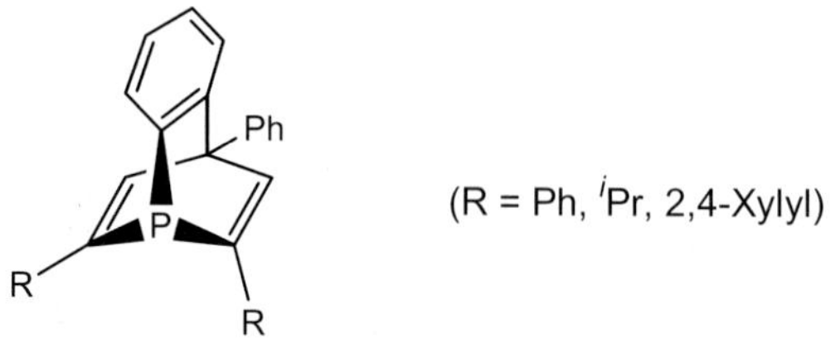

(R = Ph, ^{i}Pr, 2,4-Xylyl)

13

The kinetics of the $HRh(CO)(PPh_3)_3$-catalyzed hydroformylation of high-molecular-weight *cis*-1,4-polybutadiene in monochlorobenzene solution was studied. The hydroformylation reaction was found to be of the first order with respect to the catalyst and carbon–carbon double-bond concentrations and $CO:H_2 = 1:1$ pressure. The apparent activation energy for the hydroformylation of *cis*-1,4-polybutadiene over a temperature range of 60–80 °C was estimated to be 41 kJ mol^{-1} [37].

Various phosphane-modified rhodium catalysts were tested in the hydroformylation reaction of unsaturated esters. Pronounced temperature dependence was observed on the regioselectivity and catalytic activity for these reactions; and under the appropriate conditions, it was possible to obtain preferentially either linear or quaternary products (Equations 7.6 and 7.7) [38].

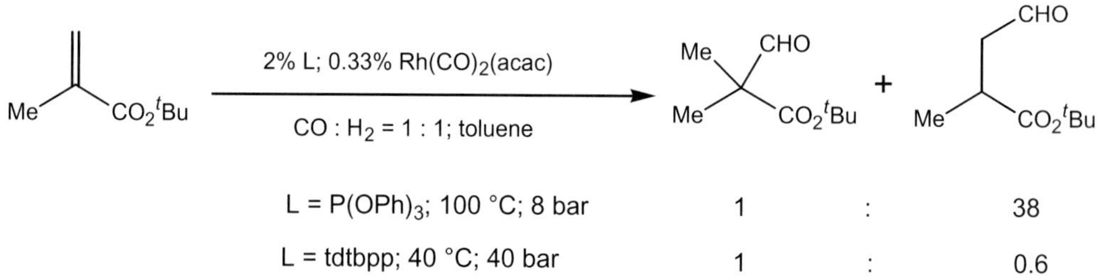

2% L; 0.33% $Rh(CO)_2(acac)$; $CO : H_2 = 1 : 1$; toluene

L	Quaternary (CHO)		Linear (CHO)
L = $P(OPh)_3$; 100 °C; 8 bar	1	:	38
L = tdtbpp; 40 °C; 40 bar	1	:	0.6

(7.6)

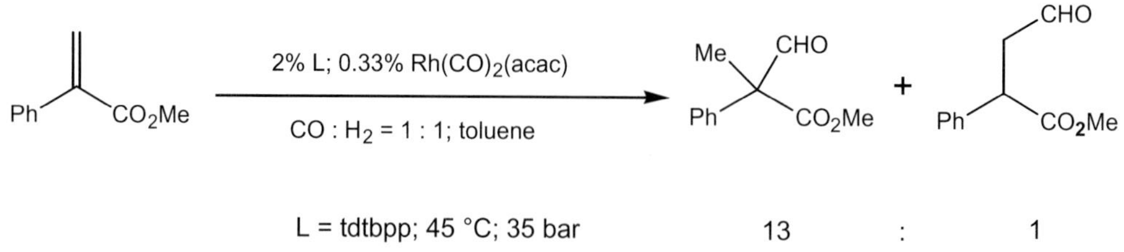

2% L; 0.33% $Rh(CO)_2(acac)$; $CO : H_2 = 1 : 1$; toluene

L	Quaternary (CHO)		Linear (CHO)
L = tdtbpp; 45 °C; 35 bar	13	:	1

tdtbpp = (2,4-di-^{t}Bu-$C_6H_3O)_3P$

(7.7)

Hydroformylation of styrene catalyzed by rhodium(III) complexes and the ethoxycarbonylation of styrene catalyzed by palladium(II) complexes containing novel phosphane ligands **14–20** were described. By using these ligands in $RhCl_2(Cp^*)$ (monodentate P ligand) complexes as catalyst precursors in styrene hydroformylation, a chemoselectivity higher than 99% was observed. A strong dependence of the regioselectivity on the P substituents was found in the case of the P heterocyclic ligands **14–17** [39].

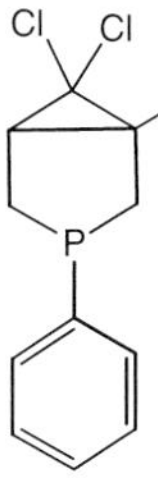

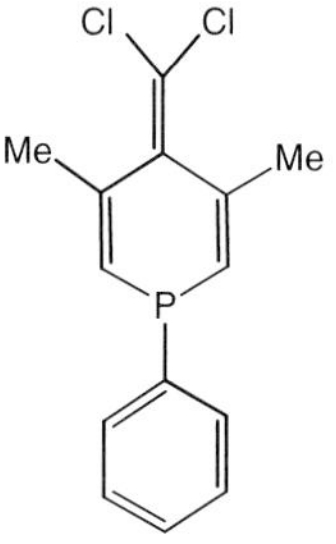

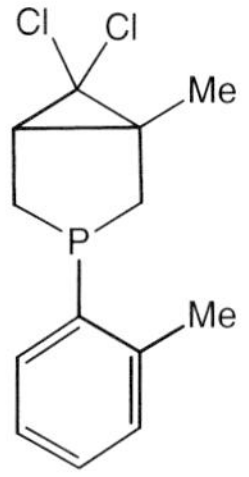

Rhodium(I) complexes composed of an anionic rhodium center containing chloride ligands and a cationic rhodium center coordinated by a diamine ligand were found to catalyze the hydroformylation of styrene and vinyl acetate under mild reaction conditions (e.g., 25 °C, 70 bar $CO : H_2 = 1 : 1$) in excellent activity and branched aldehyde selectivity [40].

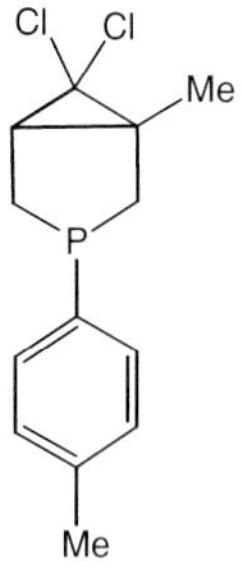

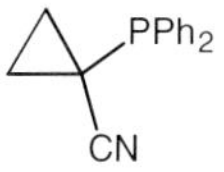

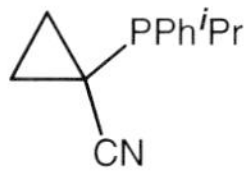

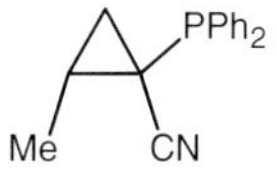

Rhodium(III) complexes of the type $RhCl(Cp^*)(L)$ (L = P heterocyclic or cyclopropane-based P ligand) were tested in styrene hydroformylation. A high activity at 40–100 °C and 100 bar $CO : H_2 = 1 : 1$, coupled with excellent chemoselectivity toward aldehydes, was observed [41].

The catalytic activity of **21** was tested in the hydroformylation of olefins. Good conversion yields and turnover frequencies were obtained in the hydroformylation of styrene and cyclohexene under mild conditions with low catalyst loading. The hydroformylation of styrene occurs with a high regioselectivity (93/7) in favor of the branched aldehyde isomer [42].

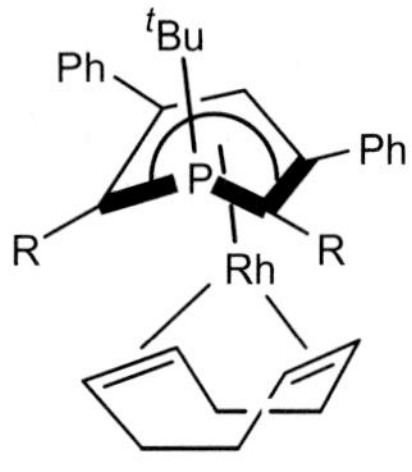

R = TMS, Ph

R = Me, Et

21 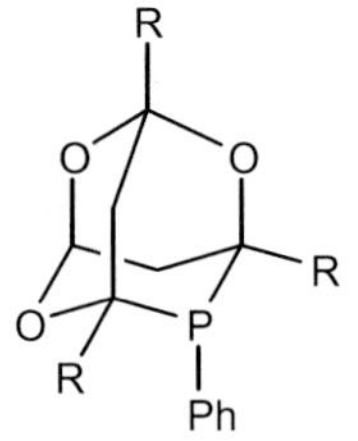 **22**

Rhodium complexes of ligand **22** were found to be similarly active in hydroformylation as catalysts derived from phosphites. The catalysts derived from **22** gave unusual low linear selectivity in the hydroformylation of hexenes. Using ligand **22** in rhodium-catalyzed hydroformylation of unsaturated esters a quaternary-selective synthetic method was developed [43].

The effects of the amounts and the type of heteropolyacids on the rhodium-catalyzed hydroformylation of styrene and 1-octene were studied. The rhodium cluster $Rh_6(CO)_{16}$ associated with the heteropolyacid $H_3PW_{12}O_{40} \cdot 25H_2O$ was found to improve the conversion of styrene and the selectivity toward the branched aldehyde [44].

A series of phenoxaphosphane- and dibenzophosphole-modified xantphos-type ligands were used for the rhodium-catalyzed hydroformylation of terminal and internal olefins. The effect of natural bite angle on hydroformylation activity and selectivity was investigated. Ligands with larger bite angle lead to more selective systems, but above 125° the regioselectivity dropped. A correlation between the selectivity for the hydroformylation of 1-octene and *trans*-2-octene was observed, suggesting that the selectivity-determining step remains unchanged between terminal and internal olefins [45].

Rhodium nanoparticles were used as catalyst precursors for the solventless hydroformylation of 1-alkenes. Linear/branched selectivities up to 25 were achieved by adding xantphos to the catalyst system [46].

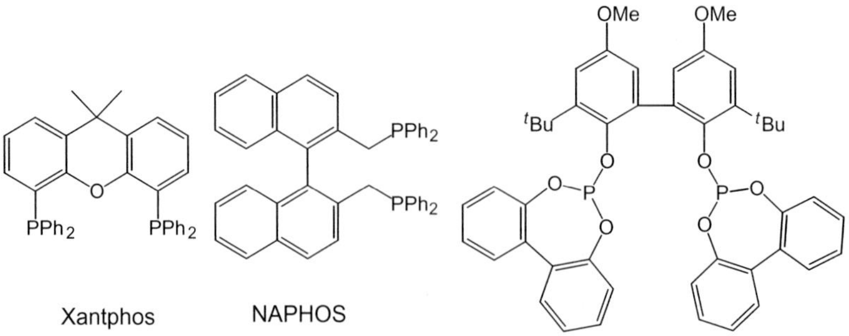

Xantphos NAPHOS BIPHEPHOS

The effect of phosphine and diphosphane ligands on the chemoselectivity and regioselectivity of the rhodium-catalyzed hydroformylation reaction of allylbenzenes and propenylbenzenes was studied. It was found that the Rh-NAPHOS system promotes the hydroformylation of allylbenzenes into linear aldehydes in near 95% selectivity and propenylbenzenes into branched aldehydes with a formyl group in β-position to the phenyl ring in near 90% selectivity, while the Rh-dppp system gives branched aldehydes with a formyl group in α-position in near 70% selectivity starting from allylbenzenes [47].

The hydroformylation of terminal and internal olefins to terminal aldehydes using a catalyst system composed of $Rh(CO)_2(acac)$ and BIPHEPHOS was investigated. Conversions up to 99% and yields up to 86% of the aldehydes (linear : branched ratio 99 : 1) were found. A very high TOF of 44 000 h^{-1} was achieved at 140 °C and 30 bar $CO : H_2 = 1 : 1$ using a rhodium-to-1-dodecene ratio of 1 : 100 000 [48]. An efficient product and catalyst separation was achieved by the application of a temperature-dependent multicomponent solvent (TMS) system in rhodium-BIPHEPHOS catalyzed hydroformylation of *trans*-4-octene to *n*-nonanal. By application of a TMS system, classical extraction process step can be omitted and catalyst leaching reduced [49]. The rhodium-BIPHEPHOS catalyzed isomerizing hydroformylation of oleic acid ester and linoleic acid ester was found to give the corresponding linear aldehyde at 115 °C and 20 bar $CO : H_2 = 1 : 1$ in up to 26 and 34% yield, respectively [50].

Preformed rhodium complexes or *in situ* systems with asymmetrical cyclic phosphite (**23**–**26**) and phosphinite (**27**) ligands bearing either trifluoromethyl or pentafluorophenyl group(s) were tested in the hydroformylation of styrene. Both systems were found to provide excellent hydroformylation activities at 100 °C. High regioselectivities toward the branched aldehyde (2-phenyl-propanal) were achieved at 40 °C [51]

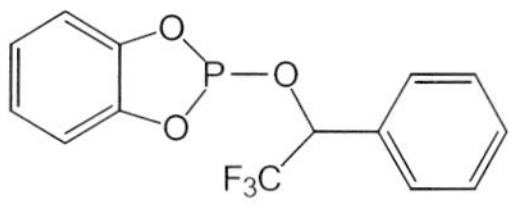

23 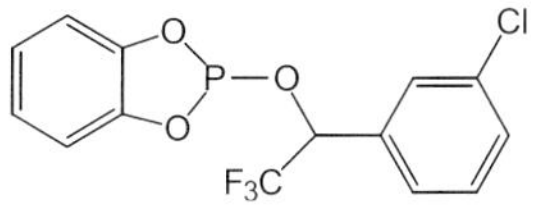 **24**

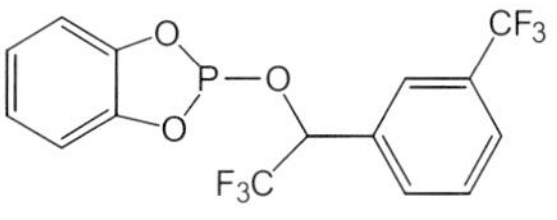

25 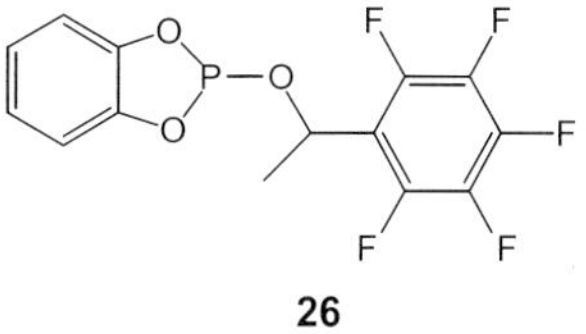 **26** 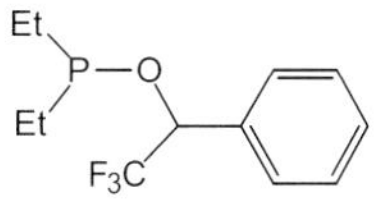 **27**

The synthesis and the catalytic activity in the rhodium-catalyzed hydroformylation of styrene as well as quantum–chemical estimation of structure–activity relationship of novel cyclic amidophosphonite ligands bearing polyfluorinated tails were reported [52].

A new rhodium complex with a nitrogen-containing bis(phosphane oxide) ligand (**28**) was applied to hydroformylation of styrene. High activity (TON up to 1443 mol aldehyde/mol catalyst) and regioselectivity toward the branched aldehyde (96.9%) were found at 30–40 °C and 20–100 bar $CO : H_2 = 1 : 1$ [53].

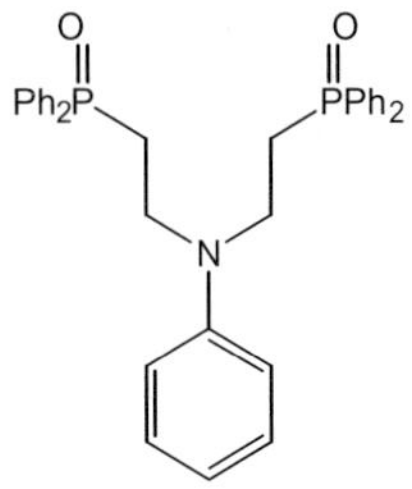

28

A new class of phosphabarrelene/rhodium catalysts was found to display very high activity toward hydroformylation of internal alkenes with unusually low tendency toward alkene isomerization (Equation 7.6) [54].

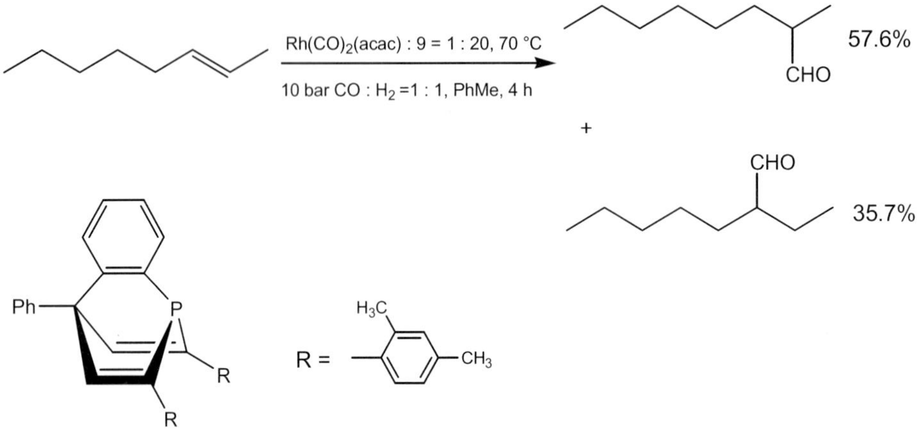

$Rh(CO)_2(acac) : 9 = 1 : 20$, 70 °C; 10 bar $CO : H_2 = 1 : 1$, PhMe, 4 h → 57.6% + 35.7%

29 R = 2,4-dimethylphenyl (H_3C, CH_3)

(7.8)

Rhodium carboxylato complexes bearing 1,3-diarylimidazol-2-ylidene ligands are catalytically active for the regioselective hydroformylation of both aryl and aliphatic alkenes with branched regioselectivities above 40 : 1 [55].

An additional gas such as dinitrogen, argon, or xenon present in the reaction medium in high concentration was found to decrease the rate of hydroformylation of cyclohexene, 1-hexene, or styrene in the presence of $RhH(CO)(PPh_3)$. This effect was attributed to a competition between the additional gas and one of the reagents, alkene, dihydrogen, or carbon monoxide for a coordinative unsaturated site available on the catalytically active intermediates [56].

Increasing the concentration of triphenylphosphane was found to accelerate the rhodium-catalyzed hydroformylation of conjugated dienes, such as isoprene and myrcene. The hydroformylation of the nonconjugated diene, limonene, followed a contrary tendency common to the most of alkenes [57,58]

The homogeneous rhodium-catalyzed hydroformylation of 3,3-dimethyl-1-butene in *n*-hexane was studied at 25 °C and variable total pressure using high-pressure *in situ* infrared spectroscopy as the analytical tool. All anticipated major and minor component spectra were reconstructed using band-target entropy minimization (BTEM) without spectral preconditioning. Beside $Rh_4(\sigma\text{-}CO)_9(\mu\text{-}CO)_3$, RC(O)Rh$(CO)_4$, and $Rh_6(CO)_{16}$, the presence of a previously unknown complex $Rh_4(\sigma\text{-}CO)_{12}$ was identified [59]. The same technique has led to the reconstruction of the pure-component spectra of $RhH(CO)_4$ and $RhD(CO)_4$ for the first time [60].

High-pressure *in situ* FTIR and polymer matrix techniques were used to study the rhodium-catalyzed hydroformylation of 1-octene, 1-butene, propene, and ethene using Rh(acac)$(CO)_2$ or Rh(acac)(CO)(PPh_3) in a polyethylene matrix as the catalyst precursor. The acyl rhodium intermediates, RC(=O)Rh$(CO)_4$ and RC(=O)Rh$(CO)_3(PPh_3)$, were observed. It was found that the acyl rhodium tetracarbonyl intermediates easily react with ethene to form acyl rhodium tricarbonyl species RC(=O)Rh$(CO)_3(C_2H_4)$ [61]. Deuterioformylation of 1-phenyl-1-(*n*-pyridyl)-ethenes in the presence of a phosphane-modified $Rh_4(CO)_{12}$ as catalyst precursor was carried out at 100 bar of CO : D_2 = 1 : 1 and 80 °C at partial substrate conversion. On basis of a direct ^{2}H NMR analysis of the crude reaction mixture, it was concluded that the branched alkyl rhodium intermediate is almost exclusively formed [62].

A variable-temperature multinuclear NMR study using enriched ^{13}CO has shown that the HRh$(CO)_2(TPP)_2$ (TPP = 1,2,5-triphenyl-1*H*-phosphole) complex, the resting state of the catalytic system for the hydroformylation of styrene by the Rh/TPP mixture, exists in solution as two stereoisomers in equilibrium. The minor isomer (10%) exhibits a geometry in which the hydride ligand occupies an equatorial position [63].

The origin of chemoselectivity in the hydroformylation of alkenes catalyzed by trialkylphosphane complexes of rhodium was investigated. In alkene hydroformylation reactions carried out in protic solvents, monophosphane acyl intermediates were found responsible for the aldehyde formation in the case of P^iPr_3 and P^iBu_3, while a disubstituted acyl intermediate caused the alcohol formation in the case of PEt_3 [64].

New mechanistic insights were obtained in the rhodium(I)/aminophosphane-catalyzed styrene hydroformylation by using D_2O as deuterium-labeling agent. H/D exchange of the rhodium-hydride complex and the reversibility of the styrene coordination were established. Based on the product compositions, protonolysis of the rhodium-acyl intermediate and a bimolecular reaction involving the rhodium-acyl and rhodium-hydride intermediates as the aldehyde-forming step were excluded [65].

7.1.3 Ruthenium Catalysts

Ruthenium carbonyl compounds were found to be suitable catalysts for the hydroformylation of alkenes using CO_2 as the source of CO along with H_2. In the multistep process, CO_2 is first reduced to CO and then used *in situ* in hydroformylation. The best results were obtained by using a $[Ru(CO)_3Cl_2]_2/Li_2CO_3$ system [66].

The homogeneous hydroformylation of 1-octene using either of the hetero-bimetallic complexes **30** or **31** as catalysts was studied. The best results were obtained with complex **31**, which showed the highest regioselectivity for the linear aldehyde [67].

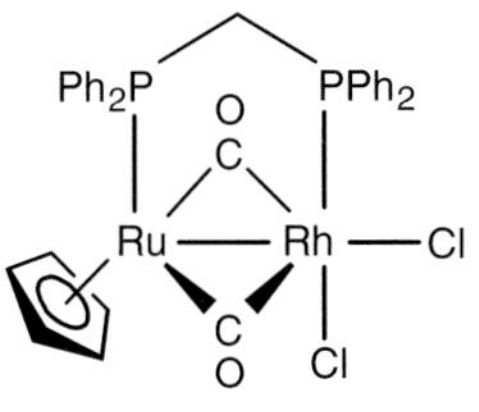

The reactivity of a series of ruthenium *ortho*-substituted triphenylphosphane complexes serving as catalysts for the hydroformylation of 1-hexene was studied. The activities were found to depend on the binding mode of the phosphane and on the strength of the ruthenium–phosphane interaction [68].

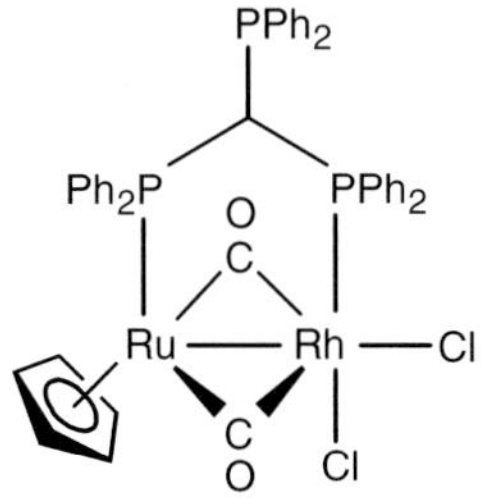

The hydroformylation of ethene with CO/H_2O to give propanal and CO_2 in the presence of acyldicarbonyl–Ru(II) complexes was found to be highly selective in diglyme containing 7% H_2O (by weight) in the presence of propionic acid/propionate. A TOF $= 20\,h^{-1}$ [mol of propanal/(mol Ru/h)] and a selectivity of >90% were obtained at 140 °C and 50 bar CO : ethene = 3 : 2 [69].

A mixed ionic liquid [bmim][Cl,NTf_2] system was successfully used as a reaction medium for ruthenium-catalyzed hydroformylation of 1-hexene with carbon dioxide in the absence of toxic CO and any volatile organic solvents. Thus, from 20 mmol 1-hexene in [bmim][Cl,NTf_2] containing 0.1 mmol $Ru_3(CO)_{12}$ 82% heptanal was obtained at 160 °C under 40 bar CO_2 and 40 bar H_2 pressure in 10 h [70].

7.1.4 Platinum–Tin Catalysts

The isolated "preformed" platinum catalyst [Pt(PP_3)($SnCl_3$)]$SnCl_3$ (PP_3 = tris[2-(diphenylphosphano)ethyl]-phosphane) from the precursor [Pt(PP_3)Cl]Cl and $SnCl_2$ was found to show high aldehyde selectivity (99%) in styrene hydroformylation at 100 °C and 100 bar $CO/H_2 = 1$ [71].

2-Benzyloxy- and 2-tosyloxystyrene were hydroformylated under various reaction conditions to obtain the corresponding linear aldehydes. The best results (up to 70% linear aldehyde at 80 °C and low pressure) were obtained by using the catalyst precursor Pt(xantphos)Cl_2 in toluene or the water-soluble catalytic system Rh$(CO)_2$(acac)/2,7-bis$(SO_3Na)_2$-xantphos in the biphasic medium water/toluene [72].

The new diphosphanes **32** and **33** were applied in the platinum/tin-catalyzed hydroformylation of 1-octene. Moderate activities and high regioselectivities (linear : branched ratio up to 45) were found [45].

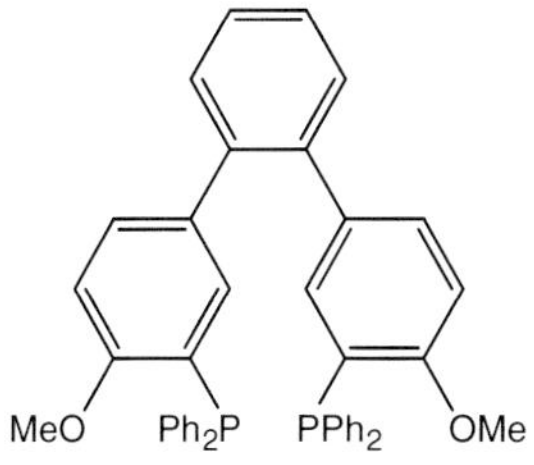

32 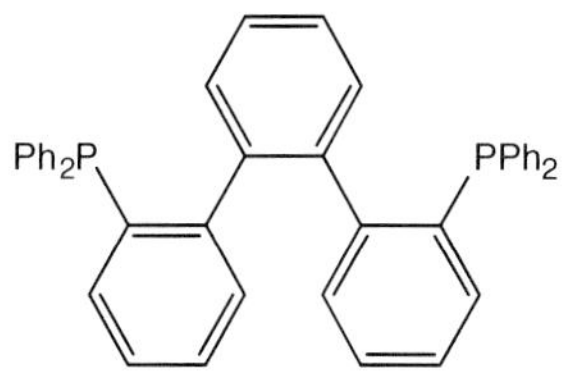 **33**

The hydroformylation of 1-octene and the *i*-octenes was performed with the precursor [Pt(sixantphos)Cl_2]. For the internal alkenes, selective tandem isomerization/hydroformylation toward *n*-nonanal was observed. *In situ* UV–Vis spectroscopic studies revealed rapid formation of the corresponding Pt–stannate complex upon reaction with $SnCl_2$, whereas high-pressure *in situ* IR spectroscopy showed the formation of a Pt–CO species and a short-lived Pt–H species under syngas, as well as a rapid evolution of aldehyde product upon addition of 1-octene to the preformed catalyst in the IR autoclave [73].

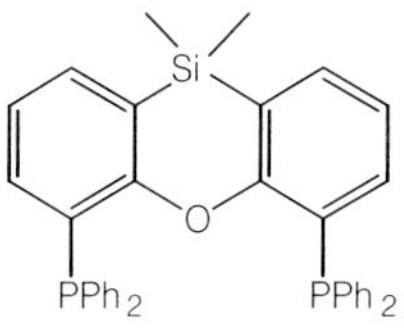

Sixantphos

7.1.5 Palladium Catalysts

Catalyst systems consisting of a palladium(II) diphosphane complex with weakly or noncoordinating counterions were found to be efficient catalysts for the hydrocarbonylation of higher olefins. Variation of ligand, anion, and/or solvent were used to steer the reaction selectively toward aldehydes/alcohols, ketones, or oligoketones. Noncoordinating anions and arylphosphane ligands were found to produce primarily oligoketones, while increasing ligand basicity shifts selectivity toward monoketones. Increasing ligand basicity and/or anion coordination strength leads to high selectivity for hydroformylation products, aldehydes, and alcohols [74].

Highly selective halide anion-promoted palladium-catalyzed hydroformylation of internal alkenes to linear alcohols was studied. A (bcope)Pd$(OTf)_2$ complex (bcope) = bis(cyclooctyl)phosphinoethane with substoichiometrically added halide anion was found to be a highly efficient homogeneous catalyst to selectively convert internal linear alkenes into predominantly linear alcohols under mild conditions (105 °C, 60 bar $CO : H_2 = 1 : 1$) [75].

7.1.6
Iridium Catalysts

The mechanism of the iridium-catalyzed hydroformylation has been studied by NMR spectroscopy. Species including iridium acyl and alkyl dihydride intermediates were detected. The results confirmed that the CO-deficient atmosphere favors hydrogenation over carbonylation [76].

The hydride complexes $IrH(CO)_2$(xantphos) and $IrH_3(CO)$(xantphos), as well as the propionyl complex $Ir(COEt)(CO)_2$(xantphos) were found to be modest catalysts for the hydroformylation of 1-hexene and styrene under mild conditions. Propionyl dihydride species $IrH_2(COEt)(CO)$(xantphos) was detected by addition of *para* hydrogen to $Ir(COEt)(CO)_2$(xantphos) [77].

7.1.7
Bimetallic Catalysts

Efficient chemoselective hydroformylation of monosubstituted alkenes was observed at room temperature under atmospheric pressure of $CO : H_2 = 1 : 1$, without affecting functional groups such as disubstituted alkene moieties, aryl and alkenyl iodide moieties, and hydroxy and carboxy groups (Equations 7.9 and 7.10) [78].

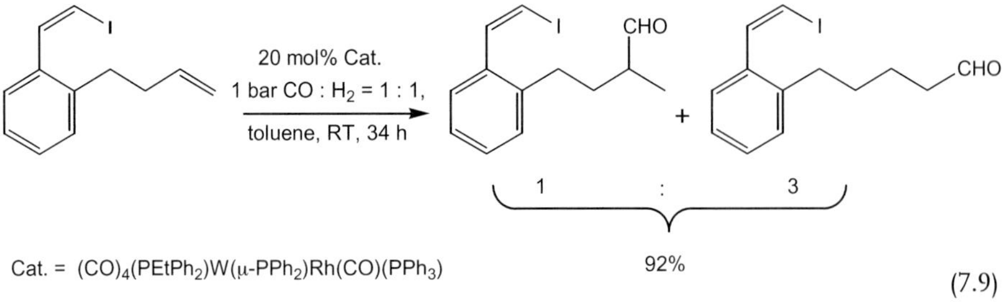

I
20 mol% Cat.
1 bar CO : H_2 = 1 : 1,
toluene, RT, 34 h
I CHO
I CHO
+
1 : 3
92%
Cat. = $(CO)_4(PEtPh_2)W(\mu\text{-}PPh_2)Rh(CO)(PPh_3)$

(7.9)

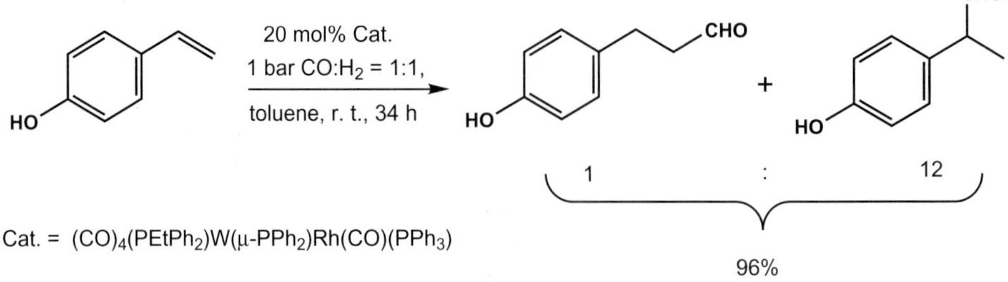

HO
20 mol% Cat.
1 bar CO:H_2 = 1:1,
toluene, r. t., 34 h
HO CHO
+
HO CHO
1 : 12
96%
Cat. = $(CO)_4(PEtPh_2)W(\mu\text{-}PPh_2)Rh(CO)(PPh_3)$

(7.10)

Adding manganese carbonyl hydride to the reaction mixture of the unmodified rhodium-catalyzed hydroformylation of 3,3-dimethyl-1-butene led to a significant increase in system activity. On the basis of the *in situ* spectroscopic information, this increase in rate of product formation was correlated with bimetallic catalytic binuclear elimination [79]. The origin of synergism in the $Rh_4(CO)_{12}$-catalyzed hydroformylation of cyclopentene promoted with $HMn(CO)_5$ was investigated by *in situ* high-pressure FTIR spectroscopy. Only four organometallic species, $Rh_4(CO)_{12}$, $C_5H_9CORh(CO)_4$, $HMn(CO)_5$, and $Mn_2(CO)_{10}$ were found in the reaction mixture. The kinetics of cyclopentanecarboxaldehyde formation suggest that the origin of synergism is the $HMn(CO)_5$ attack on the acyl species [80].

7.1.8
Supported Complexes

The concept of supported ionic liquid catalysis involves the surface of a support material that is modified with a monolayer of covalently attached ionic liquid fragments. Treating this surface with additional ionic liquid results in the formation of multiple layers of free ionic liquid on the support material. These layers serve as the reaction phase in which a homogeneous hydroformylation catalyst was dissolved. The concept of supported ionic liquid catalysis has successfully been used for hydroformylation reactions [81].

Rhodium-complexed dendrimers supported on a resin were evaluated as catalysts for the hydroformylation of aryl olefins and vinyl esters. Up to 99% yields and an outstanding selectivity for the branched aldehydes (up to 38 : 1) were obtained at room temperature and 69 bar $CO : H_2 = 1 : 1$. The dendritic catalysts were recycled by simple filtration and reused even up to the 10th cycle without any loss of activity and selectivity [82].

Neutral or cationic rhodium complexes, $[Rh(COD)(\mu\text{-}S(CH_2)_{10}CO_2H)]_2$ and $[Rh(COD)(\eta^6\text{-benzoic acid})][BF_4]$, were used as polymer-supported catalysts in hydroformylation of 4-vinylanisole and the effect of the size of the immobilizing supporters on the catalytic activity was investigated at 55 °C and 46 bar $CO : H_2 = 1 : 3$. It was found that the cobalt-ferrite nanomagnet-supported cationic rhodium catalyst had a high catalytic activity comparable with the homogeneous cationic rhodium catalyst and it could be easily recovered from the reaction mixture by magnetic decantation [83,84].

SiO_2-tethered rhodium complexes, derived from $Rh(CO)_2(acac)$ and 3-(mercapto) propyl- and 3-(1-thioureido)propyl-functionalized silica gel, were used as catalysts in the hydroformylation of various vinyl arenes and vinyl acetate. Conversions, chemoselectivity, and regioselectivity obtained with the SiO_2-tethered catalysts were comparable with those of the well-known homogeneous rhodium catalysts [85].

A rhodium(I) diphosphane complex, [Rh(**43**)(COD)]Cl, anchored on functionalized carbon support was tested as a catalyst for the hydroformylation of 1-octene. It was found that the catalyst shows an outstanding behavior, being fully active and with almost constant selectivity to the linear aldehyde in four consecutive catalytic runs [86].

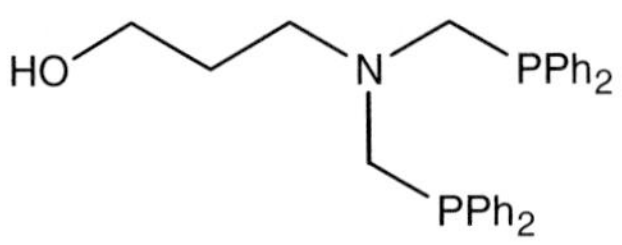

HO N PPh2 PPh2

34

Exchanging the Rh/TPPTS complex with an anion exchange resin resulted in a stable heterogenized catalyst for the hydroformylation of alkenes. The kinetics of hydroformylation of 1-hexene using this catalyst was investigated. The rate was found to be first-order dependent on the catalyst, 1-hexene concentration, and H_2 partial pressure. A maximum in the rate with increasing partial pressure of carbon monoxide was observed [87].

A "heterogenized" version of homogeneous ionic-liquid catalyst system in which the catalyst remains active, highly selective, and stable over extended periods in a continuous gas-phase hydroformylation process was described. Thus, using a rhodium–sulfoxantphos complex in [bmim][n-$C_8H_{17}OSO_3$] on a partly dehydrogenated silica support gave a >20 *normal/iso* butyraldehyde ratio, TON = 2600, TOF = 44 h^{-1} over at least 60 h [88].

Rhodium-supported catalysts were prepared by impregnating rhodium(I) and rhodium(III) complexes with and without heteropolyacids for the hydroformylation of styrene derivatives. A clear effect of the heteropolyacid $H_3PW_{12}O_{40} \cdot yH_2O$ in increasing the catalytic activity of the rhodium-supported catalyst was found [89].

The effect of $H_3PW_{12}O_{40}$ and $P(OPh)_3$ on the activity and selectivity of MCM-41-supported rhodium(I) and rhodium(III) catalysts was investigated in the hydroformylation reaction of 1-octene and styrene. The heteropolyacid-impregnated catalyst was found to enhance the catalytic activity [90]. The addition of $P(OPh)_3$ increased the selectivity of the reaction toward the branched product [91].

Nixantphos (**35**) and its modified derivatives were immobilized on soluble and solid polymer supports, such as dendrimers, polyglycerol, and polyurethanes. The new catalysts were tested in rhodium-catalyzed regioselective hydroformylation of 1-octene and proved to be selective and reusable. Polymer support was found to require spacers between the ligand and the supporting units [92].

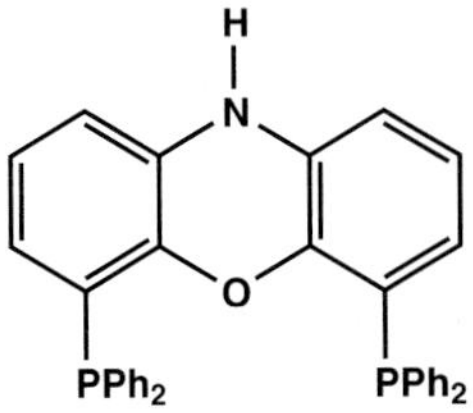

H N O PPh2 PPh2

35

7.1.9
Biphasic Systems

Improved selectivity, catalyst retention, and product separation were reported in the hydroformylation of linear terminal alkenes using rhodium-based catalysts under fluorous biphasic conditions. The best results, which compare well with those obtained in commercial systems, were obtained using Rh ($2.0\,mmol\,dm^{-3}$)/P $(4\text{-}C_6H_4C_6F_{13})_3$ ($20\,mmol\,dm^{-3}$) at $70\,°C$ and 20 bar CO/H_2 to give a linear aldehyde selectivity of 80.9% and an initial productivity of $8.8\,mol\,dm^{-3}\,h^{-1}$. Rhodium leaching into the aldehyde phase was found to be 0.05% of the rhodium charged and the phosphorus leaching is 3.3% [93].

The water-soluble complex derived from $Rh(CO)_2(acac)$ and human serum albumin (HSA) was found to be efficient in the hydroformylation of styrene and 1-octene at $60\,°C$ and 70 bar $CO/H_2 = 1$ even at very low catalyst concentrations. The chemoselectivity and regioselectivity were generally higher than those obtained by using the classic catalytic systems such as TPPTS–Rh(I) [94].

Hydroformylation of 1-dodecene using a rhodium–TPPTS catalyst in a microemulsion has been described. High activities and good selectivities toward the *n*-aldehyde were achieved at $80\,°C$ and 80 bar $CO : H_2 = 1 : 1$. The reaction rates were higher than those in a two-phase system. Technical-grade polyglycolether type surfactants were used for the preparation of the microemulsions [95].

The ionic compounds 1,2,3-trimethylimidazolium triflate and 1-ethyl-2,3-dimethylimidazolium triflate and the coordination compound (3-butylimidazole) triphenylboron were used as solvents for biphasic rhodium-catalyzed hydroformylation of 1-hexene and 1-dodecene. High conversions with varying linear/branched aldehyde ratios were observed. Compared to the conventional solvent toluene, similar turnover numbers, but a higher tendency toward isomerization and hydrogenation, were found [96].

A novel phenoxaphosphino-modified ligand **36** was used in the rhodium-catalyzed hydroformylation of 1-hexene and 1-octene in 1-butyl-3-methylimidazolium hexafluorophosphate. Very high linear selectivity and high activity were found with almost complete retention of the catalyst in the ionic phase. At $100\,°C$, high turnover frequencies ($>6200\,h^{-1}$), high linear/branched ratios (>40), and negligible catalyst loss (Rh loss $< 0.07\%$, P loss $< 0.4\%$) were observed. [97,98]

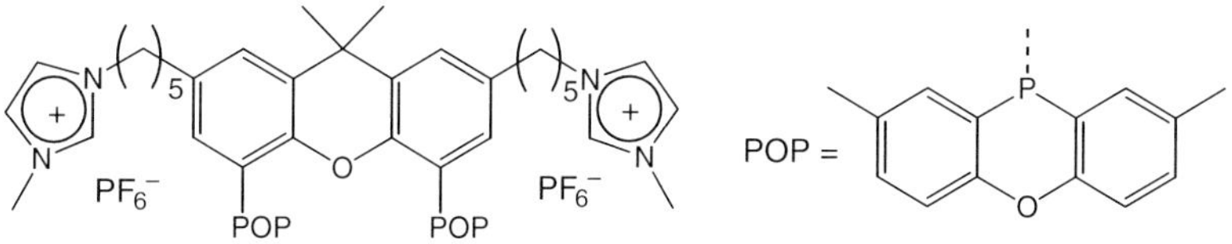

36

The rhodium-catalyzed hydroformylation of 1-dodecene was investigated with a series of sulfonated water-soluble phosphane ligands such as $Ph_2P(CH_2)_2S(CH_2)_2SO_3Na$ and $Ph_2P(CH_2)_2S(CH_2)_3SO_3Na$ at $120\,°C$ and 60 bar $CO : H_2 = 1 : 1$. The ratio of 1-dodecene/rhodium could be increased up to 10 000. Turnover numbers >50 000 were achieved without any surfactant and TONs were increased up to about 65 000 in the presence of polyoxyethylene–polyoxypropylene–polyoxyethylene triblock copolymers [99].

The *in situ* catalyst from $RhCl_3 \cdot 3H_2O$ and *N*,*N*-dipolyoxyethylene-substituted-2-diphenylphosphino)phenylamine was used in the aqueous–organic biphasic hydroformylation of 1-decene. At 120 °C and 50 bar CO : H_2 = 1 : 1, up to 99% yield of aldehydes was achieved, which remained as high as 94% after the catalyst had been recycled 20 times [100].

The micellar effect in hydroformylation of 1-octene and 1-decene using water-soluble rhodium complexes with sulfonated diphosphanes in the presence of ionic surfactants and methanol in water was studied. The hydroformylation activities using cetyltrimethylammonium hydrogen sulfate and methanol additives were found to be higher than those in experiments without these additives [101].

An improved hydroformylation of 1-hexene with CO_2 as the carbonyl carbon source using $Ru_3(CO)_{12}$ as the catalyst precursor in a biphasic system consisting of ionic liquid and organic solvent has been reported. In the presence of 2 mol% $Ru_3(CO)_{12}$ 1-hexene gave in a toluene/[bmim]Cl system at 140 °C and 80 bar CO_2 : H_2 = 1 : 1 in 30 h 84% total yield of C_7 alcohols [102].

Hydroformylation of 1-octene under aqueous two-phase conditions was studied using an *n* *N*-heterocyclic carbene rhodium catalyst immobilized to an amphiphilic, water-soluble block copolymer support. The catalyst showed high activity up to 2360 h^{-1} turnover frequency at 100 °C and 50 bar CO : H_2 = 1 : 1 pressure in four consecutive cycles [103].

The effect of methylated cyclodextrins on the $RhH(CO)(TPPTS)_3$ complex in hydroformylation conditions (50 bar, CO : H_2 = 1 : 1, and 80 °C) were investigated by high-pressure ^{31}P NMR spectroscopy. It was found that the formation of the stable inclusion complex between methylated β-cyclodextrin and TPPTS influences the TPPTS dissociation equilibrium. The methylated α-cyclodextrin does not interact with the TPPTS and the methylated γ-cyclodextrin can only weakly bind to the TPPTS. These results explain why a decrease in the normal-to-branched aldehydes ratio is always observed when cyclodextrins are used as mass-transfer agents in aqueous biphasic hydroformylation processes [104].

The two-phase hydroformylation of higher olefins with the rhodium/TPPTS catalytic system in the presence of various chemically modified α-cyclodextrins was investigated. The reaction rate and chemoselectivity were increased, but the regioselectivity, what was observed previously with chemically modified β-cyclodextrins, did not change [105]. Rhodium/TPPTS complexes noncovalently bound to cyclodextrins were tested as water-soluble supramolecular catalysts for the biphasic hydroformylation of higher olefins. The cyclodextrins greatly increased the reaction rate, the chemoselectivity, and the linear-to-branched aldehyde ratio [106]. The potential of sulfonated xantphos as ligand for a cyclodextrin-based rhodium-catalyzed 1-octene and 1-decene hydroformylation process was investigated. Activity enhancement and an increase in the linear-to-branched aldehyde ratio (up to 33) were observed [107].

The water-soluble rhodium complex $[Rh(\mu\text{-pz})(CO)(TPPTS)]_2$ (pz = pyrazolate) was used as catalyst precursor during the two-phase catalytic hydroformylation of different olefins at 100 °C, 50 bar (CO : H_2 = 1 : 1), 600 rpm, and substrate : catalyst ratio of 100 : 1. A reaction order – 1-hexene > styrene > allylbenzene > 2,3-dimethyl-1-butene > cyclohexene – was found. The experiments also showed that the binuclear catalyst precursor was resistant to possible sulfur poisons [108].

The rhodium/TPPTS-catalyzed hydroformylation of 1-octene and other higher olefins in an aqueous biphasic system was studied at pressures from 40 to 90 bar and temperatures of up to 120 °C. Nonionic amphiphiles of the alkoxyethylene type were applied to promote the contact between the reacting species by enlargement of the interfacial area. The highest reaction rates were obtained at a surfactant concentration of about 1 wt% [109].

The synthesis of *n*-hexanal from 2-pentene in water using $Rh(CO)_2(acac)$ and sulfonated NAPHOS as catalyst precursors was investigated. It was found that by lowering the partial pressure of carbon monoxide in a buffered system at pH 7 and 8, significant increase in the aldehyde yield (up to 73%) and excellent regioselectivity (*normal* : *iso* = 99 : 1) were observed. Similar results for the hydroformylation of 2-pentene were also obtained in the presence of tertiary amines such as triethylamine [110].

The rhodium/TPPTS-catalyzed hydroformylation of higher olefins in organic/aqueous biphasic system in the presence of double long-chain cationic surfactants (**37**) was studied at 100 °C and 20 bar $CO : H_2 = 1 : 1$ pressure. The reaction rate was comparable with that in homogeneous catalysis system [111].

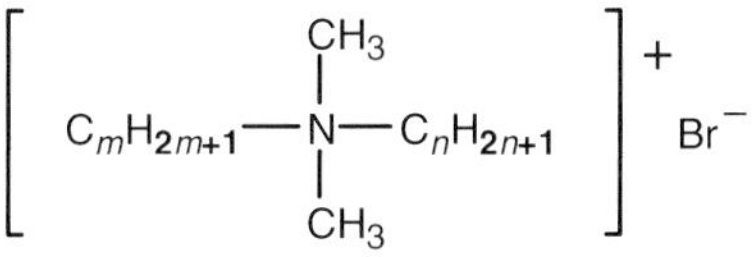

$$\left[C_mH_{2m+1}-N(CH_3)_2-C_nH_{2n+1} \right]^+ Br^-$$

$m = 16, 20$

$n = 2, 4, 8, 12, 16, 22$

37

The potential of alkyl-sulfonated diphosphane ligands associated with methylated α- and β-cyclodextrins during the reaction of rhodium-catalyzed hydroformylation of 1-decene was studied. In all cases, the presence of cyclodextrins increased the conversion and the chemoselectivity, whereas the linear-to-branched ratio of the aldehyde product decreased. The decrease in regioselectivity was attributed to the formation of low-coordinated phosphane species [112].

Amino acids and oligopeptides were used as ligands for $Rh(CO)_2(acac)$ in the aqueous biphasic hydroformylation of styrene. The water-soluble catalytic system maintained its activity practically unchanged during three recycled experiments [113].

The development of a thermoregulated polyethylene glycol (PEG) biphasic system composed of PEG-400/1,4-dioxane/*n*-heptane was reported. By applying this system in the hydroformylation of 1-dodecene catalyzed by Rh complexes modified with phosphite ligand containing PEG chains, TMPGP ($P[O(CH_2CH_2O)_nCH_3]_3$, $n = 8$), the conversion of 1-dodecene and the yield of aldehydes reached up to 96 and 94%, respectively. The catalyst, after being recycled 23 times, showed no appreciable loss of catalyst activity [114].

7.1.10
Hydroformylation in Supercritical Fluids

The cobalt-catalyzed hydroformylation of 1-octene was investigated using bis{tri(3-fluorophenyl)phosphane}hexacarbonyldicobalt and $Co_2(CO)_8$ as precatalysts in supercritical carbon dioxide. The catalytic performance in $scCO_2$ was compared to the one in toluene as a conventional solvent. Similar activities and selectivities were obtained in both reaction media. In $scCO_2$, a substantial improvement in the selectivity for aldehydes was found by using the phosphane-containing complex in comparison to the unmodified catalyst $Co_2(CO)_8$ [115]. The formation of cobalt complexes in supercritical carbon dioxide derived from $Co_2(CO)_6[P(p\text{-}CF_3C_6H_4)_3]_2$ under hydroformylation conditions has been investigated by *in situ* high-pressure NMR spectroscopy. It was found that the phosphane-modified catalyst system is stable under low carbon monoxide pressures and the hydroformylation reactions can be carried out at low pressures. ^{31}P and ^{59}Co NMR spectra of the solution show that $HCo(CO)_3[P(p\text{-}CF_3C_6H_4)_3]$ and $HCo(CO)_2[P(p\text{-}CF_3C_6H_4)_3]_2$ are the only hydrido cobalt complexes present in detectable concentrations in $Co_2(CO)_6[P(p\text{-}CF_3C_6H_4)_3]_2$-catalyzed hydroformylation reactions [116].

Simple trialkylphosphanes such as PEt_3 have been used as ligands in phosphane-modified homogeneous rhodium-catalyzed 1-hexene hydroformylation in supercritical carbon dioxide. The catalyst derived from PEt_3 is more active and slightly more selective for the linear products in $scCO_2$ than in toluene, and under the same reaction conditions. Above 200 bar total pressure at 100 °C, heptanals are obtained with 97% selectivity. Under subcritical conditions, however, heptanols from hydrogenation of the first formed heptanals are the main products [117]. An extended X-ray absorption fine structure characterization of dilute rhodium-PEt_3 catalysts of the 1-octene hydroformylation reaction was made in supercritical carbon dioxide. Significant metal clustering was detected with a Rh : P ratio of 1 : 1 [118].

The rhodium-catalyzed hydroformylation of 1-octene with the new P-donor ligands $PPh_{3-n}(OC_9H_{19})_n$ ($n = 3, 2, 1$) containing branched alkyl chains was investigated in supercritical carbon dioxide and toluene as solvents. The selectivities for aldehydes were higher in the supercritical medium than in toluene [119].

The combination of supercritical CO_2, $Rh(CO)_2(acac)$, and a fluorous polymeric phosphane **38** was found to be a highly effective catalytic system for the chemoselective hydroformylation of usually unreactive alkyl acrylates (Equation 7.11) [120].

Rh(CO)$_2$(acac)/ 4
80 °C, scCO$_2$, 1h
30 bar CO:H$_2$ = 1:1

CHO

O=C, O, (CH$_2$)$_2$, (CF$_2$)$_7$, CF$_3$; PPh$_2$; 1; 5

(7.11)

38

The effects of ligands and pressures in hydroformylation of 1-hexene catalyzed with rhodium-fluorinated phosphane complexes in supercritical carbon dioxide and in conventional organic solvents at a temperature of 333 K were investigated. The catalytic activities in supercritical carbon dioxide were found to be comparable with those in toluene [121]. Hydroformylation of styrene was studied in supercritical CO_2 using soluble rhodium catalysts bound to a fluoroacrylate copolymer backbone through phosphane ligands. Conversions up to almost 100% and branched aldehyde selectivities of 95–100% were obtained [122]. 1-Octene, 1-decene, and styrene were hydroformylated in supercritical carbon dioxide using a CO_2-philic fluorous ligand associated with a rhodium catalyst. The effect of P/Rh molar ratio, partial pressure of CO/H_2, and the total pressure of carbon dioxide were studied [123].

Two different *in situ* prepared catalysts generated from $Rh(CO)_2(acac)$ and trifluoromethyl-substituted triphenylphosphane ligands were evaluated for their activity and selectivity in the hydroformylation of 1-octene. The highest value for the turnover frequency, 9820 $mol_{1-octene}\,mol_{Rh}{}^{-1}\,h^{-1}$, was obtained in supercritical carbon dioxide using $P[C_6H_3\text{-}3,5\text{-}(CF_3)_2]_3$ as the ligand [124].

A new inverted biphasic catalysis system using supercritical CO_2 as the stationary catalyst phase and water as the continuous phase was described for rhodium-catalyzed hydroformylation of polar substrates. Product separation and catalyst recycling was possible without depressurizing the autoclave. Turnover numbers of up to 3560 were obtained in three consecutive runs and rhodium leaching into the aqueous phase was below 0.3 ppm [125]. Hydroformylation of propene was carried out in supercritical carbon dioxide + water and in supercritical propene + water mixtures using Rh(acac)$(CO)_2$ and $P(m\text{-}C_6H_4SO_3Na)_3$ as catalysts. Compared to traditional hydroformylation technology, the supercritical reactions showed better activity and selectivity [126].

The continuous rhodium-catalyzed hydroformylation of low-volatility alkenes such as 1-dodecene in supercritical fluid–ionic liquid biphasic systems was investigated. The nature of the ionic liquid is very important in achieving high rates, with 1-alkyl-3-methylimidazolium bis(trifluoromethanesulfonyl)amides giving the best activity if the alkyl chain is at least C_8. It was found that under certain process conditions, the supercritical fluid–ionic liquid system can be operated continuously for several weeks without any visible sign of catalyst degradation [127,128]. The continuous-flow hydroformylation of 1-octene catalyzed by Rh/[Rmim][$Ph_2PC_6H_4SO_3$] (Rmim = 1-propyl-, or 1-pentyl-, or 1-octyl-3-methylimidazolium) using supercritical carbon dioxide as a transport vector was studied. The rhodium leaching was low over a 12 h period at a total pressure of 125–140 bar [129].

A phenoxaphosphane-modified xanthene-type compound covalently anchored on a polysiloxane support was applied as ligand in the rhodium-catalyzed hydroformylation of 1-octene in toluene and in supercritical carbon dioxide. The hydroformylation results obtained with the heterogenized ligand were found to be competitive with those obtained with homogeneous systems employing xantphos as the ligand. The use of $scCO_2$ as reaction medium in a continuous-flow setup resulted in slightly lower regioselectivities compared to batch-wise hydroformylation reactions in toluene [130].

Hydroformylation of 1-hexene in supercritical carbon dioxide was investigated using a rhodium–phosphane catalyst tethered to a silica support. The performance of the tethered catalyst was compared with a homogeneous rhodium–phosphane catalyst and was found to be equally effective under identical reaction conditions. Initial aldehyde selectivity obtained with the heterogeneous species was also comparable to that obtained with the homogeneous catalyst, but it decreased over the course of the reaction [131]. Rhodium catalysts anchored on phosphinated silica-based supports were evaluated for their performance in the hydroformylation of 1-hexene in supercritical carbon dioxide. The surface mechanism was probed using high-pressure diffuse reflectance infrared spectroscopy. The results suggest that the homogeneous mechanism is effectively transferred to the support material [132].

Hydroformylation of 1-hexene was carried out in supercritical CO_2 and in organic solvents (toluene and ethyl acetate) using polymer-supported rhodium catalysts, which were prepared from polystyrene-bound triphenylphosphane (TPP) and dicarbonylacetylacetonato rhodium. The product distribution slightly changed with CO_2 pressure. The conversion increased appreciably as H_2 pressure was raised in $scCO_2$, but CO was found to retard the reaction. The influence of H_2 pressure in $scCO_2$ was slightly different from that in toluene. Changes of the structure of rhodium complexes on the polymer during the catalyst preparation and the reaction were investigated by diffuse reflectance FTIR. The catalyst was recyclable for the reaction in $scCO_2$ and the reaction rate and selectivity of the hydroformylation were much higher than those in the organic solvents [133].

The one-pot hydroformylation/aldol reaction sequence was applied to unsaturated ketones and ketoesters in the presence of a rhodium catalyst to afford the corresponding carbocyclic aldol adducts in good yields (Equation 7.12) [134].

CO2Et

O

0.9 mol% [Rh(COD)Cl]2; 10 mol% p-TsOH

CH2Cl2; 60 bar CO : H2 = 1 : 1; 80 °C, 20 h

CO2Et

O

72%

(7.12)

The rhodium-catalyzed one-pot enolboration/hydroformylation/aldol addition sequence was applied for the regioselective and diastereoselective formation of carbocyclic quaternary centers from acyclic olefins (Equation 7.13) [135].

O

CO2Et

Me

1.05 equiv. cHx2BCl; 1.05 equiv. Et3N; 0 °C ; 60 bar

CO : H2 = 1 : 1; 0.9 mol% Rh(CO2)(acac); 16 h

1.8 mol% Xantphos

O

Me

CO2Et

OH

89%

(7.13)

The homogeneous catalytic hydroformylation of ethylene in supercritical carbon dioxide was studied using $Ru_3(CO)_{12}$ as the catalyst precursor in the temperature range between 60 and 125 °C and in the pressure range from 224 to

408 bar. The catalyst was inactive below 70 °C. $Ru_3(CO)_{12}$ was found to be sufficiently soluble in $scCO_2$ [136].

7.2 Hydrocarboxylation

Novel heterogeneous catalysts containing a palladium complex anchored on mesoporous supports for hydrocarboxylation of aryl olefins and alcohols were found to give high regioselectivity, activity, and recyclability without leaching of palladium complex from the supports. In styrene hydrocarboxylation at 115 °C and 31 bar CO pressure, 2-phenyl-propionic acid is formed with 99% selectivity at 2600 mol styrene mol^{-1} palladium h^{-1} turnover frequency [137].

High catalytic activity (TOF = 282 h^{-1}) and selectivity to the branched product (~91%) were found in the biphasic hydrocarboxylation of vinyl aromatic compounds to the isomeric arylpropanoic acids using a novel water-soluble palladium complex, $[Pd(pyridine\text{-}2\text{-}carboxylato)(TPPTS)]^+TsO^-$, as the catalyst [138].

The polymer-supported bimetallic catalyst system PVP–$PdCl_2$–$NiCl_2$/TPPTS/PPh_3 (PVP = polyvinylpyrrolidone) was found to have good activity in the hydroxycarbonylation of styrene under aqueous–organic two-phase condition and can be reused four times with little loss of catalytic activity. The effects of temperature, CO pressure, and reaction time were studied to obtain optimum reaction conditions (Equation 7.14) [139].

PVP—$PdCl_2$—$NiCl_2$
TPPTS/PPh_3
40 bar, 100 °C
100% conversion in 6 h

CO_2H + CO_2H

63 : 36

(7.14)

The ruthenium(II) complex *fac*-$[Ru(CO)_2(H_2O)_3C(=O)C_2H_5][CF_3SO_3]$ dissolved in aqueous tetrabutylammonium hydrogen sulfate or sodium hydrogen sulfate catalyzed the hydrocarboxylation of ethylene to propionic acid under water gas shift conditions at 150 °C and 88 bar CO : C_2H_4 = 1 : 1 [140].

The catalytic potential of transition metal sulfides for abiotic carbon fixation was assayed. It was found that at 2000 bar and 250 °C, the sulfides of iron, cobalt, nickel, and zinc promote the hydrocarboxylation reaction via carbonyl insertion at a metal sulfide bound alkyl group. The results of the study support the hypothesis that transition metal sulfides may have provided useful catalytic functionality for geochemical carbon fixation in a prebiotic world [141].

Water-soluble palladium complexes with guanidinumphosphane ligands were found to be active catalyst precursors for the hydrocarboxylation of styrene in water with conversions up to 96% [142].

High selectivity in acids (up to 90%) was obtained in hydrocarboxylation of 1-octene in supercritical carbon dioxide using a Pd/P(4-CF_3-C_6H_4)$_3$ catalyst system and a perfluorinated surfactant **39** [119].

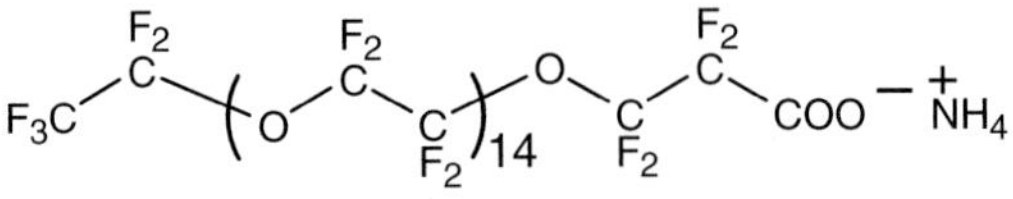
F2
C
F3C
O
F2
C
C
F2
14
O
C
F2
F2
C
COO
NH4

39

The palladium-catalyzed aqueous hydrocarboxylation reaction of styrene, 1-octene, 3-buten-1-ol, and 4-pentenoic acid was studied in acidic solutions. The catalyst employing N3P as ligand (N3P = *N*-bis(*N*,′*N*′-diethyl-2-aminoethyl)-4-aminomethyl-phenyl-diphenylphosphane) was found to show an inverted regioselectivity compared to the TPPTS system. Noncoordinating anions gave the best results in terms of activity and stability of the catalyst [143].

A mild and effective palladium(II)-catalyzed, copper(II)-mediated protocol for the arylation/carboalkoxylation of unactivated olefins with indoles was developed (Equation 7.15) [100].

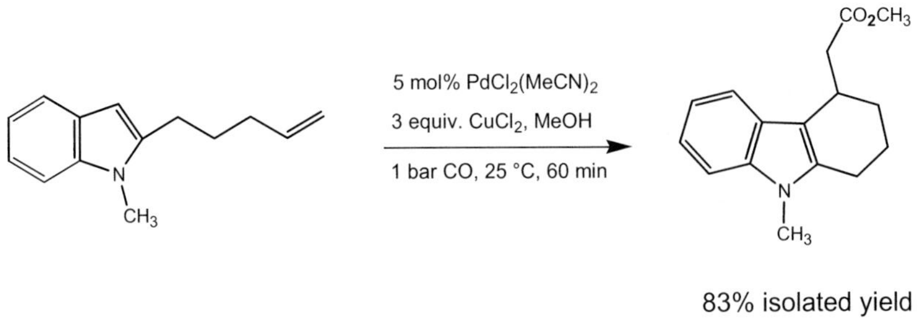
N
CH3
5 mol% PdCl2(MeCN)2
3 equiv. CuCl2, MeOH
1 bar CO, 25 °C, 60 min
CO2CH3
N
CH3
83% isolated yield

(7.15)

7.3
Hydroalkoxycarbonylation

The mechanism of the pyridine-modified cobalt-catalyzed hydromethoxycarbonylation of 1,3-butadiene was studied by high-pressure IR and NMR spectroscopy. It was found that pyridine accelerates the conversion of η^3-$C_4H_7Co(CO)_3$ to methyl-3-pentenoate and the methanolysis of the intermediate $CH_3CH{=}CHCH_2C({=}O)Co(CO)_4$ [144].

Thiourea-based ligands were evaluated for the palladium-catalyzed bis(methoxycarbonylation) of terminal olefins. The best results were obtained by using tetrasubstituted thioureas such as **40** and **41** (Equation 7.16) [145].

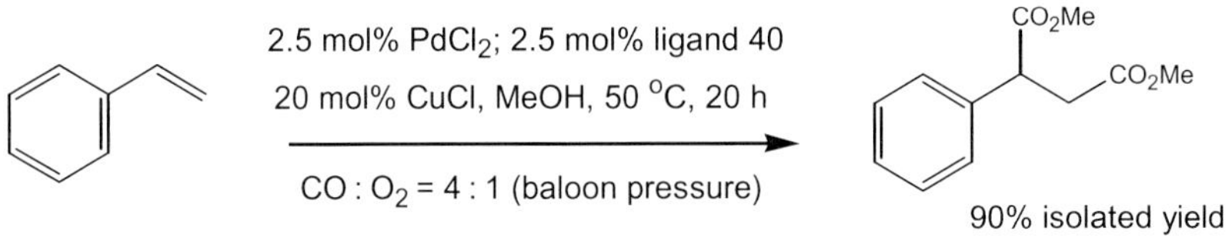

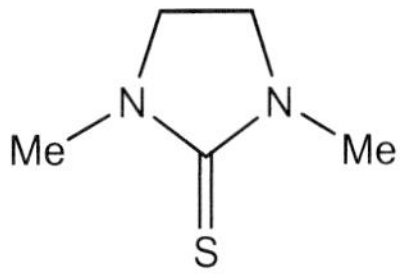

2.5 mol% $PdCl_2$; 2.5 mol% ligand 40
20 mol% CuCl, MeOH, 50 °C, 20 h
CO : O_2 = 4 : 1 (baloon pressure)
90% isolated yield

40 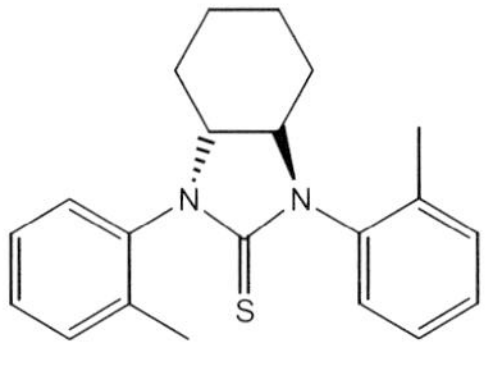 **41**

(7.16)

Palladium(II) complexes with 1,1′-bis(diphenylphosphano)ferrocene, 1,1′-bis(diphenylphosphano)octamethylferrocene, 1,1′-bis(diphenyphosphano)ruthenocene, and 1,1′-bis(diphenylphosphano)osmocene were used to catalyze the methoxycarbonylation of styrene. The regioselective (up to 85%) formation of 3-phenylpropanoate was observed. The highest turnover frequency (334 h^{-1} at 100 °C and 42 bar CO pressure) was obtained with the 1,1′-bis(diphenyphosphano)ruthenocene precursor in the presence of *p*-toluene sulfonic acid cocatalyst [146].

The mechanism for palladium-catalyzed alkoxycarbonylation of cinnamyl chloride was investigated. An associative mechanism was observed at low pressure, while an insertion mechanism was observed at high pressure or when an excess of ligand (PPh_3, P^cHx_3, or hemilabile P–N ligands **42**, respectively) was used [147].

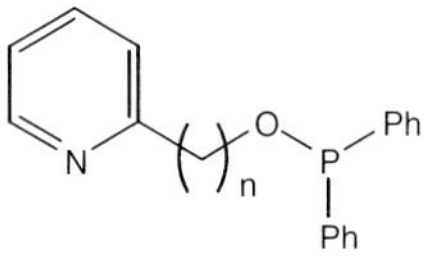

n = 1, 2, or 3

42

Highly active, selective, and recyclable palladium catalyst systems for the hydroesterification of styrene (at room temperature and 6 bar CO pressure) and vinyl acetate (at 40–60 °C and 6–10 bar CO pressure) were found by using 1,2-bis(di-*tert*-butylphosphinomethyl)benzene as ligand and polymeric sulfonic acids of limited SO_3H loadings as promoter [148].

The methoxycarbonylation of unsaturated acids or esters catalyzed by palladium complexes of 1,2-bis(di-*tert*-butylphosphinomethyl)benzene was found to produce α,ω-diesters with >95% selectivities, even if the double bond is deep in the chain or conjugated to the carbonyl group [149].

The effect of various alcohols on the regioselectivity of the palladium-catalyzed hydroalkoxycarbonylation of styrene in [bmim][BF_4] and [bmim][PF_6] ionic liquids was investigated. The highest regioselectivities toward branched ester were observed with "preformed" $PdCl_2(PPh_3)_2$ catalyst. The addition of diphosphanes favored the formation of the linear isomer [150].

Palladium complexes with a variety of *P,N*-bidentate ligands were tested as catalyst precursors in methoxycarbonylation of phenylacetylene in organic solvents and in supercritical CO_2. High activity (TOF up to 8000 in methanol) and high selectivity (>99%) toward the branched ester product were reported [151]. Imidazolium-based ionic liquids with $[PF_6]^-$, $[BF_4]^-$, and $[GaCl_4]^-$ counterions were synthesized and used in hydroethoxycarbonylation of styrene. Moderate to high regioselectivities toward branched esters were obtained with preformed $PdCl_2(PPh_3)_2$ catalyst [152].

A palladium-salicylborate-catalyzed efficient and regioselective methoxycarbonylation of terminal alkyl and aryl olefins was described. The regioselectivity, in favor of the linear ester, was found to be up to quantitative in the case of styrenes [153].

The electronic effect of diphosphanes on the regioselectivity of the palladium-catalyzed hydroesterification of styrene was studied. It was found that in the presence of electron-poor phosphanes, branched esters are produced and high activities of the catalytic systems were observed [154].

Palladium complexes of 2-pyridyldiphenylphosphane anchored on polystyrene, polymethylmethacrylate, and styrene–methylmethacrylate copolymer were found to form highly active heterogeneous catalysts for the alkoxycarbonylation of terminal alkynes with activities approaching those obtained under homogeneous conditions [155].

7.4 Tandem Carbonylation Reactions

The selective synthesis of linear amines from internal olefins or olefin mixtures was achieved through a rhodium/phosphane (**45**) catalyzed one-pot reaction at 120 °C, $P(H_2) = 50$ bar, and $P(CO) = 10$ bar, consisting of an initial olefin izomerization followed by hydroformylation and reductive amination [156].

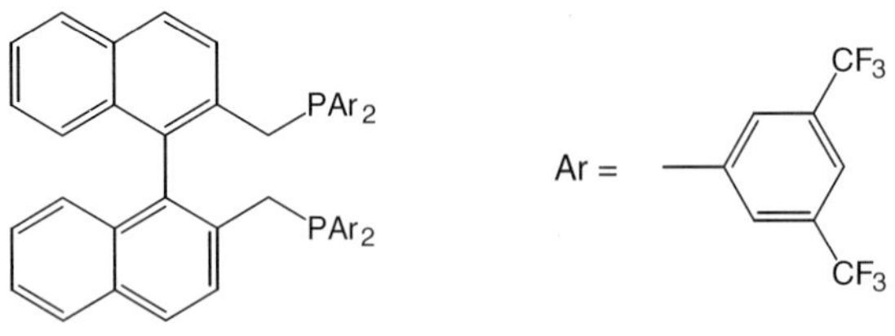

45

New steroidal derivatives of androstene and pregnene containing an α-amino moiety have been prepared in a one-pot hydroformylation–amidocarbonylation reaction using a rhodium or a rhodium–cobalt complex catalyst [157].

Carbonylation reactions of allene in alcohols and amines in the presence of $Ru_3(CO)_{12}$ precatalyst at 100 °C and 15 bar CO pressure gave methacrylates and methacrylamides, respectively, in up to 89% yield with an atom economy of 100% [158].

The one-pot hydroformylation–cyclotrimerization reaction of cyclopentene and cyclohexene was found to be catalyzed by $Rh_6(CO)_{16}$ and $H_3PW_{12}O_{40} \cdot xH_2O$ in THF at 100 °C and 40 bar $CO : H_2 = 1 : 1$ to give 2,4,6-trisubstituted-1,3,5-trioxanes (**46**) as major products along with the corresponding aldehydes (**47**) (Equation 7.17) [159].

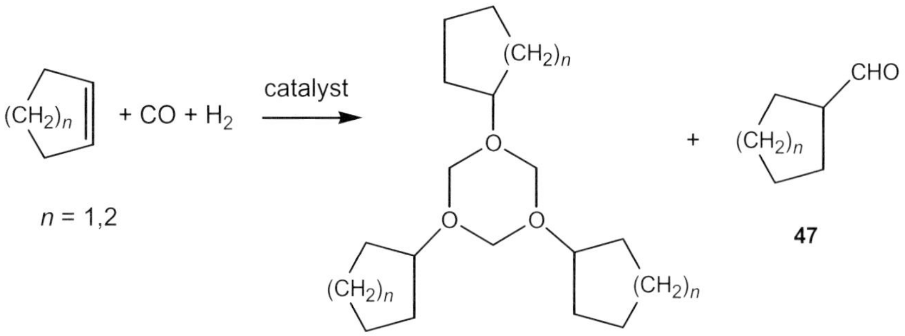

$(CH_2)_n$ + CO + H_2 —catalyst→ **46** + **47**

$n = 1,2$

(7.17)

A novel rhodium-catalyzed one-pot synthesis of indole systems via tandem hydroformylation–Fischer indole synthesis starting from olefins and arylhydrazines was described. The procedure leads directly to 3-substituted indoles if unsubstituted phenylhydrazine is used. Using *para*- or *ortho*-substituted arylhydrazines, the corresponding 3,5- and 3,7-disubstituted indoles are formed, respectively [160].

The platinum-catalyzed amidocarbonylation of aldehydes with amides and carbon monoxide was described. In contrast to precedent palladium catalysis, a remarkable ligand acceleration by phosphane was observed [161].

Hydroaminomethylation of terminal as well as internal aliphatic and aromatic olefins with various amines in the presence of [Rh(COD)(Imes)Cl] as a catalyst was described (Imes = 1,3-dimesitylimidazol-2-ylidene). Good to excellent yields and high chemoselectivities were obtained in THF at 85–105 °C using 0.1 mol% of catalyst [162].

Hydroaminomethylation of 1-dodecene with dimethylamine using a water-soluble rhodium–phosphane complex, $RhCl(CO)(TPPTS)_2$ in an aqueous–organic two-phase system in the presence of cetyltrimethylammonium bromide was investigated. High reactivity and selectivity for tertiary amine were achieved at 130 °C and 30 bar pressure [163].

The isomeric 5,6,7,8-tetrahydroindazolines **48** and **49** were obtained in a domino hydroformylation/cyclization reaction of 3-acetyl-1-pyrrole with $Rh_4(CO)_{12}$ as catalyst precursor at 140 °C and 30 bar $CO : H_2 = 1 : 1$ [72].

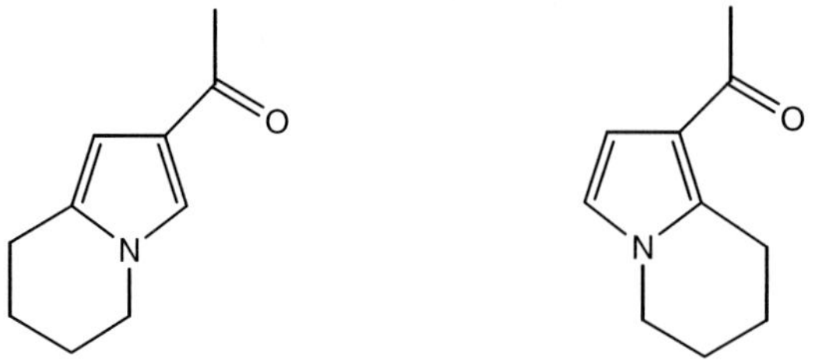

48

49

Tandem hydroformylation/hydrazone formation from amino olefins and aryl hydrazines were described (Equation 7.18) [164].

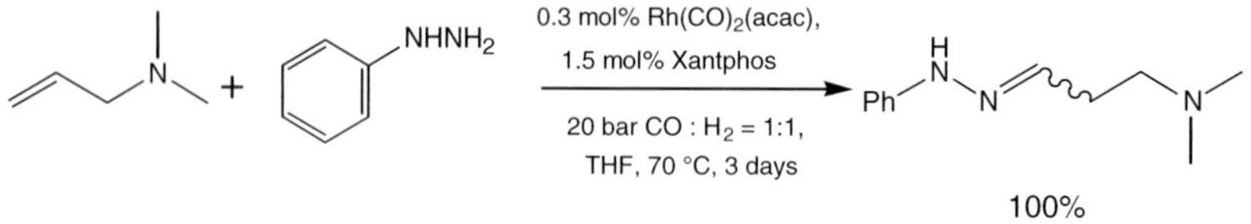

0.3 mol% $Rh(CO)_2(acac)$, 1.5 mol% Xantphos; 20 bar CO : H_2 = 1:1, THF, 70 °C, 3 days

$NHNH_2$ Ph N H N N 100%

(7.18)

The one-pot acetal formation in the rhodium-catalyzed styrene hydroformylation in alcohols as solvents was investigated. The effects of the addition of different types and amounts of phosphanes and phosphites were studied to improve the regioselectivity of the reaction [165].

Rhodium-catalyzed hydroformylation of terminal alkenes in the presence of stabilized phosphorus ylides was found to initiate a domino hydroformylation-Wittig olefination process. When monosubstituted acceptor-stabilized phosphorus ylides were employed, a hydrogenation step follows the Wittig olefination to give a domino hydroformylation-Wittig olefination hydrogenation process (Equation 7.19) [166].

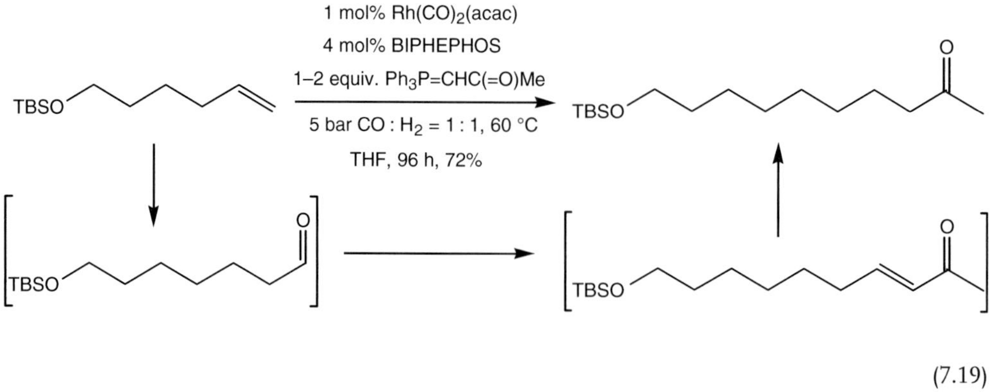

1 mol% $Rh(CO)_2(acac)$; 4 mol% BIPHEPHOS; 1–2 equiv. $Ph_3P{=}CHC({=}O)Me$; 5 bar CO : H_2 = 1 : 1, 60 °C; THF, 96 h, 72%

TBSO TBSO O TBSO O TBSO O

(7.19)

The rhodium-catalyzed hydroaminomethylation of 1-octene with morpholine was studied using temperature-dependent solvent systems consisting of propylene carbonate, an alkane, and a semipolar mediator such as *N*-octylpyrrolidone. The conversion of 1-octene and the selectivity to the corresponding amine is 92%. After the reaction, the catalyst can be easily recovered by a simple phase separation with only a negligible loss of rhodium [167].

The tandem hydroformylation-Fischer indolization protocol was used in the synthesis of 2,3-disubstituted indoles. Several olefins, bearing substituents with various functional groups, as well as cyclic olefinic systems were investigated (Equations 7.20 and 7.21) [168].

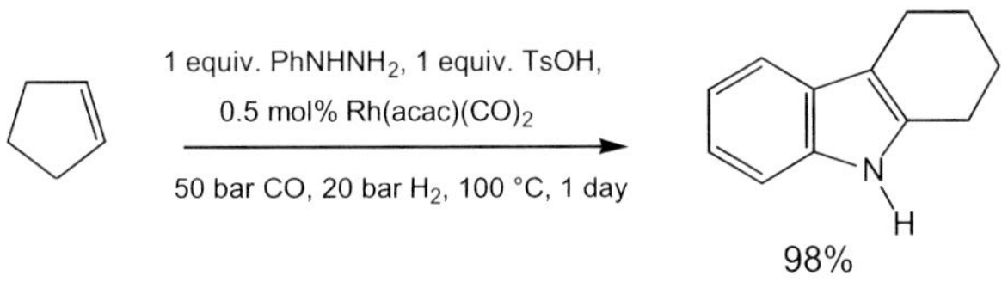

1 equiv. $PhNHNH_2$, 1 equiv. TsOH,
0.5 mol% $Rh(acac)(CO)_2$
50 bar CO, 20 bar H_2, 100 °C, 1 day
N
H
98%

(7.20)

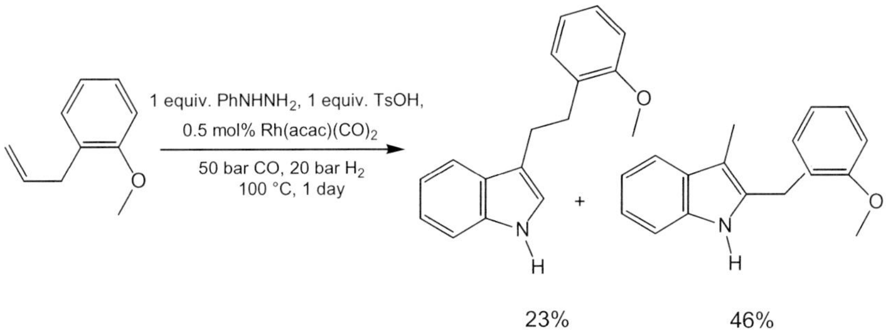

1 equiv. $PhNHNH_2$, 1 equiv. TsOH,
0.5 mol% $Rh(acac)(CO)_2$
50 bar CO, 20 bar H_2
100 °C, 1 day
O O O
N N
H H
23% 46%

(7.21)

Polyamines, structurally related to putrescines and spermidines, were obtained via rhodium-catalyzed hydroaminomethylation of methallylphthalimide with primary or secondary amines (Equations 7.22 and 7.23) [169].

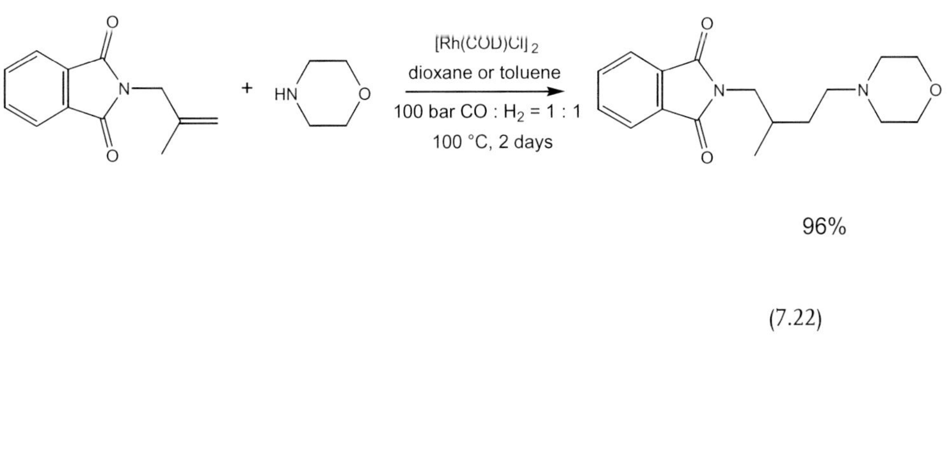

O O
N + HN O N N O
O O
$[Rh(COD)Cl]_2$
dioxane or toluene
100 bar CO : H_2 = 1 : 1
100 °C, 2 days
96%

(7.22)

$[Rh(COD)Cl]_2$
dioxane or toluene
80 bar CO : H_2 = 1 : 1
140 °C, 1 day

85%

(7.23)

Phenoxaphosphino-modified xantphos-type ligands in the rhodium-catalyzed hydroaminomethylation of internal olefins were found to give linear amines. Hydroaminomethylation and each of its individual steps were monitored by high-pressure infrared spectroscopy. The results suggest that hydroaminomethylation takes place by a sequential isomerization/hydroformylation/amination/hydrogenation pathway [170].

The hydroaminomethylation of long-chain alkenes with secondary amines in aqueous–organic two-phase system catalyzed by rhodium catalyst precursor and water-soluble diphosphane ligand BISBIS was investigated. The use of BISBIS gave higher activity and higher regioselectivity for linear amine than the monophosphane ligand TPPTS. The ratio of linear and branched amine was found to be up to 83 [171]. The hydroaminomethylation of long-chain alkenes with secondary amines was performed efficiently in ionic liquids 1-*n*-alkyl-3-methylimidazolium tosylates (alkyl *n*-butyl, octyl, dodecyl, cetyl) with Rh–BISBIS complex as catalyst [172].

References

1 Bianchi, M., Frediani, P., Piacenti, F., Rosi, L. and Salvini, A. (2002) *European Journal of Inorganic Chemistry*, 1155.

2 Allmendinger, M., Zintl, M., Eberhardt, R., Luinstra, G.L., Molnar, F. and Rieger, B. (2004) *Journal of Organometallic Chemistry*, **689**, 971–979.

3 Tannenbaum, R. and Bor, G. (2004) *Journal of Molecular Catalysis A: Chemical*, **215**, 33–43.

4 Tannenbaum, R. and Bor, G. (2004) *Journal of Physical Chemistry A*, **108**, 7105–7111.

5 Caporali, M., Frediani, P., Salvini, A. and Laurenczy, G. (2004) *Inorganica Chimica Acta*, **357**, 4537–4543.

6 Dwyer, C., Assumption, H., Coetzee, J., Crause, C., Damoense, L. and Kirk, M. (2004) *Coordination Chemical Reviews*, **248**, 653–669.

7 Polas, C., Wilton-Ely, J.D.E.T., Slawin, A.M.Z., Foster, D.F., Steynberg, P. J., Green, M.J. and Cole-Hamilton, D.J. (2003) *Dalton Transactions*, 4669–4677.

8 Godard, C., Duckett, S.B., Polas, S., Tooze, R. and Whitwood, A.C. (2005) *Journal of the American Chemical Society*, **127**, 4994–4995.

9 Klingler, R.J., Chen, M.J., Rathke, J.W. and Kramarz, K.W. (2007) *Organometallics*, **26**, 352–357.

10 Marhenke, J., Massick, S.M. and Ford, P.C. (2007) *Inorganica Chimica Acta*, **360**, 825–836.

11 He, D., Pang, D., Wei, L., Chen, Y., Wang, T., Tang, Z., Liu, J., Liu, Y. and Zhu, Q. (2002) *Catalysis Communications*, **3**, 429–433.

12 Seepersaud, M., Kettunen, M., Abu-Surrah, A.S., Repo, T., Voelter, W. and Al-Abed, Y. (2002) *Tetrahedron Letters*, **43**, 1793–1795.

13 Cossu, S., Peluso, P., Alberico, E. and Marchetti, M. (2006) *Tetrahedron Letters*, **47**, 2569–2572.

14 Clarke, M.L. and Roff, G.L. (2006) *Chemistry – A European Journal*, **12**, 7978–7986.

15 Lambers, M., Beijer, F.H., Padron, J.M., Tóth, I. and de Vries, J.G. (2002) *The Journal of Organic Chemistry*, **67**, 5022–5024.

16 Kandanarachchi, P., Guo, A. and Petrovic, Z. (2002) *Journal of Molecular Catalysis A: Chemical*, **184**, 65–71.

17 Ropartz, L., Haxton, K.J., Foster, D.F., Morris, R.E., Slawin, A.M.Z. and Cole-Hamilton, D.J. (2002) *Journal of the Chemical Society, Dalton Transactions*, 4323–4334.

18 Huang, Y.Y., Zhang, H.L., Deng, G.J., Tang, W.J., Wang, X.Y., He, Y.M. and Fan, Q.H. (2005) *Journal of Molecular Catalysis A: Chemical*, **227**, 91–96.

19 Abu-Reziq, R., Alper, H., Wang, D.S. and Post, M.L. (2006) *Journal of the American Chemical Society*, **128**, 5279–5282.

20 Liu, X., Li, H., Wang, Y. and Jin, Z. (2002) *Journal of Organometallic Chemistry*, **654**, 83–90.

21 van der Slot, S.C., Duran, J., Luten, J., Kamer, P.C.J. and van Leeuwen, P.W.N.M. (2002) *Organometallics*, **21**, 3873–3883.

22 Poyatos, M., Uriz, P., Mata, J.A., Claver, C., Fernandez, E. and Peris, E. (2003) *Organometallics*, **22**, 440–444.

23 Bortenschlager, M., Mayr, M., Nuyken, O. and Buchmeiser, M.R. (2005) *Journal of Molecular Catalysis A: Chemical*, **233**, 67–71.

24 van der Vlugt, J.I., Bonet, J.M., Mills, A.M., Spek, A.L. and Vogt, D. (2003) *Tetrahedron Letters*, **44**, 4389–4392.

25 Slagt, V.F., van Leeuwen, P.W.N.M. and Reek, J.N.H. (2003) *Chemical Communications*, 2474–2475.

26 Slagt, V.F., van Leeuwen, P.W.N.M. and Reek, J.N.H. (2003) *Angewandte Chemie-International Edition*, **42**, 5619–5623.

27 Peixoto, A., Pereira, M.M., Neves, M.G. P.M.S., Silva, A.M.S. and Cavaleiro, J.A. S. (2003) *Tetrahedron Letters*, **44**, 5593–5595.

28 Keglevich, G., Kégl, T., Chuluunbaatar, T., Dajka, B., Mátyus, P., Balogh, B. and Kollár, L. (2003) *Journal of Molecular Catalysis A: Chemical*, **200**, 131–136.

29 Breit, B. and Seiche, W. (2003) *Journal of the American Chemical Society*, **125**, 6608–6609.

30 Wang, Y., Jiang, J., Cheng, F. and Jin, Z. (2002) *Journal of Molecular Catalysis A: Chemical*, **188**, 79–83.

31 Aubry, D.A., Bridges, N.N., Ezell, K. and Stanley, G.G. (2003) *Journal of the American Chemical Society*, **125**, 11180–11181.

32 Kostas, I.D., Vallianatou, K.A., Kyritsis, P., Zednik, J. and Vohlidai, J. (2004) *Inorganica Chimica Acta*, **357**, 3084–3088.

33 Sullivan, J.T., Sadula, J., Hanson, B.E. and Rosso, R.J. (2004) *Journal of Molecular Catalysis A: Chemical*, **214**, 213–218.

34 Breit, B., Demel, P. and Gebert, A. (2004) *Chemical Communications*, 114–115.

35 Jeon, J.K., Park, Y.K. and Kim, J.M. (2004) *Chemistry Letters*, **33**, 174–175.

36 Breit, B. and Fuchs, E. (2004) *Chemical Communications*, 694–695.

37 Im-Erbsin, S., Prasassarakich, P. and Rempel, G.L. (2004) *Journal of Applied Polymer Science*, **93**, 854–869.

38 Clarke, M.L. (2004) *Tetrahedron Letters*, **45**, 4043–4045.

39 Keglevich, G., Kégl, T., Odinets, I.L., Vinogradova, N.M. and Kollár, L. (2004) *Comptes Rendus Chimie*, **7**, 779–784.

40 Kim, J.J. and Alper, H. (2005) *Chemical Communications*, 3059–3061.

41 Odinets, I.L., Vinogradova, N.M., Matveeva, E.V., Golovanov, D.D., Lyssenko, K.A., Keglevich, G., Kollár, L., Röschenthaler, G.V. and Mastryukova, T.A. (2005) *Journal of Organometallic Chemistry*, **690**, 2559–2570.

42 Moores, A., Mezailles, N., Ricard, L. and Le Floch, P. (2005) *Organometallics*, **24**, 508–513.

43 Baber, R.A., Clarke, M.L., Heslop, K.M., Marr, A.C., Orpen, A.G., Pringle, P.G., Ward, A. and Zambrano-Williams, D.E. (2005) *Dalton Transactions*, 1079–1085.

44 El Ali, B., Tijani, J., Fettouhi, M., Al-Arfaj, A. and El-Faer, M. (2005) *Applied Organometallic Chemistry*, **19**, 329–338.

45 Bronger, R.P.J., Kamer, P.C.J. and van Leeuwen, P.W.N.M. (2003) *Organometallics*, **22**, 5358–5369.

46 Bruss, A.J., Gelesky, M.A., Machado, G. and Dupont, J. (2006) *Journal of Molecular Catalysis A: Chemical*, **252**, 212–218.

47 da Silva, A.C., de Oliveira, K.C.B., Gusevskaya, E.V. and dos Santos, E.N. (2002) *Journal of Molecular Catalysis A: Chemical*, **179**, 133–141.

48 Vogl, C., Paetzold, E., Fischer, C. and Kragl, U. (2005) *Journal of Molecular Catalysis A: Chemical*, **232**, 41–44.

49 Behr, A., Henze, G., Obst, D. and Turkowski, B. (2005) *Green Chemistry*, **7**, 645–649.

50 Behr, A., Obst, D. and Westfechtel, A. (2005) *European Journal of Lipid Science and Technology*, **107**, 213–219.

51 Odinets, R., Kégl, T., Sharova, E., Artyushin, O., Goryunov, E., Molchanova, G., Lyssenko, K., Mastryukova, T., Röschenthaler, G.V., Keglevich, G. and Kollár, L. (2005) *Journal of Organometallic Chemistry*, **690**, 3456–3464.

52 Artyushin, O., Odinets, I., Goryunov, E., Fedyanin, I., Lyssenko, K., Mastryukova, T., Röschenthaler, G.-V., Kégl, T., Keglevich, G. and Kollár, L. (2006) *Journal of Organometallic Chemistry*, **691**, 5547–5559.

53 Tolis, E.I., Vallianatou, K.A., Andreadaki, F. J. and Kostas, I.D. (2006) *Applied Organometallic Chemistry*, **20**, 335–337.

54 Fuchs, E., Keller, M. and Breit, B. (2006) *Chemistry – A European Journal*, **12**, 6930–6939.

55 Praetorius, J.M., Kotyk, M.W., Webb, J.D., Wang, R. and Crudden, C.M. (2007) *Organometallics*, **26**, 1057–1061.

56 Caporali, M., Frediani, P., Piacenti, F. and Salvini, A. (2003) *Journal of Molecular Catalysis A: Chemical*, **204**, 195–200.

57 Barros, H.J.V., Guimarães, C.C., dos Santos, E.N. and Gusevskaya, E.V. (2007) *Catalysis Communications*, **8**, 747–750.

58 Barros, H.J.V., Guimarães, C.C., dos Santos, E.N. and Gusevskaya, E.V. (2007) *Organometallics*, **9**, 2211–2218.

59 Widjaja, E., Li, C. and Garland, M. (2002) *Organometallics*, **21**, 1991–1997.

60 Li, C., Widjaja, E., Chew, W. and Garland, M. (2002) *Angewandte Chemie-International Edition*, **41**, 3785–3789.

61 Zhang, J., Poliakoff, M. and George, M.W. (2003) *Organometallics*, **22**, 1612–1618.

62 Lazzaroni, R., Settambolo, R., Prota, G., Botteghi, C., Paganelli, S. and Marchetti, M. (2004) *Inorganica Chimica Acta*, **357**, 3079–3083.

63 Berounhou, C., Neibecker, D. and Mathieu, R. (2003) *Organometallics*, **22**, 782–786.

64 Cheliatsidou, P., White, D.F.S. and Cole-Hamilton, D.J. (2004) *Dalton Transactions*, 3425–3427.

65 Andrieu, J., Camus, J.M., Balan, C. and Proli, P. (2006) *European Journal of Inorganic Chemistry*, 62–68.

66 Jaaskelainen, S. and Haukka, M. (2003) *Applied Catalysis A: General*, **247**, 95–100.

67 Abou Rida, M., and Smith, A.K. (2003) *Journal of Molecular Catalysis A: Chemical*, **202**, 87–95.

68 Moreno, M.A., Haukka, M., Jaaskelainen, S., Vuoti, S., Pursiainen, J. and Pakkanen, T.A. (2005) *Journal of Organometallic Chemistry*, **690**, 3803–3814.

69 Fachinetti, G., Funaioli, T. and Marchetti, F. (2005) *Chemical Communications*, 2912–2914.

70 Tominaga, K. (2006) *Catalysis Today*, **115**, 70–72.

71 Fernández, D., García-Seijo, M.I., Kégl, T., Petőcz, G., Kollár, L. and Garća-Fernández, M.E. (2002) *Inorganic Chemistry*, **41**, 4435–4443.

72 Botteghi, C., Paganelli, S., Moratti, F., Marchetti, M., Lazzaroni, R., Settambolo, R. and Piccolo, O. (2003) *Journal of Molecular Catalysis A: Chemical*, **200**, 147–156.

73 van Duren, R., van der Vlugt, J.I., Kooijman, H., Spek, A.L. and Vogt, D. (2007) *Dalton Transactions*, 1053–1059.

74 Drent, E., Mul, W.P. and Budzelaar, P.H.M. (2002) *Comments on Inorganic Chemistry*, **23**, 127–147.

75 Kónya, D., Lenero, K.Q.A. and Drent, E. (2006) *Organometallics*, **25**, 3166–3174.

76 Godart, C., Duckett, S.B., Henry, C., Polas, S., Toose, R. and Whitwood, A.C. (2004) *Chemical Communications*, 1826–1827.

77 Fox, D.J., Duckett, S.B., Flaschenriem, C., Brennessel, W.W., Schneider, J., Gunay, A. and Eisenberg, R. (2006) *Inorganic Chemistry*, **45**, 7197–7209.

78 Yamane, M., Yukimura, N., Ishiai, H. and Narasaka, K. (2006) *Chemistry Letters*, **35**, 540–541.

79 Li, C.Z., Widjaja, E. and Garland, M. (2003) *Journal of the American Chemical Society*, **125**, 5540–5548.

80 Li, C.Z., Widjaja, E. and Garland, M. (2004) *Organometallics*, **23**, 4131–4138.

81 Mehnert, C.P., Cook, R.A., Dispenziere, N.C. and Afeworki, M. (2002) *Journal of the American Chemical Society*, **124**, 12932–12933.

82 Lu, S.M. and Alper, H. (2003) *Journal of the American Chemical Society*, **125**, 13126–13131.

83 Yoon, T.J., Kim, J.I. and Lee, J.K. (2003) *Inorganica Chimica Acta*, **345**, 228–234.

84 Yoon, T.J., Lee, W., Oh, Y.S. and Lee, J.K. (2003) *New Journal of Chemistry*, **27**, 227–229.

85 Marchetti, M., Paganelli, S. and Viel, E. (2004) *Journal of Molecular Catalysis A: Chemical*, **222**, 143–151.

86 Roman-Martinez, M.C., Diaz-Aunon, J.A., De Lecea, C.S.M. and Alper, H. (2004) *Journal of Molecular Catalysis A: Chemical*, **213**, 177–182.

87 Diwakar, M.M., Deshpande, R.M. and Chaudhari, R.V. (2005) *Journal of Molecular Catalysis A: Chemical*, **232**, 179–186.

88 Riisager, A., Fehrmann, R., Flicker, S., van Hal, R., Haumann, M. and Wasserscheid, P. (2005) *Angewandte Chemie-International Edition*, **44**, 815–819.

89 El Ali, B., Tijani, J., Fettouhi, M., El-Faer, M. and Al-Arfaj, A. (2005) *Applied Catalysis A: General*, **283**, 185–196.

90 El Ali, B., Tijani, J. and Fettouchi, M. (2006) *Journal of Molecular Catalysis A: Chemical*, **250**, 153–162.

91 El Ali, B., Tijani, J. and Fettouchi, M. (2006) *Applied Catalysis A: General*, **303**, 213–220.

92 Ricken, S., Osinski, P.W., Eilbracht, P. and Haag, R. (2006) *Journal of Molecular Catalysis A: Chemical*, **257**, 78–88.

93 Foster, D.F., Gudmunsen, D., Adams, D.J., Stuart, A.M., Hope, E.G., Cole-Hamilton, D.J., Schwarz, G.P. and Pogorzelec, P. (2002) *Tetrahedron*, **58**, 3901–3910.

94 Bertucci, C., Botteghi, C., Giunta, D., Marchetti, M. and Paganelli, S. (2002) *Advanced Synthesis & Catalysis*, **344**, 556–562.

95 Haumann, M., Koch, H., Hugo, P. and Schomäcker, R. (2002) *Applied Catalysis A: General*, **225**, 239–249.

96 Stenzel, O., Raubenheimer, H.G. and Esterhuysen, C. (2002) *Journal of the Chemical Society, Dalton Transactions*, 1132–1138.

97 Bronger, R.P.J., Silva, S.M., Kamer, P.C.J. and van Leeuwen, P.W.N.M. (2002) *Journal of the Chemical Society, Chemical Communications*, 3044–3045.

98 Bronger, R.P.J., Silva, S.M., Kamer, P.C.J. and van Leeuwen, P.W.N.M. (2004) *Dalton Transactions*, 1590–1596.

99 Paetzold, E., Oehme, G., Fischer, C. and Frank, M. (2003) *Journal of Molecular Catalysis A: Chemical*, **200**, 95–103.

100 Liu, C., Jiang, J.Y., Wang, Y.H., Cheng, F. and Jin, Z.L. (2003) *Journal of Molecular Catalysis A: Chemical*, **198**, 23–27.

101 Gimenez-Pedros, M., Aghmiz, A., Claver, C., Masdeu-Bulto, A.M. and Sinou, D. (2003) *Journal of Molecular Catalysis A: Chemical*, **200**, 157–163.

102 Tominaga, K. and Sasaki, Y. (2004) *Chemistry Letters*, **33**, 14–15.

103 Zarka, M.T., Bortenschlager, M., Wurst, K., Nuyken, O. and Weberskirch, R. (2004) *Organometallics*, **23**, 4817–4820.

104 Monflier, E., Bricout, H., Hapiot, F., Tillroy, S., Aghmiz, A. and Masdeu-Bulto, A.M. (2004) *Advanced Synthesis & Catalysis*, **346**, 425–431.

105 Leclercq, L., Sauthier, M., Castanet, Y., Mortreux, A., Bricout, H. and Monflier, E. (2005) *Advanced Synthesis & Catalysis*, **347**, 55–59.

106 Sueur, B., Leclercq, L., Sauthier, M., Castanet, Y., Mortreux, A., Bricout, H. and Tilloy, S. (2005) *Chemistry – A European Journal*, **11**, 6228–6236.

107 Leclercq, L., Hapiot, F., Tilloy, S., Ramkisoensing, K., Reek, J.N.H., van Leeuwen, P.W.N.M. and Monflier, E. (2005) *Organometallics*, **24**, 2070–2075.

108 Baricelli, P.J., Lopez-Linares, F., Bruss, A., Santos, R., Lujano, E. and Sanchez-Delgado, R.A. (2005) *Journal of Molecular Catalysis A: Chemical*, **239**, 130–137.

109 Miyagawa, C.C., Kupka, J. and Schumpe, A. (2005) *Journal of Molecular Catalysis A: Chemical*, **234**, 9–17.

110 Klein, H., Jackstell, R. and Beller, M. (2005) *Chemical Communications*, 2283–2285.

111 Fu, H.Y., Li, M., Chen, H. and Li, X.J. (2006) *Journal of Molecular Catalysis A: Chemical*, **259**, 156–160.

112 Tilloy, S., Crowyn, G., Monflier, E., van Leeuwen, P.W.N.M. and Reek, J.N.H. (2006) *New Journal of Chemistry*, **30**, 377–383.

113 Paganelli, S., Marchetti, M., Bianchin, M. and Bertucci, C. (2007) *Journal of Molecular Catalysis A: Chemical*, **269**, 234–239.

114 Yang, Y., Jiang, J., Wang, Y., Liu, C. and Jin, Z. (2007) *Journal of Molecular Catalysis A: Chemical*, **261**, 288–292.

115 Patcas, F., Maniut, C., Ionescu, C., Pitter, S. and Dinjus, E. (2007) *Applied Catalysis B: Environmental*, **70**, 630–636.

116 Chen, M.J., Klingler, R.J., Rathke, J.W. and Kramarz, K.W. (2004) *Organometallics*, **23**, 2701–2707.

117 Sellin, M.F., Bach, I., Webster, J.M., Montilla, F., Rosa, V., Aviles, T., Poliakoff, M. and Cole-Hamilton, D.J. (2002) *Journal of the Chemical Society, Dalton Transactions*, 4569–4576.

118 Fiddy, S.G., Evans, J., Nelsius, T., Sun, X.Z., Jie, Z. and George, M.W. (2004) *Chemical Communications*, 676–677.

119 Gimenez-Pedros, M., Aghmiz, A., Ruiz, N. and Masdeu-Bulto, A.M. (2006) *European Journal of Inorganic Chemistry*, 1067–1075.

120 Hu, Y., Chen, W., Banet Osuna, A.M., Iggo, J.A. and Xiao, J. (2002) *Chemical Communications*, 788–789.

121 Fujita, S.I., Fujisawa, S., Bhanage, B.M., Ikushima, Y. and Arai, M. (2002) *New Journal of Chemistry*, **26**, 1479–1484.

122 Kani, I., Flores, R., Fackler, J.P. and Akgerman, A. (2004) *Journal of Supercritical Fluids*, **31**, 287–294.

123 Pedros, M.G., Masdeu-Bulto, A.M., Bayardon, J. and Sinou, D. (2006) *Catalysis Letters*, **107**, 205–206.

124 Koeken, A.C.J., van Vliet, M.C.A., van den Broeke, L.J.P., Deelman, B.J. and Keurentjes, J.T.F. (2006) *Advanced Synthesis & Catalysis*, **348**, 1553–1559.

125 McCarthy, M., Stemmer, H. and Leitner, W. (2002) *Green Chemistry*, **4**, 501–504.

126 Zhang, J.C., Wang, H.B., Liu, H.T. and Cao, W.L. (2006) *Journal of Molecular Catalysis A: Chemical*, **260**, 95–99.

127 Webb, P.B., Sellin, M.F., Kuene, T.E., Williamson, S., Slawin, A.M.Z. and

Cole-Hamilton, D.J. (2003) *Journal of the American Chemical Society*, **125**, 15577–15588.

128 Hintermair, U., Zhao, G., Santini, C.C., Muldoon, M.J. and Cole-Hamilton, D.J. (2007) *Chemical Communications*, 1462–1464.

129 Webb, P.B. and Cole-Hamilton, D.J. (2004) *Chemical Communications*, 612–613.

130 Bronger, R.P.J., Bermon, J.P., Reek, J.N.H., Kamer, P.C.J., van Leeuwen, P.W.N.M., Carter, D.N., Licence, P. and Poliakoff, M. (2004) *Journal of Molecular Catalysis A: Chemical*, **224**, 145–152.

131 Hemminger, O., Marteel, A., Mason, M.R., Davies, J.A., Todd, A.R. and Abraham, M.A. (2002) *Green Chemistry*, **4**, 507–512.

132 Bektesevic, S., Tack, T., Mason, M.R. and Abraham, M.A. (2005) *Industrial & Engineering Chemistry Research*, **44**, 4973–4981.

133 Fujita, S., Akihara, S., Fujisawa, S. and Arai, M. (2007) *Journal of Molecular Catalysis A: Chemical*, **268**, 244–250.

134 Keränen, M.D., Kot, K., Hollmann, C. and Eilbracht, P. (2004) *Organic and Biomolecular Chemistry*, **2**, 3379–3384.

135 Keränen, M.D. and Eilbracht, P. (2004) *Organic and Biomolecular Chemistry*, **2**, 1688–1690.

136 Erkey, C., Diz, E.L., Süss-Fink, G. and Dong, X. (2002) *Catalysis Communications*, **3**, 213–219.

137 Mukhopadhyay, K., Sarkar, B.R. and Chaudhari, R.V. (2002) *Journal of the American Chemical Society*, **124**, 9692–9693.

138 Jayasree, S., Seayad, A., Sarkar, B.R. and Chaudhari, R.V. (2002) *Journal of Molecular Catalysis A: Chemical*, **181**, 221–235.

139 Li, F.W., Xu, L.W. and Xia, C.G. (2003) *Applied Catalysis A: General*, **253**, 509–514.

140 Pardey, A.J., Rivas, A.B., Longo, C., Funaioli, T. and Fachinetti, G. (2004) *Journal of Coordination Chemistry*, **57**, 871–882.

141 Cody, G.D., Boctor, N.Z., Brandes, J.A., Filley, T.R., Hazen, R.M. and Yoder, H.S. (2004) *Geochimica et Cosmochimica Acta*, **68**, 2185–2196.

142 Aghmiz, A., Gimenez-Pedros, M., Masdeu-Bulto, A.M. and Schmidtchen, F.P. (2005) *Catalysis Letters*, **103**, 191–193.

143 Karlsson, M., Ionescu, A. and Andersson, C. (2006) *Journal of Molecular Catalysis A: Chemical*, **259**, 231–237.

144 Tuba, R., Mika, L.T., Bodor, A., Pusztai, Z., Tóth, I. and Horváth, I.T. (2003) *Organometallics*, **22**, 1582–1584.

145 Dai, M.J., Wang, C.H., Dong, G.B., Xiang, J., Luo, T.P., Liang, B., Chen, J.H. and Yang, Z. (2003) *European Journal of Organic Chemistry*, 4346–4348.

146 Bianchini, C., Meli, A., Oberhauser, W., Parisel, S., Gusev, O.V., Kalsin, A.M., Vologdin, N.V. and Dolgushin, F.M. (2004) *Journal of Molecular Catalysis A: Chemical*, **224**, 35–49.

147 van Haaren, R.J., Oevering, H., Kamer, P.C.J., Goubitz, K., Fraanje, J., van Leeuwen, P.W.N.M. and van Strijdonck, G.P.F. (2004) *Journal of Organometallic Chemistry*, **689**, 3800–3805.

148 Ooka, H., Inoue, T., Itsuno, S. and Tanaka, M. (2005) *Chemical Communications*, 1173–1175.

149 Jimenez-Rodriguez, C., Eastham, G.R. and Cole-Hamilton, D.J. (2005) *Inorganic Chemistry Communications*, **8**, 878–881.

150 Rangits, G. and Kollár, L. (2005) *Journal of Molecular Catalysis A: Chemical*, **242**, 156–160.

151 de Pater, J.J.M., Maljaars, C.E.P., de Wolf, E., Lutz, M., Spek, A.L., Deelman, B.J., Elsevier, C.J. and van Koten, G. (2005) *Organometallics*, **24**, 5299–5310.

152 Rangits, G. and Kollár, L. (2006) *Journal of Molecular Catalysis A: Chemical*, **246**, 59–64.

153 Vieira, T.O., Green, M.J. and Alper, H. (2006) *Organic Letters*, **8**, 6143–6145.

154 Guiu, E., Caporali, M., Munoz, B., Muller, C., Lutz, M., Spek, A.L., Claver, C. and van Leeuwen, P.W.N.M. (2006) *Organometallics*, **25**, 31021–3104.

155 Doherty, S., Knight, J.G. and Betham, M. (2006) *Journal of the Chemical Society, Chemical Communications*, 88–90.

156 Seayad, A., Ahmed, M., Klein, H., Jackstell, R., Gross, T. and Beller, M. (2002) *Science*, **297**, 1676–1678.

157 Nagy, E., Benedek, C., Heil, B. and Tőrös, S. (2002) *Applied Organometallic Chemistry*, **16**, 628–634.

158 Zhou, D.-Y., Yoneda, E., Onitsuka, K. and Takahashi, S. (2002) *Chemical Communications*, 2868–2869.

159 El Ali, B. (2003) *Journal of Molecular Catalysis A: Chemical*, **203**, 53–58.

160 Kohling, P., Schmidt, A.M. and Eilbracht, P. (2003) *Organic Letters*, **5**, 3213–3216.

161 Sagae, T., Sugiura, M., Hagio, H. and Kobayashi, S. (2003) *Chemistry Letters*, **32**, 160–161.

162 Seayad, A.M., Selvakumar, K., Ahmed, M. and Beller, M. (2003) *Tetrahedron Letters*, **44**, 1679–1683.

163 Wang, Y.Y., Luo, M.M., Li, Y.Z., Chen, H. and Li, X.J. (2004) *Applied Catalysis A: General*, **272**, 151–155.

164 Schmidt, A.M. and Eilbracht, P. (2005) *The Journal of Organic Chemistry*, **70**, 5528–5535.

165 El Ali, B., Tijani, J. and Fettouhi, M. (2005) *Journal of Molecular Catalysis A: Chemical*, **230**, 9–16.

166 Breit, B. and Zahn, S.K. (2005) *Tetrahedron*, **61**, 6171–6179.

167 Behr, A. and Roll, R. (2005) *Journal of Molecular Catalysis A: Chemical*, **239**, 180–184.

168 Linnepe, P., Schmidt, A.M. and Eilbracht, P. (2006) *Organic and Biomolecular Chemistry*, **4**, 302–313.

169 Müller, K.S., Koc, F., Ricken, S. and Eilbracht, P. (2006) *Organic and Biomolecular Chemistry*, **4**, 826–835.

170 Ahmed, M., Bronger, R.P.J., Jackstell, R., Kamer, P.C.L., van Leeuwen, P.W.N.M. and Beller, M. (2006) *Chemistry – A European Journal*, **12**, 8979–8988.

171 Wang, Y.Y., Chen, J.H., Luo, M.M., Chen, H. and Li, X.J. (2006) *Catalysis Communications*, **7**, 979–981.

172 Wang, Y.Y., Luo, M.M., Lin, Q., Chen, H. and Li, X.J. (2006) *Green Chemistry*, **8**, 545–548.

8
Carbonylation of Diazoalkanes

Neszta Ungvári, Ferenc Ungváry

The replacement of the diazo group in diazoalkanes by carbon monoxide to afford the corresponding ketenes (Equation 8.1) might be a synthetically useful reaction since ketenes are versatile intermediates in organic synthesis [1–3].

$$XYC{=}N{=}N + CO \longrightarrow XYC{=}C{=}O + N_2 \quad (8.1)$$

(X,Y = H, alkyl, aryl, alkoxy, alkylthio, acyl, alkoxycarbonyl, alkylamino, F, Cl, trialkylsilyl, etc.)

According to theoretical calculations, there is a low-energy pathway for the direct interaction of diazomethane with carbon monoxide [4]. However, experimental results so far show only evidences for two-step pathways. These involve first the dediazotation of the diazoalkanes in thermal, photochemical, or transition metal mediated reactions [5] resulting in free carbenes (Equation 8.2) or transition metal coordinated carbenes (Equation 8.3), which under proper reaction conditions couple in the second step with carbon monoxide (Equation 8.4 or 8.5) to form ketene products.

$$XYC{=}N{=}N \xrightarrow{\Delta \text{ or } h\nu} XYC{:} + N_2 \quad (8.2)$$

$$XYC{=}N{=}N + ML_n \longrightarrow XYC{-}ML_n + N_2 \quad (8.3)$$

(ML_n = one-, two-, or polynuclear transition metal complexes with various electron-donor ligands)

$$XYC{:} + CO \longrightarrow XYC{=}C{=}O \quad (8.4)$$

Modern Carbonylation Methods. Edited by László Kollár

ISBN: 978-3-527-31896-4

$$\underset{Y}{\overset{X}{>}}C{=}ML_n + CO \longrightarrow \underset{Y}{\overset{X}{>}}C{=}C{=}O + ML_n \tag{8.5}$$

In the following discussion, we will review the known examples of the reactions depicted in Equations 8.2–8.5 with various X, Y, and ML_n groups, leading to direct or indirect experimental evidences of ketene formation. The scientific literature is covered till April 2007.

8.1 Reactions of Diazoalkanes with Carbon Monoxide in the Absence of Transition Metal Complexes

Dediazotations of diazoalkanes by heat or photochemical activation give carbenes and dinitrogen (Equation 8.2). Regardless of the kind of activation, the primary carbene products are singlet and react in solution generally as the singlet before they decay into the triplet state. The ground-state electron configuration of a carbene depends on the substituents [6]. For example, difluorcarbene is a singlet in the ground state, but methylene is a triplet in the ground state and its singlet state is higher in energy. The experimentally determined singlet–triplet splitting was found to be 8.1 ± 0.8 [7] and $9.05 \pm 0.06\,\text{kcal mol}^{-1}$ [8]. For recent calculations of the singlet–triplet splitting of various arylcarbenes see Ref. [9].

The formation of acetanilide in low yield from the reaction of aniline and the pyrolysis product of a mixture of diazomethane and carbon monoxide was regarded as the most important argument for the existence of the highly reactive and short-lived methylene intermediate that couples with carbon monoxide to give the ketene intermediate [10]. However, the heating of diphenyldiazomethane in the presence of carbon monoxide has not led to the expected diphenylketene [11].

The matrix isolation technique combined with ESR, IR, UV–vis, and fluorescence spectroscopy turned out to be very useful for studying the formation and reactivity of highly reactive carbenes [12]. Thus, photolysis of diazomethane in the presence of an 18-fold molar excess of CO in a dinitrogen matrix at 20 K gave up to 50% yield of ketene, based on *in situ* infrared spectroscopic evidence [13]. The photochemistry of diazoethane was investigated in a variety of inert gas matrices. In CO-doped matrices, methylketene was found indicating the CO trapping by singlet ethylidene [14]. Similar experiments with diazocyclopentadiene in the presence of a very large excess of CO gave exclusively the corresponding ketene. With lower molar excess or in the absence of carbon monoxide, the presence of cyclopentadienylidene and the formation of fulvalene were observed in the matrix (reaction 8.6). The formation of the ketene product was explained by the reaction of the primary photoproduct singlet-state cyclopentadienylidene with CO, which is faster than the decay of the singlet cyclopentadienylidene into the triplet state, the precursor of fulvalene [15,16].

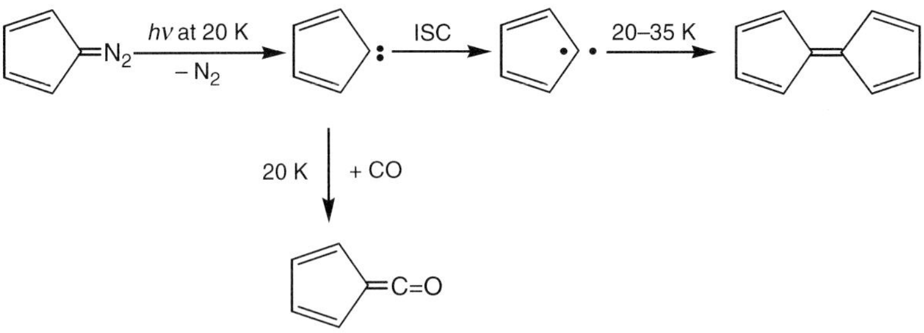

hv at 20 K, $-N_2$; ISC; 20–35 K; 20 K, + CO; C=O

(8.6)

Irradiation of diazoketones (reactions 8.7 and 8.8) [17–19], diazoesters (reaction 8.9) [20], and various other diazoalkanes (reactions 8.10 and 8.11) [21,22] as well as diazirines (reactions 8.12 and 8.13) [23,24] under matrix isolation conditions also led to the detection of the corresponding carbene intermediate, which could be trapped with carbon monoxide in the form of ketenes. For example,

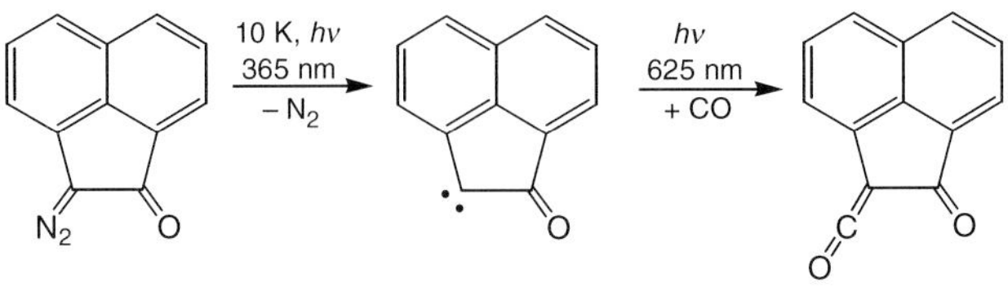

N_2, O; 10 K, hv, 365 nm, $-N_2$; O; hv, 625 nm, + CO; O, C, O

(8.7)

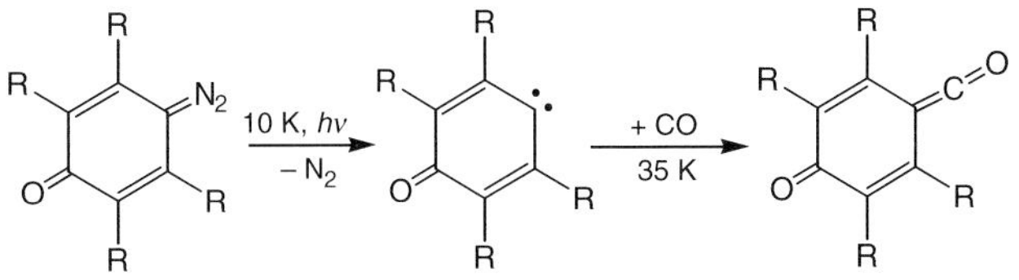

R, N_2, O; 10 K, hv, $-N_2$; + CO, 35 K; C=O

(R = H, F, Cl)

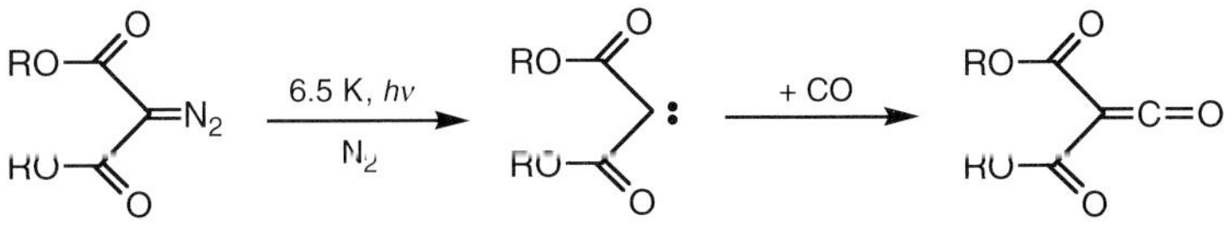

(8.8)

RO, O, N_2, RO, O; 6.5 K, hv, N_2; + CO; C=O

(R = Me, Et)

(8.9)

N, N_2; 12 K, hv, $-N_2$; + CO; C, O

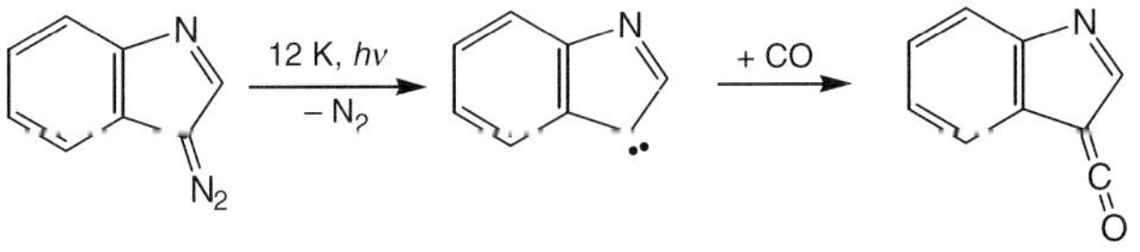

(8.10)

$$(CF_3)(Ph)C{=}N_2 \xrightarrow[-N_2]{12\ K,\ h\nu} (CF_3)(Ph)C{:} \xrightarrow[40\ K]{+CO} (CF_3)(Ph)C{=}C{=}O \tag{8.11}$$

$$(CF_3)_2C(N{=}N) \xrightarrow[-N_2]{12\ K,\ h\nu} (CF_3)_2C{:} \xrightarrow[40\ K]{+CO} (CF_3)_2C{=}C{=}O \tag{8.12}$$

$$(c\text{-}C_3H_5)_2C(N{=}N) \xrightarrow[N_2,\ 6\ K]{h\nu} (c\text{-}C_3H_5)_2C{=}N{=}N + (c\text{-}C_3H_5)_2C{:} \xrightarrow[38\ K]{+CO} (c\text{-}C_3H_5)_2C{=}C{=}O \tag{8.13}$$

Conflicting results were reported for the reactivity of stable imidazol-2-ylidenes toward carbon monoxide. The synthesis of 1,3-di-1-adamantylimidazol-2-carbonyl from 1,3-di-1-adamantylimidazol-2-ylidene in 20–30% yield at 10–15 °C in THF solutions under atmospheric pressure of carbon monoxide was described (reaction 8.14) [25].

$$\text{1,3-Ad}_2\text{-imidazol-2-ylidene} + CO \xrightarrow[THF]{10\text{–}15\ ^\circ C;\ 1\ bar} \text{1,3-Ad}_2\text{-imidazol-2-ylidene}{=}C{=}O \tag{8.14}$$

(Ad = 1-adamantyl) 20–30% yield

Attempts to duplicate the above experimental results were unsuccessful. 1,3-Di-1-adamantylimidazol-2-ylidene does not react with carbon monoxide even under pressure of excess carbon monoxide [26]. In agreement with this negative result, neither mixtures of 1,3-di-*tert*-butyl-imidazol-2-ylidene nor 1,3-di-*tert*-butyl-4,5-dihidroimidazol-2-ylidene give the corresponding ketene product upon contact with carbon monoxide [27]. Surprisingly, a smooth addition of carbon monoxide at 1 bar pressure to the stable amino-substituted acyclic carbenes and a cyclic carbene at room temperature in THF was reported most recently, which resulted in the corresponding stable ketenes in 80–82 and 65% yields, respectively (reactions 8.15 and 8.16) [28].

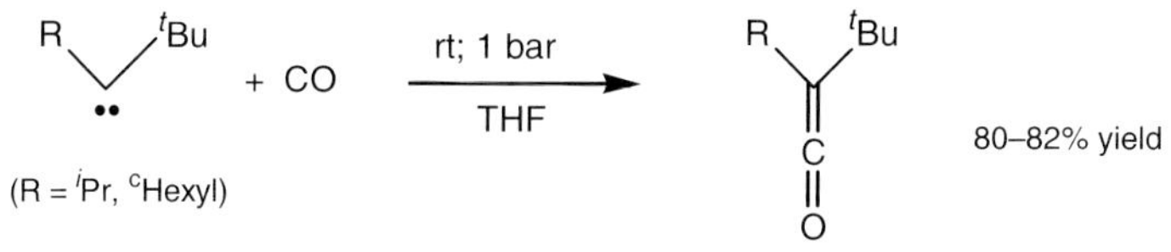

(8.15)

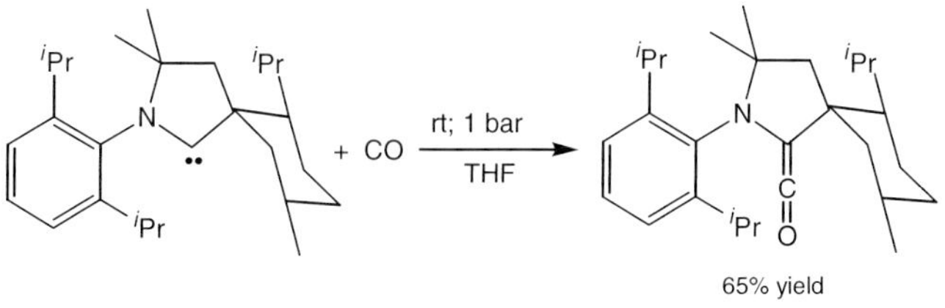

(8.16)

8.2 Reactions of Diazoalkanes with Carbon Monoxide in the Presence of Transition Metal Complexes

The transition metal complex mediated carbonylative dediazotation of diazoalkanes consists of at least two steps shown in Equations 8.3 and 8.5, which can be parts of a catalytic cycle depicted in Scheme 8.1 in a general form [29].

According to this scheme the alpha carbon of the diazoalkane coordinates to a vacant position of the metal complex L_nM. As a consequence of this coordination, dinitrogen loss and the formation of a metal carbene occur. Transfer of the electrophilic carbene to an electron-rich substrate (S:) regenerates the L_nM complex and the catalytic cycle can start again. If S: is carbon monoxide, the coupling product is a ketene. The individual steps in Scheme 8.1 are more or less explored by experimental and theoretical works in the case of a few transition metals. However, the transfer of the coordinated carbene to carbon monoxide is less explored. The known results for the different transition metals are discussed in the following sections.

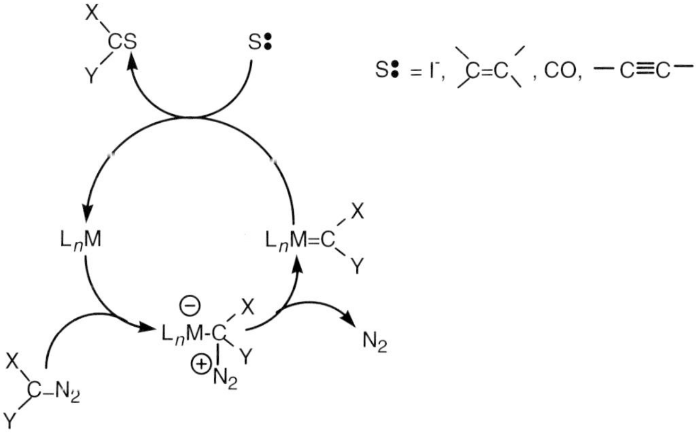

Scheme 8.1 Catalytic dediazotation of diazoalkanes and accompanying reactions with electron-rich reagents.

8.2.1
Titanium and Zirconium

Heterobinuclear complexes with μ-methylene ligands $Cp_2Ti(Cl)CH_2Pt(CH_3)L_2$ (L = $P(CH_3)_2Ph$, PCH_3Ph_2) (**1**) were found to react with carbon monoxide at atmospheric pressure rapidly to give μ-(C,O)-ketene species $Cp_2Ti(Cl)–OC(=CH_2)–Pt(CH_3)L_2$ (**2**) (reaction 8.17) [30].

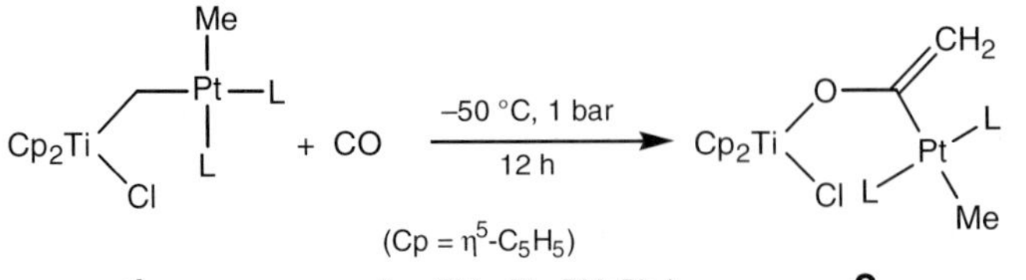

1 + CO → (−50 °C, 1 bar, 12 h) → **2** 80–95% yield

(Cp = η^5-C_5H_5)
(L = PMe_2Ph, $PMePh_2$)

(8.17)

The carbene–carbon monoxide coupling was suggested as a potential elementary C–C coupling step in catalytic carbon monoxide reduction systems based on the observation of the reaction product $\{(\eta^5\text{-}C_5(CH_3)_5)_2ZrH\}_2(\mu\text{-}OCH{=}CHO)$ [31]. The zirconoxycarbene complex **3** was found to react with carbon monoxide at room temperature to afford a zirconium-coordinated ketene complex **4** in 30% isolated yield (reaction 8.18). The X-ray structures of both **3** and **4** were determined [32].

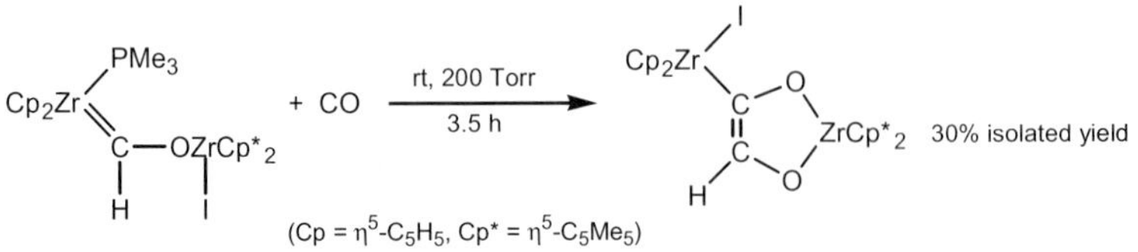

3 + CO → (rt, 200 Torr, 3.5 h) → **4** 30% isolated yield

(Cp = η^5-C_5H_5, Cp* = η^5-C_5Me_5)

(8.18)

8.2.2
Chromium, Molybdenum, and Tungsten

Carbonylation of the chromium carbene complex $(CO)_5Cr{=}C(OCH_3)Ph$ at 150 bar pressure in the presence of 1-vinyl-2-pyrrolidone resulted in $Cr(CO)_6$ and organic products that were rationalized by the intermediate formation of methoxyphenylketene [33]. The formation of a free vinylketene and vinylketene chromium complexes were found in the reaction of $(CO)_5Cr{=}C(OCH_3)Ph$ and bis(trimethylsilyl)acetylene at 50 °C (reaction 8.19) [34–36].

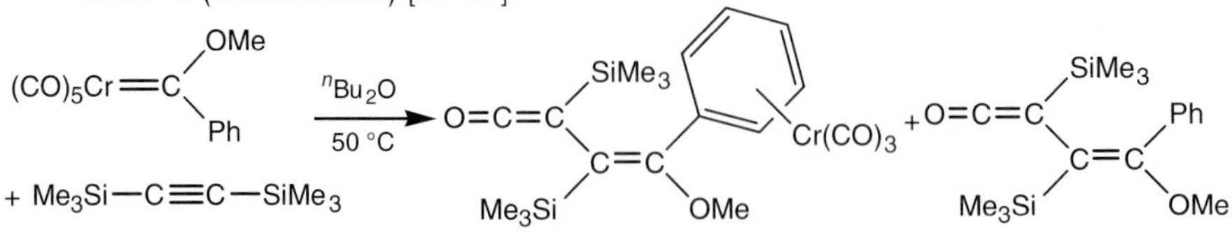

$(CO)_5Cr{=}C(OMe)Ph$ + $Me_3Si–C{\equiv}C–SiMe_3$ → ($^n Bu_2O$, 50 °C) → vinylketene chromium complex, 52% isolated yield + free vinylketene, 20% isolated yield

(8.19)

In the presence of ethanol, the reaction of $(CO)_5Cr{=}C(OCH_3)Ph$ and ethyl propiolate gives malonate derivatives [37].

$$(CO)_5Cr{=}C(OMe)Ph + H{-}C{\equiv}C{-}CO_2Et + EtOH \xrightarrow[60\,^\circ C]{THF} (Ph)(MeO)C{=}C(H)CH(CO_2Et)_2 \;(E) : (Z) = 6 : 1,\ 83\%\ \text{yield} \tag{8.20}$$

Diphenylketene was isolated in up to 70% yield from $(CO)_5Cr{=}CPh_2$ under 1 bar carbon monoxide at 50 °C. An intramolecular carbene–carbon monoxide coupling was assumed in the reaction based on the observation that no incorporation of ^{13}CO in the diphenylketene product was found under ^{13}CO at 25% conversion [38].

Ketene $Ph_3Si(EtO)C{=}C{=}O$ prepared *in situ* from the carbene complexes $(CO)_3M{=}C(OEt)SiPh_3$ (M=Cr, Mo, W) with 50 bar carbon monoxide was found to react at 80–100 °C, among others, with *N*-methylbenzimine to give the corresponding β-lactam in 90% isolated yield (reaction 8.21) [39].

$$(CO)_5M{=}C(OEt)SiPh_3 + CO + PhCH{=}NMe \xrightarrow[80-100\,^\circ C,\ 2\ \text{days}]{50\ \text{bar}} \beta\text{-lactam (3-EtO, 3-Ph}_3\text{Si, 4-Ph, N-Me)},\ 90\%\ \text{isolated yield} \quad (M = Cr, Mo, W) \tag{8.21}$$

The photolysis of chromium carbene complexes leading to short-lived species with ketenelike reactivity was applied for the synthesis of carboxylic acid derivatives, aminoacids, β-lactams, and cyclobutanones (reaction 8.22) [40].

$$(CO)_5Cr{=}C(X)R \underset{}{\overset{h\nu}{\rightleftharpoons}} [(CO)_4Cr{-}C(X)(R){=}C{=}O] \longrightarrow \begin{cases} \xrightarrow{ROH} XCH(R)C(O)OR \\ \xrightarrow{R_2NH} XCH(R)C(O)NR_2 \\ \xrightarrow{\text{imine}} \beta\text{-lactam} \\ \xrightarrow{\text{alkene}} \text{cyclobutanone} \end{cases} \quad (X = OR, NR_2;\ R = \text{alkyl}) \tag{8.22}$$

The reaction of $Cp_2Mo_2(CO)_4$ with diphenyldiazomethane in CH_2Cl_2 at room temperature was found to afford $Cp_2Mo_2(CO)_4(N_2CPh_2)$ in more than 90% yield, which can be decomposed in benzene solution at 60 °C with the loss of dinitrogen to form the stable μ-diphenylmethylene complex $Cp_2Mo_2(CO)_4(CPh_2)$ in high yield. Carbonylation of the μ-diphenylmethylene complex at 50 °C and 3 bar CO pressure yields diphenylketene and $Cp_2Mo_2(CO)_6$ [41].

The thiocarbene complex $(CO)_5W{=}C(SEt)SiPh_3$ was found to decompose at 20 °C both in the solid state and in pentane solution yielding (ethylthio)(triphenylsilyl) ketene and a tungsten complex in which this ketene is coordinated to a $(CO)_5W$ fragment[42]. The ethoxycarbene complex $(CO)_5W{=}C(OEt)SiPh_3$ was found to show much less reactivity than $(CO)_5W{=}C(SEt)SiPh_3$. The corresponding ketene product $EtO(Ph_3Si)C{=}C{=}O$ was formed from $(CO)_5W{=}C(OEt)SiPh_3$ only by carbonylation with carbon monoxide at 50 bar pressure [43].

Unusual easy coupling of carbon monoxide with carbene ligands of neutral carbene tungsten complexes $(CO)_5W{=}C(Ar)H$ (Ar = Ph, *p*-tolyl) was reported. Thus, in the presence of methanol in CH_2Cl_2 solution, $(CO)_5W{=}C(Ar)H$ gives arylacetic methyl esters $ArCH_2CO_2CH_3$ under 1 bar CO even at −20 °C. The yield of the esters increases with increasing CO pressure. External ^{13}CO was found to incorporate into the ester [44]. By the carbonylation of a ditungsten methylene complex $Cp_2(CO)_6W_2(\mu\text{-}CH_2)$ using 1 bar carbon monoxide pressure, a bridging ketene complex $Cp_2(CO)_6W_2(\mu\text{-}CH_2CO)$ with no metal–metal bond was obtained [45,46].

8.2.3
Manganese

Manganese carbonyl complexes $LMn(CO)_2(THF)$ ($L = \eta^5\text{-}CH_3C_5H_4$ or $\eta^5\text{-}C_5H_5$) were found to be effective in the dediazotation of diazoalkanes and in the formation of stable manganese carbene carbonyl complexes. See for examples reactions 8.23 and 8.24 [47,48].

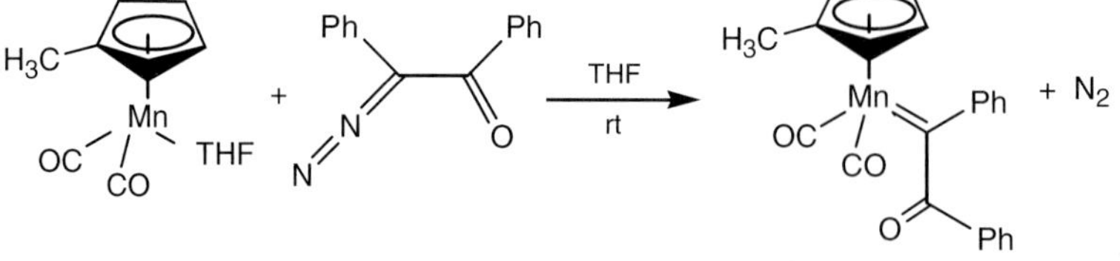

(8.23)

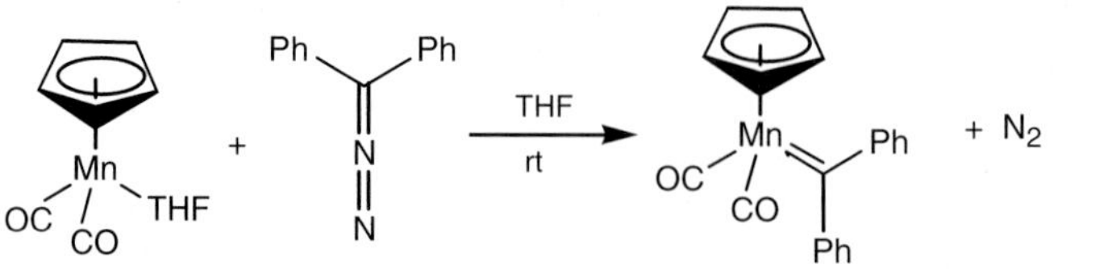

(8.24)

The high-pressure (650 bar) carbonylation of the carbene complex CpMn $(CO)_2(CPh_2)$ afforded the corresponding ketene complex $CpMn(CO)_2(O{=}C{=}CPh_2)$ at 30–50 °C in 45% isolated yield [49,50].

8.2.4 Iron, Ruthenium, and Osmium

The iron carbene complex $(\eta^5\text{-}C_5H_5)Fe(CO)(CH_2)^+$ generated by the electron impact from $(\eta^5\text{-}C_5H_5)Fe(CO)_2CH_2OCH_3$ in the gas phase was found to react with cyclohexene or *n*-donor bases such as NH_3, CH_3CN, or CD_3CDO by displacement of ketene from the iron center (reaction 8.25). Methylene–carbon monoxide coupling was assumed to occur in the coordination sphere of iron during the reaction [51].

$$(\eta^5\text{-}C_5H_5)Fe(CO)(CH_2)^+ + B \longrightarrow (\eta^5\text{-}C_5H_5)Fe(B)^+ + CH_2{=}C{=}O \quad (8.25)$$

(B = cyclohexene, NH_3, CH_3CN, CD_3CDO)

The formation of methyl acetate in up to 75% yield was observed in a stoichiometric reaction of methanol with the methylene-bridged complex $Fe_2(\mu\text{-}CH_2)(CO)_8$ (reaction 8.26). By coupling μ-methylene with carbon monoxide to form a ketene-bridged intermediate in the reaction was postulated [52].

H_2 C (OC)$_4$Fe—Fe(CO)$_4$ → [H_2C—C(=O) (OC)$_4$Fe—Fe(CO)$_4$] —(CH_3OH, 60 °C, 14 h)→ $CH_3C(=O)OCH_3$

65–75% yield (8.26)

The vinylketene complex **5** was shown to undergo reversible loss of carbon monoxide to form vinylcarbene complex **6** (reaction 8.27) [53].

5 ⇌ (60 °C; – CO / 20 °C; + CO) **6** (8.27)

Ph Ph O=C O C=C—C C=O C=C Fe (CO)$_3$ Ph Ph

Ph Ph O=C O C=C—C C=C Fe(CO)$_3$ Ph Ph

The formation of a stable (η^2-C,C) ketene compound $[(\eta^5\text{-}C_5H_5)(CO)_2Fe^+(CH_2{=}C{=}O)][PF_6^-]$ by carbonylation of an iron–methylidene complex $[(\eta^5\text{-}C_5H_5)(CO)_2Fe^+{=}CH_2)][PF_6^-]$ at 25 °C and 6 bar carbon monoxide pressure was described. Carbon-13 labeling study has shown that the electrophilic methylidene ligand picks up exogenous carbon monoxide and not a coordinated one [54].

Carbonylation of $[(\eta^5\text{-}C_5H_5)(CO)_2Ru]_2(\mu\text{-}CH_2)$ at 25 °C and 3 bar carbon monoxide pressure was found to occur readily to give the corresponding bridged ketene complex $[(\eta^5\text{-}C_5H_5)(CO)_2Ru]_2(\mu\text{-}CH_2{=}C{=}O)$ in 80% yield. This carbene–carbon monoxide coupling seemingly does not require a ruthenium–ruthenium bond [55]. Upon heating the ketene-bridged ruthenium complex **7** in toluene, rapid generation of the μ-methylene complex **8** was observed in 88% yield (reaction 8.28). This transformation could not be reversed even under 150 bar CO at 100 °C [56].

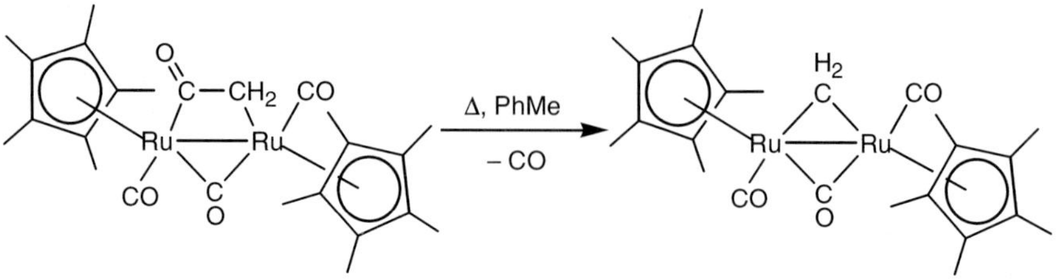

Δ, PhMe; – CO

7 → **8**

(8.28)

The interconversion of methylene and ketene ligands on a triosmium cluster was reported. It was found that a triosmium methylene complex in CH_2Cl_2 solution readily adds two moles of carbon monoxide at 22 °C to yield the ketene derivative **9** (reaction 8.29). The formation of the ketene complex **9** from the methylene complex can be made reversible by heating under reduced pressure. Labeling experiments have shown that the ketene carbonyl derives from one of the initial $Os_3(CO)_{11}(\mu\text{-}CH_2)$ carbonyl ligands and not from the added CO. This implies a preequilibrium between $Os_3(CO)_{11}(\mu\text{-}CH_2)$ and a coordinatively unsaturated ketene complex that subsequently adds two moles of carbon monoxide to give the isolated product **9** [57,58].

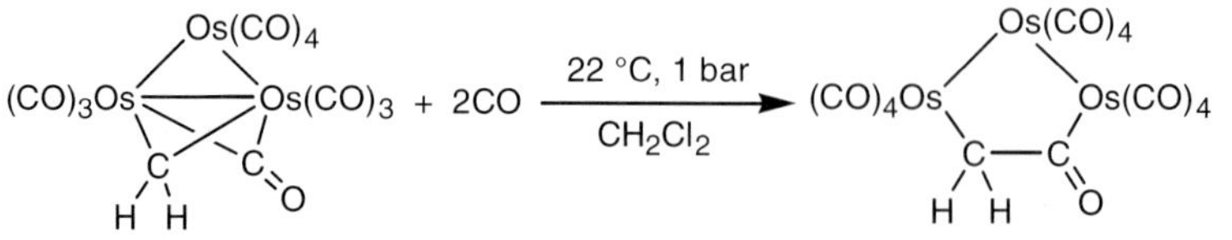

$Os_3(CO)_{11}(\mu\text{-}CH_2)$ + 2CO → (22 °C, 1 bar, CH_2Cl_2) → **9**

(8.29)

Insertion of carbon monoxide into the bridging-methylene ligands was found to be up to 100 times faster if, instead of the bridging carbonyl, a bridging halide ion is present [59,60]. The osmium methylene complex **10** was found to add carbon monoxide rapidly affording the ketene complex **11** (reaction 8.30) [61].

$$(PPh_3)_2Cl(NO)Os{=}CH_2 + CO \longrightarrow (PPh_3)_2Cl(NO)Os(CH_2{=}C{=}O) \qquad (8.30)$$

10 → **11**

8.2.5
Cobalt, Rhodium, and Iridium

In the reaction of the cobalt carbene complex **12** with 3-hexine at 25 °C, the η^4-vinylketene complex **13** was formed in 35% isolated yield (reaction 8.31) [62].

$$Ph_3Sn{-}Co(CO)_3{=}C(OMe)Ph \; (\mathbf{12}) + EtC{\equiv}CEt \xrightarrow[\text{benzene}]{25\,^\circ C,\ 48\ h} \mathbf{13} \tag{8.31}$$

In a slow dediazotation reaction of bis(trifluoromethyl)diazomethane with $(Ph_2PCH_3)_2Ir(CO)Cl$, the iridium carbene complex **14** was formed and isolated in 37% yield (reaction 8.32) [63].

$$trans\text{-}(Ph_2PMe)_2Ir(CO)Cl + (CF_3)_2CN_2 \xrightarrow[\text{benzene}]{\text{rt, 20 days}} (Ph_2MeP)_2Ir(CO)Cl{=}C(CF_3)_2 \; (\mathbf{14}) + N_2 \tag{8.32}$$

The carbene transfer from bis(trifluoromethyl)diazomethane to octacarbonyl dicobalt gives at room temperature in 28 days the μ^2-bis(trifluoromethyl)methylene complex **15** that was isolated in 54% yield (reaction 8.33) [64,65].

$$Co_2(CO)_8 + (CF_3)_2CN_2 \xrightarrow[n\text{-hexane}]{\text{rt, 28 days}} (OC)_3Co{-}Co(CO)_3(\mu\text{-}C(CF_3)_2)(\mu\text{-}CO) \; (\mathbf{15}) + N_2 \tag{8.33}$$

The dediazotation reaction of various diazoalkanes (reaction 8.34) was successfully applied in the preparation of stable dinuclear cobalt carbene complexes (**16**) containing η^5-cyclopentadienyl ligands [66–71].

$$CpCo(CO)_2 + N_2{=}CR^1R^2 \longrightarrow Cp(CO)Co{-}Co(CO)Cp(\mu\text{-}CR^1R^2) \; (\mathbf{16}) + N_2 \tag{8.34}$$

(Cp = η^5-C_5H_5
$R^1 = R^2$ = Ph; R^1 = H, R^2 = CO_2Alkyl (Alkyl = Et, ^{t}Bu)
$R^1 = R^2$ = $CO_2{}^t$Bu; R^1 = CO_2Me, R^2 = $CO_2{}^t$Bu; R^1 = CO_2Et, R^2 = $CO_2{}^t$Bu)

By the reaction of $Co_2(CO)_6$(dppm) (dppm = bis-diphenylphosphanomethane, $Ph_2PCH_2PPh_2$) with diazomethane or ethyl diazoacetate stable μ-methylene complexes **17**, **18**, **19**, and **20** were prepared (reactions 8.35–8.38). On the basis of NMR spectra obtained at different temperatures, an equilibrium transformation of μ-methylene into terminal methylene in complex **17** (reaction 8.39) [72] was shown.

Using excess diazoethane as the reagent of $Co_2(CO)_6$(dppm) or complex **17**, the complexes **21** and **22** were prepared, respectively (reactions 8.40 and 8.41) [73].

$$Co_2(CO)_6(dppm) + H_2CN_2 \xrightarrow{25\,°C} \mathbf{17} + N_2 \tag{8.35}$$

$$\mathbf{17} + H_2CN_2 \xrightarrow{35\,°C} \mathbf{18} + N_2 \tag{8.36}$$

$$Co_2(CO)_6(dppm) + EtO_2CCHN_2 \xrightarrow[CH_2Cl_2]{40\,°C,\ 1\,h} \mathbf{19} + N_2 \tag{8.37}$$

$$\mathbf{17} + EtO_2CCHN_2 \xrightarrow[CH_2Cl_2]{40\,°C} \mathbf{20} + N_2 \tag{8.38}$$

$$\mathbf{18} \rightleftharpoons (CO)_2Co(=CH_2)-Co(CO)_2(=CH_2)(dppm) \tag{8.39}$$

$$\text{Co}_2(\text{CO})_4(\mu\text{-CO})_2(\mu\text{-dppm}) + 2\,CH_3CHN_2 \longrightarrow \textbf{21} + 2\,N_2 \qquad (8.40)$$

$$\textbf{17} + CH_3CHN_2 \longrightarrow \textbf{22} + N_2 \qquad (8.41)$$

An indirect evidence of ketene formation by coupling of μ-alkylidene groups of complexes **17–22** with carbon monoxide was obtained on the basis of ^1H NMR detection of the corresponding esters if the reaction was performed at room temperature in the presence of CD_3OD in sealed NMR tubes [74]. For example,

$$18 + CO + CD_3OD \xrightarrow[\text{CDCl}_3]{\text{rt, 1 bar}} \text{Co}_2(\text{CO})_4(\mu\text{-CO})_2(\mu\text{-dppm})\ (90\%) + EtO_2CCHDCO_2CD_3\ (90\%) + CH_2DCO_2CD_3\ (95\%)$$

To show a metal-mediated coupling of a carbene with carbon monoxide, the isolation of the ketene complex $Co_2(CO)_7(CH_2{=}C{=}O)$ in 32% yield in the reaction of $Co_2(CO)_8$ with the carbene precursor CH_2Br_2 in the presence of Zn at 20 °C for 2 days in THF was taken into account. This cobalt ketene complex was found to catalyze the methoxycarbonylation or aminocarbonylation of a mixture of CH_2Br_2 and Zn under 5 bar carbon monoxide pressure at −40–20 °C in THF solution, yielding methyl acetate (385% based on $Co_2(CO)_7(CH_2{=}C{=}O)$) and *N,N*-diethylacetamide (184% based on $Co_2(CO)_7(CH_2{=}C{=}O)$) using methanol and diethylamine as the additional reagents in the reaction, respectively [75].

Octacarbonyl dicobalt was found to be an effective dediazotation reagent for ethyl diazoacetate. Under ambient conditions, ethoxycarbonylcarbene-bridged dicobalt carbonyl complexes **23** and **24** were isolated from the reaction mixtures in good yields (reactions 8.42 and 8.43).

$$Co_2(CO)_8 + EtO_2CHN_2 \xrightarrow[-CO,\ -N_2]{5\text{–}25\ °C,\ 1\ bar} \mathbf{23} \quad \text{82\% isolated yield} \tag{8.42}$$

$$\mathbf{23} + EtO_2CHN_2 \xrightarrow[-CO,\ -N_2]{5\text{–}25\ °C,\ 1\ bar} \mathbf{24} \quad \text{73\% isolated yield} \tag{8.43}$$

Under 50 bar carbon monoxide pressure in the presence of ethanol, both complexes **23** and **24** convert quantitatively to $Co_2(CO)_8$ and diethyl malonate (reactions 8.44 and 8.45).

$$\mathbf{23} + 2CO + EtOH \xrightarrow[60\ min]{25\ °C,\ 50\ bar} Co_2(CO)_8 + EtO_2CCH_2CO_2Et \tag{8.44}$$

$$\mathbf{24} + 4CO + 2EtOH \xrightarrow[60\ min]{25\ °C,\ 50\ bar} Co_2(CO)_8 + 2EtO_2CCH_2CO_2Et \tag{8.45}$$

Experiments at atmospheric pressure of carbon monoxide have shown that complex **24** converts rapidly first into **23** and diethyl malonate and in a much slower reaction, complex **23** converts further into octacarbonyl dicobalt and an other mole of diethyl malonate. In both reactions, the formation of highly reactive ethoxycarbonylketene was assumed by coupling of the ethoxycarbonyl carbene ligand with carbon monoxide, whose intermediate is scavenged by ethanol [76]. Reactions of complex **24** with ^{13}CO in the presence of excess ethanol result in diethyl malonate with natural isotopic distribution of ^{13}C [77]. The combination of these steps led to an effective one-pot procedure for the preparation of various malonic acid derivatives by catalytic carbonylation of ethyl diazoacetate in the presence of a few mol% octacarbonyl dicobalt catalyst precursor [76]. Two new example procedures illustrate the usefulness of the method.

The preparation of propanedioic acid ethyl-5-methyl-2-(1-methyl-ethyl)cyclohexyl ester [CAS: 478288-49-2] (reaction 8.46):

$$EtO_2CCHN_2 + CO + \text{(menthol)} \xrightarrow[\text{rt, 50 bar, 1 day}\ \ CH_2Cl_2]{1\ mol\%\ Co_2(CO)_8} \text{menthyl–}O_2CCH_2CO_2Et \tag{8.46}$$

Ethyl diazoacetate (184.3 mg, 1.9 mmol), (1*R*,2*S*,5*R*)-(−)-menthol (297 mg, 1.9 mmol) and methylene chloride (7.0 ml) were placed into a stainless steel autoclave (21.8 ml total capacity) containing an open glass insert loaded with octacarbonyl dicobalt (10.6 mg, 0.031 mmol) and were pressurized at room temperature with carbon monoxide to 50 bar pressure. After dissolving the precatalyst (octacarbonyl dicobalt) by turning the autoclave upside down, the autoclave was shaken at room temperature for 24 h. The pressure was slowly released in a hood. The brown solution from the autoclave was filtered and was concentrated under vacuum to about 2 ml before it was subjected to flash chromatography on alumina (Brockman I) 2.5 cm × 15 cm, using methylene chloride as the eluent. Removing the solvent from the first 150 ml eluate in vacuum gave the title compound (325 mg, 1.2 mmol, 63.2% yield) as colorless oil. $R_f = 0.83$. IR (CH_2Cl_2) ν(C=O) 1744 cm^{-1} ($\varepsilon_M = 760.2\ cm^2\ mmol^{-1}$), 1724 cm^{-1} ($\varepsilon_M = 963.0\ cm^2\ mmol^{-1}$).

Preparation of 3-(4-morpholinyl)-3-oxo-propanoic acid ethyl ester [CAS 37714-64-0] (reaction 8.47):

$$EtO_2CCHN_2 + CO + HN(CH_2CH_2)_2O \xrightarrow[\text{rt, 50 bar, 1 day}\ CH_2Cl_2]{1\ mol\%\ Co_2(CO)_8} EtO_2CCH_2C(=O){-}N(CH_2CH_2)_2O \quad (8.47)$$

Ethyl diazoacetate (184.3 mg, 1.9 mmol), morpholine (165.5 mg, 1.9 mmol), and methylene chloride (7.0 ml) were placed into a stainless steel autoclave (21.8 ml total capacity) containing an open glass insert loaded with octacarbonyl dicobalt (10.6 mg, 0.031 mmol) and were pressurized at room temperature with carbon monoxide to 50 bar pressure. After dissolving the precatalyst (octacarbonyl dicobalt) by turning the autoclave upside down, the autoclave was shaken at room temperature for 24 h. The pressure was slowly released in a hood. The light brown solution from the autoclave was exposed to air for 2 h before filtration. Extraction with pentane (30 ml) gave a colorless solution. Removing the solvent in vacuum gave the title compound (210 mg, 1.04 mmol, 55.7 % yield) as colorless oil, which solidified overnight. IR (CH_2Cl_2) ν(C=O) 1735 cm^{-1} ($\varepsilon_M = 621.5\ cm^2\ mmol^{-1}$), 1650 cm^{-1} ($\varepsilon_M = 937.7\ cm^2\ mmol^{-1}$). ^1H NMR and MS spectra were identical with those of published ones [78].

Triphenylphosphane substituted derivatives of complex **23** were found to show similar but lower catalytic activity in ethyl diazoacetate carbonylation than $Co_2(CO)_8$, or complexes **23** and **24** [79].

(Trimethylsilyl)ketene was found to form selectively in the octacarbonyl dicobalt-mediated carbonylation of (trimethylsilyl)diazomethane. In the presence of 10 mol% $Co_2(CO)_8$, solutions of (trimethylsilyl)diazomethane in *n*-heptane are carbonylated at 10 °C under an atmosphere of carbon monoxide to (trimethylsilyl)ketene to give a 92% isolated yield (reaction 8.48) [80].

$$Me_3SiCHN_2 + CO \xrightarrow[10\ °C,\ 1\ bar,\ 4\ h]{10\ mol\%\ Co_2(CO)_8} Me_3SiCH{=}C{=}O + N_2 \quad (8.48)$$

In the absence of external carbon monoxide, the coordinated carbon monoxide ligands of octacarbonyl dicobalt are consumed for the (trimethylsilyl)ketene formation (reaction 8.49).

$$8Me_3SiCHN_2 + Co_2(CO)_8 \xrightarrow[10\,°C,\ 30\,min]{Ar,\ 1\,bar} 8Me_3SiCH{=}C{=}O + 8N_2 + 2Co \quad (8.49)$$

In the electron impact induced decomposition of $Rh_2(\mu\text{-}CH_2)(CO)_2Cp_2$, the elimination of the ketene fragment was observed by mass spectrometry [81].

The dediazotation reaction of Ph_2CN_2 or $Ph(CF_3)CN_2$ using the rhodium complex **25** gives the corresponding carbene rhodium complexes **26** and **27** in up to 96% yields (reaction 8.50) [82].

$$\underset{\mathbf{25}}{Cl{-}Rh(Sb^iPr_3)_2(\eta^2\text{-}CH_2{=}CH_2)} + Ph(R)CN_2 \xrightarrow[n\text{-pentane}]{rt,\ fast} Cl{-}Rh(Sb^iPr_3)_2{=}CPh(R) + CH_2{=}CH_2 \quad (8.50)$$

R = Ph **26**
R = CF_3 **27**

Carbonylation of the rhodium diphenylcarbene complexes **26** and **28** resulted in the formation of diphenylketene in good to excellent yields (reactions 8.51 and 8.52). It was assumed that the diphenylketene is formed by carbon–carbon coupling in the coordination sphere of rhodium [83].

$$\underset{\mathbf{26}}{Cl{-}Rh(Sb^iPr_3)_2{=}CPh_2} + CO \xrightarrow[n\text{-pentane}]{rt,\ 1\,bar,\ 1\,h} Cl{-}Rh(Sb^iPr_3)_2{-}CO + Ph_2C{=}C{=}O \quad (8.51)$$

$$\underset{\mathbf{28}}{CpRh(P^iPr_3){=}CPh_2} + CO \xrightarrow[n\text{-pentane}]{rt,\ 1\,bar,\ 5\,h} CpRh(P^iPr_3)(CO) + Ph_2C{=}C{=}O \quad (8.52)$$

The fundamental interconversion of a ketene iridium complex with a carbene–CO iridium complex has been investigated by experimental [84] and theoretical [85] methods. A controlled, reversible conversion of a diphenylketene ligand to diphenylcarbene and CO ligands on a single iridium center was discovered and rationalized (reaction 8.53).

$$[Ir(Cl)(Ph_2C{=}C{=}O)(^tBu_2PCH_2P^tBu_2)] \underset{[Ph_3P{=}N{=}PPh_3]Cl,\ CD_2Cl_2,\ -50\ to\ 25°C,\ 0{,}5\,h}{\overset{AgPF_6,\ CH_2Cl_2,\ 25\,°C,\ 2\,h}{\rightleftharpoons}} [Ir(CO)({=}CPh_2)(^tBu_2PCH_2P^tBu_2)]^{\oplus} \quad (8.53)$$

8.2.6
Nickel, Platinum

In the presence of a large excess of tetracarbonyl nickel and ethanol diphenyldiazomethane, diazofluorene, bis-(4-methoxyphenyl)diazomethane, and ethyl diazoacetate give carbonylation products trapped by ethanol and processed into the corresponding carboxylic acid in 74, 38, 26, and 8.5% isolated yields, respectively, which presumably arise from a nickel carbene carbonyl intermediate that releases a substituted ketene upon decomposition at 50–66 °C. In the absence of ethanol, by refluxing a solution of 1 mol diphenylketene with 6.4 mol tetracarbonyl nickel in diethylether 35% isolated yield of diphenylketene was observed [86]. The η^2-(C,C)-ketene complex of nickel (**31**) was isolated in 17% yield from the reaction of nickelacyclobutane (**29**) with carbon monoxide (3 bar) at −50 °C (reaction 8.54) [87]. Complex **29** is believed to be in equilibrium with the nickel–carbene–olefin complex **30** [88]. The nickel–ketene complex **31** was also obtained either by direct reaction of $Ni(PPh_3)_4$ with ketene or by carbonylation of the nickel–carbene complexes presumably formed from the reaction of $Ni(PPh_3)_4$ and CH_2Br_2 in the presence of metallic zinc [89].

$(Ph_3P)_2Ni$ (**29**) ⇌ $(Ph_3P)_2Ni$ (CH_2) (**30**) —CO, 3 bar; −50 °C, 5 days→ $(Ph_3P)_2Ni(CH_2C{=}O)$ (**31**) + (olefin)

(8.54)

The η^2-(C,C)-ketene complex of platinum (**32**) was isolated in 23% yield from the reaction of $Pt(PPh_3)_4$ and CH_2Br_2 in the presence of zinc powder under 3 bar CO pressure (reaction 8.55) [90].

$$Pt(PPh_3)_4 + CH_2Br_2 + Zn + CO \xrightarrow[-50\,^\circ C,\ 10\ h]{3\ bar} (Ph_3P)_2Pt(CH_2C{=}O)\ (\mathbf{32}) + ZnBr_2 \quad (8.55)$$

8.2.7
Thorium

On the basis of ^{13}C labeling experiments, the formation of the dionediolate complex **35** from **33** was explained by the rapid dimerization of a transient ketene complex **34**, a product of coupling of the carbenelike dihaptoacyl complex **33** with carbon monoxide (reaction 8.56) [91].

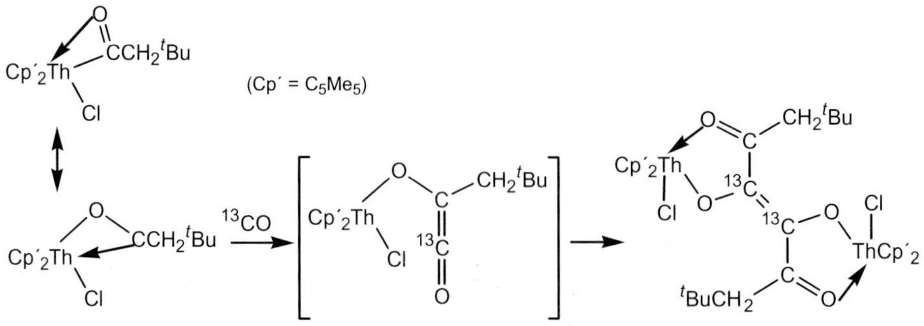

(Cp´ = C_5Me_5)

33 **34** **35**

(8.56)

8.3 Concluding Remarks

Despite the fragmentary results known to date, the dediazotation of diazoalkanes coupled with the carbene–carbon monoxide trapping reaction seems to be a promising synthetic path for the preparation of various carboxylic acid derivatives through the ketene intermediate product. Especially the highly effective and selective cobalt-catalyzed examples of the reactions are worth to explore in more detail.

References

1 Tidwell, T.T. (2006) *Ketenes*, 2nd edn, John Wiley & Sons, Inc., Hoboken, NJ.

2 Borrmann, D. (1968) *Methoden der Organischen Chemie, Houben-Weyl*, vol. 7, part 4, G. Thieme Verlag, Stuttgart.

3 Schaumann, E. and Scheiblich, S. (1993) *Methoden der Organischen Chemie, Houben-Weyl*, vol. E15, G. Thieme Verlag, Stuttgart.

4 Seres, B., Fördős, E., Ungvári, N., Ungváry, F. and Kégl, T. (2007) The mechanism of ketene formation from carbon monoxide and diazomethane: a computational assessment. *Journal of the Chemical Society, Chemical Communications*, submitted.

5 Zollinger, H. (1995) *Diazo Chemistry II. Aliphatic, Inorganic and Organometallic Compounds*, VCH, Weinheim, pp. 314–318.

6 Brückner, R. (2003) *Reaktionsmechanismen*, 2 Aufl., Spektrum Akademischer Verlag, Heidelberg, Berlin, pp. 114–116.

7 Lengel, R.K. and Zare, R.N. (1978) Experimental determination of the singlet–triplet splitting in methylene. *Journal of the American Chemical Society*, **100**, 7495–7499.

8 McKellar, A.R.W., Bunker, P.R., Sears, T.J., Evenson, K.M., Saykally, R.J. and Langhoff, S.R. (1983) Far infrared laser magnetic resonance of singlet methylene: singlet–triplet perturbations, singlet–triplet transitions, and the singlet–triplet splitting. *Journal of Chemical Physics*, **79**, 5251–5264.

9 Woodcock, H.L., Moran, D., Brooks, B.R., Schleyer, P.v.R. and Schaefer, H.F., III (2007) Carbene stabilization by aryl

substituents. Is bigger better? *Journal of the American Chemical Society*, **129**, 3763–3770.

10 Staudinger, H. and Kupfer, O. (1912) Über Reaktionen des Methylens. III. Diazomethan. *Berichte der Deutschen Chemischen Gesellschaft*, **45**, 501–509.

11 Staudinger, H., Anthes, E. and Pfenninger, F. (1916) Diphenyl-diazomethan. *Berichte der Deutschen Chemischen Gesellschaft*, **49**, 1928–1941.

12 Sander, W., Bucher, G. and Wierlacher, S. (1993) Carbenes in matrices – spectroscopy, structure, and reactivity. *Chemical Reviews*, **93**, 1583–1621.

13 Demore, W.B., Pritchard, H.O. and Davidson, N. (1959) Photochemical experiments in rigid media at low temperatures. II. The reactions of methylene, cyclopentadienylene and diphenylmethylene. *Journal of the American Chemical Society*, **81**, 5874–5879.

14 Seburg, R.A. and McMahon, R.J. (1992) Photochemistry of matrix-isolated diazoethane and methylazirine: ethylidene trapping? *Journal of the American Chemical Society*, **114**, 7183–7189.

15 Baird, M.S., Dunkin, I.R. and Poliakoff, M. (1974) Thermal dimerization and carbonylation of a carbene in low-temperature matrices. *Journal of the Chemical Society, Chemical Communications*, 904–905.

16 Baird, M.S., Dunkin, I.R., Hacker, N., Poliakoff, M. and Turner, J.J. (1981) Cyclopentadienylidene. A matrix isolation study exploiting photolysis with unpolarized and plane-polarized light. *Journal of the American Chemical Society*, **103**, 5190–5195.

17 Hayes, R.A., Hess, T.C., McMahon, R.J. and Chapman, O.L. (1983) Photochemical Wolff rearrangements of a triplet ground-state carbene. *Journal of the American Chemical Society*, **105**, 7786–7787.

18 Sander, W., Müller, W. and Sustmann, R. (1988) 4-Oxo-2,5-cyclohexadienyliden – ein Carben mit einem stabilen Triplett- und einem metastabilen Singulettzustand? *Angewandte Chemie-International Edition*, **100**, 577–579.

19 Sander, W., Hübert, R., Kraka, E., Gräfenstein, J. and Cremer, D. (2000) 4-Oxo-2,3,5,6-tetrafluorocyclohexa-2,5-dienylidene – a highly electrophilic triplet carbene. *Chemistry – A European Journal*, **6**, 4567–4579.

20 Visser, P., Zuhse, R., Wong, M.W. and Wentrup, C. (1996) Reactivity of carbenes and ketenes in low-temperature matrices. Carbene CO trapping, Wolff rearrangement, and ketene–pyridine ylide (zwitterion) observation. *Journal of the American Chemical Society*, **118**, 12598–12602.

21 Qiao, G.G., Wong, M.W. and Wentrup, C. (1996) Synthesis of *N*-confused porphyrin analogues by β-azafulvenone tetramerization. *The Journal of Organic Chemistry*, **61**, 8125–8131.

22 Sander, W. (1988) Effects of electron-withdrawing groups on carbonyl *O*-oxides trifluoroacetophenone *O*-oxide and hexafluoroacetone *O*-oxide. *The Journal of Organic Chemistry*, **53**, 121–126.

23 Malcev, A.K., Zuev, P.S. and Nefedov, O.M. (1985) Infrared spectroscopic investigation of bis(trifluoromethyl) carbene. *Izvestiya Akademii Nauk SSSR, Seriya Khimicheskaya*, 957–958 (in Russian).

24 Ammann, J.R., Subramanian, R. and Sheridan, R.S. (1992) Dicyclopropyl-carbene: direct characterization of a singlet dialkylcarbene. *Journal of the American Chemical Society*, **114**, 7592–7594.

25 Lyashchuk, S.N. and Skrypnik, Y.G. (1994) Synthesis of 1,3-di-1-adamantylimidazol-2-carbonyl from 1,3-di-1-adamantylimidazol-2-ylidene. *Tetrahedron Letters*, **35**, 5271–5274.

26 Dixon, D.A., Arduengo, A.J., III, Dobbs, K. D. and Khasnis, D.V. (1995) On the proposed existence of a ketene derived from carbon monoxide and 1,3-di-1-adamantylimidazol-2ylidene. *Tetrahedron Letters*, **36**, 645–648.

27 Denk, M.K., Rodezno, J.M., Gupta, S. and Lough, A.J. (2001) Synthesis and reactivity of subvalent compounds. Part 11. Oxidation, hydrogenation and hydrolysis of stable diamino carbenes. *Journal of Organometallic Chemistry*, **617–618**, 242–253.

28 Lavallo, V., Canac, Y., Donnadieu, B., Schoeller, W.W. and Bertrand, G. (2006) CO fixation to stable acyclic and cyclic alkyl amino carbenes: stable amino ketenes with a small HOMO–LUMO gap. *Angewandte Chemie-International Edition*, **118**, 3568–3571; *Angewandte Chemie-International Edition*, 2006, **45**, 3488–3491.

29 Doyle, M.P., McKervey, M.A. and Ye, T. (1998) *Modern Catalytic Methods for Organic Synthesis with Diazo Compounds*, John Wiley & Sons, Inc., New York, pp. 62–66.

30 Ozawa, F., Park, J.W., Mackenzie, P.B., Schaefer, W.P., Henling, L.M. and Grubbs, R.H. (1989) Structure and reactivity of titanium/platinum and palladium heterobinuclear complexes with μ-methylene ligands. *Journal of the American Chemical Society*, **111**, 1319–1327.

31 Wolczanski, P.T. and Bercaw, J.E. (1980) On the mechanism of carbon monoxide reduction with zirconium hydrides. *Accounts of Chemical Research*, **13**, 121–127.

32 Barger, P.T., Santarsiero, B.D., Armentraut, J. and Bercaw, J.E. (1984) Carbene complexes of zirconium. Synthesis, structure, and reactivity with carbon monoxide to afford coordinated ketene. *Journal of the American Chemical Society*, **106**, 5178–5186.

33 Dorrer, B. and Fischer, E.O. (1974) Reaktion von Pentacarbonyl(methoxyphenylcarben)-chrom(0) mit Alkenyl-pyrrolidonen unter CO-Einschiebung über ein Keten zu Cyclobutanonen und Enaminoketon-Analogen. *Chemische Berichte*, **107**, 2683–2690.

34 Dötz, K.H. (1979) Stabile Vinylketene durch metallkomplex-induzierte Olefinierung und Carbonylierung von Alkinen. *Angewandte Chemie-International Edition*, **91**, 1021–1022;*Angewandte Chemie-International Edition in English*, 1979, **18**, 954–955.

35 Dötz, K.H. and Fügen-Köster, B. (1980) Stabile silylsubstituierte Vinylketene. *Chemische Berichte*, **113**, 1449–1457.

36 Dötz, K.H. and Mühlemeier, J. (1982) Bis [bis(trimethylsilyl)acetylen] dicarbonylchrom. *Angewandte Chemie-International Edition*, **94**, 936; *Angewandte Chemie Supplement*, 1982, 2023–2029.

37 Yamashita, A. and Scahill, T.A. (1982) Reaction of aryl chromium carbene complexes with ethyl propiolate. A versatile vinyl ether formation. *Tetrahedron Letters*, **23**, 3765–3768.

38 Fischer, H. (1983) Intramolekulare Carben-Carbonmonoxid-Kupplung in Chrom- und Wolfram Komplexen. *Angewandte Chemie-International Edition*, **95**, 913–914.

39 Kron, J. and Schubert, U. (1989) Siliciumhaltige Carben-Komplexe. *Journal of Organometallic Chemistry*, **373**, 203–219.

40 Hegedus, L.S. (1995) Synthesis of amino acids and peptides using chromium carbene complex photochemistry. *Accounts of Chemical Research*, **28**, 299–305, and references therein.

41 Messerle, L. and Curtis, M.D. (1980) Reaction of diaryldiazomethanes with a metal–metal triple bond: synthesis, structural characterizations and reactivity of novel bridging diazoalkane and alkylidene complexes. *Journal of the American Chemical Society*, **102**, 7789–7791.

42 Hörnig, H., Walther, E. and Schubert, U. (1985) Silicon-containing carbene complexes. 5. (Ethylthio)(triphenylsilyl) ketene by thermal decomposition of $(CO)_5WC(SEt)SiPh_3$. *Organometallics*, **4**, 1905–1906.

43 Schubert, U., Kron, J. and Hörnig, H. (1988) Siliciumhaltige Carben-Komplexe X. Ethoxy- und Ethyl(thio(triphenylsilyl)keten aus Carben-Komplexen $(CO)_5WC(XEt)SiPh_3$ ($X = O, S$). *Journal of Organometallic Chemistry*, **355**, 243–256.

44 Fischer, H., Jungklaus, H. and Schmid, J. (1989) Kinetische und mechanistische Untersuchungen von Übergangsmetall-Komplex-Reaktionen XXV. Ungewöhnlich leicht ablaufende intermolekulare Kupplung von CO und dem Carben-Liganden bei neutralen Carben-Komplexen des Typs $(CO)_5W{=}C(Aryl)H$. *Journal of Organometallic Chemistry*, **368**, 193–198.

45 Chen, M.C., Tsai, Y.J., Chen, C.T., Lin, Y.C., Tseng, T.W., Lee, G.H. and Wang, Y. (1991) Transformation of a bridging ketene to a metal-substituted acetylene in dinuclear tungsten complexes: crystal and molecular structure of a dinuclear tungsten complex containing a bridging σ-acetylene ligand. *Organometallics*, **10**, 378–380.

46 Yang, Y.L., Wang, L.J.J., Lin, Y.C., Huang, S. L., Chen, M.C., Lee, G.H. and Wang, Y. (1997) Chemistry of bridging ketene from facile carbonylation of a ditungsten methylene complex with no metal–metal bond. *Organometallics*, **16**, 1573–1580.

47 Herrmann, W.A. (1974) Ein neues Verfahren zur Darstellung von Übergangsmetall-Carben-Komplexen. *Angewandte Chemie-International Edition*, **86**, 556–557.

48 Herrmann, W.A. (1975) Reaktionen aliphatischer Diazoverbindungen mit thermostabilen Mangan-Komplexen. *Chemische Berichte*, **108**, 486–499.

49 Herrmann, W.A. and Plank, J. (1978) Hochdruckcarbonylierung metallkoordinierter Carbene und Hydrogenolyse der Keten-Komplexe. *Angewandte Chemie-International Edition*, **90**, 555–556; *Angewandte Chemie-International Edition in English*, 1978, **17**, 525–526.

50 Herrmann, W.A., Plank, J., Kriechbaum, G. W., Ziegler, M.L., Pfisterer, H., Atwood, J.L. and Rogers, R.D. (1984) Komplexchemie reaktiver organischer Verbindungen XLVII Synthese, Strukturchemie und Druckcarbonylierung von Metallcarben-Komplexen. *Journal of Organometallic Chemistry*, **264**, 327–352.

51 Stevens, A.E. and Beauchamp, J.L. (1978) Metal carbene chemistry. Formation and reactions of $(\eta^5\text{-}C_5H_5)Fe(CO)_n(CH_2)^+$ ($n = 1, 2$) in the gas phase by ion cyclotron resonance spectroscopy. *Journal of the American Chemical Society*, **100**, 2584–2585.

52 Röper, M., Strutz, H. and Keim, W. (1981) Alkyl acetates by stoichiometric reaction of alcohols with the methylene bridged complex $[Fe_2(\mu\text{-}CH_2)(CO)_8$. *Journal of Organometallic Chemistry*, **219**, C5–C8.

53 Klimes, J. and Weiss, E. (1982) Cyclopropene als Komplexliganden: $Fe_2(CO)_9$-induzierte Ringöffnung eines Spirocyclopropens und reversible CO-Addition an den Vinylcarben-Liganden. *Angewandte Chemie-International Edition*, **94**, 207; *Angewandte Chemie-International Edition in English*, 1982, **21**, 205.

54 Bodnar, T.W. and Cutler, A.R. (1983) Formation of a stable (η^2-C,C) ketene compound $(C_5H_5)Fe(CO)_2(CH_2CO)^+PF_6^-$ by carbonylation of an iron–methylidene complex. A novel entry into CO-derived C_2 chemistry. *Journal of the American Chemical Society*, **105**, 5926–5928.

55 Lin, Y.C., Calabrese, J.C. and Wreford, S.S. (1983) Preparation and reactivity of a dimeric ruthenium *m*-methylene complex with no metal–metal bond: crystal and molecular structure of $[(\eta^5\text{-}C_5H_5)Ru(CO)_2]_2(\mu\text{-}CH_2)$. *Journal of the American Chemical Society*, **105**, 1679–1680.

56 Doherty, N.M., Fildes, M.J., Forrow, N.J., Knox, S.A.R., Macpherson, K.A. and Orpen, A.G. (1986) Transformation of vinylidene to ketene at a diruthenium centre: crystal structure of $[Ru_2(CO)_2(\mu\text{-}CO)\{\mu\text{-}C(O)CH_2)\}(\eta\text{-}C_5Me_5)_2]$. *Journal of the Chemical Society, Chemical Communications*, 1355–1357.

57 Morrison, E.D., Steinmetz, G.R., Geoffroy, G.L., Fultz, W.C. and Rheingold, A.L. (1983) Interconversion of methylene and ketene ligands on a triosmium cluster. Crystal and molecular structure of the ketene complex $Os_3(CO)_{12}(\eta^2\text{-}(C,C),\mu\text{-}CH_2CO)$. *Journal of the American Chemical Society*, **105**, 4104–4105.

58 Morrison, E.D., Steinmetz, G.R., Geoffroy, G.L., Fultz, W.C. and Rheingold, A.L. (1984) Trinuclear osmium clusters as models for intermediates in CO reduction chemistry. 2. Conversion of a methylene into a ketene ligand on a triosmium cluster face. *Journal of the American Chemical Society*, **106**, 4783–4789.

59 Morrison, E.D., Geoffroy, G.L. and Rheingold, A.L. (1985) Halide-promoted insertion of CO into bridging-methylene ligands in triosmium clusters. *Journal of the American Chemical Society*, **107**, 254–255.

60 Morrison, E.D. and Geoffroy, G.L. (1985) Halide-promoted insertion of carbon monoxide into osmium–μ-methylene bonds in triosmium cluster. *Journal of the American Chemical Society*, **107**, 3541–3545.

61 W.R. Roper and A.H. Wright, unpublished results cited on p. 168 in Gallop, M.A. and Roper, W.R. Carbene and carbyne complexes of Ru, Os, and Ir. *Advances in Organometallic Chemistry*, **25**, 121–198.

62 Wulff, W.D., Gilbertson, S.R. and Springer, J.P. (1986) Reactions of cobalt carbene complexes with alkynes – η^4-vinylketene complex intermediates and a novel synthesis of bovolide. *Journal of the American Chemical Society*, **108**, 520–522.

63 Cooke, J., Cullen, W.R., Green, M. and Stone, F.G.A. (1969) Reactions of bis (trifluoromethyl)diazomethane with transition-metal complexes. *Journal of the Chemical Society A*, 1872–1874.

64 Cooke, J., Cullen, W.R., Green, M. and Stone, F.G.A. (1968) Reactions of bis (trifluoromethyl)diazomethane with transition-metal complexes. *Journal of the Chemical Society, Chemical Communications*, 170–171.

65 Cooke, J., Cullen, W.R., Green, M. and Stone, F.G.A. (1969) Reactions of bis (trifluoromethyl)diazomethane with transition-metal complexes. *Journal of the Chemical Society A*, 1872–1874.

66 Herrmann, W.A. (1978) Übergangsmetall-Methylen-Komplexe. IV. Neue Methylen-Cobalt-Komplexe. *Chemische Berichte*, **111**, 1077–1082.

67 Herrmann, W.A. and Schweizer, I. (1978) Komplexchemie reaktiver organischer Verbindungen XXV. Stabilisierung von Diphenylmethylen und Diphenyldiazomethan in Carbonylcobalt-Komplexen. *Zeitschrift Fur Naturforschung*, **33B**, 911–914.

68 Herrmann, W.A., Steffl, I., Ziegler, M.L. and Weidenhammer, K. (1979) Fünfgliedrige Cobaltacyclen durch Carben-Addition an Dicarbonyl-(η^5-cyclopentadienyl)cobalt. *Chemische Berichte*, **112**, 1731–1742.

69 Herrmann, W.A., Schweizer, I., Creswick, M. and Bernal, I. (1979) The preparation and unusual rearrangement of triply bridged μ-heteromethylenecobalt complexes. Molecular structure of μ-ethoxycarbonylmethylenebis[carbonyl-(η^5-cyclopentadienyl)cobalt] (Co(Co). *Journal of Organometallic Chemistry*, **165**, C17–C20.

70 Herrmann, W.A., Huggins, J.M., Reiter, B. and Bauer, C. (1981) Übergangsmetall-Methylen-Komplexe XXIV. Carben-Addition an eine Co=Co-Bindung. *Journal of Organometallic Chemistry*, **214**, C19–C24.

71 Herrmann, W.A., Huggins, J.M., Bauer, C., Ziegler, M.L. and Pfisterer, H. (1984) Übergangsmetall-methylen-komplexe LIII. Synthese und Struktur eines dreifach verbrückten Cobalt-Komplexes der μ-Alkyliden-Reihe. *Journal of Organometallic Chemistry*, **262**, 253–262.

72 Laws, W.J. and Puddephatt, R.J. (1983) New μ-methylene–dicobalt complexes and evidence for μ-methylene to terminal methylene transformations. *Journal of the Chemical Society, Chemical Communications*, 1020–1021.

73 Laws, W.J. and Puddephatt, R.J. (1984) Easy coupling of μ-methylene groups in di(μ-methylene)–dicobalt complexes: model reactions for the Fischer–Tropsch synthesis, *Journal of the Chemical Society, Chemical Communications*, 116–117.

74 Laws, W.J. and Puddephatt, R.J. (1986) Formation of ketenes from μ-alkylidenedicobalt complexes: model reactions for the Fischer–Tropsch synthesis, *Inorganica Chimica Acta*, L23–L24.

75 Miyashita, A., Nomura, K., Kaji, S. and Nohira, H. (1989) $Co_2(CH_2{=}C{=}O)(CO)_7$ as an active intermediate for cobalt-catalyzed alkoxycarbonylation of CH_2Br_2. *Chemistry Letters*, 1983–1986.

76 Tuba, R. and Ungváry, F. (2003) Octacarbonyl dicobalt-catalyzed selective transformation of ethyl diazoacetate into organic products containing the ethoxycarbonyl carbene building block. *Journal of Molecular Catalysis A: Chemical*, **203**, 59–67.

77 Fördős, E., Ungvári, N., Kégl, T. and Ungváry, F. (2006) Reactions of ^{13}CO with ethoxycarbonylcarbene-bridged dicobalt carbonyl complexes: [μ_2-{ethoxycarbonyl (methylene)}-μ_2-(carbonyl)-bis(tricarbonyl-cobalt) (Co–Co)] and [di-μ_2-{ethoxycarbonyl (methylene)}-bis(tricarbonyl-cobalt) (Co–Co)]. *European Journal of Inorganic Chemistry*, 1875–1880.

78 Bartsch, H. and Erker, T. (1989) Synthese und Reaktivität des 1,5-Benzoxazepin-2,4 (3*H*,5*H*)-dions. *Annalen Der Chemie-Justus Liebig*, 177–179.

79 Tuba, R., Fördős, E. and Ungváry, F. (2005) Preparation of triphenylphosphane substituted ethoxycarbonylcarbene-bridged dicobalt carbonyl complexes and their application as catalyst precursors in the carbonylation of ethyl diazoacetate to diethyl malonate. *Journal of Molecular Catalysis A: Chemical*, **236**, 113–118.

80 Ungvári, N., Kégl, T. and Ungváry, F. (2004) Octacarbonyl dicobalt-catalyzed selective carbonylation of (trimethylsilyl) diazomethane to obtain (trimethylsilyl) ketene. *Journal of Molecular Catalysis A: Chemical*, **219**, 7–11.

81 Meyer, K.K. and Herrmann, W.A. (1979) Übergangsmetall-Methylen-Komplexe VIII. Massenspectrometrische Untersuchungen an μ-Methylen-Komplexen von Cobalt, Rhodium, Mangan, und Eisen. *Journal of Organometallic Chemistry*, **182**, 361–374.

82 Schwab, P., Mahr, N., Wolf, J. and Werner, H. (1993) Carbenrhodium-Komplexe mit Diaryl- und Alkyl(aryl)carbenen als Liganden: Das fehlende Glied in der Reihe der Doppebindungssysteme *trans*-[RhCl {=C(=C)$_n$RR'}(L)$_2$] $n = 0$, 1 und 2. *Angewandte Chemie-International Edition*, **105**, 1498–1500; *Angewandte Chemie-International Edition in English*, 1993, **32**, 1480–1482.

83 Werner, H., Schwab, P., Bleuel, E., Mahr, N., Windmüller, B. and Wolf, J. (2000) Carbenerhodium complexes of the half-sandwich-type: synthesis, substitution, and addition reactions. *Chemistry – A European Journal*, **6**, 4461–4470.

84 Grotjahn, D.B., Bikzhanova, G.A., Collins, L. S.B., Concolino, T., Lam, K.C. and Rheingold, A.L. (2000) Controlled, reversible conversion of a ketene ligand to carbene and CO on a single metal center. *Journal of the American Chemical Society*, **122**, 5222–5223.

85 Urtel, H., Bikzhanova, G.A., Grotjahn, D.B. and Hofmann, P. (2001) Reversible carbon–carbon double bond cleavage of a ketene ligand at a single iridium(I) center: a theoretical study. *Organometallics*, **20**, 3938–3949.

86 Rüchard, C. and Schrauzer, G.N. (1960) Über die Carbonylierung von Carbenen und die katalytische Zersetzung von Diazoalkanen mit Nickelcarbonyl. *Chemische Berichte*, **93**, 1840–1848.

87 Miyashita, A., Shitara, H. and Nohira, H. (1985) Isolation of η^2-(C,C) ketene complexes of nickel from the reactions of nickelacyclobutane complexes with carbon monoxide. *Journal of the Chemical Society, Chemical Communications*, 850–851.

88 Miyashita, A. and Grubbs, R.H. (1981) Reactions of nickel–carbene complexes generated from nickelacycle complexes. *Tetrahedron Letters*, **22**, 1255–1256.

89 Miyashita, A., Shitara, H. and Nohira, H. (1985) Isolation of η^2-(C,C) ketene complexes of nickel from the reactions of nickelacyclobutane complexes with carbon monoxide. *Journal of the Chemical Society, Chemical Communications*, 850–851.

90 Miyashita, A., Shitara, H. and Nohira, H. (1985) Preparation and properties of platinum ketene complexes. Facile C–C bond cleavage of coordinated ketene. *Organometallics*, **4**, 1463–1464.

91 Moloy, K.G., Marks, T.J. and Day, V.W. (1983) Carbon monoxide activation by organoactinides. η^2-Acyl–CO coupling and the formation of metal-bound ketenes. *Journal of the American Chemical Society*, **105**, 5696–5698.

9
Carbonylation of Enolizable Ketones (Enol Triflates) and Iodoalkenes

Antonio Arcadi

9.1
Introduction

In recent years, there has been a growing interest in the chemistry of enol triflates and iodoalkenes **1**. These derivatives undergo, in an initial step, facile oxidative addition to Pd(0) to form alkenylpalladium intermediates **2** that have a Pd–C σ-bond. Then, complexes **2** can give a large variety of transformations accomplishing the construction of carbon–carbon as well as carbon–heteroatom bonds. The importance of the palladium-catalyzed processes of enol triflates and iodoalkenes as powerful and versatile tools available to synthetic organic chemists [1,2] has been documented excellently. Their popularity stems, in part, from the fact that they are easily available and are more reactive than the corresponding vinyl bromides, chlorides, and aryl halides/triflates [3]. Under carbonylation conditions, the organopalladium derivatives **2** are converted to α,β-unsaturated acylpalladium complexes **3**, which allows them to be employed in a wide range of applications by reaction with nucleophiles [4], organometallic reagents/related carbon nucleophiles [5], and different π-bond systems (Scheme 9.1).

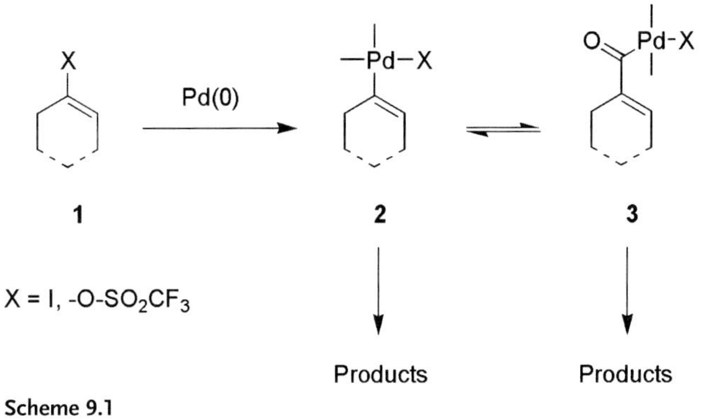

X

Pd(0)

–Pd–X

O Pd-X

1

2

3

X = I, -O-SO_2CF_3

Products

Products

Scheme 9.1

Modern Carbonylation Methods. Edited by László Kollár

ISBN: 978-3-527-31896-4

Synthetic applications of palladium-catalyzed carbonylation reactions of organic halides and pseudohalides have recently been reviewed [6]. We here focus only on selected applications with particular emphasis on the pioneering as well as some of the most recent examples of palladium-catalyzed carbonylation reactions of enol triflates and iodoalkenes.

9.2 Reactions of α,β-Unsaturated Acylpalladium Complexes with Nucleophiles

9.2.1 Introduction

A general mechanism for the reaction of the *in situ* generated α,β-unsaturated acylpalladium **3** complexes with nucleophiles is depicted in Scheme 9.2.

9.2.2 Alkoxy- and Aminocarbonylation of Enol Triflates and Iodoalkenes

Generally, successful syntheses of α,β-unsaturated esters and amides [7] can occur through the reaction of vinyl halides [8,9], triflates [10], carbon monoxide, and alcohols or primary and secondary amines in the presence of a palladium-based catalytic system. In this context, the catalytic systems could be distinguished as the ones using a preformed palladium(0) catalyst and those based on a Pd(0) catalyst formed *in situ* from a palladium(II) species [11–13]. The nucleophilic attack of the alcohol and the amine on the acylpalladium complex **3** furnishes the ester and amide product, respectively. Usually, a large excess of alcohol or amine was needed. The fact that the reaction rates are faster with amines than with alcohols means that the rate-determining step is nucleophilic attack of the alcohol/amine on the acylpalladium intermediate **3**. A base is generally added to neutralize the acid formed in the reaction. A tertiary amine and other bases such as KOAc, TlOAc, and K_2CO_3 have been used. When the reactant primary or secondary amine was strong enough as a base, the extra base was not required. DMF and CH_3CN were the most common solvents. The palladium-catalyzed carbonylative amidation with *cis* and *trans* vinylic halides is highly stereospecific, producing amides with retained configuration. However, Heck

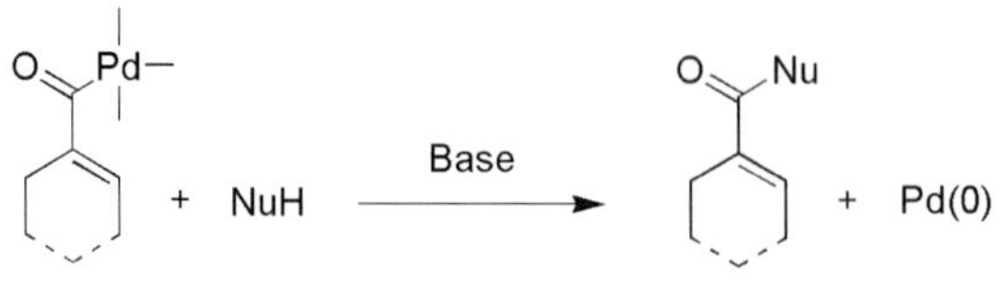

Scheme 9.2

and coworkers [8] found that the carboxylation of terminal vinylic iodides with alcohols can suffer from isomerization under the reaction conditions. For example, *cis*-$CH_3(CH_2)_3CH{=}CHI$ yields *cis*-$CH_3(CH_2)_3CH{=}CHCO_2Bu$ and *trans*-$CH_3(CH_2)_3CH{=}CHCO_2Bu$ in 79 and 6% yield, respectively. In contrast, (*E*)- and (*Z*)-1,2-difluoro-1-iodoalkenes and (*E*)- and (*Z*)-α,β-difluoro-β-iodostyrenes give corresponding esters in the presence of catalytic $Cl_2Pd(PPh_3)_2$, alcohol, trialkylamine, and carbon monoxide in excellent yields with complete retention of configuration. Fluorinated iodides require an elevated carbon monoxide pressure (80–180 psi) to react at reasonable rates [14]. Carbonylation of β-fluoroenol triflates [15] with carbon monoxide in methanol gave the corresponding methyl β-fluoro enoate. Homologation of trifluoroacetimidoyl iodides and the related perfluoro compounds by palladium catalyst under CO (1 atm) atmosphere in the presence of alcohols gives α-imino perfluoroalkanoates, which are transformed to α-imino perfluoroalkanoic acids [16]. The palladium-mediated carbonylation of vinyl triflates of *N*-substituted lactams [17] and subsequent catalytic asymmetric hydrogenation allowed the synthesis of (*S*)-pipecolic acid to be accomplished in high yield and high ee [18]. A comparison between the reactivity of the readily available enol triflates of tropinone **5** [19] and the 3-iodo-2-tropene **6** [20] shows that different reaction conditions are needed for the success of their palladium-catalyzed alkoxycarbonylation reaction (Scheme 9.3). An optimum yield of methyl trop-2-ene-3-carboxylate **7** starting from the enol triflate **5** (65% yield after workup and bulb-to-bulb distillation) could be achieved under a CO (g) atmosphere (1 atm) in methanol/DMF, with $Pd(OAc)_2$ (3 mol%) and 1,2′-bis(diphenylphosphino)ferrocene (dppf) at room temperature. The feature of the palladium ligand is prominent for the success of the alkoxycarbonylation reaction. One limitation of the reaction is the concentration of alcohol necessary for complete conversion (30-fold excess). A lower ratio of substrate/nucleophile = 25 accomplished higher yields of the 3-alcoxycarbonyl-3-tropene derivatives by heating the vinyl iodide **6** under carbon monoxide atmosphere at 60 °C in the presence of Et_3N, $Pd(OAc)_2$, and PPh_3.

Pd-catalyzed carbonylation of enol triflates can be successfully applied to the one-carbon elongation on the keto-carbonyl carbon in the total syntheses of various natural products and biologically active derivatives [21–26]. The functionalization of various steroidal skeletons by Pd-catalyzed carbonylations of vinyl iodides/triflates played a relevant role [27]. The strength of the palladium-catalyzed homogeneous carbonylation reaction can be shown by the fact that even the hindered positions 12 of hecogenin derivative **9** can be functionalized to give ester or amide derivative **10** in moderate-to-good yield without any side reaction of the further functionalities (Scheme 9.4) [28].

9.2.3
Double Carbonylation Reactions

Alkenyl bromides or iodides having phenyl substituent(s) on the vinyl group can be doubly carbonylated to the corresponding α-ketoamides **13** [29]. This latter derivative is formed as a consequence of the reductive elimination of the palladium complex **12**

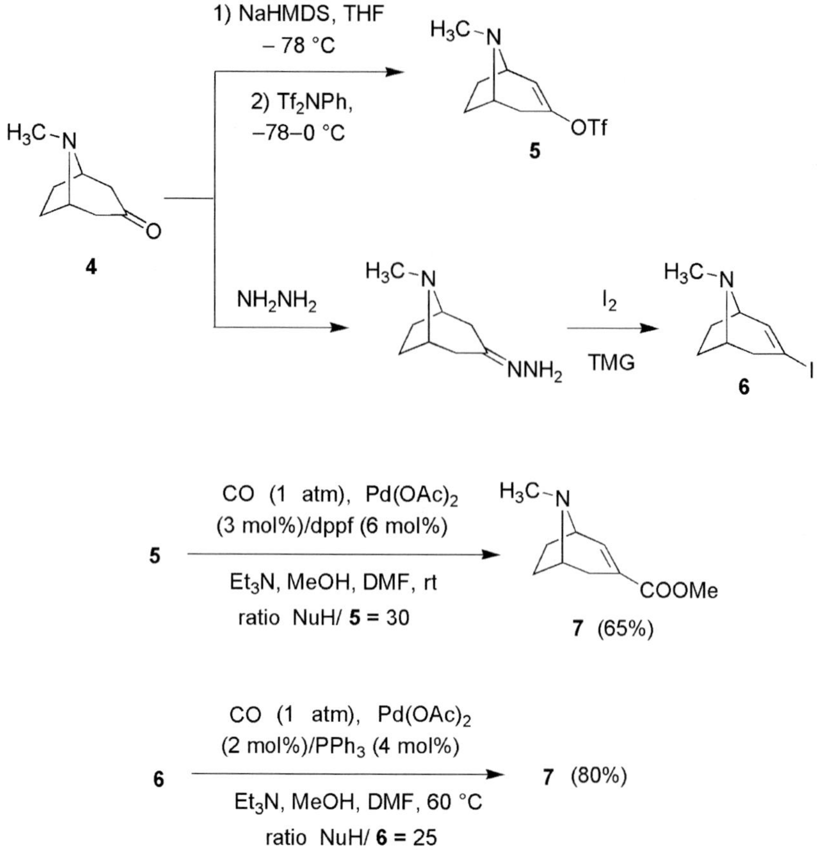

1) NaHMDS, THF
– 78 °C
2) Tf_2NPh,
–78–0 °C
H_3C–N
OTf
5
H_3C–N
O
4
NH_2NH_2
H_3C–N
NNH_2
I_2
TMG
H_3C–N
I
6

5
CO (1 atm), $Pd(OAc)_2$
(3 mol%)/dppf (6 mol%)
Et_3N, MeOH, DMF, rt
ratio NuH/ **5** = 30
H_3C–N
COOMe
7 (65%)

6
CO (1 atm), $Pd(OAc)_2$
(2 mol%)/PPh_3 (4 mol%)
Et_3N, MeOH, DMF, 60 °C
ratio NuH/ **6** = 25
7 (80%)

Scheme 9.3

derived from the nucleophilic attack of a CO ligand coordinated to the Pd(II) derivative **11** by amines (Scheme 9.5).

9.2.4
Ammonia Equivalent for the Palladium-Catalyzed Preparation of *N*-Unsubstituted α,β-Unsaturated Amides

A route to α,β-unsaturated primary amides **14** can be allowed through the palladium-catalyzed reaction of vinyl iodides/triflates with carbon monoxide in the presence of hexamethyldisilazane followed by hydrolytic workup (Scheme 9.6) [30]. To suppress the formation of by-products, the choice of the solvent played a pivotal role and satisfactory yields of the target products **14** were obtained with 1,3-dimethyl-3,4,5,6-tetrahydro-2(1*H*)-pyrimidone (DMPU) at 80–100 °C. The use of 4 mol of PPh_3 per mole of palladium was found necessary to prolong the lifetime of the catalyst.

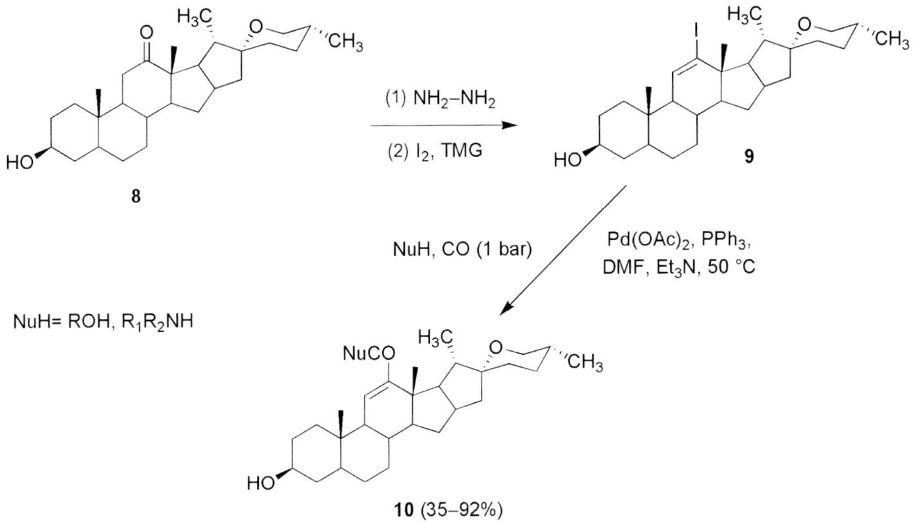

(1) NH_2–NH_2
(2) I_2, TMG
NuH, CO (1 bar)
Pd(OAc)$_2$, PPh$_3$, DMF, Et$_3$N, 50 °C
NuH= ROH, R_1R_2NH
8 **9** **10** (35–92%)

Scheme 9.4

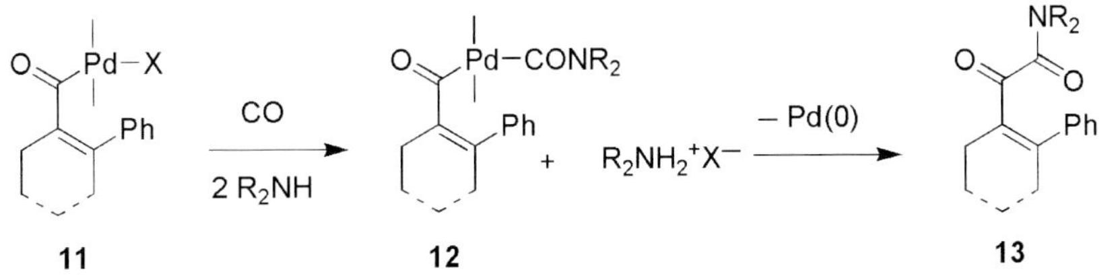

CO, 2 R_2NH; + $R_2NH_2^+X^-$; – Pd(0)
11 **12** **13**

Scheme 9.5

1 (1) Pd(0), CO, HN(SiMe$_3$)$_2$ (2) MeOH/H_3O^+
14 (40–94%)

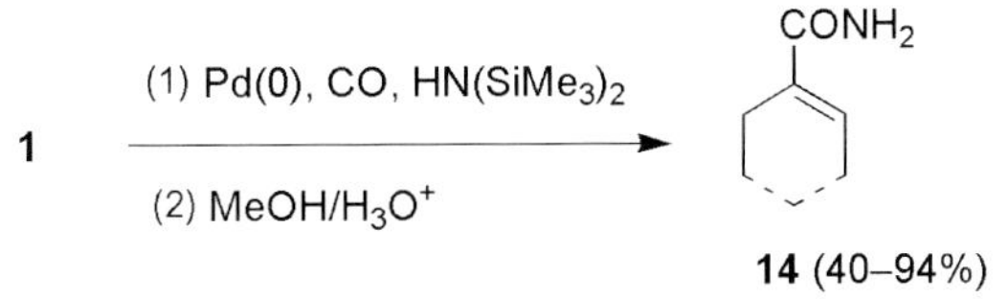

Scheme 9.6

9.2.5
Dipeptide Isosteres via Carbonylation of Enol Triflates

Interestingly, the reaction of the triflate **15** with various α-amino acid esters, generated *in situ* from their HCl salts **16**, allowed for rapid construction of a variety of peptidomimetic compounds **17** by using ~1 equiv. of the α-amino acid derivatives (Scheme 9.7) [31].

Amino acid methyl esters in a slight excess were also used as amine nucleophiles in palladium-catalyzed aminocarbonylation of iodoalkenes under conventional organic

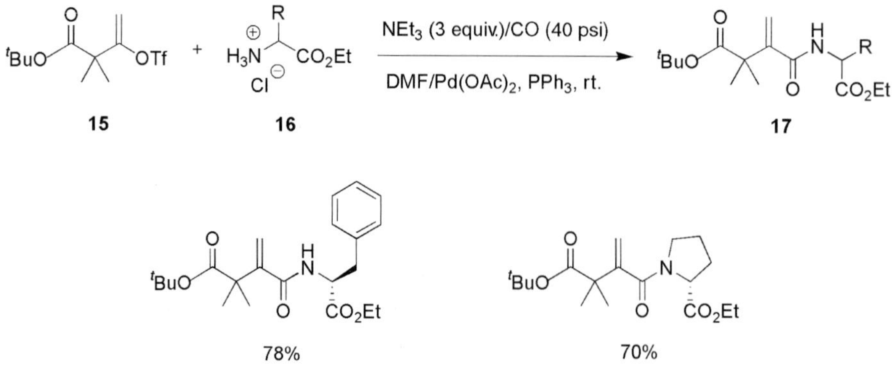

Scheme 9.7

solvents and in ionic liquid [32]. Recovery of the catalyst, particularly where high quantities are required, is a pressing problem, which can be approached by the use of ionic liquids. Consequently, homogeneous palladium-catalyzed amino- and hydroxycarbonylation of alkenyl halides in ionic liquids as solvents have been investigated [33,34]. The product can be smoothly separated from the solvent/catalyst mix, which can be successfully recycled. The conversion strongly depends on both the properties of the ionic liquid and those of the catalyst.

9.2.6
Carbonylation Reactions of Enol Triflates and Iodoalkenes with Bidentate Nucleophile

Several triflates, treated with CO in the presence of $Pd(PPh_3)_4$ and chiral aminoalcohols **18**, gave the corresponding hydroxy amides **19** easily converted into the oxazolines **20** in good yield (Scheme 9.8) [35].

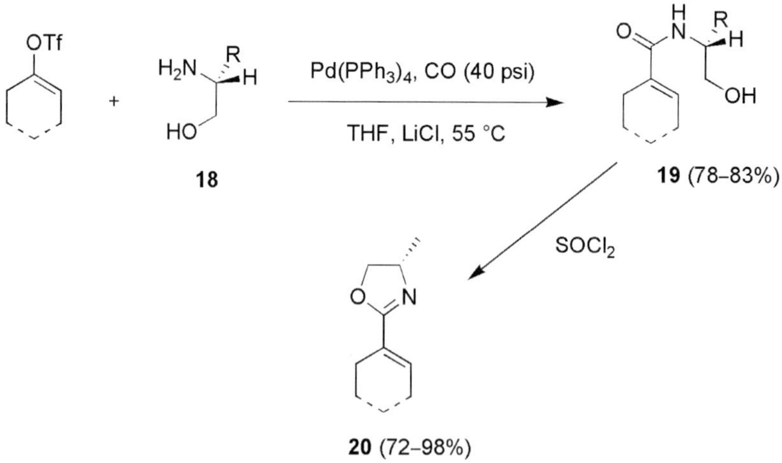

Scheme 9.8

Hydrazides **21–25** were synthesized under mild reaction conditions with high yields in carbonylation reactions with substituted hydrazines as reagents (Scheme 9.9) [36]. Homogeneous catalytic hydrazinocarbonylation of some steroid derivatives possessing iodoalkenyl moiety was carried out in the presence of a palladium catalyst, Et_3N, and mono- or disubstituted hydrazines as the nucleophilic reagent. The *N*-acetamido-/*N*-benzamido-carbamoyl steroidal derivatives **23**, obtained in high yields, served for the synthesis of steroidal 1,3,4-oxadiazole compounds **26** [37]. All the reactions were highly regioselective. The regioselectivity of palladium-catalyzed hydrazinocarbonylation seemed to be controlled by the substituent(s) of the hydrazine reagent, resulting in a substantial difference between the basicities of the two hydrogen atoms. For comparison, the hydrazinocarboxylation of some steroidal derivatives containing the triflate leaving group was also carried out.

The use of the hydroxamic acid derivatives as nucleophiles can lead to the *N*-**27** or *O*-acylated **28** derivatives: the site of acylation was determined by various factors (i.e., the structure of the reagent, the structure of the substrate, and the solvent) (Scheme 9.10) [38].

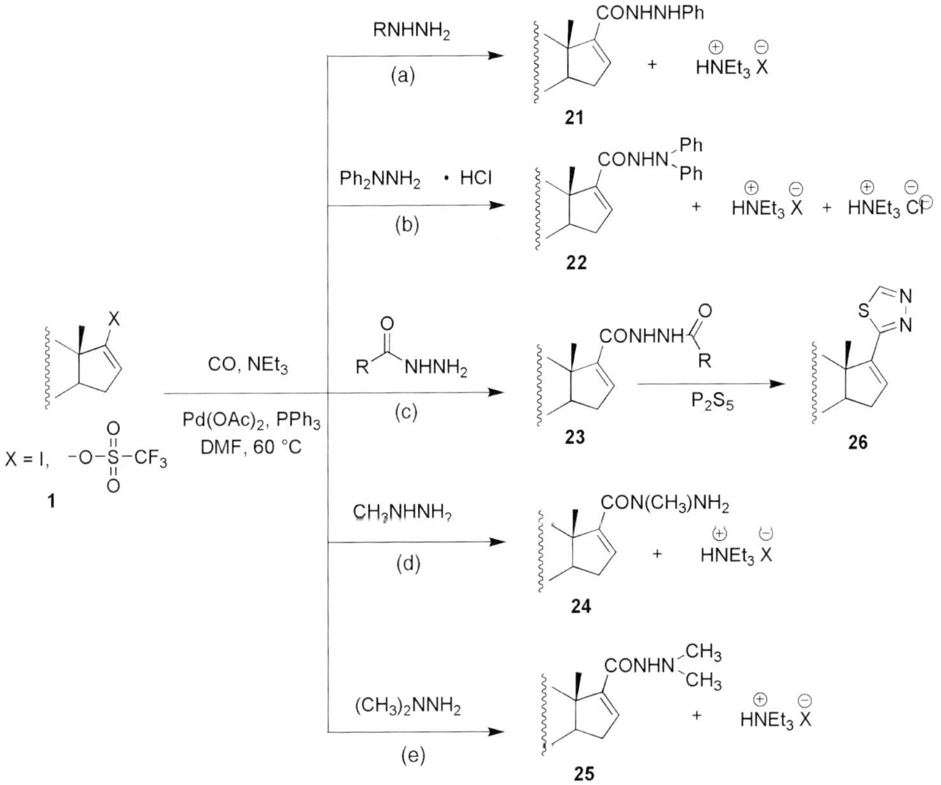

Scheme 9.9

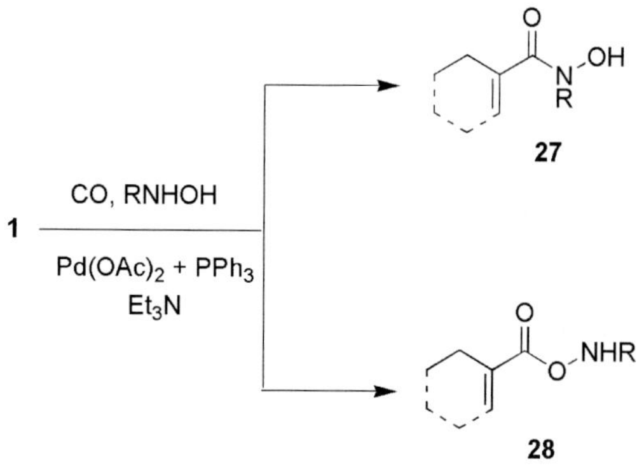

Scheme 9.10

9.2.7
Chemoselective Carbonylation Reactions of Enol Triflates and Iodoalkenes

Selective insertion of Pd(0) into vinyl iodides/triflates can take place in preference to the corresponding aryl derivatives. Indeed, the palladium-catalyzed carbonylation of a mixture of iodophenols **29** and vinyl triflates in anhydrous CH_3CN at room temperature under an atmosphere of CO, in the presence of a catalytic amount of $Pd(PPh_3)_4$ and potassium carbonate as base, afforded *o*-iodophenyl α,β-unsaturated esters **30** in high yield. These derivatives can undergo a following regioselective 5-*exo-trig* palladium-catalyzed cyclization to afford 3-spiro-fused benzofuran-2(3*H*)-ones **31** [39]. Moreover, the formation of 3-spiro-2(3*H*)-benzofuranones as well as of the 3-spiro-2-oxyndoles can be achieved in a one-pot process by the optimal planning of reaction conditions, which allow ter- and tetramolecular queuing sequences (Scheme 9.11) [40].

A different sequence afforded the biologically active 2-vinyl-4*H*-3,1-benzoxazin-4-ones **33** through the palladium-catalyzed reaction of *o*-iodoaniline **32** with unsaturated halides or triflates in the presence of K_2CO_3 and a catalytic amount of $Pd(PPh_3)_4$ under an atmosphere of carbon monoxide (Scheme 9.12) [41,42].

Chemoselective carbonylation of a vinyl iodide **34** with alcohol containing a vinyl bromide moiety **35** has been successfully employed for the solid-phase synthesis of a macrosphelide precursor **36** [43]. After the 4-methoxyphenylmethyl (MPM) group was removed, the palladium-catalyzed carbonylative macrolactonization of the vinyl bromide **37** achieved the synthesis of the macrosphelide-supported derivative **38** (Scheme 9.13). The combinatorial synthesis of a 122-member macrosphelide library has been performed by the three-component strategy based on the palladium-catalyzed chemoselective carbonylation/macrolactonization reaction.

9.2.8
Heterocyclization Reactions Through Intramolecular Carbonylative Lactonization and Lactamization

Intramolecular cyclization processes of suitable enol triflates and iodoalkenes bearing proximate oxygen or nitrogen nucleophiles **39** via palladium-catalyzed carbonylative

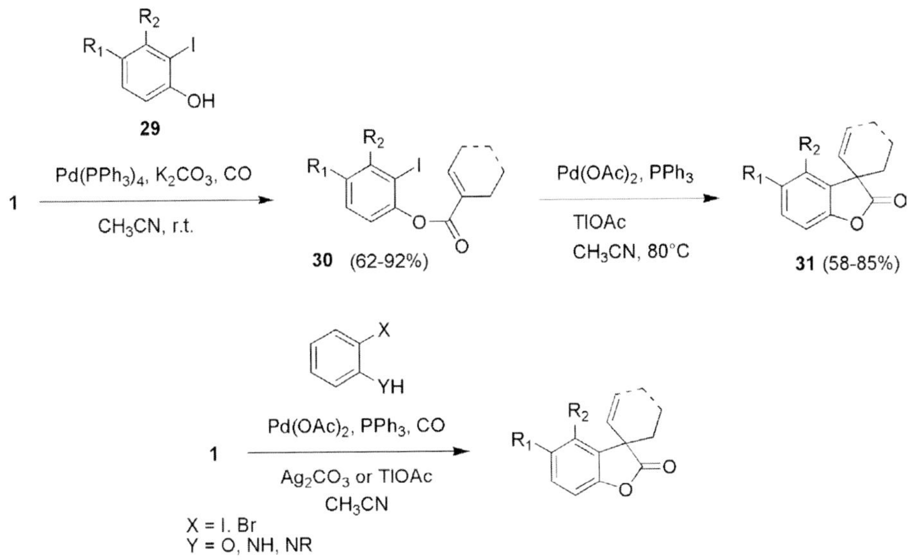

Scheme 9.11

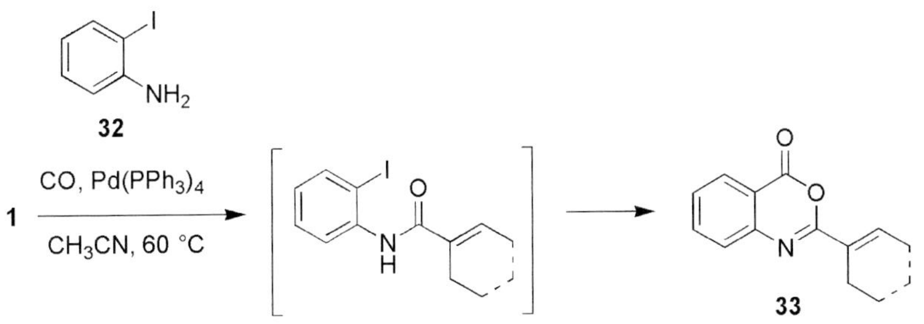

Scheme 9.12

lactonization and lactamization to give the corresponding derivative **40** have been extensively explored (Scheme 9.14) [44–50]. The most common ring sizes prepared by this method are four, five, six, and seven-membered *N*-heterocycles [51]. When five- and six-membered ring closures are in direct competition, the five-membered ring cyclization mode is slightly faster.

9.2.9
Carbon Monoxide Free Aminocarbonylation of Iodoalkenes

Many efforts have been devoted to the development of innovative ways for transferring a carbonyl group by using substitutes for carbon monoxide [52]. Thus, the palladium-catalyzed reaction of iodoalkene **41** with DMF in the presence of $POCl_3$ afforded the corresponding amide **42**. Very likely, the reaction proceeded via the

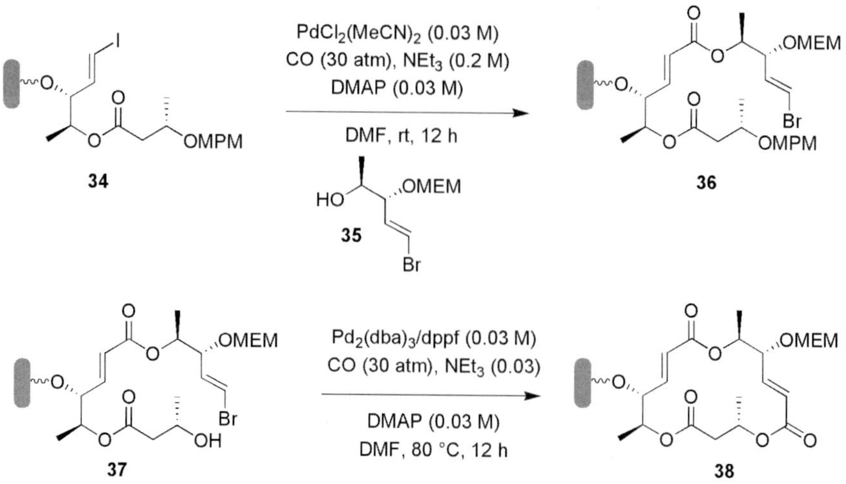

$PdCl_2(MeCN)_2$ (0.03 M), CO (30 atm), NEt_3 (0.2 M), DMAP (0.03 M); DMF, rt, 12 h

34, **35**, **36**

$Pd_2(dba)_3$/dppf (0.03 M), CO (30 atm), NEt_3 (0.03); DMAP (0.03 M), DMF, 80 °C, 12 h

37, **38**

Scheme 9.13

Pd(0), CO

39, **40**

X = I, $-O-SO_2-CF_3$

Y = O, NH, NR

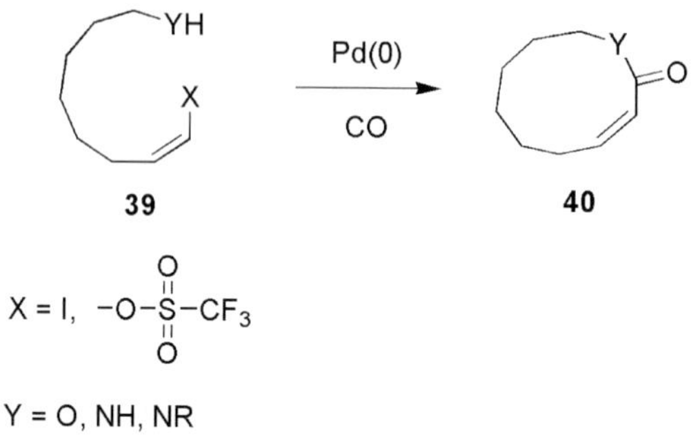

Scheme 9.14

Heck-type addition of the vinylpalladium complex to the Vilsmeier iminium species **43**, generated from DMF and $POCl_3$ (Scheme 9.15) [53].

Carbamoylsilane also led to the CO-free aminocarbonylation of alkenyl halides through catalysis by palladium complexes [54].

9.2.10 Hydroxycarbonylation of Enol Triflates and Iodoalkenes

Hydroxycarbonylation of vinyl triflates/halides has received considerable attention. The formation of carboxylic acids from palladium-catalyzed carbonylation of vinyl triflates is presumed to occur through the intermediacy of an unstable mixed anhydride. Hydroxycarbonylation of vinyl triflates has been reported to occur at room temperature in DMF under a CO balloon in the presence of $Pd(OAc)_2(PPh_3)_2$ and potassium acetate [55]. The crucial role of the acetate anion is supported when the starting triflate is recovered after it has been reacted under usual conditions omitting potassium acetate. Acrylic acids have been obtained by palladium-catalyzed

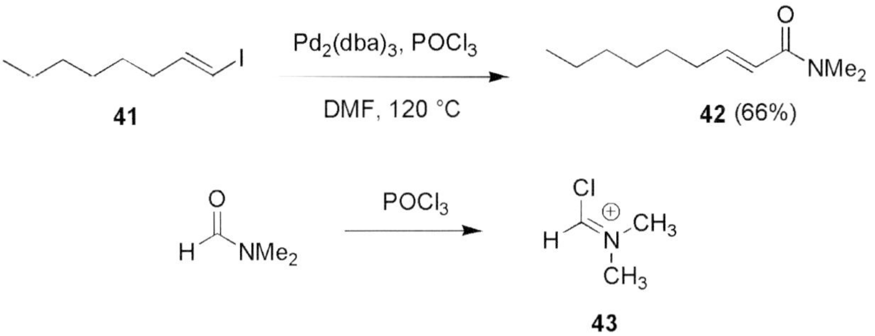

Scheme 9.15

carbonylation of enol triflates by using ammonium formate as nucleophile [56]. Considering that the introduction of an α,β-unsaturated carboxylic acid moiety by palladium-catalyzed carbonylation reaction is a key step in many synthetic protocols, particularly in the synthesis of biologically active compounds [57] and in the production of new substances for high-throughput screening in pharmaceutical industry [58], the development of techniques where carbon monoxide is gradually generated *in situ* is the target of current interest [59]. The palladium-catalyzed reaction of vinyl halides or triflates in the presence of acetic anhydride and lithium formate as a condensed source of carbon monoxide through the formation of the mixed anhydride **44** provides an efficient route to the corresponding carboxylic acids **45** (Scheme 9.16) [60]. The employment of the commercially available $H^{13}COONa$ allows for an easy way of introducing a labeled carbonyl into the products and shows that the formate anion can be used as an alternative source of the carbonyl group.

Interestingly, the procedure of hydroxycarbonylation by using lithium formate and acetic anhydride as internal condensed sources of carbon monoxide can be carried out in the presence of a recoverable and reusable phosphine-free palladium–carbon aerogel catalyst [61]. To support high-speed chemistry and automated organic synthesis, an operationally simple and environmentally safe hydroxycarbonylation of vinyl triflates can

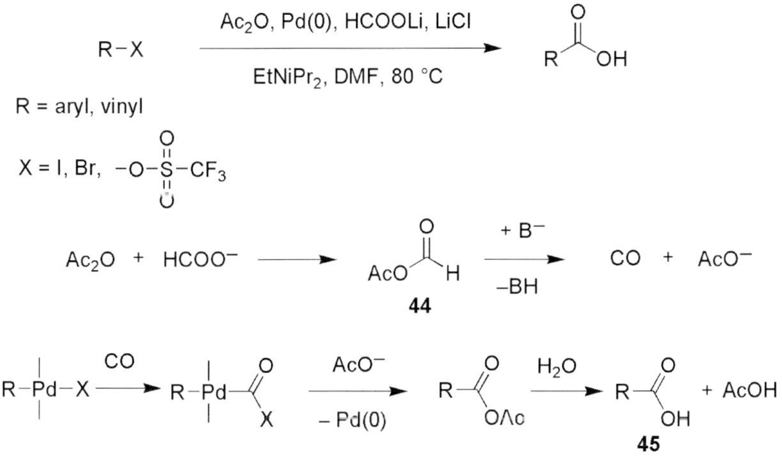

Scheme 9.16

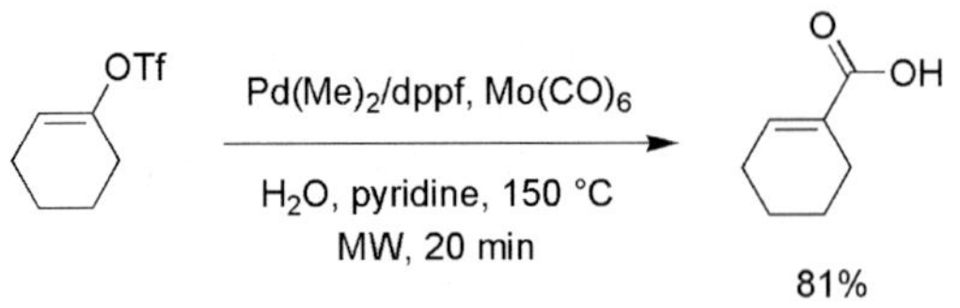

OTf
$Pd(Me)_2$/dppf, $Mo(CO)_6$
H_2O, pyridine, 150 °C
MW, 20 min
OH
81%

Scheme 9.17

be performed in water under microwave irradiation using $Pd(OAc)_2$ and $Mo(CO)_6$ [62] as commercially available, stable and solid carbon monoxide sources (Scheme 9.17) [63].

Moreover, the stereoselective palladium-catalyzed hydroxycarbonylation of vinyl halides to α,β-unsaturated carboxylic acids has been accomplished in ionic liquids as an alternative clean reaction media [64].

9.2.11
Palladium-Catalyzed Formylation of Enol Triflates and Iodoalkenes

Aldehydes are particularly desirable as functional group for the synthesis of fine and pharmaceutical chemicals. Therefore, palladium-catalyzed reductive carbonylation of halides/triflates has been extensively investigated [65–67]. Vinyl iodides were reported [68] to react under 80–100 bar pressure of 1 : 1 carbon monoxide–hydrogen in the presence of a basic tertiary amine and $PdX_2(PPh_3)_2$ at 100 °C to form α,β-unsaturated aldehydes **46** in good yields (Scheme 9.18a).

Formate salts [69] or formic acid under conditions converting them electrochemically into formate [70,71], silyl, and tin hydrides have been also explored to perform formylations of organic halides/triflates under lower pressure of carbon monoxide. Acetic formic anhydride has been used as a source of carbon monoxide for the conversion of aryl iodides in conjunction with R_3SiH [72]. The palladium-catalyzed formylation of a wide variety of vinyl iodides and triflates with tin hydride and carbon

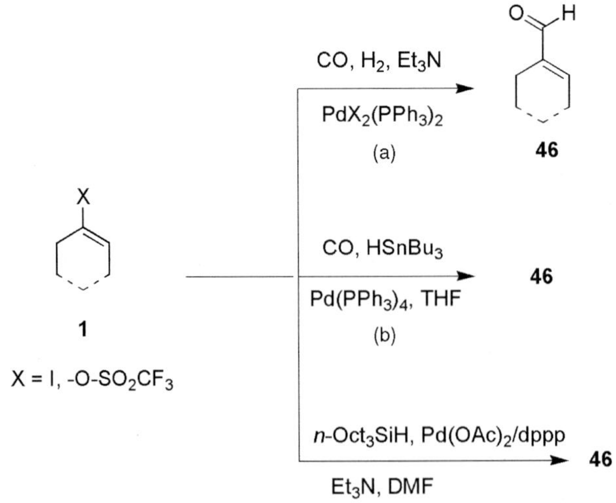

X
1
X = I, -O-SO_2CF_3

CO, H_2, Et_3N
$PdX_2(PPh_3)_2$
(a)
O H
46

CO, $HSnBu_3$
$Pd(PPh_3)_4$, THF
(b)
46

n-Oct_3SiH, $Pd(OAc)_2$/dppp
Et_3N, DMF
46

Scheme 9.18

monoxide gave good yields of α,β-unsaturated aldehydes under mild conditions (50 °C, 1–3 atm CO, and 2.5–3.5 h) and tolerated a number of functional groups (Scheme 9.18b) [73]. A competitive side reaction, the direct reduction of the halide or triflate, could be minimized by the slow addition of tributyltin hydride and higher pressure of carbon monoxide. The best conversions and selectivities can also be determined by making a suitable choice of the phosphine: the application of flexible chelating phosphines such as dppb and dppp proved to be superior to both rigid bidentate diphosphine and monodentate phosphine [74]. For the preparation of α,β-unsaturated aldehydes, vinyl triflates were preferred to vinyl iodides. Although the reaction required the addition of lithium chloride, vinyl triflates as well as vinyl iodides were formylated, yet triflates offered the advantages of greater stability, a regiospecific control during their preparation. Application of the formylation reaction with tributyl hydride has been extended to the incorporation of unsaturated aldehyde moiety into carbohydrates via an enol triflate [75] and to the preparation compounds of medicinal interest [76–78]. However, contamination with Sn is highly undesirable for pharmaceutical products, and other groups found R_3SiH to be equally effective. The conversion of enol triflates to the corresponding α,β-unsaturated aldehydes was optimized using n-Oct_3SiH and other silanes under 1 atm of CO in DMF at 70 °C with $Pd(OAc)_2$/1,3-bis(diphenylphosphino)-propane as catalyst (Scheme 9.18c) [79].

9.2.12
Trapping of α,β-Unsaturated Acylpalladium with Active C–H Compounds

Intramolecular [80] and intermolecular [81,82] trapping of acylpalladium intermediates with enolates has been studied. Intermolecular versions can involve trapping with both C- and O-enolates. The Pd-catalyzed carbonylation reactions of alkenyl iodides in the presence of various ketone enolate precursors displayed an interesting dichotomy: the expected O-enolate trapping product **47** may undergo cyclization to give six-membered lactone **48**, and the product's composition critically depends on the amount of a base and the structure of ketones (Scheme 9.19).

9.2.13
Sequential Carbopalladation/Carbonylation Reactions of Enol Triflates and Iodoalkenes

Acylpalladium complexes can be generated through an intramolecular carbopalladation/carbonylation sequence of vinyl iodides/triflates bearing a proximate alkene or

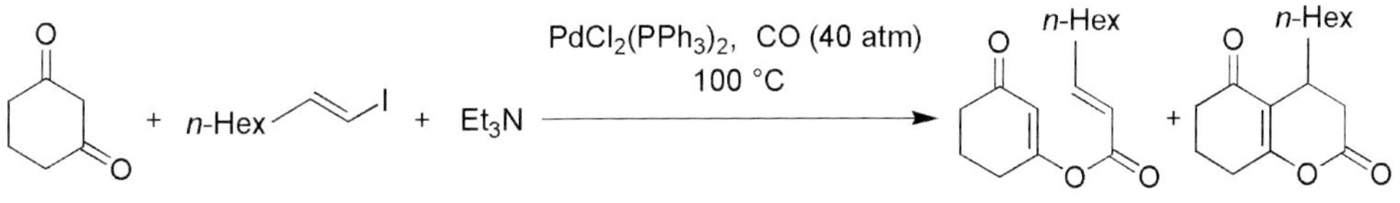

+ n-Hex + Et_3N → ($PdCl_2(PPh_3)_2$, CO (40 atm), 100 °C) **47** + **48**

1,3-cyclohexanedione		n-Hex vinyl iodide		Et_3N	**47**	**48**
1	:	1	:	1.2	85%	<5%
1	:	1	:	2	<5%	78%

Scheme 9.19

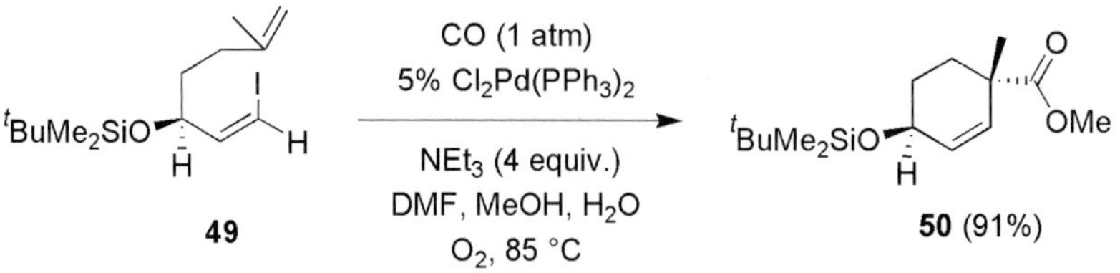

CO (1 atm)
5% $Cl_2Pd(PPh_3)_2$
NEt_3 (4 equiv.)
DMF, MeOH, H_2O
O_2, 85 °C

^{t}BuMe$_2$SiO ... 49 → ^{t}BuMe$_2$SiO ... OMe 50 (91%)

Scheme 9.20

alkyne moiety [83]. Indeed, Pd-catalyzed reaction of iododienes **49** under 1 atm of CO and a small amount of O_2 in the presence of NEt_3, as well as CH_3OH in DMF, gave the corresponding ester-containing cyclization products **50** in 91% yield through a highly diastereoselective cyclic carbopalladation–carbonylative esterification sequential process (Scheme 9.20) [84].

Termination of cyclic carbopalladation of alkynes via carbonylative lactonization/lactamization has been achieved with alkenyl iodides containing an ω oxydryl or amido/amino moiety [85].

9.3 Reactions of α,β-Unsaturated Acylpalladium Complexes with Organometals and Related Carbon Nucleophiles

9.3.1 Introduction

Carbonylative cross-coupling reactions of organometallic compounds with organic halides have been extensively studied and reported to provide excellent methods for the synthesis of unsymmetrical ketones [86]. The general catalytic cycle for this carbonylative coupling reaction is analogous to that of nucleophile except for the transmetalation step (Scheme 9.21).

9.3.2 Synthesis of Divinyl Ketones

Vinyl iodides/triflates have been reported to undergo carbonylative palladium-catalyzed coupling reaction with organostannanes, organoboron derivatives, and metal acetylide. The palladium-catalyzed three-component coupling reaction of vinyl iodides with trimethyl- or tributylvinylstannanes gives unsymmetrical divinyl

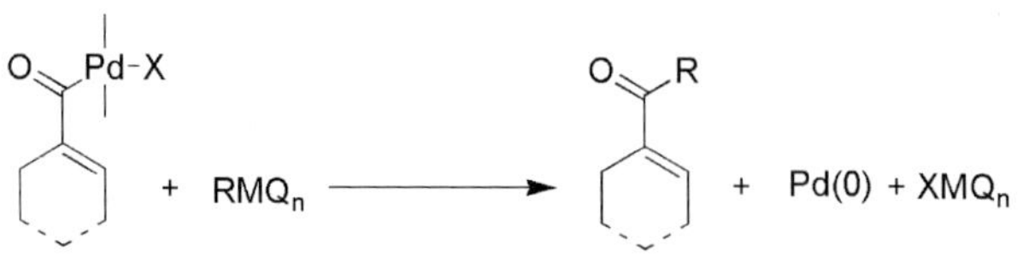

O Pd-X (3) + RMQ_n → O R + Pd(0) + XMQ_n

Scheme 9.21

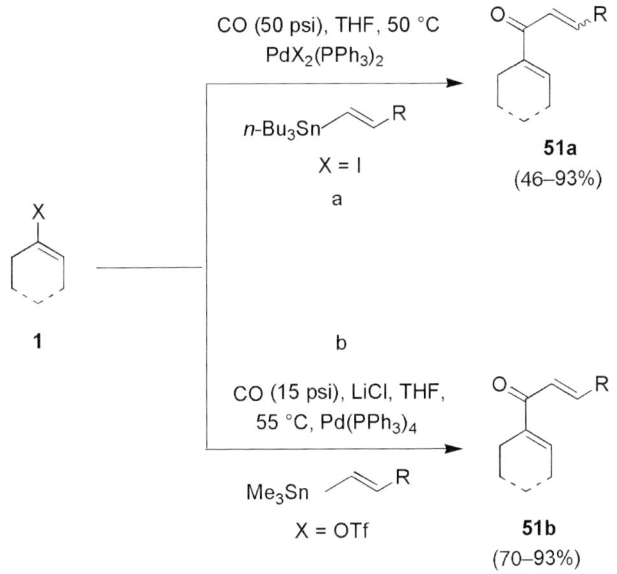

Scheme 9.22

ketones **51a** in good yields (Scheme 9.22a) [87]. The reaction is highly catalytic, requiring only 1–2 mol% of palladium(II) catalyst. The conditions are neutral and mild enough (40–50 °C, 3 atm CO) so that other functional groups in either coupling partner can be brought unaltered into the coupled product. Cycloalkenyl iodides can afford entry into functionalized bicyclo[*n*.3.0] ring systems by a sequence that concludes with a Nazarov cyclization. The vinyl iodides also entered into the carbonylative cross-coupling reaction with phenyl and acetylenic tin reagents, giving moderate yields of phenyl vinyl and acetylenic vinyl ketones, respectively. The formation of the product of carbonylative coupling depended on the rates of carbon monoxide insertion and transmetalation. The carbonylative cross-coupling reaction, when carried out under 50 psi of carbon monoxide, occurred without competition from a direct coupling of the vinyl iodide and vinyltin partners. The palladium-catalyzed carbonylative coupling of vinyl triflates with alkyl, vinyl, allyl, and arylstannanes gives good yields of the cross-coupled ketone products **51b** (Scheme 9.22b) [88]. The presence of an excess of LiCl is essential for carbonylation reaction to take place. Temperature and the features of the reaction medium are other important factors [89]. Regioselectively formed vinyl triflates can be used to produce divinyl ketones as regioisomerically pure compounds. In sharp contrast with vinyl iodides, the reaction with vinyl triflates may not accompany the isomerization of the products and the geometry of the vinylstannane is maintained during the reaction.

The methodology was applied to total synthesis of natural products [90]. The presence of a doubly unsaturated ketone in α-carboalkoxy-α,α′-dienone scaffold **54**, recognized within the skeleton of important cytotoxic natural products of the eleustide family, suggested the application of the palladium-catalyzed carbonylative coupling between the electron-poor (carboalkoxy-substituted) unsaturated triflate **52**

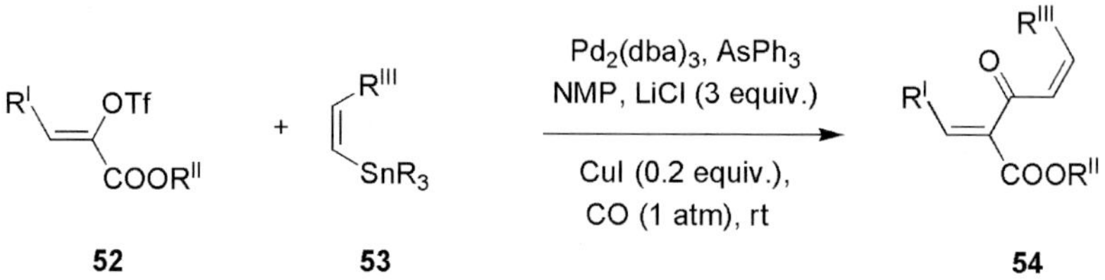

Scheme 9.23

and vinyl stannane **53** for their synthesis [91]. Acceleration of the carbon monoxide insertion over the transmetalation step has been obtained through the use of a soft ligand ($AsPh_3$) and of CuI, thus allowing the carbonylative coupling to take place at room temperature and atmospheric pressure of carbon monoxide even on electron-poor systems, without formation of the direct coupling products (Scheme 9.23).

Furthermore, the inclusion of a copper salt can significantly improve the efficiency of the carbonylative coupling between hindered alkenylstannane and enol triflate partners. Coupling of the bicyclic enol triflate **55** with vinylstannane **56** provided the corresponding dienone **57** in excellent yield (Scheme 9.24) [92].

The palladium-catalyzed three-component cross-coupling reaction between organoboron compounds, carbon monoxide, and electrophiles represents a straightforward route to the synthesis of unsymmetrical ketones [93]. Recently, new catalytic systems have opened up new opportunities [94]. However, there are only sporadic results for the applications of this protocol to the synthesis of alkyl- and aryl-1-alkenyl ketones [95]. Alkyl-1-alkenyl ketones have been synthesized by the reaction of 9-alkyl-9-BBN with 1-alkenyl iodides in the presence of $Pd(PPh_3)_4$ and K_3PO_4 as a base. Intramolecular reaction afforded cyclic ketones. The triethyl(1-methylindol-2-yl)borate, generated *in situ* from 2-lithio-1-methylindole and triethylborane, underwent palladium-catalyzed carbonylative cross-coupling reaction with vinyl iodides/triflates to form indol-2-yl ketones [96]. With both vinyl iodides and triflates, the carbonylation proceeded in good yields under carbon monoxide (15 atm) at 60 °C in THF. Steroidal phenyl ketones were synthesized in high yields by palladium-catalyzed carbonylation reactions of 17-iodo-androst-16-ene derivatives in the presence of $NaBPh_4$ in toluene [97]. The use of a catalytic amount of $Pd(PPh_3)_4$ together with an excess of Et_3N made it possible to use borate/steroid ratios lower than 1. Enol triflates gave lower yields in the same reaction. 1-Iodocyclohexene and phenylboronic acid were reacted under carbon monoxide (1 bar) atmosphere in the presence of *in situ* formed palladium(0) catalysts. The detailed analysis of the reaction mixture revealed that a great variety of side products may be formed (Scheme 9.25) [98].

The reaction showed a strong dependence on the solvent. By using $Pd(OAc)_2$–dppf as a catalytic system, the complete ketone selectivity was obtained with DMF. A further interesting feature noticed was that the increased carbon monoxide

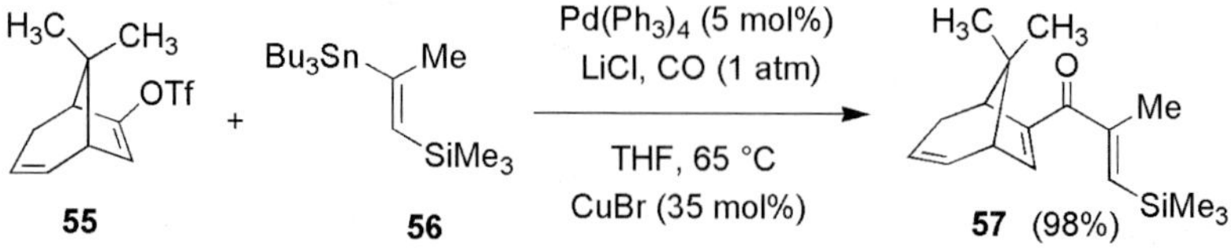

Scheme 9.24

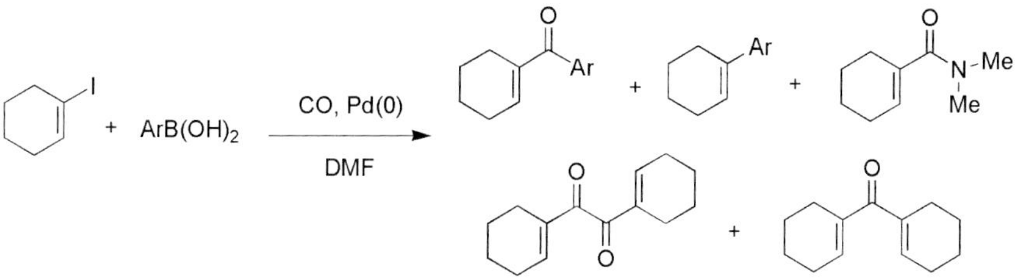

+ $ArB(OH)_2$ CO, Pd(0) DMF

Scheme 9.25

pressure had a negative influence on the selectivity toward the formation of the carbonylative cross-coupling derivative.

9.3.3 Synthesis of α,β-Alkynyl Ketones

The synthesis of α,β-alkynyl ketones attracts considerable interest because of their appearance in a wide variety of biologically active molecules and their key synthetic intermediates [99]. An easy route to prepare these derivatives should involve the transition metal catalyzed coupling of terminal alkynes or the metalated derivatives with organic halides in the presence of carbon monoxide [100]. Besides the carbonylative cross-coupling of alkynylstannanes with vinyl iodides/triflates [87,88] and of alkynylsilanes with aryl iodides [101], the selective carbonylative cross-coupling of terminal alkynes with alkenyl halides has been carried out at 120 °C and 20 atm in the presence of triethylamine and a catalytic amount of palladium(II) complex [102]. Conversely, the palladium-catalyzed cross-coupling reaction of vinyl triflates with 1-alkynes in the presence of carbon monoxide to give the α,β-alkynyl ketones **58** can take place under milder conditions (60 °C under a CO balloon) (Scheme 9.26) [99,103].

9.4 Reactions of α,β-Unsaturated Acylpalladium Complexes with π-Bond Systems

9.4.1 Introduction

The "living" nature of acylpalladium permits a series of interconversions that can also occur in cyclic manners, and these reactions are the subject of this section.

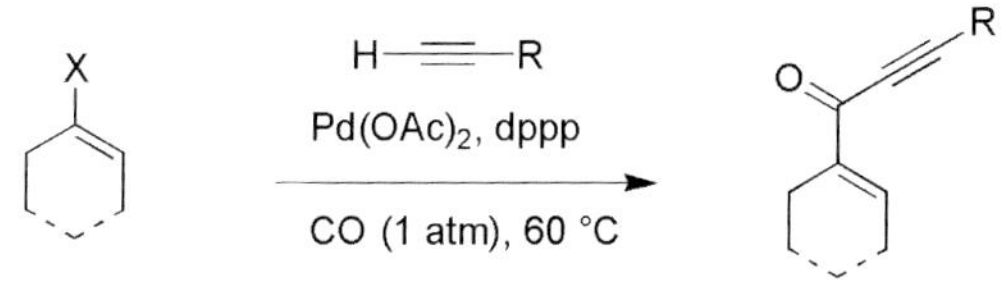

1 + H—≡—R, $Pd(OAc)_2$, dppp, CO (1 atm), 60 °C → **58** (53–83%)

X = $-O-SO_2CF_3$

Scheme 9.26

9.4.2
Intramolecular Acylpalladium Reactions with Alkenes, Alkynes, and Related Unsaturated Compounds

Palladium-catalyzed carbonylative reactions of vinyl iodides/triflates normally begin with migratory insertion of CO to produce the corresponding α,β-unsaturated acylpalladium derivatives. Their intramolecular cyclic acylpalladation reactions into carbon–carbon multiple bonds have been extensively explored and some interesting variants of propagation processes have been developed. "Living" carbopalladation has been linked together to devise cascading processes. Some such processes involving the formation of three or more carbon–carbon bonds have been adapted for the synthesis of complex organic molecules (Scheme 9.27) [104,105].

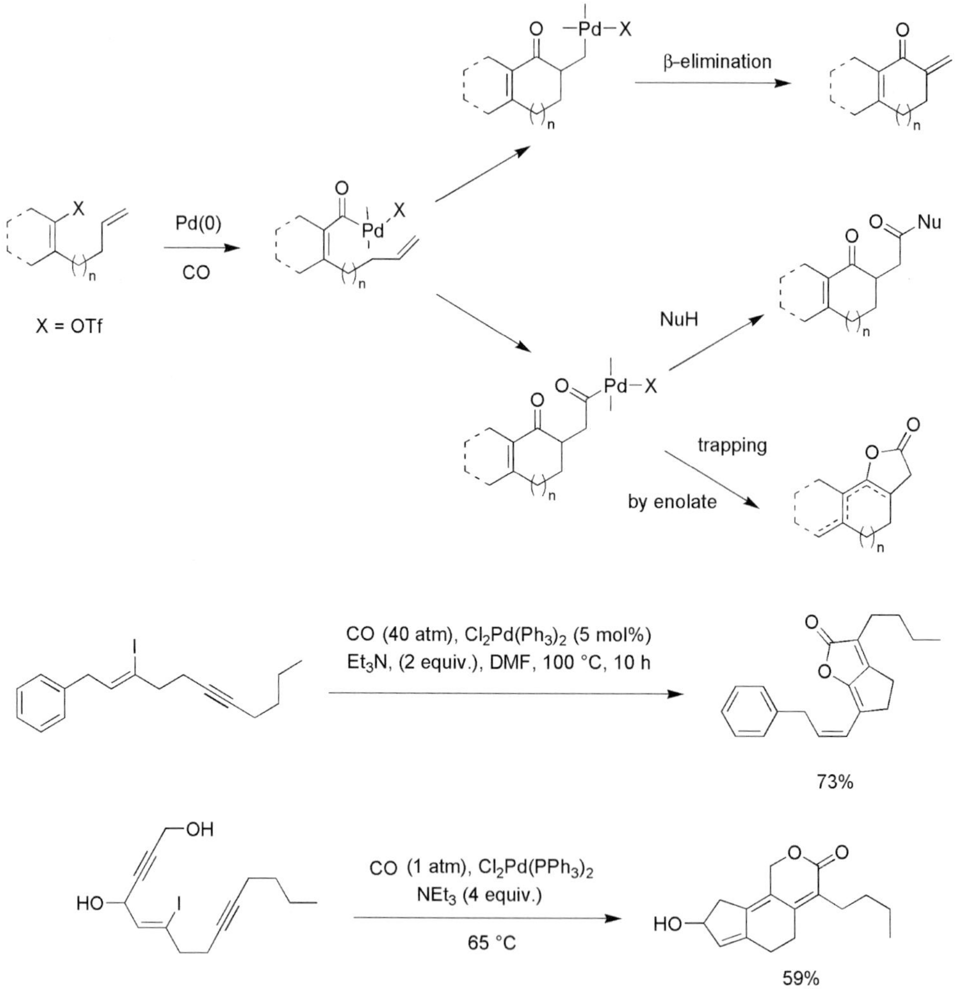

Pd(0)
CO
X = OTf
β-elimination
NuH
trapping
by enolate
CO (40 atm), $Cl_2Pd(Ph_3)_2$ (5 mol%)
Et_3N, (2 equiv.), DMF, 100 °C, 10 h
73%
CO (1 atm), $Cl_2Pd(PPh_3)_2$
NEt_3 (4 equiv.)
65 °C
59%

Scheme 9.27

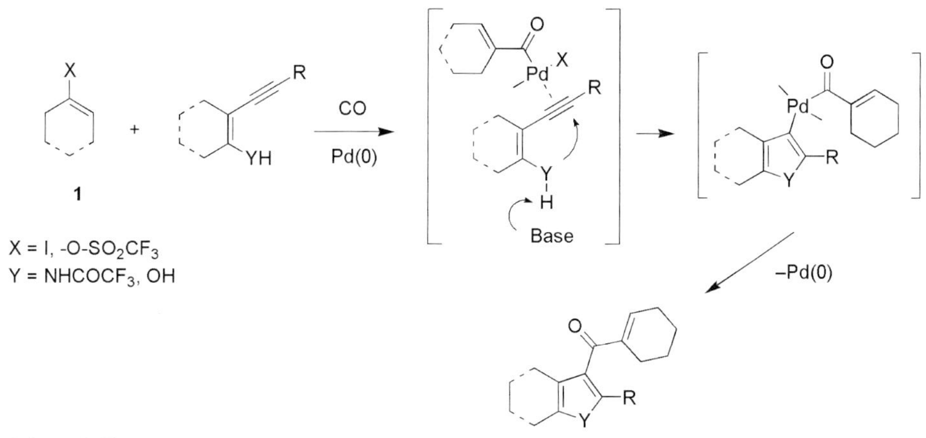

X = I, -O-SO_2CF_3
Y = $NHCOCF_3$, OH

Scheme 9.28

9.4.3
Intermolecular Acylpalladium Reactions with Alkynes Bearing Proximate Nucleophiles

The ability of the *in situ* generated acylpalladium complexes to activate carbon–carbon triple bonds toward intramolecular nucleophilic attack emerged as a valuable tool for the synthesis of functionalized heterocycle derivatives (Scheme 9.28). The reaction is considered to proceed through the intramolecular regioselective nucleophilic attack across the carbon–carbon triple bond activated by coordination to the acylpalladium complex, followed by a reductive elimination step that produces the functionalized cyclic derivative and regenerates the catalytic species.

2-Substituted-3-acylindoles have been prepared through the palladium-catalyzed carbonylative cyclization of 2-alkynyltrifluoroacetanylides with vinyl triflates [102,106]. Analogously, the palladium-catalyzed cyclocarbonylation of bis(*o*-trifluoroacetamidophenyl)acetylene **59** with vinyl halides and triflates afforded 12-acylindolo[1,2-*c*] quinazoline derivatives in high yields. The palladium-catalyzed reaction of **59** with the vinyl triflates **60** at 50 °C under 5 bar of carbon monoxide allowed the synthesis of the corresponding quinazoline derivative **61** in 98% yield (Scheme 9.29) [107].

The Pd-catalyzed reaction of 2-propynyl-1,3-dicarbonyls **62** with vinyl triflates in CH_3CN at 60 °C under 2 atm of CO in the presence of $Pd(PPh_3)_4$ and K_2CO_3 provided acyl furans **63** containing a vinyl fragment (Scheme 9.30) [108].

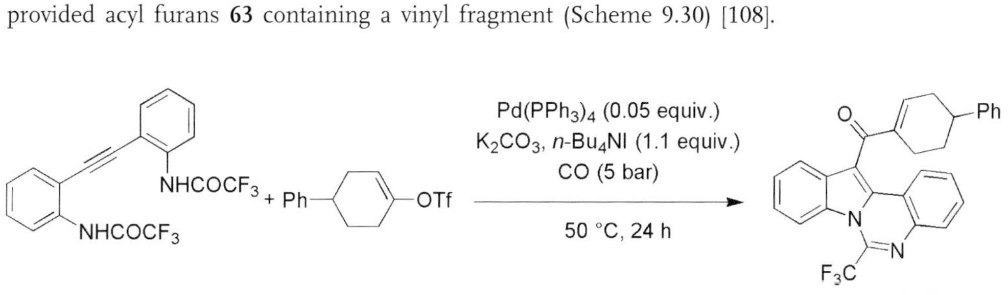

Scheme 9.29

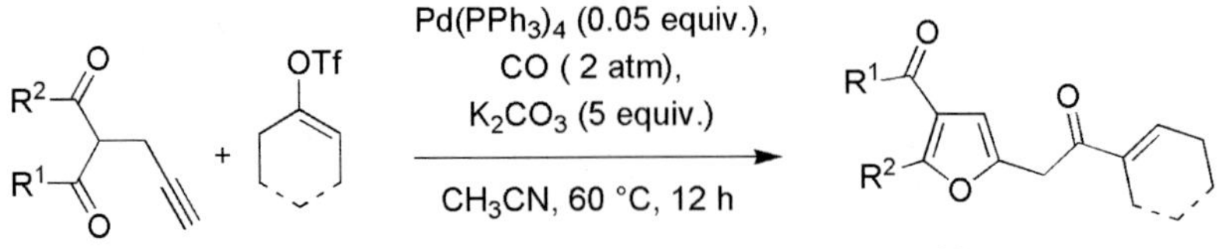

Pd(PPh$_3$)$_4$ (0.05 equiv.), CO (2 atm), K$_2$CO$_3$ (5 equiv.)

CH$_3$CN, 60 °C, 12 h

62 **63**

Scheme 9.30

A divergent behavior has been shown by *o*-arylethynyl-phenols [109] and *o*-ethynyl-phenols [110] under their Pd-catalyzed carbonylative reactions with vinyl triflates. For example, treatment of *o*-(phenylethynyl)phenol **64** and 3,3,5,5-cyclohex-1-en-1-yl triflate **65** under usual carbonylative conditions (Scheme 9.31) gives the 3-acylbenzo[*b*]furan **66** along with the ester **67**.

In contrast, *o*-ethynyl-phenols **68** undergo a palladium-catalyzed cyclocarbonylation to afford 3-alkylidene-2-coumaranone **69** (Scheme 9.32). The reaction is envisioned to involve the key intermediate **70**. The carbonylpalladium fragment of **70** adds intramolecularly to the triple bond. The addition proceeds with *syn* stereochemistry. The resulting σ-vinylpalladium intermediate **71** undergoes the reductive elimination of Pd (0) to give the 2-coumaranone derivative and the active catalyst. Thermal isomerization of the *Z* isomer can occur to some extent under the reaction conditions.

The same heteroannulative coupling involving the styrenyl iodide **74** has been suggested to accomplish the formation of 3-alkykidenebenzo[*b*]furanone **75** through a four-component, one-pot palladium-catalyzed approach (Scheme 9.33) [111].

9.5
Concluding Remarks

To summarize, palladium-catalyzed carbonylation reactions of enol triflates and iodoalkenes readily available from the corresponding ketones have gained great importance as powerful tools in organic synthesis. The number of applications for the synthesis of complex and pharmacologically interesting substrates is steadily increasing because of the breakthrough in the development of chemo- and regioselective processes as well as in the replacement of conventional multistep

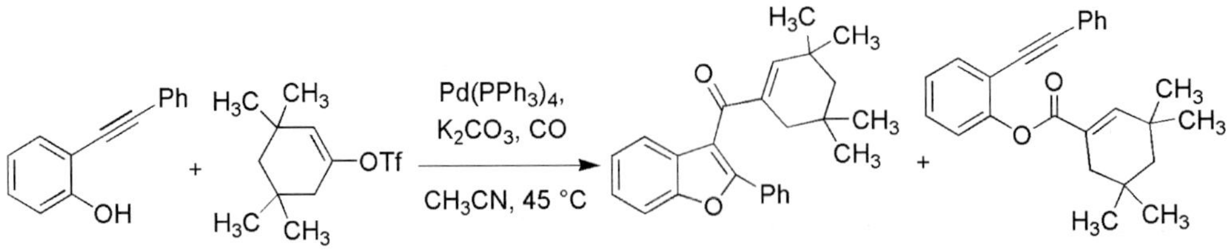

Pd(PPh$_3$)$_4$, K$_2$CO$_3$, CO

CH$_3$CN, 45 °C

64 **65** **66** (54%) **67** (21%)

Scheme 9.31

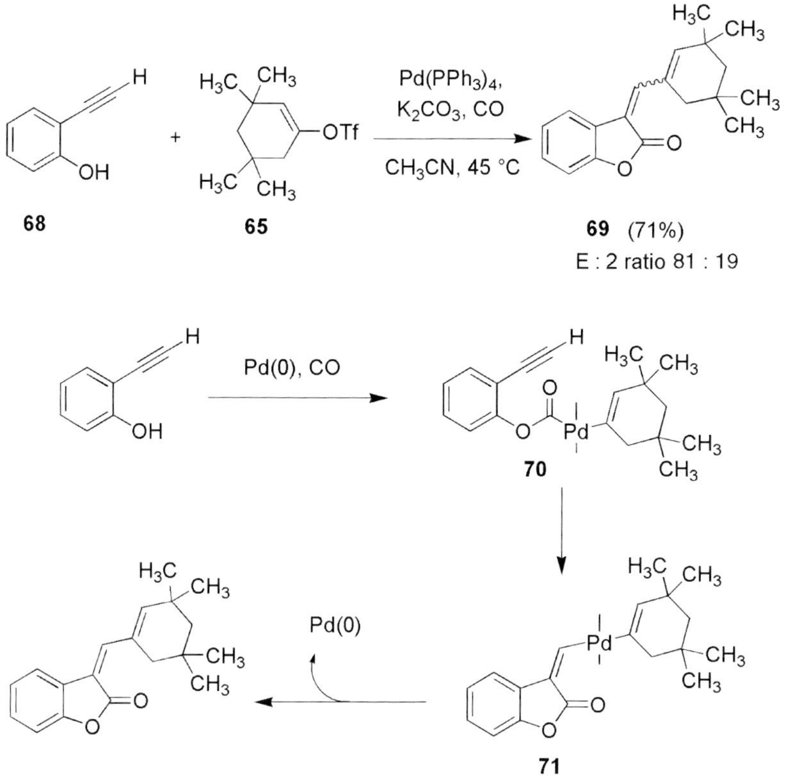

68 + 65 → (Pd(PPh$_3$)$_4$, K$_2$CO$_3$, CO; CH$_3$CN, 45 °C) → **69** (71%), E : 2 ratio 81 : 19

Pd(0), CO → **70** → **71** → Pd(0)

Scheme 9.32

synthesis of a target compound with selective one-pot catalytic process. The wide use of automated combinatorial chemistry has helped overcome the trouble of using gaseous carbon monoxide in carbonylation by applying suitable precursors.

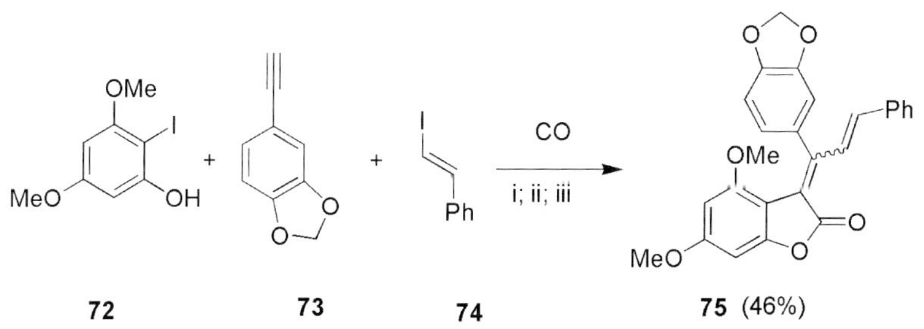

72 + **73** + **74** → (CO; i; ii; iii) → **75** (46%)

i= **72**, **73**, 2× MeMgCl$_2$ in THF, 0 °C; ii= Pd(PPh$_3$)$_2$Cl$_2$, 65 °C, 1–2 h;
iii= CO (g) was introduced upon
addition of **74** and DMSO and the reaction heated to 80–95 °C for 20 h.

Scheme 9.33

References

1 Tsuji, J. (2003) *Palladium Reagents and Catalysts: New Perspectives for the 21st Century*, John Wiley & Sons, Inc., New York.

2 Tsuji, J. (2002) *Transition Metal Reagents and Catalysts: Innovation in Organic Synthesis*, Wiley-VCH Verlag GmbH, Weinheim, Germany.

3 Ritter, K. (1993) Synthetic transformations of vinyl and aryl triflates. *Synthesis*, 735–762.

4 Mori, M. (2002) Palladium-catalyzed carbonylation of aryl and vinylic halides, in *Handbook of Organopalladium Chemistry for Organic Synthesis*, vol. II (ed. E. Negishi), Wiley–Interscience, New York. pp. 2313–2332.

5 Tamaru, Y. and Kimura, M. (2002) Reactions of acylpalladium derivatives with organometals and related carbon nucleophiles, in *Handbook of Organopalladium Chemistry for Organic Synthesis*, vol. II (ed. E. Negishi), Wiley–Interscience, New York. pp. 2425–2454.

6 Skoda-Földes, R. and Kollár, L. (2002) Synthetic applications of palladium catalysed carbonylation of organic halides. *Current Organic Chemistry*, **6**, 1097–1119.

7 Cook, G.R. (2005) Amides: synthesis from amines by carbonylation. *Science of Synthesis*, **21**, 114–130.

8 Schoenberg, A., Bartoletti, I. and Heck, R. F. (1974) Palladium-catalyzed carboxylation of aryl, benzyl, and vinylic halides. *Journal of Organic Chemistry*, **39**, 3318–3326.

9 Schoenberg, A.R. and Heck, F. (1974) Palladium-catalyzed amidation of aryl, heterocyclic, and vinylic halides. *Journal of Organic Chemistry*, **39**, 3327–3331.

10 Cacchi, S., Morera, E. and Ortar, G. (1985) Palladium-catalyzed carbonylation of enol triflates. A novel method for one-carbon homologation of ketones to α,β-unsaturated carboxylic acid derivatives. *Tetrahedron Letters*, **26**, 1109–1112.

11 Amatore, C. and Jutand, A. (2000) Anionic Pd(0) and Pd(II) intermediates in palladium-catalyzed Heck and cross-coupling reactions. *Accounts of Chemical Research*, **33**, 314–321.

12 Christmann, U. and Vilar, R. (2005) Monoligated palladium species as catalysts in cross-coupling reactions. *Angewandte Chemie-International Edition*, **44**, 366–374.

13 Dupont, J., Consorti, C.S. and Spencer, J. (2005) The potential of palladacycles: more than just precatalyst. *Chemical Reviews*, **105**, 2527–2571.

14 Wesolowski, C.A. and Burton, D.J. (1999) Palladium-catalysed stereospecific carboalkoxylation of 1,2-difluoro-1-iodoalkenes and α,β-difluoro-β-iodostyrenes. *Tetrahedron Letters*, **40**, 2243–2246.

15 Hossain, M.A. (1997) β-Fluorenol triflates: synthesis and some palladium catalysed reactions. *Tetrahedron Letters*, **38**, 49–52.

16 Watanabe, H., Hashizume, Y. and Uneyama, K. (1992) Homologation of trifluoroacetimidoyl iodides by palladium-catalyzed carbonylation. An approach to α-imino perfluoroalkanoic acids. *Tetrahedron Letters*, **33**, 4333–4336.

17 Luker, T., Hiemstra, H. and Speckamp, W. N. (1996) Synthesis and reactivity of pyrrolidinone- and piperidone-derived enol triflates. *Tetrahedron Letters*, **37**, 8257–8260.

18 Foti, C.J. and Comins, D.L. (1995) Synthesis and reactions of α-(trifluoromethanesulfonyloxy) enecarbamates prepared from *N*-acyllactams. *Journal of Organic Chemistry*, **60**, 2656–2657.

19 Cheng, J., Moore, Z., Stevens, E.D. and Trudell, M.L. (2002) Stereoselective synthesis of the three isomers of ethylene glycol bis(tropane-3-carboxylate). *Journal of Organic Chemistry*, **67**, 5433–5436.

20 Horvát, L., Berente, Z. and Kollár, L. (2005) High-yielding synthesis of 3-alkoxycarbonyl- and 3-carboxyamido-3-tropene derivatives in

homogeneous carbonylation reactions of 3-iodo-2-tropene. *Letters in Organic Chemistry*, **2**, 54–56.

21 Rizzo, C.J. and Smith, A.B., III (1991) Aphidicolin synthetic studies: a stereocontrol end game. *Journal of the Chemical Society, Perkin Transactions I*, 969–979.

22 Smith, A.B., III, Sulikowski, G.A., Sulikowski, M.M. and Fujimoto, K. (1992) Applications of an asymmetric [2+2]-photocycloaddition: total synthesis of (−)-echinosporin. Construction of an advanced 11-deoxyprostaglandin intermediate. *Journal of the American Chemical Society*, **114**, 2567–2576.

23 Crisp, G.T. and Meyer, A.G. (1995) Synthesis of optically active α-methylene γ-butyrolactones and (+)-mintlactone. *Tetrahedron*, **51**, 5831–5846.

24 Snider, B.B., Vo, N.H., O'Neil, S.V. and Foxman, B.M. (1996) Synthesis of (±)-allocyathin B_2 and (+)-erinacine A. *Journal of the American Chemical Society*, **118**, 7644–7645.

25 Konaklieva, M.I., Shi, H. and Turos, E. (1997) Palladium-promoted derivatizations of novel C-fused penem ring systems. *Tetrahedron Letters*, **38**, 8647–8650.

26 Sun, H., Yang, J., Amaral, K.E. and Horenstein, B.A. (2001) Synthesis of a new transition-state analogue of the sialyl donor. Inhibition of sialyltransferases. *Tetrahedron Letters*, **42**, 2451–2553.

27 Skoda-Földes, R. and Kollár, L. (2003) Transition-metal-catalyzed reactions in steroid synthesis. *Chemical Reviews*, **103**, 4095–4129.

28 Ács, P., Müller, E., Czira, G., Mahó, S., Perreira, M. and Kollár, L. (2006) Facile synthesis of 12-carboxamido-11-spirostenes via palladium-catalyzed carbonylation reactions. *Steroids*, **71**, 875–879.

29 Yamamoto, A., Ozawa, F. and Yamamoto, T. (1986) *Fundamental Research in Homogeneous Catalysis* (ed. A.E. Shilov), Gordon & Breach Science Publishers, London, pp. 181.

30 Morera, E. and Ortar, G. (1988) A palladium-catalyzed carbonylative route to primary amides. *Tetrahedron Letters*, **39**, 2835–2838.

31 Freskos, J.N., Ripin, D.H. and Reilly, M.L. (1993) Synthesis of succinate containing dipeptide isosteres via carbonylation of enol triflates. *Tetrahedron Letters*, **34**, 255–256.

32 Müller, E., Péczely, G., Skoda-Főldes, R., Takács, E., Kovotos, G., Bellis, E. and Kollár, L. (2005) Homogenous catalytic aminocarbonylation of iodoalkenes and iodobenzene with amino acid esters under conventional conditions and ionic liquids. *Tetrahedron*, **61**, 797–802.

33 Skoda-Főldes, R., Takács, E., Horvát, J., Tuba, Z. and Kollár, L. (2003) Palladium-catalysed aminocarbonylation of steroidal 17-iodo derivatives in *N,N'*-dialkyl-imidazolium-type ionic liquids. *Green Chemistry*, **5**, 643–645.

34 Zhao, X., Alper, H. and Yu, Z. (2006) Stereoselective hydroxycarbonylation of vinyl bromides to α,β-unsaturated carboxylic acids in the ionic liquid [BMIM] PF_6. *Journal of Organic Chemistry*, **71**, 3988–3990.

35 Meyers, A.I., Robichaud, A.J. and McKennon, M.J. (1992) The synthesis of chiral α,β-unsaturated and aryl oxazolines from ketones and arols via their triflates and Pd-catalyzed CO and amino alcohol coupling. *Tetrahedron Letters*, **33**, 1181–1184.

36 Skoda-Főldes, R., Szarka, Z., Kollár, L., Dinya, Z., Horváth, J. and Tuba, Z. (1999) Synthesis of *N*-substituted steroidal hydrazides in homogeneous catalytic hydrazinocarbonylation reaction. *Journal of Organic Chemistry*, **64**, 2134–2136.

37 Szarka, Z., Skoda-Főldes, R., Horváth, J., Tuba, Z. and Kollár, L. (2002) Synthesis of steroidal diacyl hydrazines and their 1,3,4-oxadiazole derivatives. *Steroids*, **67**, 581–586.

38 Szarka, Z., Skoda-Főldes, R., Kollár, L., Berente, Z., Horváth, J. and Tuba, Z. (2000)

Highly efficient synthesis of steroidal hydroxamic acid derivatives via homogeneous catalytic carbonylation reaction. *Tetrahedron*, **56**, 5253–5257.

39 Anacardio, R., Arcadi, A., D'Anniballe, G. and Marinelli, F. (1995) Palladium-catalyzed selective carbonylation of vinyl triflates in the presence of 2-iodophenols: a new route to 3-spiro-fused benzofuran-2(3*H*)-ones. *Synthesis*, 831–836.

40 Grigg, R., Putnikovic, B. and Urch, C.J. (1996) Palladium-catalysed ter- and tetra-molecular queuing processes. One pot routes to 3-spiro-2-oxindoles and 3-spiro-2(3*H*)-benzofuranones. *Tetrahedron Letters*, **37**, 695–698.

41 Cacchi, S., Fabrizi, G. and Marinelli, F. (1996) A novel one-pot palladium-catalysed synthesis of 2-substituted-4*H*-3,1-benzoxazin-4-ones. *Synlett*, 997–998.

42 Arcadi, A., Asti, C., Caselli, G., Marinelli, F. and Ruggirei, V. (1999) Synthesis and *in vitro* and *in vivo* evaluation of the 2-(6′-methoxy-3′,4′-dihydro-1′-naphthyl)-4*H*-3,1-benzoxazin-4-one as a new potent substrate inhibitor of human leucocyte elastase. *Biorganic & Medicinal Chemistry Letters*, **9**, 1291–1294.

43 Takahashi, T., Kusaka, S., Doi, T., Sunazuka, T. and Ōmura, S. (2003) A combinatorial synthesis of a macrosphelide library utilizing a palladium-catalyzed carbonylation on a polymer support. *Angewandte Chemie-International Edition*, **42**, 5230–5234.

44 Martin, L.D. and Stille, J.K. (1982) Palladium-catalyzed carbonylation of vinyl halides: a route to the synthesis of α-methylene lactones. *Journal of the American Chemical Society*, **47**, 3630–3633.

45 Zhang, C. and Lu, X. (1997) A convenient synthesis of 3-iodohomoallylic alcohols and the further transformation to α,β-unsaturated γ-lactones. *Tetrahedron Letters*, **38**, 4831–4834.

46 Crisp, G.T. and Meyer, A.G. (1992) Palladium-catalyzed, carbonylative intramolecular coupling of hydroxyl vinyl triflates. Synthesis of substituted α,β-butenolides. *Journal of Organic Chemistry*, **57**, 6972–6975.

47 Suzuki, T., Uozumi, Y. and Shibasaki, M. (1991) A Catalytic asymmetric synthesis of α-methylene lactones by the palladium-catalysed carbonylation of prochiral alkenyl halides. *Journal of the Chemical Society, Chemical Communications*, 1593–1595.

48 Copéret, C., Sugihara, T., Wu, G., Shimoyama, I. and Negishi, E. (1995) Acylpalladation of internal alkynes and palladium-catalyzed carbonylation of (*Z*)-β-iodoenones and related derivatives producing γ-lactones and γ-lactams. *Journal of the American Chemical Society*, **117**, 3422–3431.

49 Crisp, G.T. and Meyer, A.G. (1995) Synthesis of α,β-unsaturated lactams by palladium-catalysed intramolecular carbonylative coupling. *Tetrahedron*, **51**, 5585–5596.

50 Aoyagi, S., Hasegawa, S., Hirashima, S. and Kibayashi, C. (1998) Total synthesis of (+)-homopumiliotoxin 223G. *Tetrahedron Letters*, **39**, 2149–2152.

51 Mori, M., Chiba, K. and Nan, Y. (1978) Reactions and syntheses with organometallic compounds. 7. Synthesis of benzolactams by palladium-catalyzed amidation. *Journal of Organic Chemistry*, **43**, 1684–1687.

52 Morimoto, T. and Kakiuchi, K. (2004) Evolution of carbonylation catalysis: no need for carbon monoxide. *Angewandte Chemie-International Edition*, **43**, 5580–5588.

53 Hosoi, K., Nozaky, K. and Hiyama, T. (2002) Carbon monoxide free aminocarbonylation of aryl and alkenyl iodides using DMF as an amide source. *Organic Letters*, **4**, 2849–2851.

54 Cunico, R.F. and Maity, B.C. (2003) Direct carbamoylation of alkenyl halides. *Organic Letters*, **5**, 4947–4949.

55 Cacchi, S. and Lupi, A. (1992) Palladium-catalysed hydroxycarbonylation of vinyl and aryl triflates: synthesis of α,β-unsaturated and aromatic carboxylic acids. *Tetrahedron Letters*, **25**, 3939–3942.

56 Freskos, J.N., Laneman, S.A., Reilly, M.L. and Ripin, D.H. (1994) Synthesis of chiral succinates via Pd(0) catalyzed carbonylation/asymmetric hydrogenation sequence. *Tetrahedron Letters*, **27**, 835–838.

57 Murai, A., Tanimoto, N., Sakamoto, N. and Masamune, T. (1988) Total synthesis of glycinoeclepin A. *Journal of the American Chemical Society*, **110**, 1985–1986.

58 Blaszczak, L.C., Brown, R.F., Cook, G.K., Hornback, W.J., Hoying, R.C., Indelicato, J. M., Jordan, C.L., Katner, A.S., Kinnick, M. D., McDonald, J.H., III, Morin, J.M., Jr, Munroe, J.E. and Pasini, C.E. (1990) Comparative reactivity of 1-carba-1-dethiacephalosporins with cephalosporins. *Journal of Medicinal Chemistry*, **33**, 1656–1662.

59 Berger, P., Bessmernykh, A., Caille, J.C. and Mignonac, S. (2006) Palladium-catalyzed hydroxycarbonylation of aryl and vinyl bromides by mixed acetic formic anhydride. *Synthesis*, 3106–3110.

60 Cacchi, S., Fabrizi, G. and Goggiamani, A. (2003) Palladium-catalyzed hydroxycarbonylation of aryl and vinyl halides or triflates by acetic anhydride and formate anions. *Organic Letters*, **5**, 4269–4272.

61 Cacchi, S., Cotet, C.L., Fabrizi, G., Forte, G., Goggiamani, A., Martínez, S., Molins, E., Moreno-Mañas, M., Petrucci, F., Roig, A. and Vallribera, A. (2007) Efficient hydroxycarbonylation of aryl iodides using recoverable and reusable carbon aerogels doped with palladium nanoparticles as catalyst. *Tetrahedron*, **63**, 2519–2523.

62 Lesma, G., Sacchetti, A. and Silvani, A. (2006) Palladium-catalyzed hydroxycarbonylation of aryl and vinyl triflates by *in situ* generated carbon monoxide under microwave irradiation. *Synthesis*, 594–596.

63 Kaiser, N.F.K., Hallberg, A. and Larhed, M. (2002) *In situ* generation of carbon monoxide from solid molybdenum hexacarbonyl. A convenient route to palladium-catalyzed carbonylation reactions. *Journal of Combinatorial Chemistry*, **4**, 109–111.

64 Zhao, X., Alper, H. and Yu, Z. (2006) Stereoselective hydroxycarbonylation of vinyl bromides to α,β-unsaturated carboxylic acids in the ionic liquid [BMIM] PF_6. *Journal of Organic Chemistry*, **71**, 3988–3990.

65 Larsen, R.D. and King, A.O. (2002) Synthesis of aldehydes via hydrogenolysis of acylpalladium derivatives, in *Handbook of Organopalladium Chemistry for Organic Synthesis*, vol. II (ed. E. Negishi), Wiley–Interscience, New York, pp. 2473–2504.

66 Ashfield, L. and Barnard, C.F.J. (2007) Reductive carbonylation – an efficient and practical catalytic route for the conversion of aryl halides to aldehydes. *Organic Process Research & Development*, **11**, 39–43.

67 Klaus, S., Neumann, H., Zapf, A., Strübing, D., Hübner, S., Almena, J., Riermeier, T., Groβ, P., Sarich, M., Krahnert, W.R., Rossen, K. and Beller, M. (2006) A general and efficient method for the formylation of aryl and heteroaryl bromides. *Angewandte Chemie-International Edition*, **45**, 154–158.

68 Schoenberg, A. and Heck, R.F. (1974) Palladium-catalyzed formylation of aryl, heterocyclic, and vinylic halides. *Journal of the American Chemical Society*, **96**, 7761–7764.

69 Okano, T., Harada, N. and Kiji, J. (1994) Formylation of aryl halides with carbon monoxide and sodium formate in the presence of palladium catalyst. *Bulletin of the Chemical Society of Japan*, **67**, 2329–2332.

70 Carelli, I., Chiarotto, I., Cacchi, S., Pace, P., Amatore, C., Jutand, A. and Meyer, G. (1999) Electrosynthesis of aromatic aldehydes by palladium-catalyzed carbonylation of aryl iodides in the presence of formic acid. *European Journal of Organic Chemistry*, 1471–1473.

71 Chiarotto, I., Carelli, I., Carnicelli, V., Marinelli, F. and Arcadi, A. (1996) Electrochemical behaviour of

$Pd^{II}(PPh_3)_2Cl_2$ in the presence of carbon monoxide and its use in the palladium-catalyzed electrochemical formylation of iodoanisole. *Electrochimica Acta*, **41**, 2503–2509.

72 Cacchi, S., Fabrizi, G. and Goggiamani, A. (2004) Palladium-catalyzed synthesis of aldehydes from aryl iodides and acetic formic anhydride. *Journal of Combinatorial Chemistry*, **6**, 692–694.

73 Baillargeon, V.P. and Stille, J.K. (1986) Palladium-catalyzed formylation of organic halides with carbon monoxide and tin hydride. *Journal of the American Chemical Society*, **108**, 452–461.

74 Petz, A., Horváth, J., Tuba, Z., Pintér, Z. and Kollár, L. (2002) Facile synthesis of 17-formyl steroids via palladium-catalyzed homogeneous carbonylation reaction. *Steroids*, **67**, 777–781.

75 Al-Abed, Y. and Voelter, W. (1997) An expeditious methodology for the incorporation of unsaturated systems into carbohydrates via enol triflate. *Tetrahedron Letters*, **38**, 7303–7306.

76 Johnson, C.R., Golebiowski, A., Schoffers, E., Sundram, H. and Braun, P.M. (1955) Chemoenzymatic synthesis of azasugars: D-*talo*- and D-*manno*-1-deoxynojirimycin. *Synlett*, 313–314.

77 Kaneko, S., Nakajima, N., Shikano, M., Katoh, T. and Terashima, S. (1998) Synthetic studies of huperzine A and its fluorinated analogues. 2. Synthesis and acetylcholinesterase inhibitory activity of novel fluorinated huperzine A analogues. *Tetrahedron*, **54**, 5485–5506.

78 Almstead, J.I., Demuth, T.P. and Ledoussal, B. (1998) An investigation into the total synthesis of clerocidin: stereoselective synthesis of clerodane intermediate. *Tetrahedron: Asymmetry*, **9**, 3179–3183.

79 Kotsuki, H., Datta, P.K. and Suenaga, H. (1996) An efficient procedure for palladium-catalyzed hydroformylation of aryl/enol triflates. *Synthesis*, 470–472.

80 Negishi, E. and Tour, J.M. (1986) Complete reversal of regiochemistry in cyclic acylpalladium. Novel synthesis of quinones. *Tetrahedron Letters*, **27**, 4869–4872.

81 Kobayashi, T. and Tanaka, M. (1986) Acylation of active methylene compounds via palladium complex-catalyzed carbonylative cross-coupling of organic halides. *Tetrahedron Letters*, **27**, 4745–4748.

82 Negishi, E., Liou, S.Y., Xu, C., Shimoyama, I. and Makabe, H. (1993) Intermolecular trapping of acylpalladium and related acylmetal derivatives with active C–H compounds. *Journal of Molecular Catalysis A: Chemical*, **143**, 279–286.

83 Mouriño, A., Torneiro, M., Vitale, C., Fernández, S., Pérez-Sestelo, J., Anné, S. and Gregorio, C. (1997) Efficient and versatile synthesis of A-ring precursors of 1α,25-dihydroxy-vitamin D_3 and analogues. Application to the synthesis of Lythgoe–Roche phosphine oxide. *Tetrahedron Letters*, **38**, 4713–4716.

84 Copéret, C. and Negishi, E. (1999) Palladium-catalyzed highly diastereoselective cyclic carbopalladation–carbonylative esterification tandem reaction of iododienes and iodoarylalkenes. *Organic Letters*, **1**, 165–167.

85 Copéret, C., Ma, S., Sugihara, T. and Negishi, E. (1996) Cyclic carbopalladation of alkynes terminated by carbonylative amidation. *Tetrahedron*, **52**, 11529–11544.

86 Diederich, F. and Stang, P. (1998) *Metal-Catalyzed Cross-Coupling Reactions*, Wiley-VCH Verlag GmbH, Weinheim, Germany.

87 Goure, W.F., Wright, M.E., Davis, P.D., Labadie, S.S. and Stille, J.K. (1984) Palladium-catalyzed cross-coupling of vinyl iodides with organostannanes: synthesis of unsymmetrical divinyl ketones. *Journal of the American Chemical Society*, **106**, 6417–6422.

88 Crisp, G.T., Scott, W.J. and Stille, J.K. (1984) Palladium-catalyzed carbonylative coupling of vinyl triflates with organostannanes. A total synthesis of $(\pm)\Delta^{9(12)}$-capnellene. *Journal of the American Chemical Society*, **106**, 7500–7506.

89 Skoda-Főldes, R., Kollár, L., Marinelli, F. and Arcadi, A. (1994) Direct and carbonylative vinylation of steroidal triflates in the presence of homogeneous palladium catalysts. *Steroids*, **59**, 691–695.

90 Nicolau, K.C., Bulger, P.G. and Sarlah, D. (2005) Palladium-catalyzed cross-coupling reactions in total synthesis. *Angewandte Chemie-International Edition*, **44**, 4442–4489.

91 Ceccarelli, S., Piarulli, U. and Gennari, C. (2000) Effect of ligands and additives on the palladium-promoted carbonylative coupling of vinyl stannanes and electron-poor enol triflates. *Journal of Organic Chemistry*, **65**, 6254–6256.

92 Mazzola, R.D., Glese, S., Jr, Benson, C.L. and West, F.G. (2004) Improved yields with added copper(I) salts in carbonylative Stille couplings of sterically hindered vinylstannanes. *Journal of Organic Chemistry*, **69**, 220–223.

93 Ishiyama, T., Kizaki, H., Hayashi, T., Suzuki, A. and Miyaura, N. (1998) Palladium-catalyzed carbonylative cross-coupling reaction of arylboronic acids with aryl electrophiles: synthesis of biaryl ketones. *Journal of Organic Chemistry*, **63**, 4726–4731.

94 Mingji, D., Liang, B., Wang, C., You, Z., Xiang, J., Dong, G., Chen, J. and Yang, Z. (2004) A novel thiourea ligand applied in the Pd-catalyzed Heck, Suzuki and Suzuki carbonylative reactions. *Advanced Synthesis & Catalysis*, **346**, 1669–1673.

95 Miyaura, N. and Suzuki, A. (1995) Palladium-catalyzed cross-coupling reactions of organoboron compounds. *Chemical Reviews*, **95**, 2457–2483.

96 Ishikura, M. and Tarashima, M. (1994) Palladium-catalyzed carbonylative cross-coupling reaction with triethyl(1-methylindol-2-yl)borate: a simple route to 1-methylindol-2-yl ketones. *Journal of Organic Chemistry*, **59**, 2634–2637.

97 Skoda-Főldes, R., Székvlgyi, Z., Kollár, L., Berente, Z., Horváth, J. and Tuba, Z. (2000) Facile synthesis of steroidal phenyl ketones via homogeneous catalytic carbonylation. *Tetrahedron*, **56**, 3415–3418.

98 Petz, A., Péczely, G., Pintér, Z. and Kollár, L. (2006) Carbonylative and direct Suzuki–Miyaura cross-coupling reactions with 1-iodo-cyclohexene. *Journal of Molecular Catalysis A: Chemical*, **255**, 97–102.

99 Abbiati, G., Arcadi, A., Marinelli, F., Rossi, E. and Verdecchia, M. (2006) Rh-catalyzed hydroarylation/hydrovinylation–heterocyclization of β-(2-aminophenyl)-α,β-ynones with organoboron derivatives: a new approach to functionalized quinolines. *Synlett*, 3218–3224.

100 Ahmed, M.S.M. and Mori, A. (2003) Carbonylative Sonogashira coupling of terminal alkynes with aqueous ammonia. *Organic Letters*, **5**, 3057–3060.

101 Arcadi, A., Cacchi, S., Marinelli, F., Pace, P. and Sanzi, G. (1995) The palladium-catalysed carbonylative coupling of 5-(trimethylsilylethynyl)-3′,5′-di-*O*-acetyl-2′-deoxyuridine and 1-alkynes with aryl iodides. *Synlett*, 823–824.

102 Kobayashi, T. and Tanaka, M. (1981) Carbonylation of organic halides in the presence of terminal acetylenes; novel acetylenic ketone synthesis. *Journal of the Chemical Society, Chemical Communications*, 333–334.

103 Ciattini, P.G., Morera, E. and Ortar, G. (1991) A new pathway to alkynyl ketones via palladium-catalyzed carbonylative coupling of vinyl triflates with 1-alkynes. *Tetrahedron Letters*, **32**, 6449–6452.

104 Negishi, E., Ma, S., Amanfu, J., Copéret, C., Miller, J.A. and Tour, J.M. (1996) Palladium-catalyzed cyclization of 1-iodo-substituted 1,4-, 1,5-, and 1,6-dienes as well as of 5-iodo-1,5-dienes in the presence of carbon monoxide. *Journal of the American Chemical Society*, **118**, 5919–5931.

105 Negishi, E., Wang, G. and Zhu, G. (2006) Palladium-catalyzed cyclization via carbopalladation and acylpalladation. *Topics in Organometallic Chemistry*, **19**, 1–48.

106 Arcadi, A., Cacchi, S., Carnicelli, V. and Marinelli, F. (1994) 2-Substituted-3-acyindoles through the palladium-catalysed carbonylative cyclization of 2-alkynyltrifluoroacetanylides with aryl halides and vinyl triflates. *Tetrahedron*, **50**, 437–452.

107 Battistuzi, G., Cacchi, S., Fabrizi, G., Marinelli, F. and Parisi, L.M. (2002) 12-Acylindolo[1,2-*c*]quinazolines by palladium-catalyzed cyclocarbonylation of *o*-alkynyltrifluoroacetanilides. *Organic Letters*, **4**, 1355–1358.

108 Arcadi, A., Cacchi, S., Fabrizi, G., Marinelli, F. and Parisi, L.M. (2003) Highly substituted furans from 2-propynyl-1,3-dicarbonyls and organic halides or triflates via oxypalladation-reductive elimination domino reaction. *Tetrahedron*, **59**, 4661–4671.

109 Arcadi, A., Cacchi, S., Del Rosario, M., Fabrizi, G. and Marinelli, F. (1996) Palladium-catalyzed reaction of *o*-ethynylphenols, *o*-((trimethylsilyl) ethynyl)phenyl acetates and o-alkynylphenols with unsaturated triflates or halides: a route to 2-substituted-2,3-disubstituted- and 2-substituted-3-acylbenzo[*b*]furans. *Journal of Organic Chemistry*, **61**, 9280–9288.

110 Arcadi, A., Cacchi, S., Fabrizi, G. and Moro, L. (1999) Palladium-catalyzed cyclocarbonylation of *o*-ethynylphenols and vinyl triflates to form 3-alkylidene-2-coumaranones. *European Journal of Organic Chemistry*, 1137–1141.

111 Chaplin, J.H. and Flynn, B. (2001) A multi-component coupling approach to benzo[*b*] furans and indoles. *Journal of the Chemical Society, Chemical Communications*, 1594–1595.

10
Recent Developments in Alkyne Carbonylation

Simon Doherty, Julian G. Knight, Catherine H. Smyth

10.1
Introduction

Since its discovery in the late 1930s, the carbonylation of alkynes in the presence of a protic nucleophile has evolved into a powerful and highly versatile transformation for the synthesis of a range of α,β-unsaturated carboxylic acids and their derivatives, depending on the choice of nucleophilic partner. While the original Reppe carbonylation was based on using a nickel catalyst and has been used for the commercial-scale production of acryclic acids and esters [1], more recent developments have extended this methodology to include higher alkynes for the synthesis of other important monomers and building blocks such as methyl methacrylate (MMA) and acrylamides [2]. Although complexes of many of the late transition metals catalyze the carbonylation of alkynes, the most widely used are based on palladium, most likely because they are highly active and generally afford excellent levels of regio- and stereocontrol under mild conditions. Intramolecular versions of this transformation with acetylenic alcohols and ketones have been used to prepare a range of carbonylated heterocycles such as five- and six-membered unsaturated lactones, a subunit that appears in a number of biologically important natural and nonnatural compounds as well as α-methylene butyrolactones [3]. In addition, under oxidative conditions, carbonylation of terminal alkynes affords 2-alkynoates and 2-ynamides via monocarbonylation, maleic and fumaric acid derivatives via dicarbonylation, and products of tricarbonylation depending on the reaction conditions [4]. Moreover, intramolecular versions of oxidative carbonylation with acetylenic alcohols and amines complement their non-oxidative counterparts and afford a range of functionalized and substituted heterocycles such as β-lactones and lactams, substituted 2(5*H*)-furanones and their pyrrole counterparts, as well as a host of other carbonylated and noncarbonylated heterocycles, details of which have been the subject of a number of thorough and comprehensive reviews [5]. Inter- and intramolecular palladium-catalyzed carbonylative cyclization and annulation of alkynes have proven to be extremely useful strategies for the synthesis of a range of carbonylated heterocycles, and several recent review articles

Modern Carbonylation Methods. Edited by László Kollár

ISBN: 978-3-527-31896-4

cover the relevant literature [6]. More recent developments involving alkyne carbonylation include the use of multicomponent and tandem carbonylation reactions as an elegant means for the construction of complex molecular architectures from readily available building blocks. The key developments that define the reaction chemistry of alkynes with carbon monoxide and a selection of some recent noteworthy advances that illustrate the scope of alkyne carbonylation in synthesis will be presented.

10.2 Hydrochalcogenocarbonylation and Dichalcogenocarbonylations

10.2.1 Terminal Alkynes

The carbonylation of alkynes in the presence of oxygen-based nucleophiles has been thoroughly investigated during the past two decades and provides convenient access to a wide range of acyclic and cyclic α,β-unsaturated oxygenates [7]. In contrast, the corresponding transformation with sulfur-based substrates such as thiols and disulfides has not received a similar level of alteration, probably because of a widespread prejudice that organosulfur compounds poison thiophilic late transition metal catalysts and render them inactive. However, several recent reports on the use of transition metal complexes to catalyze transformations with organosulfur substrates, examples of which include bisthiolation, hydrothiolation, thioboration, and hydrothiolation of unsaturates [8], provide compelling evidence that sulfur-containing compounds are compatible with the late transition metals. In this section, we review the reactions of alkynes with chalcogen-containing substrates in the presence of CO, which demonstrate that transition metal complexes affect the catalytic thiocarbonylation of alkynes, in many cases with quite remarkable efficiency and with excellent levels of chemo-, regio-, and stereoselectivity. Although this is an area in its infancy, the utility of transition metals for the synthesis of organosulfur compounds can now be considered as well established with immense potential for the synthesis of useful reagents and intermediates.

Transition metal catalyzed thioformylation was first reported by Sonoda [9] and involved the rhodium-catalyzed addition of CO and benzenethiol to 1-octyne to give the corresponding β-thio-α,β-unsaturated aldehyde (**1**) in good yield and moderate regioselectivity (Equation 10.1). Although the highest yields were obtained with RhH(CO)$(PPh_3)_3$ in THF, the highest regioselectivity was achieved in CH_3CN, which was ultimately adopted as the solvent of choice. The reaction was functional group tolerant within the limited number of substrates examined. The *Z*-isomer was shown to slowly isomerize to the thermodynamically more stable *E*-isomer under the reaction conditions (120 °C, CH_3CN, 5 h). On the basis of stoichiometric reactions between RhH(CO)$(PPh_3)_3$ and PhSH, a tentative mechanism was presented that involved regioselective thiorhodation of the alkyne to generate a rhodium vinyl intermediate, insertion of CO to give the corresponding α,β-unsaturated acyl complex, oxidative addition of PhSH, and reductive elimination to liberate the β-thio-α,β-unsaturated aldehyde (Scheme 10.1a). In contrast, when the same reaction

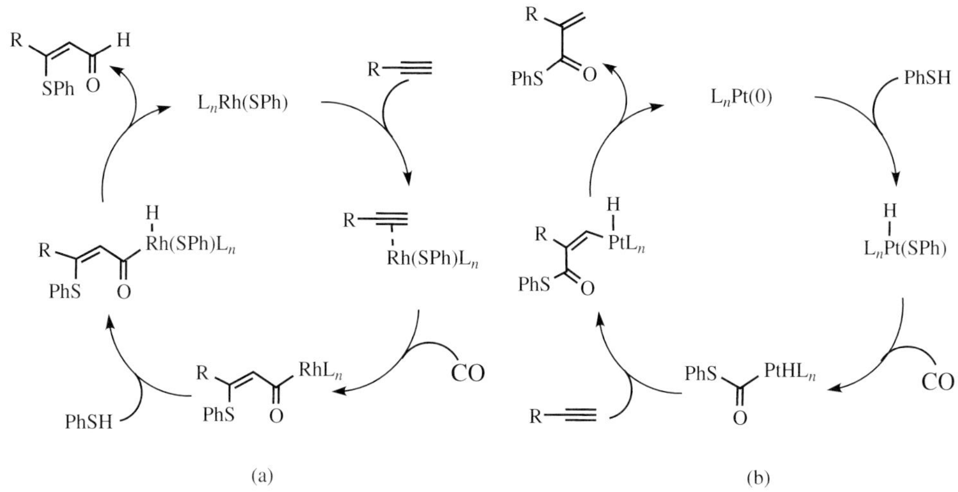

Scheme 10.1

was catalyzed by $Pt(PPh_3)_4$, highly regioselective hydrothiocarbonylation of 1-octyne afforded a mixture of α,β-unsaturated thioester **2**, together with its conjugate addition product **3** (Equation 10.2) [10]. A stoichiometric reaction between $Pt(PPh_3)_4$ and PhSH in CH_3CN resulted in oxidative addition to afford *trans*-$[Pt(H)SPh(PPh_3)_2]$, which was unequivocally identified by NMR spectroscopy. A solution of this hydride in CH_3CN catalyzed the hydrothiocarbonylation of 1-octyne, which was consistent with a mechanism involving initial oxidative addition of PhSH to Pt(0), insertion of CO into the Pt–S bond, thioalkoxycarbonylplatination of the alkyne, and reductive elimination from the resulting platinum vinyl hydride (Scheme 10.1b).

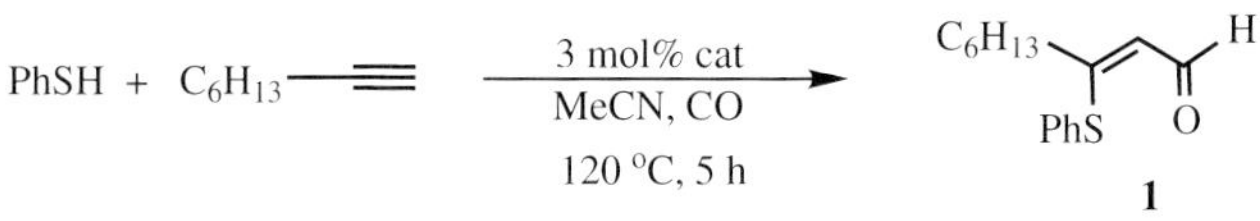

PhSH + C_6H_{13}–≡ → (3 mol% cat; MeCN, CO; 120 °C, 5 h) **1** (10.1)

PhSH + C_6H_{13}–≡ → ($Pt(PPh_3)_4$; MeCN, CO; 120 °C, 5 h) **2** + **3** (10.2)

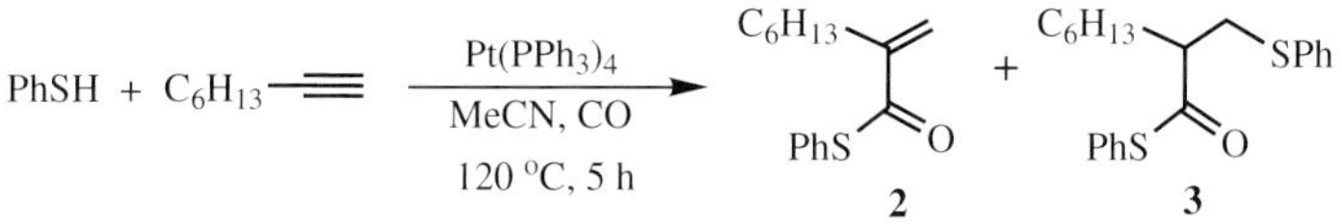

More recent studies have shown that the regioselectivity of alkyne thiocarbonylation depends on both the metal–ligand combination and the solvent with $Pd(OAc)_2$/dppb in THF affording the linear α,β-unsaturated thioester **4** as the dominant product, whereas $Pd(OAc)_2$/dppp in CH_2Cl_2 was highly regioselective for the branched isomer **5** (Equation 10.3) [11].

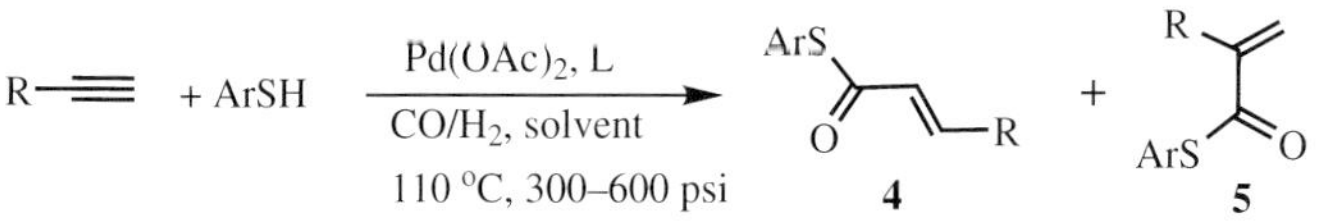

R–≡ + ArSH → ($Pd(OAc)_2$, L; CO/H_2, solvent; 110 °C, 300–600 psi) **4** + **5** (10.3)

The thiocarbonylation reaction has been extended to include the chemo- and regioselective addition to 1,3-conjugated enynes bearing a terminal triple bond [12]. The model reaction between 1-ethynylcyclohexene, thiophenol, and CO was catalyzed by a mixture of $Pd(OAc)_2$ and dppp in tetrahydrofuran at 100 °C to afford 2-(phenylthiocarbonyl)-1,3-diene **6** together with **7**, the latter resulting from a subsequent 1,4-addition of thiophenol (Equation 10.4). Although $Pd(PPh_3)_4$ is an efficient catalyst for the thiocarbonylation of alkynes and allenes, it gave a poor yield with 1,3-enynes. The thiocarbonylation of a series of acyclic and cyclic enynes was examined, and in each case the transformation occurred with high chemo- and regioselectivity, addition of the thiocarbonyl group occurring exclusively at the internal carbon atom of the alkyne. The mechanism proposed to account for this transformation involved oxidative addition of PhSH to Pd(0), carbonyl insertion into the Pd–SPh bond followed by chemo- and regioselective intramolecular acylpalladation of the coordinated enyne, and reductive elimination of the resulting palladium vinyl hydride (Scheme 10.2).

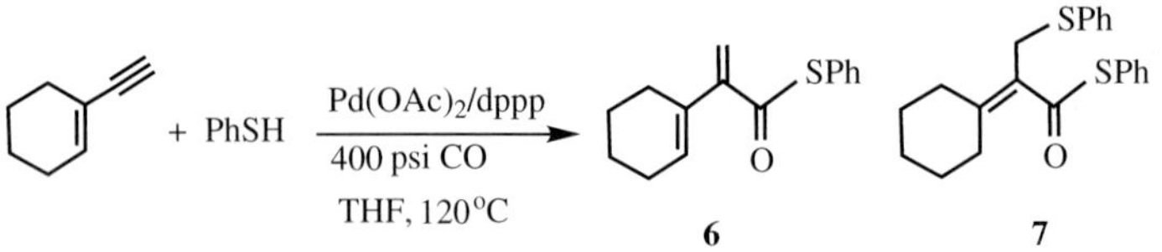

+ PhSH; $Pd(OAc)_2$/dppp, 400 psi CO, THF, 120 °C → **6** + **7** (10.4)

Diarylchalcogenides have also been used as a source of ArS and PhSe in the carbonylation of alkynes [13]. Having established that $Pd(PPh_3)_4$ catalyzes the stereoselective addition of diaryl disulfides and diselenides to terminal alkynes to afford (*Z*)-1,2-bis(arylthio)-1-alkenes and (*Z*)-1,2-bis(arylseleno)-1-alkenes, respectively, the reaction was conducted under carbonylative conditions to afford the corresponding (*Z*)-1,3-bis(arylthio)-2-alken-1-ones and (*Z*)-1,3-bis(arylseleno)-2-alken-1-ones **8** with excellent levels of regio- and stereocontrol such that CO was incorporated at the terminal carbon of the alkyne and the *Z*-isomer favored with selectivities of 94–100% (Equation 10.5). Two pathways were considered on the basis of the regio- and stereoselectivity of carbonylation: one involving the insertion

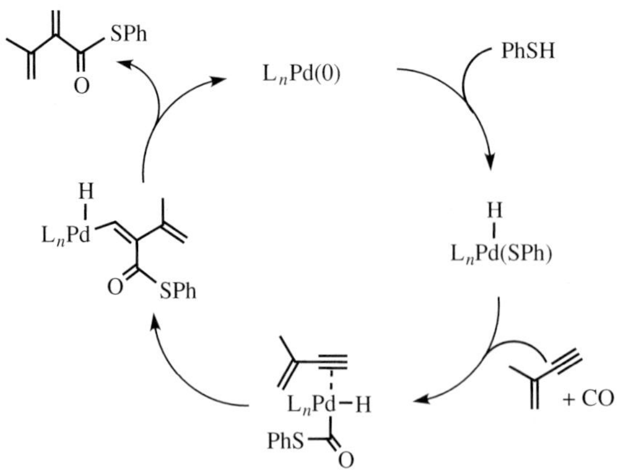

Scheme 10.2

of CO into a palladium vinyl intermediate and the other involving initial insertion of CO into a Pd–SR bond of a palladium thiolate followed by 1,2-insertion of alkyne and reductive elimination.

$$\text{R}{-}{\equiv} + (\text{ArY})_2 \xrightarrow[\text{CO, solvent, } 80\,^{\circ}\text{C}]{\text{cat-Pd(PPh}_3)_4} \text{R(ArY)C=CH–C(=O)YAr} \quad \mathbf{8} \tag{10.5}$$

On the basis that alkyl radicals, generated by the addition of an arylthiol radical to an alkene, could be effectively trapped by molecular oxygen, Yoshida and Isoe reasoned that such radicals could also be trapped with CO to give carbonylation products and showed that the thiocarbonylation of alkynes could be catalyzed by a radical initiator under an atmosphere of carbon monoxide (30 mol% AIBN, 80 atm CO, benzene, 100 °C) to afford β-alkylthio-α,β-unsaturated aldehydes **9** with high selectivity for the *E*-stereoisomer (Equation 10.6) [14]. A pathway involving carbonylation of a β-alkylthio alkenyl radical followed by hydrogen abstraction from the thiol was suggested. Interestingly, the β-alkylthio alkenyl radical appeared to exist as an equilibrium mixture of *E* and *Z* isomers, as evidenced by the isolation of an *E*/*Z* mixture of alkenyl sulfide by-product, which ultimately generated the more stable thermodynamic product.

$$\text{R}^1\text{SH} + \text{R}^2{-}{\equiv} \xrightarrow[\text{80 atm CO, benzene}]{\text{AIBN}} \text{R}^1\text{S–CH=C(R}^2\text{)–CHO} \quad \mathbf{9} \tag{10.6}$$

10.2.2 Propargyl Alcohols and Their Derivatives

By analogy with the alkoxycarbonylation of propargyl alcohols, the first example of a palladium-catalyzed thiocarbonylation of propargyl alcohols with thiols and CO gave a mixture of thiofuranones **10**, thioesters **11**, and dithioesters **12** (Equation 10.7), the distribution of which could be optimized by varying the reaction conditions [15]. Optimum selectivity for the β,γ-unsaturated thioester was obtained in THF with catalytic amounts of *p*-toluenesulfonic acid, while the dithiocarbonylation product was selectively formed in toluene under more forcing conditions (110 °C, 600 psi CO). The thiolactonization was assumed to be initiated by oxidative addition of thiophenol to Pd(0) followed by thioalkoxypalladation of the triple bond with elimination of dihydrogen (possibly involving a σ-bond metathesis-type process), insertion of carbon monoxide into either the palladium–carbon or palladium–oxygen bond followed by reductive elimination to afford the β-phenylthio-γ-lactone.

$$\text{R}^1{-}{\equiv}{-}\text{C(OH)(R}^2\text{)(R}^3\text{)} + \text{R}^4\text{S-H} \xrightarrow[\text{solvent 80–110 °C}]{\text{Pd cat, 400–600 psi CO}} \mathbf{10} + \mathbf{11} + \mathbf{12} \tag{10.7}$$

In the belief that propargyl alcohols were not the ideal substrates for dithiocarbonylation, the reaction of propargyl mesylates with thiols and carbon monoxide was investigated which resulted in a highly stereoselective dithiocarbonylation to afford dithioester **13** with *E*-stereochemistry (Equation 10.8) [16]. Among the catalysts examined, $Pd(PPh_3)_4$ exhibited the highest activity and gave yields up to 88% (3 mol %, 400 psi CO, 120 °C, 24 h). Temperature and solvent both influenced catalyst efficiency with reactions conducted in THF at 90 °C giving optimum yields. Under these conditions, the reaction was highly versatile in terms of substrate scope with arenethiols and alkylthiols both giving good yields with primary, secondary, and tertiary mesylates. Moreover, propargyl mesylates with an internal triple bond reacted with similar efficiency. A plausible mechanism was suggested to involve initial oxidative addition of the propargyl mesylate to Pd(0), substitution of mesylate for thiolate followed by CO insertion, and reductive elimination to liberate an allenyl thioester intermediate that would be susceptible to a Michael-type addition of Pd(0) to the allenyl sp carbon and addition of thiophenol to generate a vinylpalladiumthiolate that would insert CO and liberate the dithioester via reductive elimination from the resulting acyl-thiolate (Scheme 10.3).

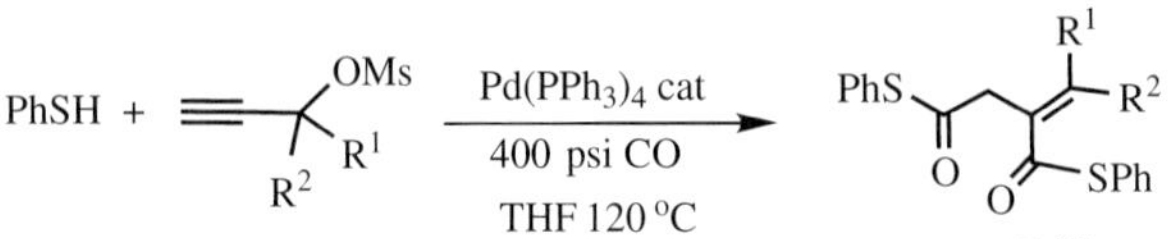

PhSH + propargyl mesylate —[$Pd(PPh_3)_4$ cat, 400 psi CO, THF 120 °C]→ *E*-**13** (10.8)

Having developed a highly selective hydrothiocarbonylation of terminal alkynes, Ogawa and coworkers investigated the corresponding reaction with acetylenic alcohols with the aim of preparing lactones via carbonylative cyclization. Under the same conditions as those used for the intermolecular hydrothiocarbonylation of alkynes (3 mol% $Pt(PPh_3)_4$, CH_3CN, 120 °C, 4 h), 5-hydroxy-1-pentyne undergoes carbonylative lactonization in the presence of benzenethiol to afford α-((phenylthio)methyl)-δ-lactone

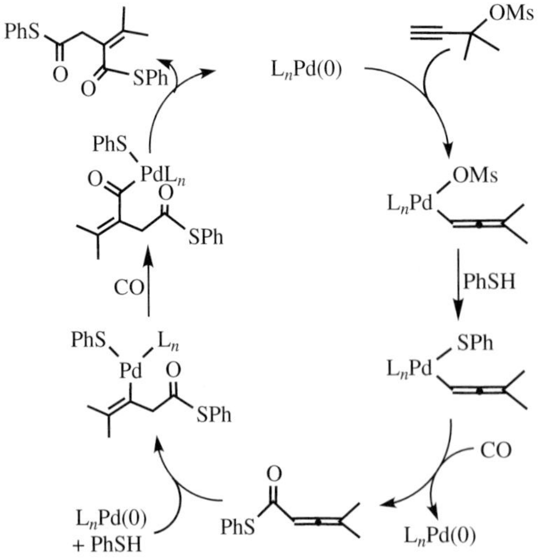

Scheme 10.3

14, which is believed to result from Michael addition of PhSH to an intermediate α-methylene-δ-lactone **15** (Equation 10.9) [17]. This was confirmed by performing the reaction in the presence of a catalytic amount of arenethiols, which resulted in carbonylative lactonization to afford α-methylene-δ-lactone as the major product in good yield. Isolation of the acyclic thiocarbonylation product **16** (Equation 10.10) from the reaction with the sterically demanding 2,6-dimethylbenzenethiol was taken as evidence that the carbonylative lactonization occurred via hydrothiocarbonylation. Under similar conditions, the palladium-catalyzed reaction between propargyl alcohols, diarylchalcogenides, and CO results in a novel one-pot carbonylative lactonization to afford α,β-unsaturated–β-chalcogenolactones **17** in moderate to good yields (Equation 10.11), whereas reaction with homopropargylic alcohols gave the corresponding δ-lactones [18]. In the case of α,α′-disubstituted propargyl alcohols, lactonization with diphenyldisulfide was not as efficient and instead the β-thio-α,β-unsaturated thioester **18** was isolated as the major product, although interestingly much higher yields of the desired lactone were obtained with diphenyl diselenide. A mechanism initiated by oxidative addition of the dichalcogenide to Pd(0), regioselective thiopalladation of the triple bond, CO insertion, *E*–*Z* double-bond isomerization involving a ketene intermediate, and intramolecular cyclization accounts for the observed product (Scheme 10.4).

HO + PhSH → [$Pt(PPh_3)_4$ cat, 30 atm CO, MeCN, 120 °C] **14** + **15** (10.9)

HO + 2,6-dimethylbenzenethiol (SH) → [$Pt(PPh_3)_4$ cat, 30 atm CO, MeCN, 120 °C] **16** (10.10)

+ $(ArY)_2$ → [$Pd(PPh_3)_4$ cat, 60 atm CO, toluene 100 °C] **17** + **18** (10.11)

10.2.3 Thiocarbamoylation of Terminal Alkynes

Sulfenamides have proven to be useful reagents for the introduction of R^1S and R^2R^3N groups in transition metal catalyzed organic synthesis, for example, the azathiolation of CO to afford thiocarbamates, the regioselective azathiolation of alkynes, and most recently the palladium-catalyzed regio- and stereoselective synthesis of β-sulfenyl- and selenyl acrylamides from an alkyne and carbon monoxide. In one case, Kuniyasu showed that a range of palladium complexes catalyzed the intermolecular thiocarbamoylation of terminal alkynes with a sulfenamide and CO [19]; the optimum catalyst was identified as $PdCl_2(PPh_3)_2$ in combination with PPh_3 and n-Bu_4NCl as additives (Equation 10.12). A range of sulfenamide substrates were examined, and in the majority of cases, the β-sulfenyl acrylamide **19** was generated as an *E*/*Z* mixture of stereoisomers in low yield and poor to moderate selectivity, with the

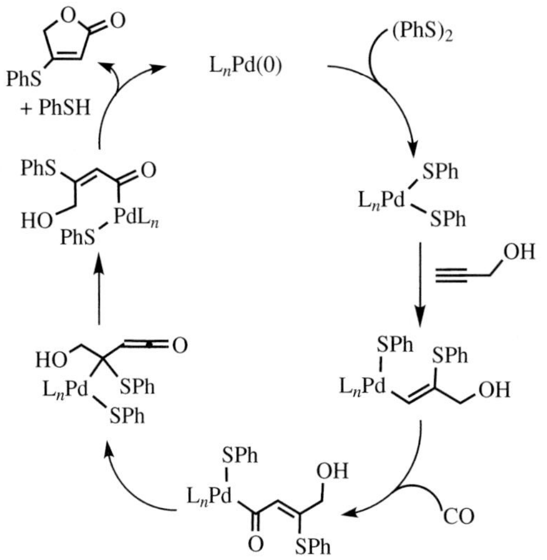

Scheme 10.4

exception of 2,4,5-$C_6H_2Cl_3$-$SNEt_2$, which was highly selective for the *Z*-stereoisomer (98 : 2). Meyer and coworker have developed a practical one-pot synthesis of (*Z*)-β-selenyl acrylamides **20** via a highly regio- and stereoselective palladium-catalyzed four-component coupling between a sulfenamide ($PhSNR^1R^2$), a terminal alkyne, CO, and diphenyl diselenide (Equation 10.13) [20]. Moderate selectivity (4 : 1) for the formation of the β-selenyl acrylamide was obtained over its sulfenyl counterpart. This transformation is reasonably versatile in that it tolerates a range of functional groups on the sulfenamide and alkyne components, although internal alkynes, phenyl acetylene, and alkynes bearing hydroxy substituents are unreactive. The *Z*-β-selenyl acrylamide products were obtained in high yields and as a single regio- and stereoisomer. A number of minor by-products arising from two- and three-component coupling were identified, including products resulting from the dichalcogenation of the alkyne, azaselenolation and azathiolation of CO, and carbonylative addition of diaryl dichalcogenides to the alkyne. On the basis of the analysis of the by-products and their distribution and role as possible intermediates, a mechanism was proposed that involved oxidative addition of diphenyldiselenide to Pd(0), stereoselective *cis*-selenopalladation, insertion of CO into the Pd–C(vinyl) bond, and σ-bond metathesis with the sulfenamide to liberate the product. The catalytic cycle would continue either by reductive elimination of diaryldichalcogenide or by direct insertion of alkyne into the

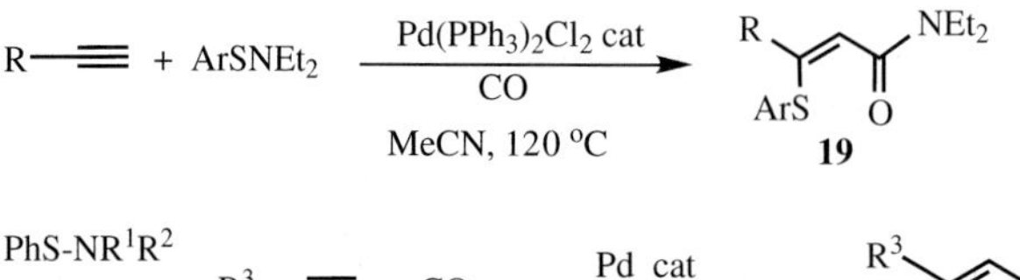

$$R{-}{\equiv} + ArSNEt_2 \xrightarrow[\text{CO, MeCN, 120 °C}]{Pd(PPh_3)_2Cl_2\ cat} \mathbf{19} \qquad (10.12)$$

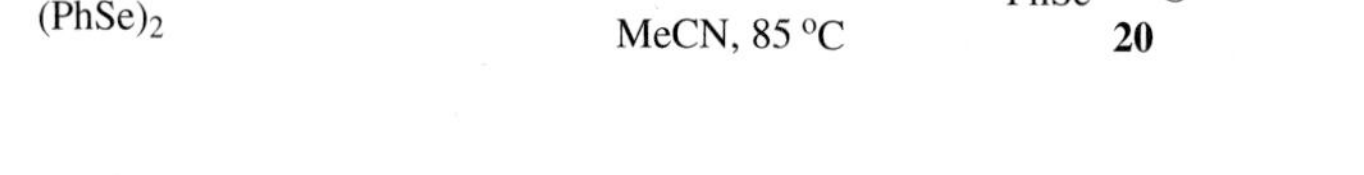

$$PhS\text{-}NR^1R^2 + (PhSe)_2 + R^3{-}{\equiv} + CO \xrightarrow[\text{CO, MeCN, 85 °C}]{Pd\ cat} \mathbf{20} \qquad (10.13)$$

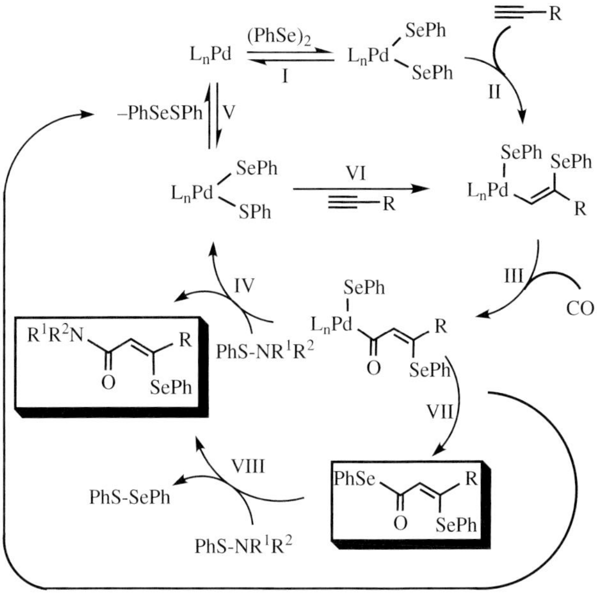

Scheme 10.5

palladium–chalcogen bond with significant by-product formation occurring via steps VII and VIII (Scheme 10.5).

10.3 Nonoxidative Hydroxy- and Alkoxycarbonylation of Alkynes

10.3.1 Terminal Alkynes

Early studies aimed at identifying alkyne carbonylation catalysts based on group 10 metal complexes modified with various tertiary phosphines met with partial successes in that good selectivities for the branched regioisomer were obtained, but in all cases reaction rates were far too low to be commercially viable. In the mid-1990s, Drent made a dramatic breakthrough in this area with the discovery that catalyst mixtures based on a palladium(II) precursor, 2-pyridyldiphenylphosphine (2-pyPPh$_2$), and a weak acid of a noncoordinating anion were highly active and selective for the methoxycarbonylation of alkynes. Initial efforts focused on the methoxycarbonylation of propyne (Equation 10.14) that generated methyl methacrylate with a regioselectivity of 99.95% and a turnover frequency of 50 000 mol product per mol catalyst per hour [21]. Experimental evidence supported a catalytic cycle involving insertion of propyne into a cationic palladium methoxycarbonyl species followed by rate-limiting protonolysis of the resulting palladium alkenyl complex to liberate the product and regenerate the palladium methoxy initiator. The 2-pyridyldiphenylphosphine was suggested to play a key role in the rate-determining protonolysis step by facilitating transfer of the proton to the palladium–alkenyl bond. Shortly after this discovery, Scrivanti applied the same catalyst system to the hydroxycarbonylation of (6-methoxy-2-naphthyl)ethyne (THF, 30 atm CO, 50 °C, substrate/acid/Pd 900/80/1) to afford 2-(6-methoxy-2-naphthyl)propenoic acid **21**, a precursor to naproxen (Equation 10.15) [22]. The same catalyst system has also been used for the hydroxycarbonylation of 5-ethynyl-1,1,2,3,3-pentamethylindane to afford the corresponding α-aryl propenoic acid **22**, in near quantitative yield, as part of the synthesis of the olfactorally active stereoisomers of the musk odorant Galaxolide (Equation 10.16).

Me—≡—H → [$Pd(OAc)_2$, 2-$pyPPh_2$; CO, $MeSO_3H$, MeOH] → Branched methyl methacrylate and/or Linear methyl crotonoate (10.14)

(6-methoxy-2-naphthyl)ethyne → [$Pd(OAc)_2$, 2-$pyPPh_2$; CO, $MeSO_3H$, THF, water] → **21** (CO_2H) (10.15)

5-ethynyl-1,1,2,3,3-pentamethylindane → [$Pd(OAc)_2$, 2-$pyPPh_2$; CO, $MeSO_3H$, THF, water] → **22** (CO_2H) (10.16)

Scrivanti replaced 2-pyridyldiphenylphosphine with 2-furyl-substituted phosphines and found that these ligands formed highly selective catalysts for the alkoxycarbonylation of phenylacetylene, generating methyl atropate exclusively, although in all cases the activities were markedly lower than those obtained with 2-pyridyldiphenylphosphine [23]. Not surprisingly, significantly higher activities were obtained by introducing a 2-pyridyl substituent, and catalyst mixtures based on (2-furyl)phenyl(2-pyridyl)phosphine and $Pd(OAc)_2$ gave selectivities in excess of 99% and conversions that matched those obtained with the original Drent system. Interestingly, at low H^+ : Pd ratios, the catalyst based on (2-furyl)phenyl(2-pyridyl) phosphine was markedly more efficient than its 2-$pyPPh_2$ counterpart, suggesting a possible role for this ligand in the alkoxycarbonylation of acid-sensitive substrates [23]. Well-defined palladium(0)–alkene complexes of iminophosphine ligands also catalyze the regioselective alkoxycarbonylation of phenylacetylene to afford the branched α,β-unsaturated ester, although the average turnover frequencies of 50 mol product per mol catalyst per hour were markedly lower than that obtained with 2-$pyPPh_2$, as were the regioselectivities of 80–87% and the chemoselectivity, as evidenced by the production of significant quantities of acetophenone [24]. The reaction rates were comparable to those obtained with the tris(2-furyl)phosphine-based system. In a related study, Reetz and coworkers investigated the coordination chemistry of 2-pyrimidyldiphenylphosphine relevant to palladium-catalyzed alkoxycarbonylation [25] and showed that this ligand coordinates in a monodentate manner through the phosphorus atom in cis-[Pd{($C_4N_2H_3$)PPh_2}$_2Cl_2$] and as a bidentate P,N chelate in the cation [Pd{($C_4N_2H_3$)PPh_2}$_2$]$^{2+}$. This is particularly relevant to the earlier studies of Drent in which the active catalyst formed from palladium(II) and 2-pyPPh_2 was proposed to contain two molecules of 2-pyPPh_2, one coordinated in a monodentate manner and the other as a four-membered PN chelate. Solutions of 2-pyrimidyldiphenylphosphine, Pd$(OAc)_2$, and CH_3SO_3H catalyze the highly regioselective alkoxycarbonylation of a range of terminal alkynes with TOFs up to 7000 mol product per mol catalyst per hour. The methoxycarbonylation of fluorinated alkynes has been examined as a potential route to fluorinated acrylates for the synthesis of fluorinated homo- and copolymers that are widely used in the coatings industry because of their refractory properties and resistance to weathering and UV rays. However, in stark contrast to propyne, the palladium-catalyzed (Pd$(OAc)_2$/(6-methylpyrid-2-yl)diphenylphosphine/CH_3SO_3H) alkoxycarbonylation of 3,3,3-trifluoropropyne gave markedly lower conversions and lower chemoselectivity [26] as evidenced by the identification of four products (Scheme 10.6), two of which were the expected regioisomeric methyl esters of 2-(trifluoromethyl)propenoic acid **23** and *E*-4,4,4-trifluorobut-2-enoic acid **24**. The remaining two products were methyl 2-trifluoromethyl-3-methoxypropanoate **25** and the methyl ester of *Z*-4,4,4-trifluorobut-2-enoic acid **26**, the former arising from addition of methanol to the double bond of the branched α,β-unsaturated ester **23**, whereas the latter corresponds to a formal *trans* addition of H and CO_2R to 3,3,3-trifluoropropyne. In addition, the alkoxycarbonylation of 3,3,3-trifluoropropyne also exhibited a dramatic dependence of the chemoselectivity on the CO pressure with the branched isomer favored at lower pressures of CO. Optimization studies showed that regioselectivities in excess of 90% could be obtained at high ligand/Pd and H^+/Pd ratios in a mixed solvent system of dichloromethane and NMP (*N*-methylpyrrolidone). Interestingly, high pressures (80 atm) and a low ligand/Pd and high H^+/Pd ratio favored the linear regioisomer (85%). The hydroesterification of terminal alkynes with a range of aliphatic alcohols has been catalyzed by Pd$(dba)_2$/PPh_3/TsOH to afford branched esters with high regioselectivity, whereas the same catalyst system gave good yields but poor regioselectivities with internal alkynes [27].

The discovery that palladium(II) complexes of 2-pyPPh_2 are highly active and selective for the alkoxycarbonylation of alkynes generated considerable interest both in its application to synthesis and the mechanism of this highly efficient process. In this regard, Scrivanti *et al.* have studied the mechanism using a combination of deuterium-labeling studies and stoichiometric reactions [28]. Methoxycarbonylation of but-2-yne established the *syn* addition of H and CO_2CH_3, and unequivocal

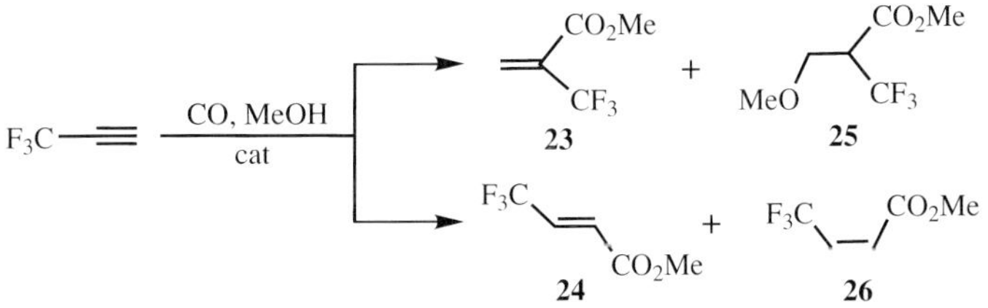

F_3C—≡ (CO, MeOH; cat) → **23** ($CH_2=C(CF_3)CO_2Me$) + **25** ($MeOCH_2CH(CF_3)CO_2Me$); **24** (F_3C–CH=CH–CO_2Me, *E*) + **26** (F_3C–CH=CH–CO_2Me, *Z*)

Scheme 10.6

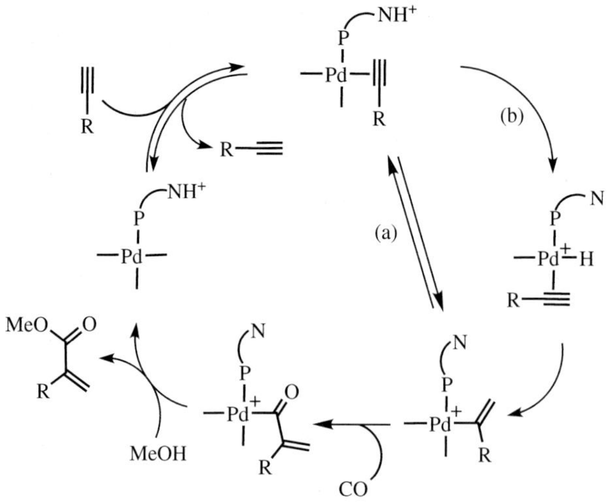

Scheme 10.7

spectroscopic evidence was obtained for the existence of a palladium vinyl intermediate (Pd–C(Ph)=CH_2). On the basis of deuterium-labeling studies, H/D exchange in the vinyl intermediate and a large kinetic isotope effect for the carbonylation of but-2-yne with a 1 : 1 mixture of CH_3OH and CH_3OD formation of the vinyl intermediate was proposed to involve rate-determining protonation of an η^2-coordinated alkyne from the protonated pyridine (Scheme 10.7). A compelling argument was presented in favor of direct protonation (pathway a) of the coordinated alkyne rather than transfer to palladium followed by hydropalladation (pathway b).

The palladium-based coordination chemistry of 2-pyPPh$_2$ has been the subject of a number of recent publications. In one of these, Edwards reported the synthesis of a range of zerovalent palladium complexes of 2-pyPPh$_2$ together with a study of their reaction chemistry relevant to the alkoxycarbonylation of alkynes. In particular, the α-phenyl vinyl complexes [Pd(2-pyPPh$_2$)$_2$(PhC=CH_2)](CF_3CO_2), prepared by oxidative addition of PhC≡CH/trifluoroacetic acid to [Pd(2-pyPPh$_2$)$_3$], catalyzed the alkoxycarbonylation of phenylacetylene to afford methyl 2-phenylpropenoate [29]. The identity of the α-vinyl complex was unequivocally established by a single-crystal X-ray structure determination and supports the spectroscopic observation of a similar species by Scrivanti. In an earlier study, the same researchers showed that [Pd(2-pyPPh$_2$)$_2$(CO_2CH_3)(CF_3CO_2)] did not react with PhC≡CH via insertion but formed the σ-alkynyl complex [Pd(2-pyPPh$_2$)$_2$(C≡CPh)(CF_3CO_2)] via metathesis and elimination of methyl formate [29]. The same complex formed an active catalyst for the regioselective production of methyl 2-phenylpropenoate when combined with phenylacetylene and methanol in d_6-benzene under 30 psi pressure of CO, although there was no catalysis at atmospheric pressure. Even though the active species was not identified, this study suggests that the methoxycarbonylation of phenylacetylene does not occur via insertion into a methoxycarbonyl palladium bond. Edwards and coworkers also examined the potential of a series of OPN donors **27–30** to act as hemilabile ligands in homogeneous catalysis [30], and in all cases O-coordination was shown to be unfavorable with the exception of **30** that appeared to adopt a terdentate OPN coordination mode in $[Pd(OPN)(THF)]^{2+}$. Catalyst mixtures containing $Pd(OAc)_2$, **30**, and CH_3SO_3H were active for the methoxycarbonylation of propyne and produced 15.2 g of methyl methacrylate in 3 h at 50 °C with 10 bar CO and 15 g of propyne. Although the TOF of 354 mol MMA per mol Pd per hour obtained using phosphine **30** was lower than that of 890 mol MMA per mol Pd per hour for the 2-pyPPh$_2$-based system, the optimum ligand : Pd ratio of 2.5 : 1 was significantly lower than that of 20–40 : 1 required to achieve high activities with 2-pyPPh$_2$. This was the first catalyst system for the carbonylation of alkynes that did not require a large excess of phosphine ligand to maintain catalyst stability.

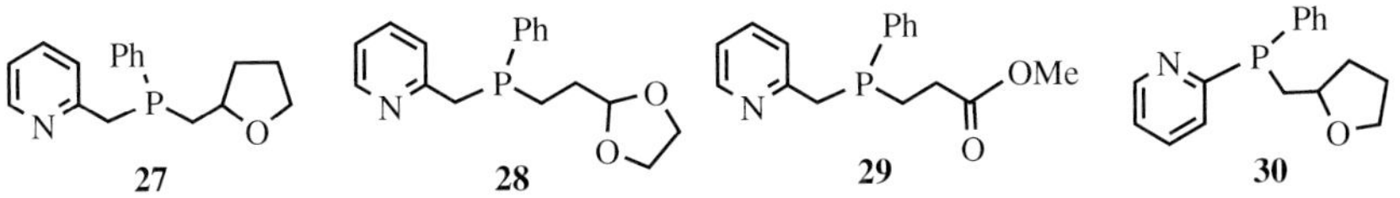

27 **28** **29** **30**

In the vast majority of reports, the palladium-catalyzed alkoxycarbonylation of terminal alkynes occurs with high regioselectivity for the branched isomer. However, regioselective alkoxycarbonylation in favor of the linear α,β-unsaturated ester can be achieved with palladium catalysts based on dppf (dppf = 1,1′-bis(diphenylphosphino)ferrocene) in methanol/CH_3CN at 80–120 °C under 40 atm of carbon monoxide [31]. Under these conditions, the linear ester could be obtained as the exclusive product while other solvents such as toluene, dichloromethane, and DMF gave lower selectivities. Hydroesterification of 1,7-octadiyne with $Pd(OAc)_2/PPh_3$/TsOH resulted in cyclization to afford methyl (*E*)-2-methylenecyclohexylideneethanoate **31** as the dominant product together with minor amounts of noncyclized diester **32** (Equation 10.17). The authors suggested that the carbonylative cyclization involved hydropalladation of one of the triple bonds, intramolecular insertion of the remaining alkyne into the resulting Pd–C (vinyl) bond to form a six-membered ring followed by CO insertion and methanolysis. Another example of selective production of the linear acrylate appeared in 1988 in the first report of a metal-catalyzed reaction between a formate ester and an alkyne [32], which described the palladium(0)/dppb-catalyzed regio- and stereoselective hydroesterifications of terminal alkynes (toluene, 80 atm CO, 100 °C, 2–3 days) to afford the *E*-α,β-unsaturated ester **33** in moderate yield (Equation 10.18). The use of dppb was critical, as catalysts based on shorter chain diphosphines and PPh_3 were ineffective. Regioselective hydroesterification of alkynes and alkynols can also be performed with formate esters using a catalyst based on $Pd(OAc)_2$/dppb/PPh_3/*p*-TsOH in THF (Equation 10.19) [33]. For terminal alkynes regioselectivity was determined by steric factors with increasing bulk favoring the linear regioisomer, whereas internal unsymmetrical alkynes suffered from low regioselectivities. Interestingly, a catalyst based on a 2 : 1 mixture of PPh_3 and dppb was more active than those containing either phosphines alone. On the basis of ^{13}C labeling experiments, the formate ester was suggested to generate a palladium alkoxy hydride initiator, via oxidative addition and deinsertion, which underwent hydropalladation of the alkyne, insertion of CO, and reductive elimination from the resulting acyl-palladium-alkoxide to liberate the corresponding α,β-unsaturated ester. High regioselectivity for the linear ester was also obtained from the hydroesterification of terminal alkynes catalyzed by $Pd(OAc)_2$/dppb in dichloromethane under syngas [33].

$Pd(OAc)_2/PPh_3/TsOH$; CO, MeOH → CO_2Me **31** + CO_2Me, CO_2Me **32** (10.17)

R–≡ + HCOOR′ — $Pd(dba)_2$, dppb; 80 atm CO, toluene → R, CO_2R' **33** (10.18)

OH + $HCOOC_4H_9$ — $Pd(OAc)_2$/dppb/PPh_3; THF, p-TsOH; 100 C, 20 atm CO → $CO_2C_4H_9$ (10.19)

Doherty and coworkers have recently anchored 2-pyPPh_2 onto a variety of different polymer supports for use in the methoxycarbonylation of terminal alkynes using Pd $(OAc)_2$ and methane sulfonic acid as cocatalysts. In the case of phenylacetylene, the activity and selectivity of catalyst generated with 2-pyPPh_2, anchored on polystyrene **34** and polymethylmethacrylate **35**, compared favorably with its homogeneous counterpart, whereas the activities obtained with catalyst supported by styrene methylmethacrylate copolymer **36** were slightly lower [34]. With propyne as the substrate, only polymer **37** gave activities comparable to the homogeneous system. Exceptionally high selectivities were obtained with polyvinylpyridine-supported catalyst, but this was at the expense of activity. With the aim of exploring the potential benefits of performing the palladium-catalyzed alkoxycarbonylation of alkynes in a fluorous biphasic system and supercritical carbon dioxide (scCO_2), Elsevier and coworkers prepared a fluorous 2-[bis(4-aryl)phosphine]pyridine ligand, **37** and several derived palladium complexes [35]. A comparison of the performance of palladium catalysts for the methoxycarbonylation of phenylacetylene in a 1 : 1 mixture of methanol and α,α′,α″-trifluorotoluene revealed that catalyst generated *in situ* from $Pd(OAc)_2$ and 2-[bis{4-((2-(perfluorohexyl)ethyl)dimethylsilyl)phenyl}phosphino] pyridine **37** gave a slightly higher activity than its 2-pyPPh_2 counterpart in this solvent, with TONs of 3400 and 3000, respectively, whereas both catalysts had similar activities in methanol. The beneficial effect of the perfluoroalkyl group was more evident for reactions conducted in supercritical CO_2, because the presence of a perfluoroalkyl chain in the palladium catalyst gave a highly active homogeneous system (TON 2000, 50 min, 50% conversion), whereas that based on 2-pyPPh_2 formed a suspension and gave low conversions (TON 120, 50 min, 3% conversion).

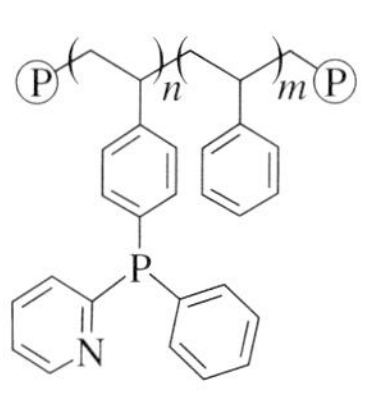

34 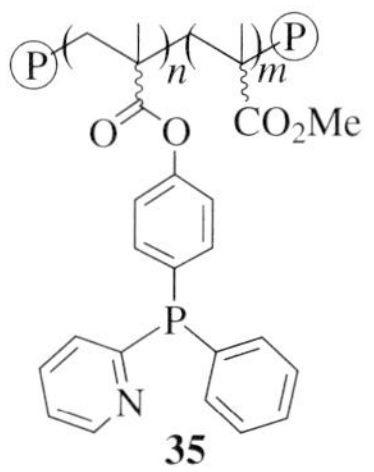 35 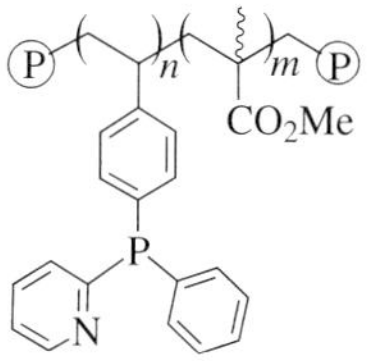 36

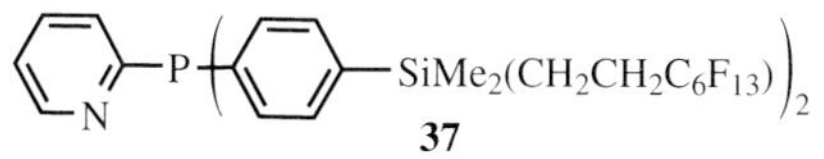

37

The benefit of using a well-defined Pd(0) precursor of known composition was also highlighted because the use of a palladium(0)–alkene complex of **37** gave a TON of 8000 mol product per mol catalyst, approximately twice of that obtained with *in situ* prepared catalyst.

Having demonstrated that the use of formic acid for the palladium-catalyzed carbonylation of terminal alkynes under oxidative conditions ($CuCl_2/O_2$) results in dicarbonylation to afford maleic and fumaric acids, Alper and coworkers investigated the corresponding reaction under nonoxidative conditions and found that the major product was a mixture of linear and branched unsaturated acids, the distribution of which was determined by a combination of electronic and steric factors [36a]. Branched acids were favored for phenylacetylene and alkynes substituted with linear alkyl groups, whereas bulkier alkynes favored the linear isomer such that it was formed as the exclusive product with trimethylsilylacetylene. Internal alkynes also undergo hydroxycarbonylation with the same catalyst systems to afford *E*-alkenoic acid, and although the regioselectivity varied with the alkyne substituent, they were generally lower than those obtained with terminal alkynes. For selected substrates, catalyst based on a combination of dppb and PPh_3 ($Pd(OAc)_2 : PPh_3 : dppb = 1:4:2$) gave markedly better conversions than those formed from a single phosphine, and this synergism was most apparent in the case of internal alkynes. Deuterium labeling studies with terminal alkynes demonstrated that the products result from a formal *cis* and *trans* addition of H and CO_2H, the latter most likely arises from a concerted *cis* addition followed by isomerization at the C=C double bond of a coordinated alkenyl intermediate. On the basis of the precedent literature and experimental observations, a tentative mechanism was suggested to involve alkyne coordination and protonation at the electron-rich Pd(0), hydropalladation of the triple bond followed by CO insertion and cleavage of the resulting palladium–acyl bond to liberate the product as the mixed anhydride (Scheme 10.8). The regioselective hydrocarboxylation of terminal alkynes with either formic acid or oleic acid can also be catalyzed by a combination of Pd/C and dppb (6–40 atm CO, 150 °C, 24 h). The authors suggested that the active catalyst could form by leaching of the metal out of the solid phase by the dppb, although a surface-bound colloidal palladium-phosphine species was also considered possible [36b].

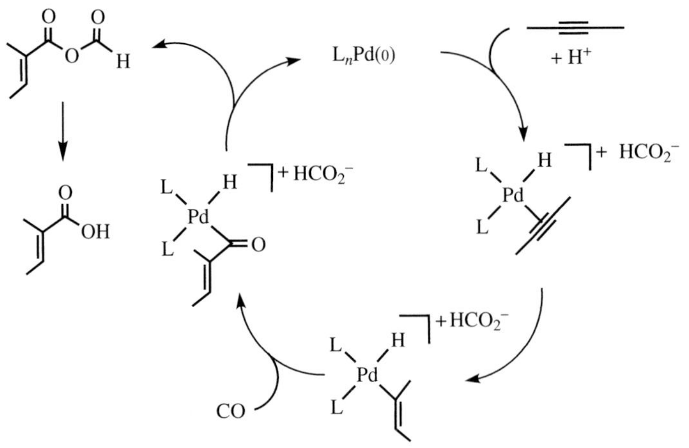

Scheme 10.8

10.3.2
Propargyl Alcohols

Having demonstrated that nickel cyanide catalyzes the direct carbonylation of allyl alcohols to acids and the regiospecific hydroxycarbonylation of alkynes to unsaturated acids, Alper and coworkers reasoned that alkynols could be converted into diacids **39**/**40** by hydroxycarbonylation of the alkyne component followed by carbonylation of the resulting allylic alcohol **38** [37]. The carbonylation of a range of alkynols, using catalytic amounts of $Ni(CN)_2$, toluene as the organic phase, 5 N sodium hydroxide as the aqueous phase, and tetrabutyl ammonium chloride as the phase transfer agent (8 h, 95 °C, 1 atm), gave an *E*/*Z* mixture of the diacid (Scheme 10.9). Interestingly, the stereoselectivity showed a marked dependence on the phase-transfer reagent, with cetyltrimethylammonium bromide and Bu_4NHSO_4 giving the highest *E*/*Z* ratios in favor of the *Z*-stereoisomer. Alper reported one of the first examples of the direct synthesis of butenolides **41** via the palladium-catalyzed cyclocarbonylation of terminal propargyl alcohols using a mixture of $Pd(dba)_2$ and dppb, in DME at 150 °C (Scheme 10.10) [38]. This transformation was later extended to include internal alkynols bearing alkyl, phenyl, and vinyl substituents by using a similar catalyst mixture under a hydrogen atmosphere (600 psi CO and 200 psi H_2). Although the precise role of hydrogen was not established, insertion of Pd(0) into the C–O bond, rearrangement of the resulting propargyl palladium hydroxide to the corresponding allenyl tautomer, CO insertion, and reductive elimination with C–O bond formation would generate the 2(5*H*)-furanone product (Scheme 10.10). In a

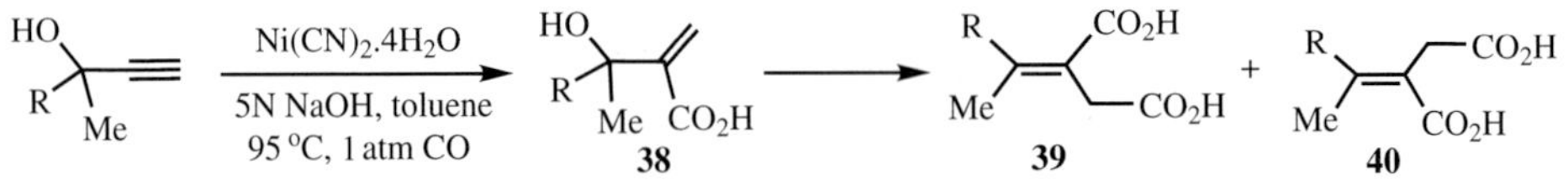

Scheme 10.9

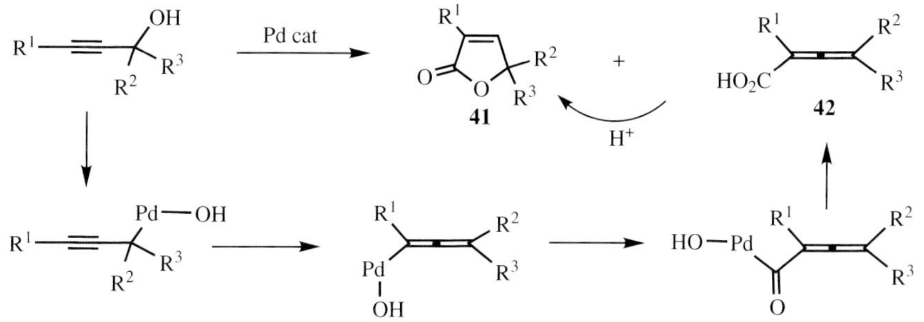

Scheme 10.10

closely related study, the cyclocarbonylation of tertiary propargylic alcohols was catalyzed by $[Pd(CH_3CN)_2(PPh_3)_2](BF_4)_2$ to give a mixture of the 2(5*H*)-furanone **41** and 2,3-dienoic acid **42**, the distribution of which depended on R^1, the substituent directly attached to the alkynyl carbon atom (Scheme 10.10) [39]. Rapid and quantitative conversion of the dienoic acid into the 2(5*H*)-furanone occurred in the presence of trace amounts of *p*-toluenesulfonic acid.

10.3.3 Propargyl Halides

The palladium-catalyzed carbonylation of propargyl halides, phosphates, and esters in the presence of protic nitrogen- and oxygen-based nucleophiles to afford products of mono- and dicarbonylation has been thoroughly investigated, details of which appear in a recent and comprehensive review [40]. Nickel cyanide also catalyzes the highly regio- and stereoselective carbonylation of propargyl (Equation 10.20) and allenyl halides (Equation 10.21) under phase-transfer conditions (4-methyl-2-pentanone, 5 N NaOH, R_4NX) to afford the allenic monoacid **43** that underwent a second much slower carbonylation to the unsaturated diacid **44** [41]. Carbonylation of the propargyl halide was suggested to occur via S_N2' nucleophilic attack of the tricarbonylnickelate $[Ni(CO)_3(CN)]^-$ on the terminal alkynyl carbon atom, whereas nucleophilic attack on the allenyl halide was considered to be more difficult, and for this substrate an oxidative addition/reductive elimination pathway assisted by excess cyanide was suggested to be more likely in this case.

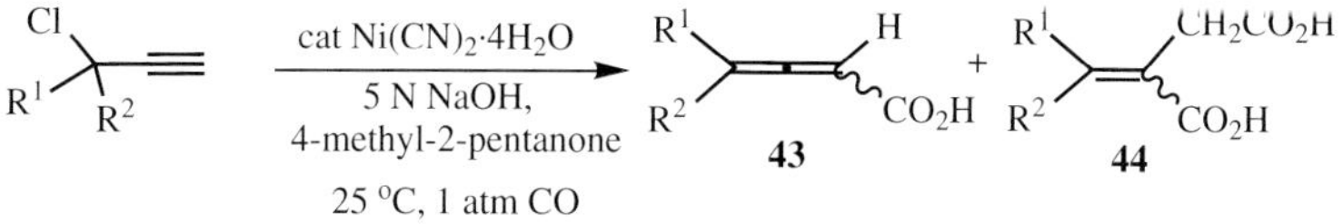

cat $Ni(CN)_2 \cdot 4H_2O$, 5 N NaOH, 4-methyl-2-pentanone, 25 °C, 1 atm CO → **43** + **44** (10.20)

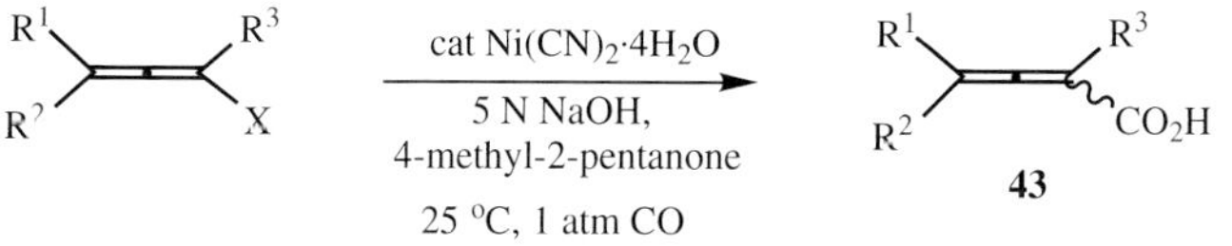

cat $Ni(CN)_2 \cdot 4H_2O$, 5 N NaOH, 4-methyl-2-pentanone, 25 °C, 1 atm CO → **43** (10.21)

10.3.4
Carbonylation of α-Ketoalkynes

Under similar conditions of phase-transfer carbonylation, α-keto alkynes gave either unsaturated hydroxybutyrolactones **45** (Equation 10.22) or 2-alkylidene 3-keto carboxylic acid **46** (Scheme 10.11), depending on whether hydrogen atoms were attached

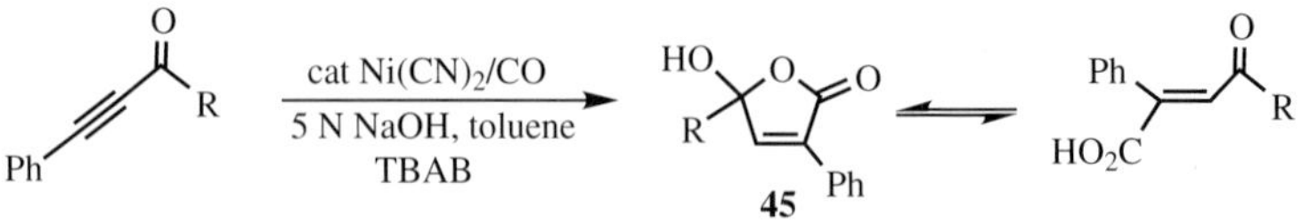

(10.22)

to the carbon atom α to the alkynyl group [42]. In both cases, a mechanism involving initial attack of the nucleophilic catalyst at the alkynyl carbon was proposed with α-hydrogen abstraction and rearrangement accounting for the formation of the unsaturated keto acid.

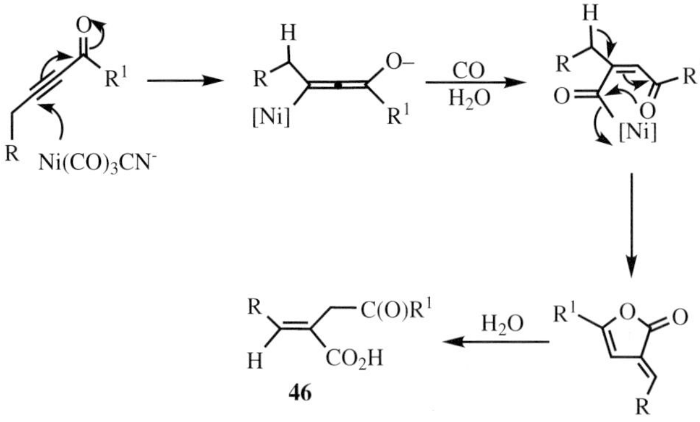

Scheme 10.11

Iron-catalyzed aqueous phase cyanohydroxycarbonylation of α-ketoalkynes with CO and KCN gave moderate to high yields of carboxy-substituted α,β-unsaturated lactams **47** under mild conditions (Equation 10.23) [43]. On the basis of IR spectroscopic evidence and previous literature precedent the active catalyst was suggested to be $[Fe(CN)_4(CO)_2]^{2-}$, generated *in situ* by ligand exchange between $K_4[Fe(CN)_6]$ and carbon monoxide.

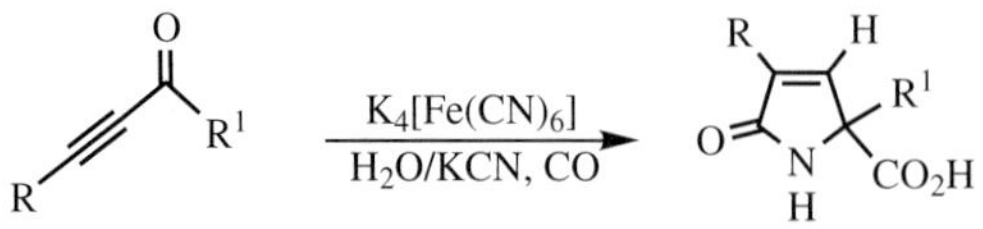

(10.23)

Alkynones are a particularly interesting class of substrate because cyclocarbonylation would generate a new stereocenter in the resulting furanone. Moreover, alkynones can be prepared from inexpensive and readily available starting materials, for example, by palladium-catalyzed cross-coupling of an acid chloride with a terminal alkyne. In this regard, α-ketoalkynes can be cyclocarbonylated with high chemo- and regioselectivity

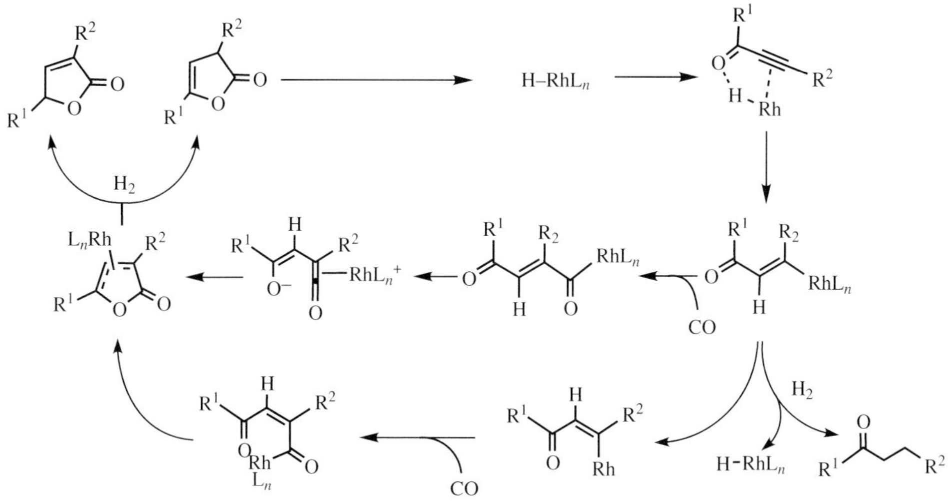

Scheme 10.12

using catalyst mixtures containing zwitterionic $(\eta^6\text{-}C_6H_5BPh_3)^-Rh^+(1,5\text{-COD})$, P$(OPh)_3$, CO, and H_2 to afford 2(3*H*)- and 2(5*H*)-furanones (Equation 10.24) [44]. Hydrogenated alkynone **50** was also produced even after the optimization of the reaction conditions. The substituent α to the keto group (R^1) has a marked influence on the ratio of 2(3*H*)-furanone and hydrogenated alkynone with primary alkyl-substituted substrates, giving the highest selectivity for 2(3*H*)-furanone **48**. High selectivity for the 2(5*H*)-furanone **49** was obtained when $R^2 = $ Ph, vinyl, and methoxymethyl, although in the latter case demethoxylation occurs under the conditions of reaction, transforming the methoxymethyl to a methyl group. Two possible mechanisms were proposed, both initiated by a regioselective *syn* hydrorhodation of the triple bond followed either by carbonylation, rearrangement to a zwitterionic ketene intermediate, followed by cyclization, or an isomerization–carbonylation–cyclization sequence (Scheme 10.12).

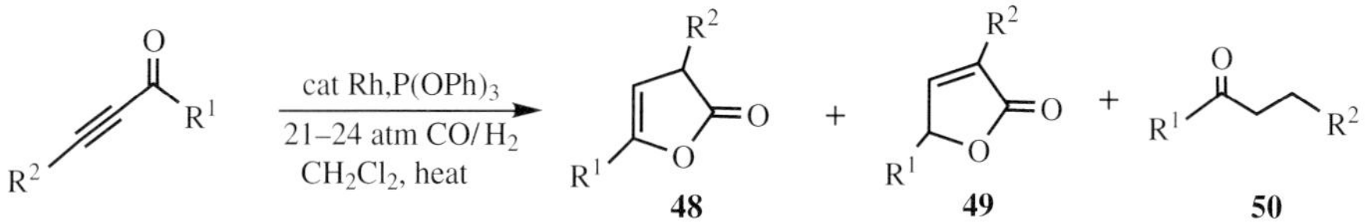

cat Rh,P$(OPh)_3$; 21–24 atm CO/H_2; CH_2Cl_2, heat → **48** + **49** + **50**

(10.24)

10.3.5
Carbonylation of Internal Alkynes

Some time ago Takahashi and coworkers demonstrated that enynes could be converted into cyclic enones in a rhodium-catalyzed hydroformylation [45] and in the course of these studies discovered that the use of water instead of hydrogen resulted in cyclocarbonylation to afford 2(5*H*)-furanones **51**/**52** (Equation 10.25). This reaction was subsequently applied to a range of internal alkynes to afford the corresponding 3,4-disubstituted 2(5*H*)-furanones, albeit with poor to moderate regioselectivity [45]. The use of water gas shift reaction conditions was critical for the reaction as no furanone was obtained when the water was replaced with a hydrogen source. Two mechanisms were considered, one involving a μ-η^1-furanone intermediate and the other a σ-furanoyl complex.

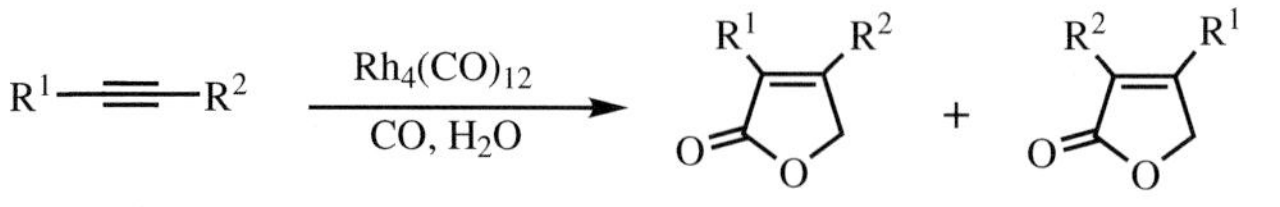

$$R^1\text{—}\equiv\text{—}R^2 \xrightarrow[\text{CO, } H_2O]{Rh_4(CO)_{12}} \mathbf{51} + \mathbf{52} \qquad (10.25)$$

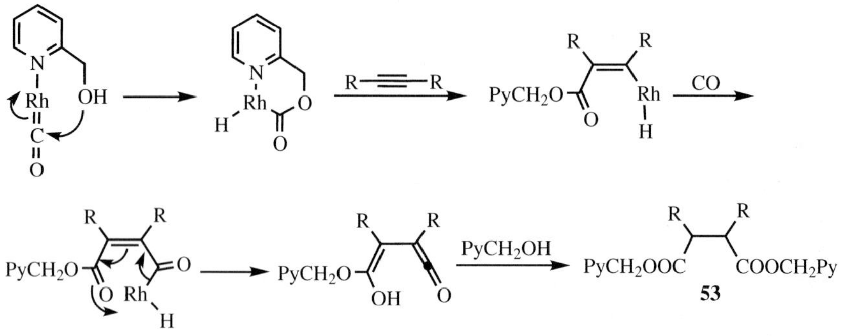

Scheme 10.13

While the oxidative carbonylation of terminal alkynes has been reported to afford maleic and fumaric acids and esters (Section 10.1.4.2), there is only a single report of the double hydroesterification of an alkyne to give a 1,4-dicarboxylate ester, although the rhodium-catalyzed carbonylation of diphenylacetylene in ethanol has been reported to give the corresponding 2(5*H*)-furanone as the major product together with a small amount of 2,3-diphenylsuccinic acid diethyl ester [46]. To this end, Chatani recently reported that internal alkynes undergo a double hydroesterification in the presence of pyridine-2-methanol and $Rh_4(CO)_{12}$ to give 1,4-dicarboxylate esters **53** as a mixture of *meso* and DL diastereoisomers in good to excellent yield [46]. Control experiments excluded a mechanism involving two consecutive steps in favor of a chelation-assisted transformation in which coordination of the pyridine nitrogen atom with rhodium facilitates intramolecular nucleophilic attack of alcohol on a coordinated CO to afford a cyclic alkoxycarbonylrhodium hydride; subsequent insertion of alkyne and CO generates a ketene intermediate that undergoes addition of a pyridine-2-methanol to afford the 1,4-dicarboxylate ester (Scheme 10.13).

Model stoichiometric reactions of $[PdCH_3(CO)(Pr^iDAB)]^+[B\{3,5\text{-}(CF_3)_2C_6H_3\}_4]^-$ (Pr^iDAB = 1,4-diisopropyl-1,4-diaza-1,3-butadiene) with alkynes and carbon monoxide have been investigated by NMR spectroscopy and DFT studies to identify the putative intermediates involved in the cyclocarbonylation of alkynes [47]. Addition of but-2-yne (R = CH_3) or 1-phenylpropyne (R = Ph) results in regioselective insertion into the Pd–acyl bond to afford a five-membered palladacycle **54** that undergoes rapid cyclocarbonylation at low temperature to afford a palladium-coordinated, η^3-allylic lactone **55**. The α,β-unsaturated γ-lactone could be liberated either by proton abstraction with a stoichiometric amount of $Na[BEt_3H]$ or by nucleophilic addition

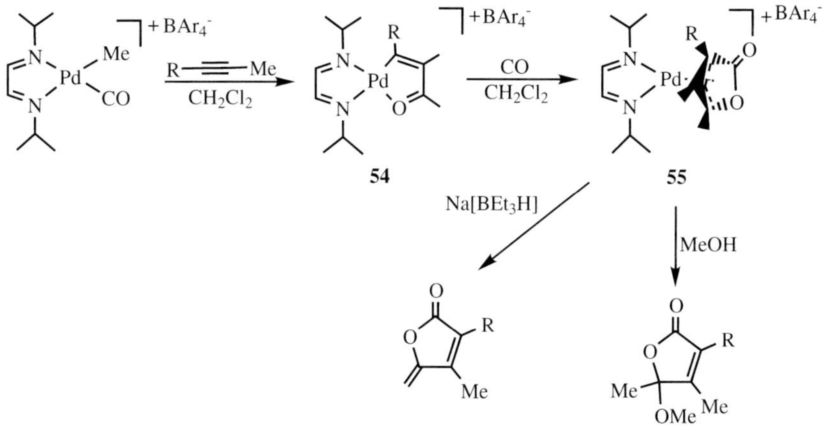

Scheme 10.14

with methanol (Scheme 10.14). Chiral palladium complexes based on BIOX react with alkynes (1-phenylpropyne, but-2-yne, hex-3-yne, and 1-phenylbutyne) and CO in a similar manner to afford the corresponding η^3-lactone complexes as a near equal mixture of diastereoisomers, which epimerize at room temperature and eventually reach an equilibrium diastereoisomeric excess of 86–94% [47].

Carbonylation of alkyne complexes of the early transition metals has received limited attention, which is somewhat surprising because this class of compound could react via different pathways and thus provide access to new carbonylation products. In this regard, Takahashi has shown that zirconocene alkyne complexes undergo double carbonylation at 0 °C to afford 4-hydroxycyclobuten-1-one derivatives **56** after hydrolysis [48]. The reaction temperature was crucial for achieving dicarbonylation, as cyclopentenone derivatives were formed at room temperature. Deuterium-labeling studies led the authors to propose a mechanism involving reductive coupling of alkyne and CO to afford a zirconacyclobutenone, insertion of a second molecule of CO to generate a maleoylzirconium complex, rearrangement to an oxazirconacyclopropane, and ultimately hydrolysis to liberate the product (Scheme 10.15). Other particularly significant and noteworthy developments in alkyne carbonylation with complexes of the group 4 metals include the titanium-catalyzed intramolecular Pauson–Khand

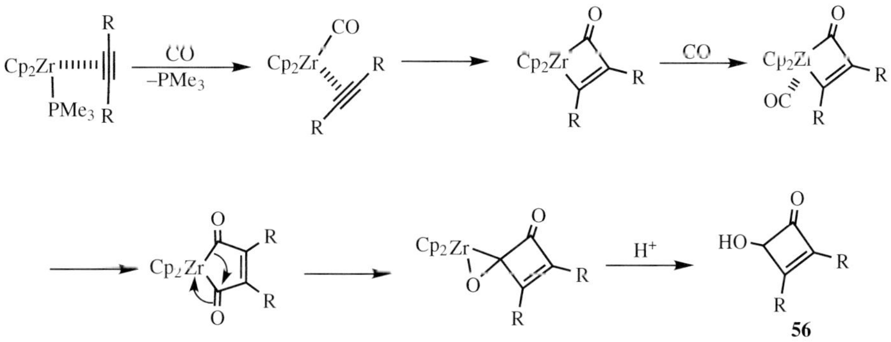

Scheme 10.15

reaction, a topic that has been exhaustively reviewed [49] and will be discussed in greater detail in other chapters.

10.3.6 Cyclocarbonylation of Alkynols

Unsaturated lactones are versatile building blocks as well as subunits for synthesis in a number of natural and nonnatural products, and thus their synthesis via cyclocarbonylation methodology has been the subject of a number of investigations. In this regard, the cyclocarbonylation of 3-butyn-1-ols is well documented and generally affords the α-alkylidene lactone as the major product. Recently though, Inoue has reported that cationic complexes of the type $[PdL_2(CH_3CN)_2](BF_4)_2$ (L_2 = dppb or $(PPh_3)_2$) catalyze the cyclocarbonylation of 3-butyn-1-ols to afford either five- or six-membered lactones depending on the reaction conditions and phosphine [50]. Six-membered ring lactones **57** were favored by catalysts based on a chelating diphosphine, such as dppb or dppf, acetonitrile as solvent, and a cationic catalyst precursor, whereas the use of $[Pd(PPh_3)_2(CH_3CN)_2](BF_4)_2$ in DMF gave the five-membered α-alkylidene lactone **58** as the sole product (Scheme 10.16). Although the formation of the α-alkylidene lactone could be accounted for by an alkoxycarbonyl-based pathway (a), the six-membered lactone would require a *trans* addition to the alkyne, which was suggested to occur via *syn* hydropalladation followed by isomerization at the resulting palladium–alkenyl double bond (b).Dupont and coworkers have developed a recyclable catalytic system for the highly regioselective intramolecular carbonylation of alkynols using $Pd(OAc)_2$/2-pyPPh$_2$ in organic solvent and 1-*n*-butyl-3-methyl imidazolium based ionic liquids (Equation 10.26) [51].Carbonylation of 1-alkyn-3-ols

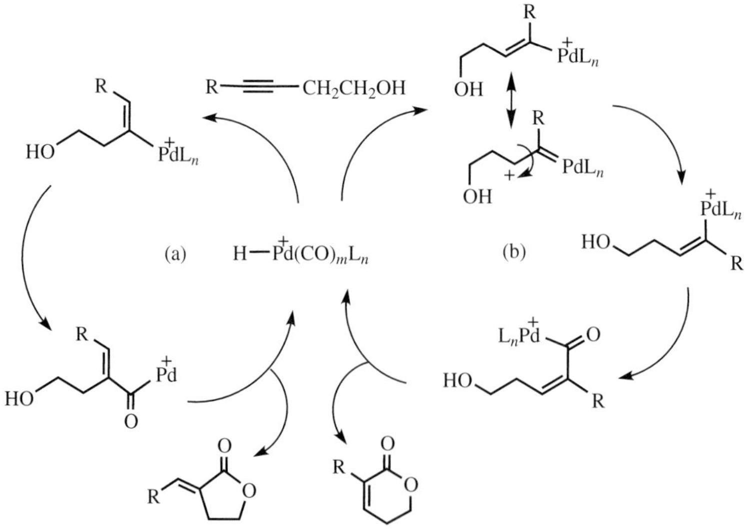

R—≡—CH₂CH₂OH

H—P̊d(CO)ₘLₙ

(a)

(b)

58

57

Scheme 10.16

and 1-alkyn-4-ols gave the corresponding *exo*-α-methylene γ- and δ-lactones, **59** and **60**, respectively, in near quantitative yields. In contrast, cyclocarbonylation of internal alkynes occurred much more slowly to afford the corresponding lactone in low yield. Ionic liquid solutions of the catalyst could be recycled after extraction of the product, although yields of lactone were significantly lower in successive reactions.

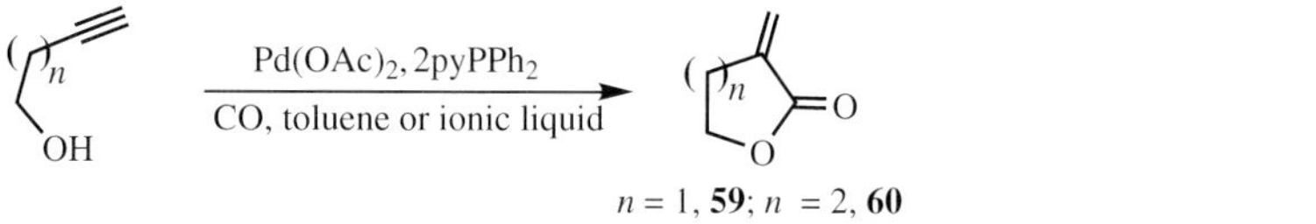

$Pd(OAc)_2$, $2pyPPh_2$; CO, toluene or ionic liquid (10.26)

$n = 1$, **59**; $n = 2$, **60**

Alper has recently extended his studies on the palladium-catalyzed thiocarbonylation of unsaturated substrates to include enynols that undergo a tandem cyclocarbonylation–thiocarbonylation sequence to afford thioester-substituted α,β-unsaturated six-membered ring lactones **61** and **62** in excellent selectivity and moderate to good yields (Equation 10.27) [52]. A screening of the catalyst revealed that triphenylphosphine-based systems gave the highest selectivity for the dicarbonylated lactone, whereas bidentate-based systems gave a mixture of mono- and dicarbonylation products, **61** and **62**, respectively. Two plausible mechanisms were considered; one involving an alkoxycarbonyl initiator and the other hydropalladation of the alkyne (Scheme 10.17).

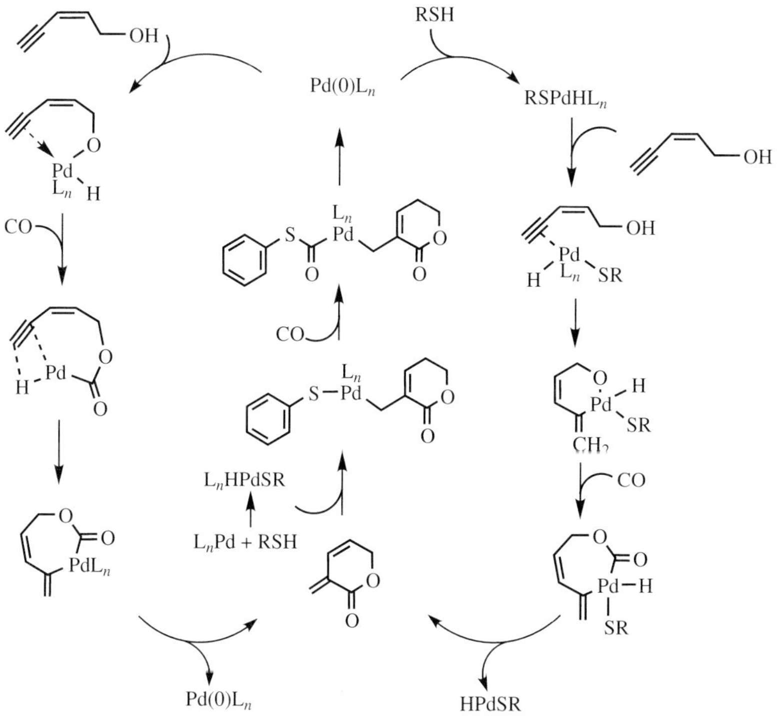

Scheme 10.17

$$\text{enynol-OH} + \text{PhS-H} \xrightarrow[\text{THF, CO, 110 °C}]{\text{Pd cat, ligand}} \mathbf{61} + \mathbf{62} \tag{10.27}$$

10.4 Aminocarbonylation of Terminal Alkynes

Acrylamides are important starting materials for a wide range of organic transformations and are also widely used in the synthesis of polymeric materials. Conventional methods for the synthesis of acrylamides involve the hydration of acrylonitrile, whereas their N-substituted counterparts are prepared in a stepwise manner starting from acrylic acid or its esters. Recently, the aminocarbonylation of terminal alkynes has been attracting interest as an alternative, more direct, atom economical and cleaner synthesis of acrylamides (Equation 10.28). In this regard, the palladium-catalyzed aminocarbonylation of phenylacetylene with $Pd(OAc)_2$/2-$pyPPh_2$/CH_3SO_3H in the presence of *n*-butylamine affords the branched acrylamide in low yield with a maximum regioselectivity of 90% [53]. Higher yields and exclusive regioselectivity for the branched isomer were obtained with the less basic aniline. Complete chemo- and regioselectivity for the branched amide was obtained in NMP and further optimization studies revealed that the highest rates were obtained at relatively low pressures (20 atm) and that catalyst activity increased with increasing acid/palladium ratio reaching a maximum at 30. The acid component of the aminocarbonylation catalyst makes the process corrosive and is incompatible with the amine, which may be the cause of the low yields obtained in some of the early studies. With the aim of developing an aminocarbonylation that did not require an acid cocatalyst or an activity booster, Alper investigated the palladium-catalyzed aminocarbonylation of alkynes (Equation 10.28) in ionic liquids, and under optimum conditions of $Pd(OAc)_2$/dppb in [bmim][NTf_2] (200 psi, 110 °C, 22 h), a range of terminal alkynes reacted to afford the branched acrylamide in moderate to excellent yields with excellent regioselectivities [54]. The high solubility of $Pd(OAc)_2$ and dppb in the ionic liquid enabled the catalyst to be recycled five times without any loss in activity.

$$R{-}{\equiv} + 2R'NH_2 \xrightarrow[\text{CO, solvent}]{\text{Pd cat}} \underset{\text{Branched}}{CH_2{=}C(R)CONHR'} + \underset{\text{Linear}}{RCH{=}CHCONHR'} \tag{10.28}$$

The palladium-catalyzed aminocarbonylation of alkyl-substituted terminal alkynes with aniline affords either the branched or linear acrylamide, depending on the catalyst composition and reaction conditions [55]. Optimum conditions for the production of the branched unsaturated amide were identified as $Pd(OAc)_2$/dppp/TsOH in THF, whereas $Pd(OAc)_2$/dppb/CO/H_2 in dichloromethane favored the formation of the linear amide with *E*-stereochemistry; in the latter case, yields increased markedly with increasing pressure. The former conditions also proved to be optimum for the carbonylative amination of terminal alkynes with secondary aryl amines to afford the tertiary unsaturated amide in high yield and with excellent selectivity for the branched regioisomer. Similarly, aminocarbonylation of terminal aryl-substituted alkynes with aniline derivatives was catalyzed by $Pd(OAc)_2$/dppb under syngas (CO/H_2) in toluene to afford the corresponding branched unsaturated amides in good yields with excellent regioselectivities [56]. Substitution of aniline for acetanilide resulted in amidocarbonylation to give the corresponding branched *N*-acyl acrylamide in good yield and with excellent regioselectivity. The same system also catalyzed the aminocarbonylation of internal alkynes, but the yields and regioselectivities were lower. The aminocarbonylation of terminal alkynes can also be conducted in the presence of a cocatalyst such as an organic iodide or ammonium iodide salt with palladium catalysts based on PPh_3 or dppb giving branched acylamides with excellent regioselectivities [57].

Radical carbonylation is proving to be an exceptionally powerful tool in the synthesis of carbonyl compounds and has recently been used to prepare branched acrylamides in a highly regioselective aminocarbonylation of terminal alkynes [58]. Minor amounts of the corresponding α-stannylmethylene amide were formed as the by-product but could be converted into the desired acrylamide by protodestannylation. Deuterium-labeling studies and DFT calculations support the tin radical catalyzed hybrid radical/ionic mechanism in which a tributyltinvinyl radical undergoes carbonylation to generate an α-ketenyl radical that is trapped by amine to afford a 1-hydroxyallyl radical. The acrylamide is then generated via a 1,4-*H* migration and β-fission (Scheme 10.18).

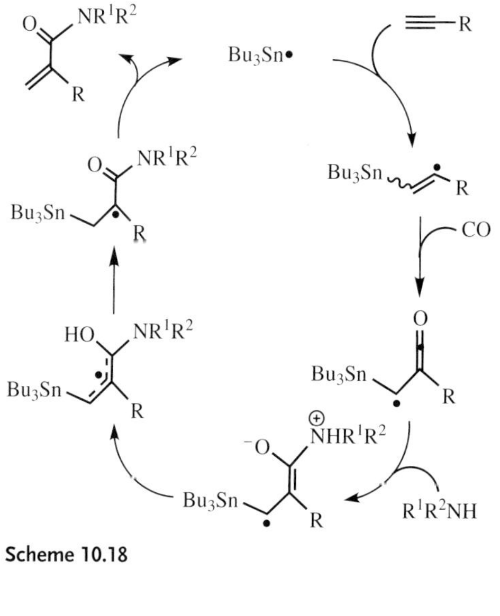

Scheme 10.18

10.5
Oxidative Carbonylations

10.5.1
Oxidative Hydroxy-, Alkoxy-, and Aminocarbonylation of Terminal Alkynes

Alkynoates are a highly versatile class of compound that has been used in the synthesis of biologically important molecules such as butenolides and carbapenem intermediates. Although these compounds can be prepared in a stoichiometric manner, there are advantages to the use of catalytic methods and, in this regard, the carbonylation of terminal alkynes has been receiving increasing attention, with an emphasis on using oxygen as the oxidant. The palladium-catalyzed oxidative alkoxycarbonylation of alkynes provides a convenient route to 2-alkynoates, and the first transformations of this type used copper(II) salts in combination with sodium acetate as an oxidant. More recently, alternative oxidants such as quinones have been used in the transition metal catalyzed carbonylation of phenylacetylene. As a result of a search for an alternative oxidizing agent to improve on existing processes, Yamamoto and Shimizu found that $Pd(OAc)_2/PPh_3$ in DMF catalyzes the oxidative carbonylation of a range of 1-alkynes to afford the corresponding 2-alkynoates **63** in moderate to good yield under 20 atm of a 4:1 mixture of CO and oxygen (Equation 10.29) [59]. The presence of triphenylphosphine was essential to obtain high yields, whereas bidentate diphosphines were less effective. Stoichiometric reactions with model compounds were undertaken to establish the putative elementary steps of the catalytic cycle. For example, a methoxycarbonyl palladium complex reacted with phenylacetylene in the presence of a base to afford methyl 3-phenyl-2-propynoate and the treatment of a phenylethynylpalladium chloride complex with methanol and NaOAc under an atmosphere of CO liberated methyl 3-phenyl-2-propynoate. A number of mechanisms were considered and, on the basis of the stoichiometric reactions, the most likely mechanism was suggested to involve a methoxycarbonyl palladium initiator, generated from a Pd(II) precursor, CO, and methanol, which reacts with a terminal alkyne to afford a palladium(II) intermediate with coordinated meth-oxycarbonyl and alkynyl groups. Reductive elimination from this species would liberate methyl alkynoate and generate Pd(0) that is reoxidized by oxygen via an η^2-O_2-Pd complex and a hydroxide-bridged dimer (Scheme 10.19).

$$R{-}{\equiv}{-}H \xrightarrow[\text{CO, [O], MeOH}]{\text{Pd cat, solvent}} R{-}{\equiv}{-}C(=O)OMe \quad \mathbf{63} \qquad (10.29)$$

In a similar transformation, oxidative aminocarbonylation of alkyl- and aryl-substituted 1-alkynes is catalyzed by PdI_2/KI under relatively mild conditions to afford 2-ynamides **64** in good yield (Scheme 10.20) [60]. Nucleophilic secondary amines were required as amines of low basicity were unreactive and primary amines gave complex reaction mixtures. The key intermediate in the mono-aminocarbonylation was proposed to be an alkynyl palladium iodide that undergoes CO insertion followed by nucleophilic abstraction by amine to generate the product and Pd(0), which is reoxidized by iodine ($2HI + 1/2O_2 = I_2 + H_2O$). Small amounts of diaminocarbonylation product (maleic

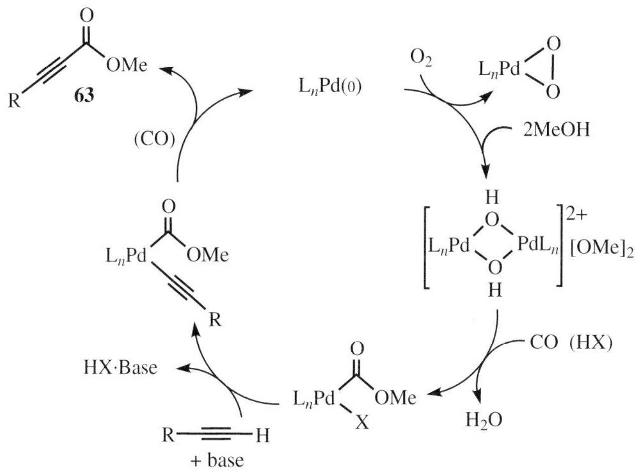

Scheme 10.19

bis-amide) were also obtained with alkyl-substituted alkynes, which were suggested to result from competitive *syn* addition of the carbamoylpalladium initiator to the triple bond, CO insertion, and nucleophilic abstraction. In contrast, oxidative carbonylation with alcohols as the nucleophilic partner resulted in dicarbonylation to afford maleic acid derivatives as the sole product (Section 10.1.4.2).

In an extension of their work on triple catalytic systems, Ishii and coworkers used mixtures of $Pd(OAc)_2$, chlorohydroquinone, molybdovanadophosphate (NPMoV), and methane sulfonic acid under O_2/CO in methanol to catalyze the oxidative alkoxycarbonylation of aryl- and alkyl-substituted terminal alkynes to afford the corresponding 2-alkynoates **63** [61]. Lower yields were obtained at higher pressures of CO/O_2, no reaction occurred in the absence of dioxygen, and reaction was slow in the absence of acid. Although mechanistic studies were not reported, the authors suggested that the key intermediate was a vinyl carboxyl palladium(II) species. However, generation of the 2-alkynoate from this intermediate would require a challenging β-hydride elimination of H-Pd-OAc involving a vinylic sp^2 carbon. High yields of 2-alkynoates have also been obtained by oxidative alkoxycarbonylation of terminal alkynes with catalyst systems based on $PdBr_2/CuBr_2$ and a base in alcohol under 1 atm of CO [62]. Electrochemical reoxidation of Pd(0) to Pd(II) (+0.4 V versus

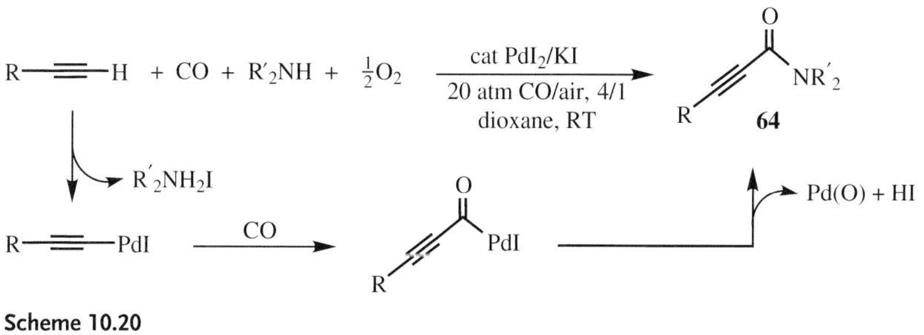

Scheme 10.20

SCE) has been investigated as an alternative to the use of the $PdCl_2/CuCl_2$ and Pd(II)/HQ-Cl/NPMoV systems [63]. The electrocarbonylation of 1-alkynes was suggested to occur via a mechanism similar to that proposed by Heck and the base was considered essential for the deprotonation of the vinylic β-H^+ to overcome the intrinsic problem of β-hydride elimination from the product of *syn* addition of L_n(OAc)PdC(O)OR to the triple bond. Catalytic oxidative carbonylation of alkynes can also be performed in the absence of an oxidant by coupling it with reductive carbonylation of the same alkyne (Equations 10.30 and 10.31) [64]. The composition of the catalyst is critical to obtaining good conversions and yields and a PdI_2–thiourea system with a thiourea/Pd ratio of 3.5 proved to be optimal. The amount of oxidative carbonylation product **65/66** balanced that obtained from reductive coupling **67/68**, and any differences were attributed to partial reduction of the palladium complex. The products of oxidative carbonylation were proposed to form via a common intermediate generated by sequential insertion of alkyne and CO into a palladium(II) alkoxycarbonyl initiator, while the pathway for reductive carbonylation was suggested to involve hydride transfer between two α,β-unsaturated acyl chains in the same molecule.

R—≡—H —(cat $PdCl_2$ / CO, MeOH)→ MeO_2C(R)C=CH(CO_2Me) *E/Z*-**65** + **66** (10.30)

R—≡—H —(cat Pd(0) / CO, MeOH)→ **67** + MeO_2C–C(R)=CH–CH$(OMe)_2$ **68** (10.31)

3-Chloroacrylate esters are valuable intermediates in organic synthesis and there are numerous methods available for their preparation, the majority of which are selective for the *E*-stereoisomer. However, the stereoselective synthesis of *Z*-chloroacrylate esters is less well developed and remains of considerable interest to the synthetic chemist. In this regard, Jiang and coworkers have recently reported that the palladium-catalyzed oxidative monocarbonylation of terminal alkynes with aliphatic alcohols and CO in the presence of excess $CuCl_2$ results in highly regio- and stereoselective chloroalkoxycarbonylation to afford the corresponding (*Z*)-3-chloroacrylates **69** (Equation 10.32) [65]. The copper(II) chloride acts both as an oxidant and as the source of chlorine. Although the authors favored a pathway involving chloropalladation, CO insertion followed by nucleophilic abstraction in preference to the alternative of an acylpalladation-reductive elimination sequence, it has since been suggested that *syn* addition of $CuCl_2$ across the alkyne and transmetalation is a more likely pathway for the formation of the β-chlorovinylpalladium intermediate.

R—≡—H + R^1OH + CO —($PdCl_2$, $CuCl_2$ / CO/O_2, 1 atm / THF, RT)→ Cl(R)C=CH(CO_2R^1) **69** (10.32)

10.5.2
Oxidative Di- and Tricarbonylation

In addition to monocarbonylation, selective oxidative dicarbonylation of terminal alkynes is of interest because the resulting maleic anhydrides are important intermediates for the synthesis of biologically active molecules or monomers for polymerization. Following his early studies of the dicarbonylation of alkynes in alcohols, Alper demonstrated that the oxidative dicarbonylation of terminal alkynes catalyzed by $PdCl_2/CuCl_2/CO/O_2$ in formic acid affords a mixture of maleic and fumaric acid **70** and maleic anhydride **71** (Equation 10.33), the composition of which depends on the steric bulk of the alkyne, with less bulky groups giving more of the fumaric acid [66]. The presence of a phosphine ligand completely inhibited the reaction. While the distribution of these products varied dramatically depending on the reaction conditions, the highest total yield and proportion of anhydride was consistently obtained in formic acid, although the precise role of the acid was not established. The initiator was suggested to be a palladium hydroxycarbonyl species Pd-CO_2H formed by nucleophilic attack of water on coordinated CO. A regio- and stereoselective *syn* insertion of the alkyne would generate a palladium vinyl intermediate that would liberate maleic acid by two possible pathways, either by insertion of CO followed by nucleophilic abstraction of the resulting α,β-unsaturated acyl with water or reductive elimination from a vinyl hydroxycarbonyl species. As the maleic and fumaric acid derivatives did not interconvert significantly under the conditions of the catalytic reaction, a number of possible isomerization pathways were considered with a four-centered concerted process and a dipolar mechanism considered to most likely. The reduced palladium was suggested to be oxidized via a hydroperoxo palladium complex that results from the insertion of O_2 into the Pd–H bond, generated from Pd(0) and HCl. In contrast, oxidative carbonylation of terminal alkynes in CH_3OH : DMF (1 : 1) in the presence of an iodide salt such as NaI or NEt_4I resulted in highly selective tricarbonylation to give **72** and **73** (Equation 10.34), whereas the use of a lower proportion of DMF and/or a lower CO pressure gave a mixture of mono- and tricarbonylation products [67]. Two possible mechanisms for the mono- and tricarbonylation were discussed, both of which were suggested to involve a common methoxycarbonyl Pd(II) intermediate.

R—≡—H + $HCOOH/H_2O$ —[$PdCl_2/CuCl_2$; CO/O_2, 1atm; THF, RT]→ **70** (R, H; HO_2C, CO_2H) + **71**

(10.33)

Ph—≡—H —[$Pd(OAc)_2$, PPh_3; CO, MeOH/DMF; 60 °C, NaI]→ Ph—≡—CO_2Me (**63**) + **72** (R, CO_2Me; MeO_2C, CO_2Me) + **73** (MeO, MeO, R, CO_2Me)

(10.34)

Although there have been a number of reports of the palladium-catalyzed synthesis of maleic anhydrides and acids, many such syntheses require the use of $CuCl_2$ cocatalyst as the oxidizing agent. Gabriele has recently reported an easy and practical palladium-catalyzed dicarbonylation of terminal alkynes with O_2 as the oxidant that is selective for the maleic anhydride or its acid depending on the reaction conditions [68]. For example, oxidative dicarbonylation of terminal alkynes in aqueous dioxane (PdI_2/KI, 4 : 1 : 10 mixture CO : air : CO_2, 60–80 °C) gave the maleic anhydride **71**, whereas reaction in the presence of a large excess of water (3 : 1 DME : H_2O) generated maleic acid **70** selectively and with an unprecedented efficiency. In the former case ^{13}C labeling studies were used to show that carbon dioxide acted as a promoter and not as a reagent. In contrast, the production of maleic acid in DME/H_2O was inhibited by CO_2, which was accounted for by competitive binding between CO_2 and the substrate. In the absence of O_2, carbonylation selectivity generated 3-substituted 2(5*H*)-furanones [68]. Regardless of the product, carbonylation was initiated via a hydroxycarbonyl palladium species, formed by attack of water on coordinated CO, *syn* insertion of alkyne into the Pd–C bond and either tautomerization/β-H elimination to afford the anhydride or nucleophilic abstraction by water to generate the maleic acid (Scheme 10.21). In an earlier study, Gabriele reported that the palladium-catalyzed (PdI_2/KI) oxidative carbonylation of terminal alkynes in water or aqueous solvents under moderate pressures of CO and O_2 occurs with high stereoselectivity [69], providing an efficient route to maleic acids, anhydrides, and their ring chain tautomers with turnover numbers up to 4000 mol product per mol of catalyst (Equation 10.35). The mechanism proposed involved insertion of alkyne into a palladium alkoxy- or hydroxycarbonyl initiator followed by insertion of CO to generate an α,β-unsaturated acyl intermediate that either undergoes nucleophilic abstraction with alcohol/water to afford the maleic acid/ester or intramolecular cyclization to generate the ring chain tautomer.

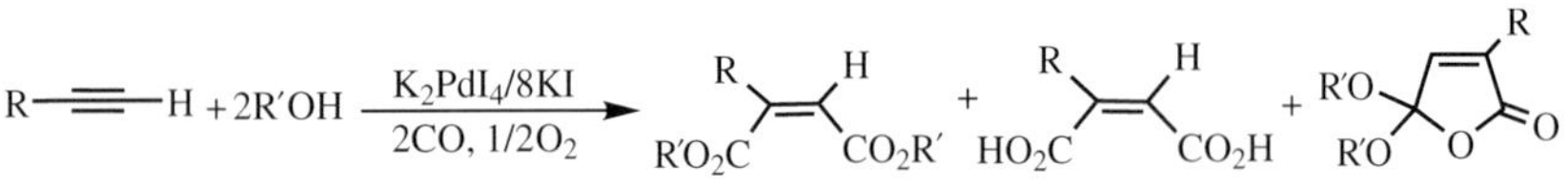
R
H + 2R′OH
K_2PdI_4/8KI
2CO, 1/2O_2
R
H
R′O_2C
CO_2R′
+
R
H
HO_2C
CO_2H
+
R′O
R′O
O
O
R

(10.35)

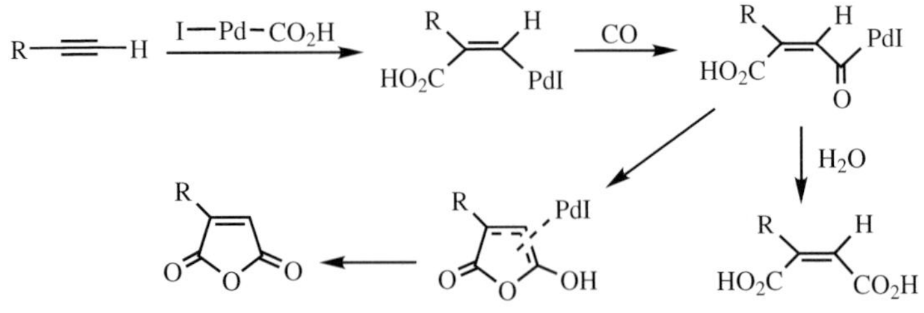
R
H
I−Pd−CO_2H
R
H
HO_2C
PdI
CO
R
H
PdI
HO_2C
O
H_2O
R
PdI
O
O
OH
R
O
O
O
71
R
H
HO_2C
CO_2H
70

Scheme 10.21

10.5.3
Oxidative Alkoxy- and Aminocarbonylation of Propargyl Alcohols, Amines and Acetates, Ynols, and Ynones

The palladium-catalyzed oxidative aminocarbonylation of terminal alkynes has been extended to include propargyl alcohols (Scheme 10.22; Y = O), which results in a direct one-pot oxidative carbonylation–conjugate addition–lactonization to afford 4-dialkylamino-5*H*-furan-2-ones 73 [70]. Molecules incorporating this fragment have shown interesting pharmacological properties. On the basis of low-conversion experiments and the reactivity of isolated intermediates, the author favored a pathway involving the palladium-catalyzed oxidative monoaminocarbonylation of the triple bond to give the 4-hydroxy-2-ynamide **72**, stereoselective conjugate addition to the corresponding *E*-enamide, and intramolecular alcoholysis of the amide to liberate furanone **73** (Scheme 10.22). Varying amounts of 4-hydroxy-2-ynamide and maleic bis(amide) were also formed, presumably via pathways similar to those shown in Schemes 10.20 and 10.21, respectively. The same authors extended this methodology to include the synthesis of 4-dialkylamino-1,5-dihydropyrrol-2-ones via oxidative aminocarbonylation of propargylamines (Y = NBn) [71]. In contrast, the corresponding reaction with propynyl alcohols (α,α-disubstituted or α-monosubstituted) and but-3-yn-1-ols under CO in methanol gave *Z*-α-(alkoxycarbonyl)methylene β-lactones **74** and the corresponding γ-lactones, respectively, in good yields and with excellent stereoselectivity (Equation 10.36) [72]. The highest selectivities for the β-lactone were obtained with α,α-disubstituted and α-monosubstituted propynyl alcohols containing sterically demanding substituents, whereas only low yields of lactone were obtained with unsubstituted propynyl alcohol and α-substituted derivatives with small alkyl groups; in such cases, the maleic ester and its cyclic tautomer were the major products. A number of pathways were presented to account for the composition and distribution of products, each of which was initiated by an alkoxycarbonyl species derived either from ROH or the acetylenic alcohol.

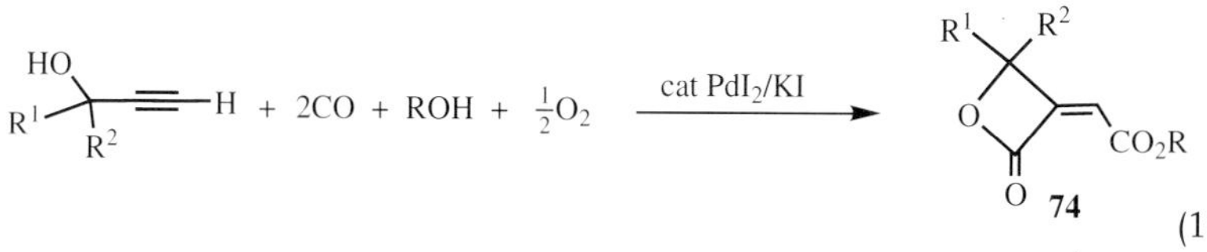

HO, R^1, R^2 ... ≡—H + 2CO + ROH + $\frac{1}{2}O_2$ —cat PdI_2/KI→ **74** (R^1, R^2, O, O, CO_2R) (10.36)

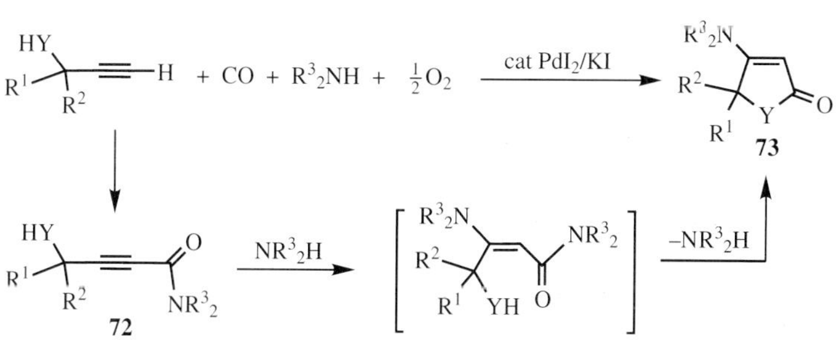

HY, R^1, R^2 ... ≡—H + CO + $R^3{}_2NH$ + $\frac{1}{2}O_2$ —cat PdI_2/KI→ **73** ($R^3{}_2N$, R^2, R^1, Y, O)

72 (HY, R^1, R^2, O, $NR^3{}_2$) —$NR^3{}_2H$→ [$R^3{}_2N$, R^2, R^1, YH, O, $NR^3{}_2$] —$-NR^3{}_2H$→ **73**

Scheme 10.22

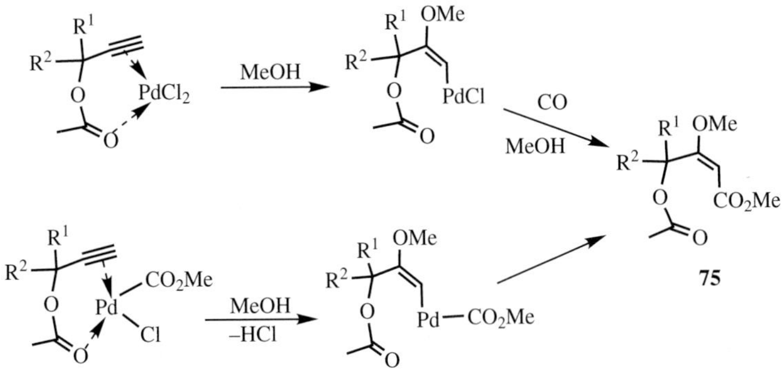

Scheme 10.23

The palladium-catalyzed intramolecular oxidative cyclocarbonylation of acetylenic substrates such as 4-yn-1-ols, 4-yn-1-ones, 4-yn-1-als, and propargylic acetates has evolved into a useful strategy for the synthesis of acyclic and heterocyclic β-alkoxyacrylates that are present as fragments in a range of natural and nonnatural antibiotics and steroids. For example, propargyl acetates can be converted into γ-acetoxy-β-methoxy-α,β-unsaturated esters **75** via palladium-catalyzed oxidative carbonylation under 1 atm of CO in methanol [73]. Overall, this transformation amounts to alkoxy-alkoxycarbonylation for which two pathways were suggested, both of which involved a common palladium vinyl intermediate that either undergoes CO insertion–nucleophilic abstraction or reductive elimination from the corresponding alkoxycarbonyl intermediate (Scheme 10.23). Propargylic acetates have also been shown to undergo an unusual highly stereoselective oxidative cyclocarbonylation ($PdCl_2(CH_3CN)_2$, *p*-benzoquinone, CH_3OH, 0 °C–RT, 1 atm CO) to form (*E*)-cyclic-orthoesters **76** in good yields (Scheme 10.24) [74]. In the case of R^1 = aryl, a dramatic dependence of the reaction yield on the *para* substituent was considered strong evidence that the carbonyl group played an important role in initiating the reaction. In combination with the NMR evidence, cyclization of propargylic acetates was suggested to be initiated by nucleophilic attack of the carbonyl oxygen on the palladium-coordinated alkyne to generate a vinyl palladium intermediate that undergoes addition of CH_3OH to the carbon atom of the carbonyl (Scheme 10.24). The synthetic utility of these transformations has been demonstrated by a number of applications to the synthesis of β-methoxyacrylate antibiotics, antifungal agents, and steroid derivatives.

Under oxidative conditions, acyclic [75a,b] and cyclic [75c] 4-yn-1-ols react via a cyclization–methoxycarbonylation pathway to afford (*E*)-cyclic-β-alkoxyacrylates (2*E*-[(methoxycarbonyl)methylene]tetrahydrofurans in good yields under mild conditions (Equations 10.37 and 10.38). The stereochemical outcome is entirely consistent with a reaction initiated by intramolecular nucleophilic attack of the hydroxy group on the palladium-coordinated triple bond to generate a vinylpalladiumiodide intermediate that undergoes methoxycarbonylation to afford the β-alkoxyacrylate. The ketopyranose subunit can also be constructed via a palladium-catalyzed oxidative cyclocarbonylation of substituted 5-yn-1-ols, which occurs with excellent stereose-

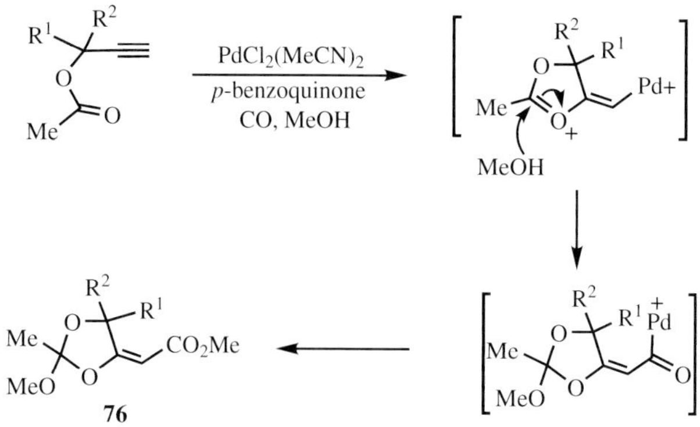

R^1 R^2 O Me O — $PdCl_2(MeCN)_2$, *p*-benzoquinone, CO, MeOH → [R^2 R^1 O Me O+ Pd+ MeOH] → [R^2 R^1 Pd+ O Me MeO O O] → R^2 R^1 O Me MeO O CO_2Me **76**

Scheme 10.24

lectivity [76]. The synthetic utility of this transformation was illustrated with the preparation of ketopyranosides of the type found in cytotoxic polyketide natural products, for example, apoptoldin. Under similar conditions of oxidative carbonylation, acyclic and cyclic 4-yn-1-ones afford cyclic ketals **77** in high yield, as a single diastereoisomer for substrates bearing a quaternary carbon atom α to the keto group (R^1, $R^2 \neq H$) and as a mixture of diastereoisomers for substrates bearing a tertiary carbon atom adjacent to the keto group (Scheme 10.25) [77]. The proposed mechanism accounted for the stereochemical outcome and involved nucleophilic addition of the carbonyl oxygen to the palladium-coordinated alkyne to generate an

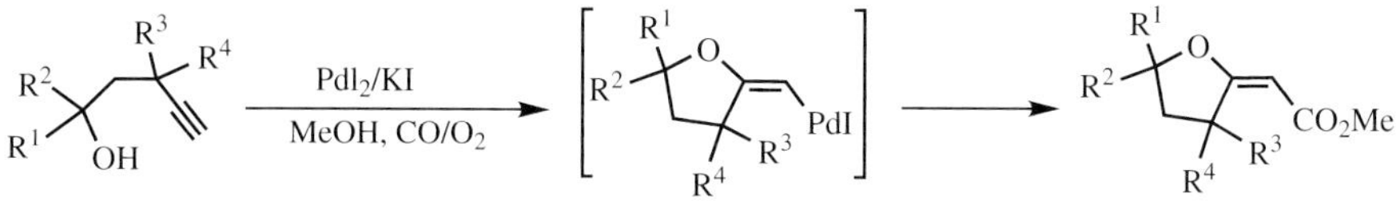

R^3 R^4 R^2 R^1 OH — PdI_2/KI, MeOH, CO/O_2 → [R^1 O R^2 PdI R^3 R^4] → R^1 O R^2 CO_2Me R^3 R^4

(10.37)

oxonium-type intermediate followed by addition of methanol to the carbon atom of the cationic carbonyl group from the side opposite to the methyl ester.

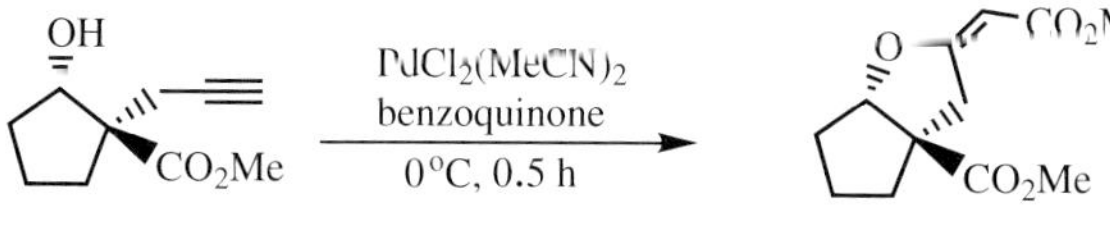

OH CO_2Me — $PdCl_2(MeCN)_2$, benzoquinone, 0°C, 0.5 h → O CO_2Me CO_2Me

(10.38)

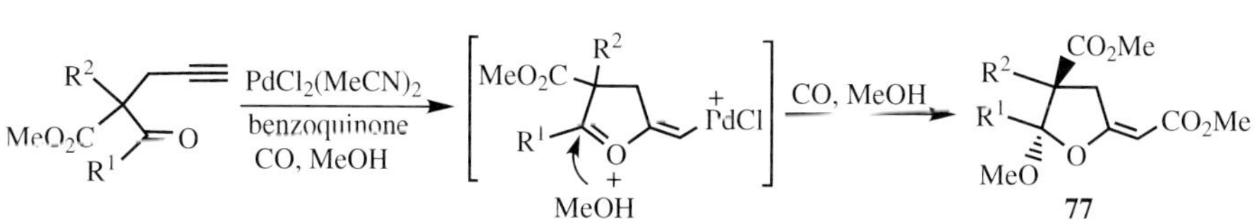

R^2 MeO_2C O R^1 — $PdCl_2(MeCN)_2$, benzoquinone, CO, MeOH → [MeO_2C R^2 R^1 O+ PdCl+ MeOH] — CO, MeOH → CO_2Me R^2 R^1 MeO O CO_2Me **77**

Scheme 10.25

10.6
Carbonylative Annulation of Alkynes

10.6.1
Intermolecular Carbonylative Annulation of Internal Alkynes

Larock investigated the reaction of 2-iodophenols with internal alkynes in the presence of carbon monoxide, reasoning that this three-component reaction would result in annulation to afford coumarins or chromones depending on the order of insertion of CO and alkyne into the palladium–aryl bond (Equation 10.39) [78]. Surprisingly, this reaction resulted in preferential insertion of alkyne into the palladium–aryl bond to give coumarins with no evidence for any product arising from the initial insertion of CO. Optimum conditions were identified as 1 atm CO, 5 mol% $Pd(OAc)_2$, 2 equiv of pyridine, and 1 equiv of n-Bu_4NCl in DMF at 120 °C. A study of the scope and limitation of this annulation showed that it was highly versatile and could be applied to a range of alkynes, including those with functional groups such as alkoxy, silyl, acyl, and esters, as well as various functionalized iodophenols including heterocyclic analogues of 2-iodophenol. Exclusive formation of the coumarins was unexpected since it requires insertion of an internal alkyne into an aryl–palladium bond in preference to CO insertion and previous examples of intermolecular reactions between alkynes, aryl iodides, and CO report that CO insertion precedes alkyne insertion even at 1 atm of CO. Intramolecular trapping experiments revealed that CO does insert into the palladium–aryl bond but that subsequent insertion of alkyne into the resulting palladium–acyl bond is slow such that decarbonylation ultimately results in alkyne insertion and carbonylation to afford the observed product (Scheme 10.26). Attempts to extend this methodology to include o-iodoaniline derivatives failed to give any of the expected 3,4-disubstituted

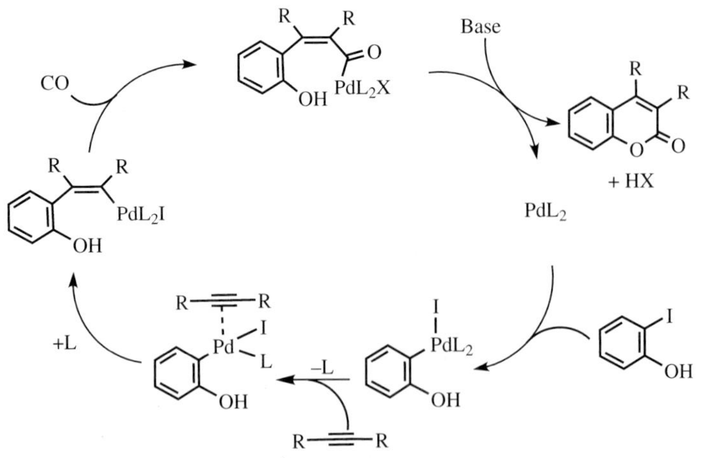

Scheme 10.26

2-quinolines. Reasoning that the failure of this reaction was associated with the high nucleophilicity of the amino group, the use of N-protected anilines gave the expected 2-quinolines as the sole product with no evidence for the isomeric 4-quinoline that would result from preferential insertion of CO into the palladium–aryl bond [79]. The highest yields were obtained with alkoxycarbonyl, *p*-tolylsulfonyl, and trifluoroacetyl protecting groups. This methodology has been applied to a range of internal diaryl, dialkyl, aryl–alkyl, and heteroaryl–alkyl alkynes with various electron-rich and electron-poor alkyl *N*-(2-iodophenyl)carbamates. Although poor to moderate regioselectivities are a limitation of this reaction, the regioisomers can be separated by column chromatography. The mechanism of this transformation was assumed to parallel to that for the synthesis of coumarins shown in Scheme 10.26. This synthetic approach represents a significant advance in the palladium-catalyzed synthesis of 2-quinolines as it utilizes readily available starting materials, is carried out under mild reaction conditions, and tolerates a variety of substituents and functional groups. Moreover, there are still relatively few examples of the palladium-catalyzed synthesis of 2-quinolines.

I, OH + R—≡—R —cat Pd(0), CO→ Coumarins + Chromones (10.39)

Under the same conditions, *o*-iodophenol and aniline derivatives also undergo carbonylative annulation with terminal alkynes in the presence of carbon monoxide to form the corresponding coumarins and 2-quinolines, respectively [80]. This is a particularly interesting discovery because analogous palladium-catalyzed reactions between a terminal alkyne, *o*-iodophenol, and CO have previously been reported to afford either aurones or chromones, whereas the corresponding reaction with *o*-iodoanilines generates six-membered ring 4-quinolines.

10.6.2 Intramolecular Carbonylative Annulation of Internal Alkynes

A range of 3-substituted 4-aroylisoquinolines **78** have been prepared by the palladium-catalyzed three-component coupling of *N*-*tert*-butyl-2-(1-alkynyl)benzaldimines, aryl halides, and CO (Equation 10.40) [81]. Minor amounts of noncarbonylated isoquinoline by-products **79** and **80** were also isolated, although their formation could be limited by the use of an organic base such as NEt_3 and N^nBu_3. High yields were obtained with a wide range of electron-rich and electron-poor aromatic iodides as well as substituted 2-(1-alkynyl)benzaldimines. By analogy with the palladium-catalyzed synthesis of related heterocycles, the mechanism was suggested to involve oxidative addition of Ar–X to Pd(0), CO insertion, intramolecular nucleophilic addition of the imine nitrogen atom to the coordinated alkyne, reductive elimination with C–C bond formation between the acyl carbon and the isoquinoline ring, and cleavage of the *tert*-butyl group.

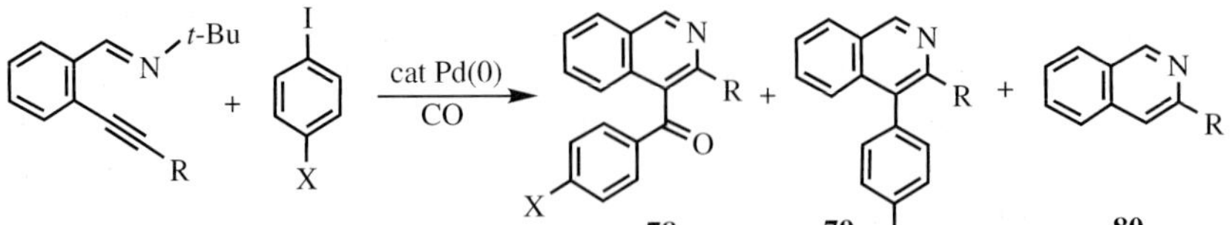

(10.40)

The palladium-catalyzed cascade carbonylative annulation of *o*-hydroxyarylalkynes in the presence of carbon monoxide results in multiple-bond formation to afford benzo[*b*]furan-3-carboxylic esters (R = alkyl) and their acids (R = H) **81** (Equation 10.41) depending on the reaction conditions; formation of the ester requires PdI_2, thiourea, CO, Cs_2CO_3 in CH_3CN, while the highest yield of acid was obtained with a catalyst mixture based on $PdCl_2(CH_3CN)_2$, AgOTs, 2-pyPPh$_2$, and CsOAc at 50 °C. In the latter case, the presence of silver salt was crucial to limit the formation of *o*-acetyloxyarylacetylene by-product [82].

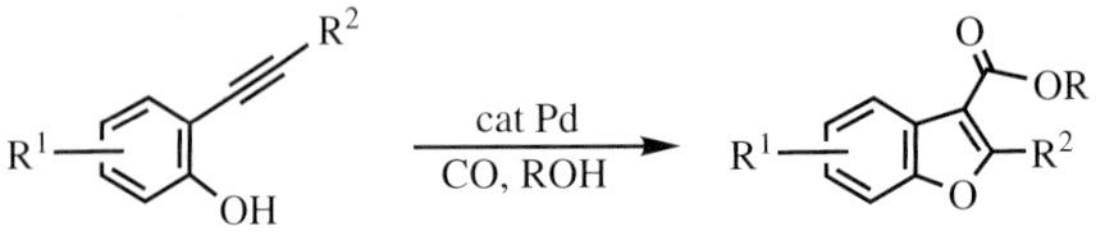

(10.41)

During a series of investigations into palladium-catalyzed cyclocarbonylation of *o*-ethynylphenols, Arcadi found that the outcome of the three-component reaction of this substrate with vinyl triflates and carbon monoxide gave either 3-alkylidene-2-coumaranones or 3-acylbenzo[*b*]furans, depending on the alkyne substituent, R (Equation 10.42) [83]. For example, the palladium catalyzed ($Pd(PPh_3)_4$) reaction between of *o*-ethynylphenol (R = H) and 3,3,5,5-tetramethylcyclohex-1-en-1yl triflate under 1 atm of carbon monoxide resulted in annulation to afford 3-(3,3,5,5-tetramethylcyclohex-1-en-1-yl)methylidene-2-coumaranone. In striking contrast, under the same conditions, the reaction between *o*-(arylethynyl)phenol (R = Ar) and 3,3,5,5-tetramethylcyclohex-1-en-1yl triflate generated 3-acylbenzo[*b*]furan with no evidence of the corresponding coumaranone.

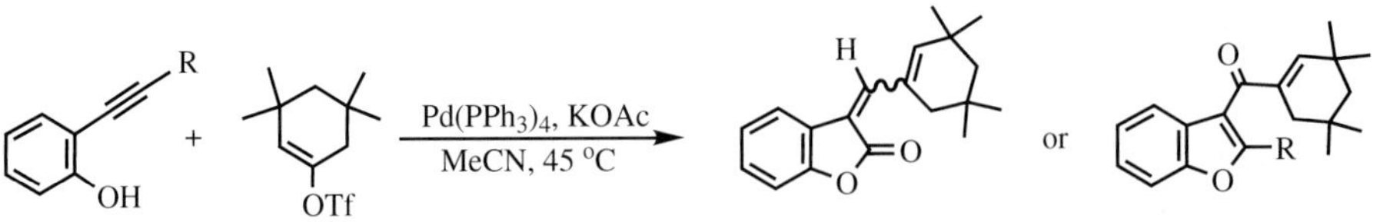

3-alkylidene-2-coumaranone or 3-acyl-2-arylbenzo[*b*]furan

(10.42)

10.7 Summary and Outlook

Alkyne carbonylation in the presence of an alcohol or amine is now a well-established and highly versatile method for the synthesis of a range of important intermediates and monomers for the industrial-scale preparation of polymeric materials. Intramolecular versions of this transformation provide access to unsaturated carbonylated heterocycles and have been used to prepare polycyclic lactones with complex molecular architectures. Other noteworthy developments include the carbonylative annulation of internal alkynes to afford carbonylated nitrogen- and oxygen-based heterocycles. Oxidative carbonylation of terminal alkynes can give 2-alkynoates and amides, maleic acids and their derivatives, and furanones as well as products of tricarbonylation, depending on the reaction conditions, whereas intramolecular versions afford a range of heterocyclic derivatives. The use of sulfur-based substrates in transition metal catalyzed alkyne carbonylation to introduce a thiocarbonyl unit has begun to increase the diversity of substrates and reaction products, and future studies will aim to advance carbonylation technology by further expanding this diversity. With the exception of a single study involving stoichiometric reactions, the stereoselective synthesis of butenolides via alkyne carbonylation has not been explored, which presents an exciting opportunity for future investigations in this area. There have been relatively few reports of the immobilization of alkyne carbonylation catalysts, and current strategies include the use of ionic liquids to effect catalyst recovery and recycling, covalent attachment to polymeric supports, and the use of fluorous biphasic systems and supercritical CO_2. Further investigations in this area will be required to develop catalyst systems suitable for use in semicontinuous processes for scale-up or in microfluidic reactors. Furthermore, the recent development of multicomponent and tandem reactions involving alkyne carbonylation serves to demonstrate the potential versatility of this chemistry.

References

1 (a) Reppe, W. (1939) Ger. Patent, 885 **110**. (b) Reppe, W. (1948) *Annals of Chemistry*, **560**, 1.

2 Kiss, G. (2001) *Chemical Reviews*, **101**, 3435–3456.

3 El Ali, B. and Alper, H. (2000) *Synlett*, 161–171.

4 Gabriele, B., Salerno, G., Costa, M. and Chiusoli, G.P. (2004) *Current Organic Chemistry*, **8**, 919–946.

5 Gabriele, B., Salerno, G. and Costa, M. (2004) *Synlett*, 2469–2483.

6 (a) Zeni, G. and Larock, R.C. (2006) *Chemical Reviews*, **106**, 4644–4680. (b) Zeni, G. and Larock, R.C. (2004) *Chemical Reviews*, **104**, 2285–2309.(c) Nakamura, I. and Yamamoto, Y. (2004) *Chemical Reviews*, **104**, 2127–2198.

7 (a) El Ali, B. and Alper, H. (2004) *Transition Metals for Organic Syntheses: Building Blocks and Fine Chemicals* (eds. M. Beller and C. Bolm) Wiley-VCH Verlag GmbH, Weinheim, Germany. (b) Falbe, J. (1980) *New Syntheses with Carbon Monoxide*, Springer-Verlag, Berlin. (c) Colquhoun, H.M., Thomson, D.J. and Twigg, M.V. (1991) *Carbonylation: Direct Synthesis of Carbonyl Compounds*, Plenum Press, Weinheim, New York.

8 (a) Ogawa, A., Kawakami, J., Sonoda, N. and Hirao, T. (1996) *Journal of Organic Chemistry*, **61**, 4161. (b) Miyaura, N. and Suzuki, A. (1995) *Chemical Reviews*, **95**, 2457. (c) Han, L. B. and Tanaka, M. (1999) *Chemistry Letters*, 863. (d) Sugoh, K., Kuniyasu, H., Sungae, T., Ohtaka, A., Takai, Y., Tanaka, A., Machino, C.,

Kambe, N. and Kurosawa, H. (2001) *Journal of the American Chemical Society*, **123**, 5108.

9 (a) Ogawa, A., Takeba, M., Kawakami, J., Ryu, I., Kambe, N. and Sonoda, N. (1995) *Journal of the American Chemical Society*, **117**, 7564–7565. (b) Kawakami, J., Takeba, M., Kamiya, I., Sonoda, N. and Ogawa, A. (2003) *Tetrahedron: Asymmetry*, **59**, 6559–6567.

10 (a) Ogawa, A., Kawakami, I., Mihara, M., Ikeda, T., Sonoda, N. and Hirao, T. (1997) *Journal of the American Chemical Society*, **119**, 12380–12381. (b) Kawakami, J., Mihara, M., Kamiya, I., Takeba, M., Ogawa, A. and Sonoda, N. (2003) *Tetrahedron: Asymmetry*, **59**, 3521–3526.

11 El Ali, B., Tijani, J., El-Ghanam, A. and Fettouhi, M. (2001) *Tetrahedron: Asymmetry*, **42**, 1567–1570.

12 Xiao, W.-J., Vasapollo, G. and Alper, H. (1999) *Journal of Organic Chemistry*, **64**, 2080–2084.

13 Kuniyasu, H., Ogawa, A., Miyazaki, S.-I., Ryu, I., Kambe, N. and Sonoda, N. (1991) *Journal of the American Chemical Society*, **113**, 9796–9803.

14 Nakatani, S., Yoshida, J. and Isoe, S. (1992) *Journal of the Chemical Society, Chemical Communications*, 880–881.

15 Xiao, W.-J. and Alper, H. (1997) *Journal of Organic Chemistry*, **62**, 3422–3423.

16 Xiao, W.-J. and Alper, H. (2005) *Journal of Organic Chemistry*, **70**, 1802–1807.

17 Ogawa, A., Kawabe, K., Kawakami, J., Mihara, M., Hirao, T. and Sonoda, N. (1998) *Organometallics*, **17**, 3111–3114.

18 Ogawa, A., Kuniyasu, H., Sonoda, N. and Hirao, T. (1997) *Journal of Organic Chemistry*, **62**, 8361–8365.

19 Kuniyasu, H., Kato, T., Asano, S., Ye, J.-H., Ohmori, T., Morita, M., Hitaike, H., Fujiwara, S., Terao, J., Kurosawa, H. and Kambe, N. (2006) *Tetrahedron Letters*, **47**, 1141.

20 (a) Knapton, D.J. and Mayer, T.Y. (2005) *Journal of Organic Chemistry*, **70**, 785–796. (b) Knapton, D.J. and Mayer, T.Y. (2004) *Organic Letters*, **6**, 687–689.

21 (a) Drent, E., Arnoldy, P. and Budzelaar, P.H.M. (1994) *Journal of Organometallic Chemistry*, **475**, 57–63. (b) Drent, E., Arnoldy, P. and Budzelaar, P.H.M. (1993) *Journal of Organometallic Chemistry*, **455**, 247–253.

22 (a) Scrivanti, A. and Matteoli, U. (1995) *Tetrahedron Letters*, **36**, 9015–9018. (b) Ciappa, A., Matteoli, U. and Scrivanti, A. (2002) *Tetrahedron: Asymmetry*, **13**, 2193–2195.

23 (a) Scrivanti, A., Beghetto, V., Zanato, M. and Matteoli, U. (2000) *Journal of Molecular Catalysis A: Chemical*, **160**, 331–336. (b) Scrivanti, A., Beghetto, V., Campagna, E. and Matteoli, U. (2001) *Journal of Molecular Catalysis A: Chemical*, **168**, 75–80.

24 Scrivanti, A., Matteoli, U., Beghetto, V., Antonaroli, S., Scarpelli, R. and Crociani, B. (2001) *Journal of Molecular Catalysis A: Chemical*, **170**, 51–56.

25 Reetz, M.T., Demuth, R. and Goddard, R. (1998) *Tetrahedron Letters*, **39**, 7089–7092.

26 (a) Scrivanti, A., Beghetto, V. and Matteoli, U. (2002) *Advanced Synthesis & Catalysis*, **244**, 543–547. (b) Matteoli, U., Botteghi, C., Sbrogio, F., Beghetto, V., Paganelli, S. and Scrivanti, A. (1999) *Journal of Molecular Catalysis A: Chemical*, **143**, 287–295.

27 Kushino, Y., Itoh, K., Miura, M. and Nomura, M. (1994) *Journal of Molecular Catalysis*, **59**, 151–158.

28 Scrivanti, A., Beghetto, V., Campagna, E., Zanato, M. and Matteoli, U. (1998) *Organometallics*, **17**, 630–635.

29 (a) Dervisi, A., Edwards, P.G., Newman, P.D. and Tooze, R.P. (2000) *Journal of the Chemical Society, Dalton Transactions*, 523–528. (b) Dervisi, A., Edwards, P.G., Newman, P.D., Tooze, R.P., Coles, S.J. and Hursthouse, M. B. (1999) *Journal of the Chemical Society, Dalton Transactions*, 1113–1120.

30 Green, M.J., Cavell, K.J., Edwards, P.G., Tooze, R.P., Skelton, B.W. and White, A.H. (2004) *Journal of the Chemical Society, Dalton Transactions*, 3251–3260.

31 Akato, M., Sugawara, S., Amino, K. and Inoue, Y. (2000) *Journal of Molecular Catalysis A: Chemical*, **157**, 117–122.

32 Alper, H., Saldana-Maldonado, M. and Lin, I.J.B. (1988) *Journal of Molecular Catalysis*, **49**, L27–L30.

33 (a) El Ali, B. and Alper, H. (1995) *Journal of Molecular Catalysis A: Chemical*, **96**, 197–201.(b) El Ali, B., Tijani, J. and El-Ghanam, A.M. (2001) *Tetrahedron Letters*, **42**, 2385–2387.

34 Doherty, S., Knight, J.G. and Betham, M. (2006) *Journal of the Chemical Society, Chemical Communications*, 88–90.

35 de Pater, J.J.M., Maljaars, C.E.P., de Wolf, E., Lutz, M., Spek, A.L., Deelman, B.-J., Elsevier, C.J. and van Koten, G. (2005) *Organometallics*, **24**, 5299–5310.

36 (a) Zargarian, D. and Alper, H. (1993) *Organometallics*, **12**, 712–724. (b) El Ali, B., Vasapollo, G. and Alper, H. (1993) *Journal of Organic Chemistry*, **58**, 4739–4741.

37 Satyanarayana, N. and Alper, H. (1991) *Organometallics*, **10**, 804–807.

38 (a) El Ali, B. and Alper, H. (1991) *Journal of Organic Chemistry*, **56**, 5357–5360. (b) Yu, W.-Y. and Alper, H. (1997) *Journal of Organic Chemistry*, **62**, 5684–5687.

39 Matsushita, K., Komori, T., Oi, S. and Inoue, Y. (1994) *Tetrahedron Letters*, **35**, 5889–5890.

40 Tsuji, J. and Mandai, T. (1995) *Angewandte Chemie-International Edition*, **34**, 2589–2612.

41 (a) Arzoumanian, H., Cochini, F., Nuel, D., Petrignani, J.F. and Rosas, N. (1992) *Organometallics*, **11**, 493–495. (b) Arzoumanian, H., Cochini, F., Nuel, D. and Rosas, N. (1993) *Organometallics*, **12**, 1871–1875.

42 Arzoumanian, H., Jean, M., Nuel, D., Cabrera, A., Garcia, J.L. and Rosas, N. (1995) *Organometallics*, **14**, 5438–5441.

43 Roasa, N., Sharma, P., Arellano, I., Ramírez, M., Pérez, D., Hernández, S. and Cabrera, A. (2005) *Organometallics*, **24**, 4893–4895.

44 Van den Hoven, B.G., El Ali, B. and Alper, H. (2000) *Journal of Organic Chemistry*, **65**, 4131–4137.

45 (a) Doyama, K., Joh, T., Onitsuka, K., Shiohara, T. and Takahashi, S. (1987) *Journal of the Chemical Society, Chemical Communications*, 7649–650. (b) Joh, T., Doyama, K., Onitsuka, K., Shiohara, T. and Takahashi, S. (1991) *Organometallics*, **10**, 2493–2498.

46 (a) Mise, T., Hong, P. and Yamazaki, H. (1983) *Journal of Organic Chemistry*, **48**, 238–242. (b) Inoue, S., Yokota, K., Tatamidani, H., Fukumoto, Y. and Chatani, N. (2006) *Organic Letters*, **8**, 2519–2522.

47 (a) Carfagna, C., Gatti, G., Mosca, L., Paoli, P. and Guerri, A. (2003) *Organometallics*, **12**, 3967–3970. (b) Carfagna, C., Gatti, G., Mosca, L., Paoli, P. and Guerri, A. (2005) *Chemistry – A European Journal*, **11**, 3268–3278.

48 Mito, S. and Takahashi, T. (2005) *Journal of the Chemical Society, Chemical Communications*, 2495–2497.

49 Shibata, T. (2006) *Advanced Synthesis & Catalysis*, **348**, 2328–2336.

50 Tezuka, K., Ishizaki, Y. and Inoue, Y. (1998) *Journal of Molecular Catalysis A: Chemical*, **129**, 199–206.

51 Consorti, C.S., Ebeling, G. and Dupont, J. (2002) *Tetrahedron Letters*, **43**, 753–755.

52 Cao, H., Xiao, W.-J. and Alper, H. (2006) *Advanced Synthesis & Catalysis*, **348**, 1807–1812.

53 Matteoli, U., Scrivanti, A. and Beghetto, V. (2004) *Journal of Molecular Catalysis A: Chemical*, **213**, 183–186.

54 Li, Y., Alper, H. and Yu, Z. (2006) *Organic Letters*, **8**, 5199–5201.

55 (a) El Ali, B., Tujani, J. and El-Ghanam, A.M. (2002) *Journal of Molecular Catalysis A: Chemical*, **187**, 17–33. (b) El Ali, B. and Tijani, J. (2003) *Applied Organometallic Chemistry*, **17**, 921–931.

56 (a) El Ali, B., Tijani, J. and El-Ghanam, A. M. (2002) *Applied Organometallic Chemistry* **16**, 369–376. (b) El Ali, B., El-Ghanam, A., Fettouhi, M. and Tijani, J. (2000) *Tetrahedron Letters*, **41**, 5761–5674.

57 Torii, S., Okumoto, H., Sadakane, M. and He Xu, L. (1991) *Chemistry Letters*, 1675–1676.

58 Uenoyama, Y., Fukuyama, T., Nobuta, O., Matsubara, H. and Ryu, I. (2005) *Angewandte Chemie-International Edition*, **44**, 1075–1078.

59 Izawa, Y., Shimizu, I. and Yamamoto, A. (2004) *Bulletin of the Chemical Society of Japan*, **77**, 2033–2045.

60 Gabriele, B., Salerno, G., Veltri, L. and Costa, M. (2001) *Journal of Organometallic Chemistry*, **622**, 84–88.

61 Sakurai, Y., Sakaguchi, S. and Ishii, Y. (1999) *Tetrahedron Letters*, **40**, 1701–1704.

62 Li, J., Jiang, H. and Chen, M. (2001) *Synthetic Communications*, **31**, 199–202.

63 Chiarotto, I. and Carelli, I. (2002) *Synthetic Communications*, **32**, 881–886.

64 Gabriele, B., Salerno, G., Costa, M. and Chiusoli, G.P. (1995) *Journal of Organometallic Chemistry*, **303**, 21–28.

65 Li, J., Jiang, H., Feng, A. and Jia, L. (1999) *Journal of Organic Chemistry*, **64**, 5984–5987.

66 Zargarian, D. and Alper, H. (1991) *Organometallics*, **10**, 2914–2921.

67 Izawa, Y., Shimizu, I. and Yamamoto, A. (2005) *Chemistry Letters*, **34**, 1060–1063.

68 (a) Gabriele, B., Veltri, L., Salerno, G., Costa, M. and Chiusoli, G.P. (2003) *European Journal of Organic Chemistry*, 1722–1728. (b) Gabriele, B., Salerno, G., Costa, M. and Chiusoli, G.P. (1999) *Tetrahedron Letters*, **40**, 989–990.

69 Gabriele, B., Costa, M., Salerno, G. and Chissoli, G.P. (1994) *Journal of the Chemical Society, Perkin Transactions*, 83–87.

70 Gabriele, B., Salerno, G., Plastina, P., Costa, M. and Crispini, A. (2004) *Advanced Synthesis & Catalysis*, **346**, 351–358.

71 Gabriele, B., Plastina, P., Salerno, G. and Costa, M. (2005) *Synlett*, 935–938.

72 Gabriele, B., Salerno, G., De Pascali, F., Costa, M. and Chiusoli, G.P. (1997) *Journal of the Chemical Society, Perkin Transactions*, 147–151.

73 Okumoto, H., Nishihara, S., Nakagawa, H. and Suzuki, A. (2000) *Synlett*, 217–218.

74 (a) Kato, K., Yamamoto, Y. and Akita, H. (2002) *Tetrahedron Letters*, **43**, 6587–6590. (b) Kato, K., Nouchi, H., Ishikura, K., Takaishi, S., Motodate, S., Tanaka, H., Okudaira, K., Mochida, T., Nishigaki, R., Shigenobu, K. and Akita, H. (2006) *Tetrahedron: Asymmetry*, **62**, 2545–2554.

75 (a) Gabriele, B., Salerno, G., De Pascali, F., Costa, M. and Chiusoli, G.P. (2000) *Journal of Organometallic Chemistry*, **593–594**, 409–415. (b) Kato, K., Nishimura, A., Yamamoto, Y. and Akita, H. (2002) *Tetrahedron Letters*, **43**, 643–645. (c) Kato, K., Nishimura, A., Yamamoto, Y. and Akita, H. (2001) *Tetrahedron Letters*, **42**, 4203–4205.

76 Marshall, J.A. and Yanik, M.M. (2000) *Tetrahedron Letters*, **41**, 4717–4721.

77 Kato, K., Yamamoto, Y. and Akita, H. (2002) *Tetrahedron Letters*, **43**, 4915–4917.

78 (a) Kadnikov, D.V. and Larock, R.C. (2000) *Organic Letters*, **2**, 3643–3636. (b) Kadnikov, D.V. and Larock, R.C. (2002) *Journal of Organic Chemistry*, **68**, 9423–9432.

79 Kadnikov, D.V. and Larock, R.C. (2004) *Journal of Organic Chemistry*, **69**, 6772–6780.

80 Kadnikov, D.V. and Larock, R.C. (2003) *Journal of Organometallic Chemistry*, **687**, 425–435.

81 (a) Dai, G. and Larock, R.C. (2002) *Organic Letters*, **4**, 193–196. (b) Dai, G. and Larock, R.C. (2002) *Journal of Organic Chemistry*, **67**, 7042–7047.

82 (a) Nan, Y., Miao, H. and Yang, Z. (2000) *Organic Letters*, **2**, 297–299. (b) Liao, Y., Reitman, M., Zhang, Y., Fathi, R. and Yang, Z. (2002) *Organic Letters*, **4**, 2607–2609. (c) Liao, Y., Smith, J., Fathi, R. and Yang, Z. (2005) *Organic Letters*, **7**, 2707–2709.

83 (a) Arcadi, A., Cacchi, S., Fabrizi, G. and Moro, L. (1999) *European Journal of Organic Chemistry*, 1137–1141. (b) Arcadi, A., Cacchi, S., Del Rosario, M., Fabrizi, G. and Marinelli, F. (1996) *Journal of Organic Chemistry*, **61**, 9280–9288.

11
Carbonylation of Allenes

Akihiro Nomoto, Akiya Ogawa

Allene has a cumulated diene structure with remarkable features such as axial chirality of the elongated tetrahedron and a higher reactivity than usual C–C double bonds. Considering the addition of organic molecules to allenes, regioselectivity is of great importance because the desired allylic and/or vinylic compounds can be synthesized selectively [1]. Although until 1980s the examples of catalytic carbonylations of allenes were sparse, their discovery substantially expanded in the 1990s [2]. Recently, many examples of the transition metal catalyzed carbonylation of allenes with carbon monoxide have been reported. On the basis of the reaction mechanisms, these carbonylation reactions of allenes can be classified into four types of processes: (A) *anti*-addition process by the attack of nucleophile (Nu) to the unsaturated bond coordinated by metal; (B) process involving in the formation of vinylidenyl π-allyl metal complexes; (C) hydrometalation or heteroatom-metalation process by H-M-YR (Y = heteroatom) formed *in situ* via oxidative addition of RY-H to the metal complex; and (D) carbometalation process by R-MX generated via oxidative addition of H-YR to metal complex.

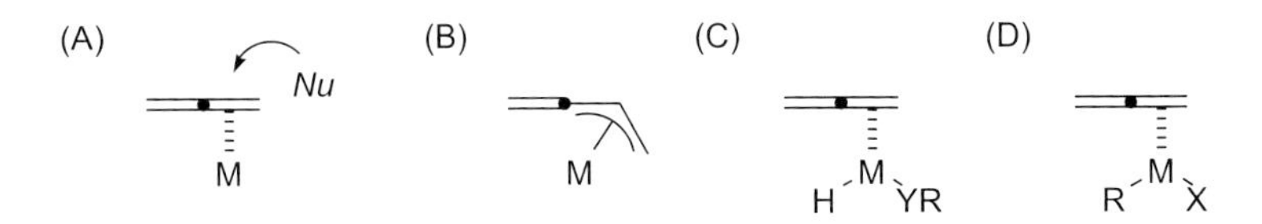

This chapter deals with an overview of the catalytic carbonylation procedures of allenes with carbon monoxide for organic synthesis.

11.1
Anti-Addition Process

As one of the pioneering works concerning allene carbonylation, Alper *et al.* report palladium-catalyzed alkoxy-alkoxycarbonylation of allenes with carbon monoxide and methanol in the presence of $CuCl_2/HCl/O_2$ (Scheme 11.1) [3].

Modern Carbonylation Methods. Edited by László Kollár

ISBN: 978-3-527-31896-4

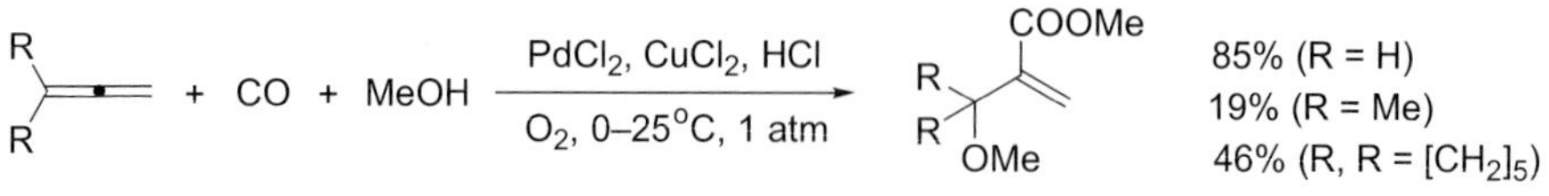

Scheme 11.1

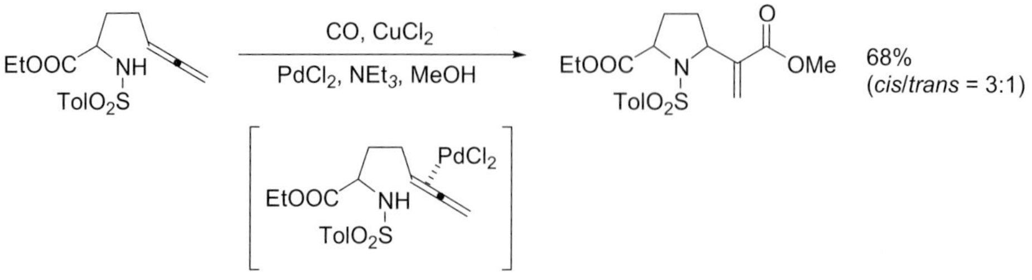

Scheme 11.2

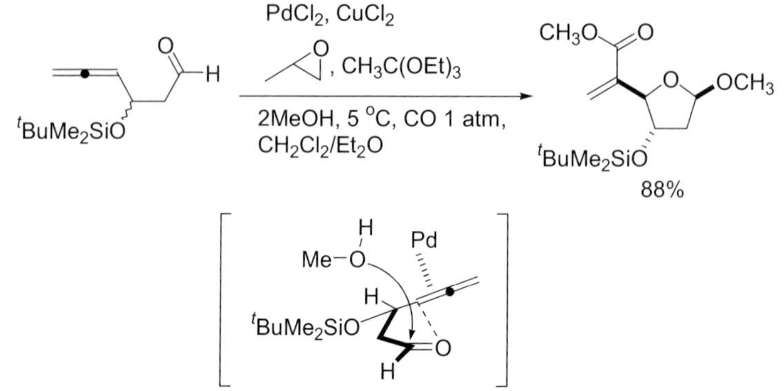

Scheme 11.3

Palladium-catalyzed carbonylation of allenes can be applied to the synthesis of a series of cyclic compounds, which proceeds via nucleophilic attack of the amino group at a suitable position to the allenyl double bond coordinated by palladium (Scheme 11.2) [4].

Similarly, aldehyde carbonyl group can be employed as the nucleophilic group, providing the corresponding furanyl acrylate derivatives in good yields. This reaction proceeds via intramolecular oxypalladation with high stereoselectivity. Alternative pathway may involve the addition of methanol to the formyl group, followed by intramolecular oxypalladation. In this reaction, propylene oxide and triethyl orthoacetate act as acid sink and water scavenger, respectively (Scheme 11.3) [5].

11.2 Vinylidenyl π-Allyl Metal Formation Process

In the presence of $Pd(PPh_3)_4$ (3 mol%), allenyl methyl carbonates react with carbon monoxide and methanol to afford the dienylcarboxylate in excellent yields, although the *E*/*Z* selectivities are moderate (Scheme 11.4) [6].

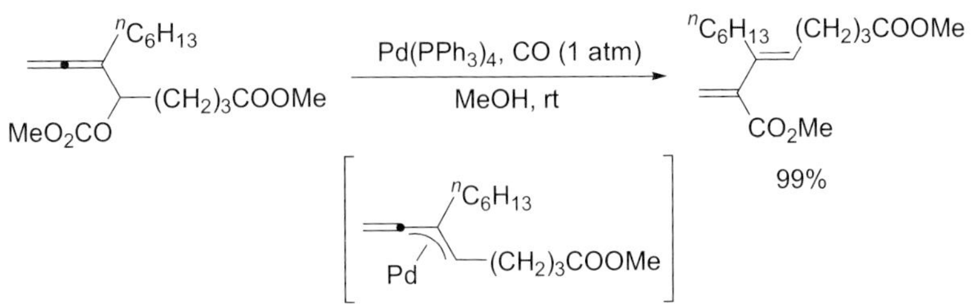

Scheme 11.4

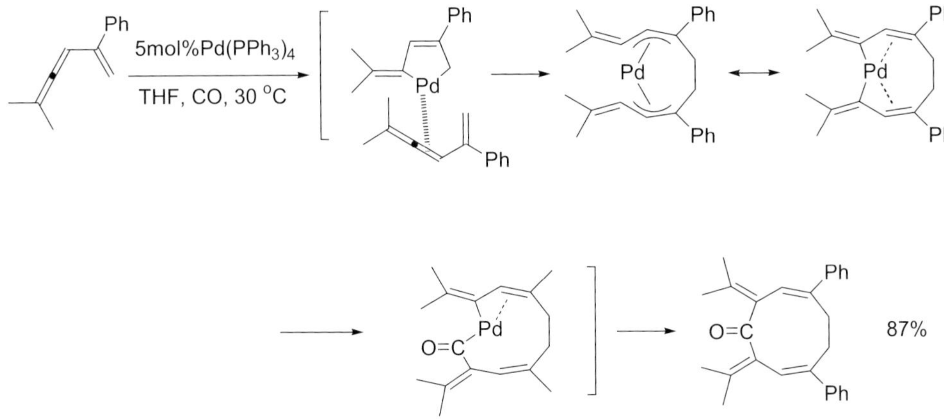

Scheme 11.5

This carbonylation of allenes includes the formation of vinylidenyl π-allylpalladium complexes as the key intermediate after the elimination of the carbonate group [7]. Furthermore, nine-membered cyclic compounds are prepared successfully with alkylidenyl π-allylpalladium complex as the key intermediate (Scheme 11.5) [8].

11.3 Hydrometalation or Heteroatom-Metalation Process

As organosulfur compounds have been widely believed to be catalyst poisons, examples of the transition metal catalyzed reaction of these sulfur compounds have been limited. After the development of transition metal catalyzed addition of organosulfur compounds such as disulfides and thiols to carbon–carbon triple bonds [9], many types of transition metal catalyzed addition reactions of organosulfur compounds have been developed. As to allenes, for example, the addition of thiols to terminal allenes successfully proceeds regioselectively at the internal double bonds of the allenes by the action of palladium acetate catalyst [10a,10b], while the disulfide addition to terminal allenes takes place at the terminal double bond in the presence of tetrakis (triphenylphosphine) palladium catalyst (Scheme 11.6) [10c].

When the $Pd(PPh_3)_4$-catalyzed addition reaction of diphenyl disulfide to terminal allenes is conducted under the pressure of carbon monoxide, a novel carbonylative

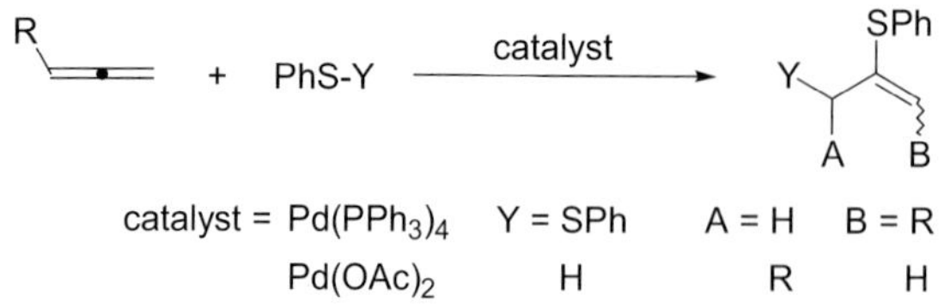

R-allene + PhS-Y —catalyst→ Y, SPh, A, B product

catalyst = $Pd(PPh_3)_4$ Y = SPh A = H B = R
$Pd(OAc)_2$ H R H

Scheme 11.6

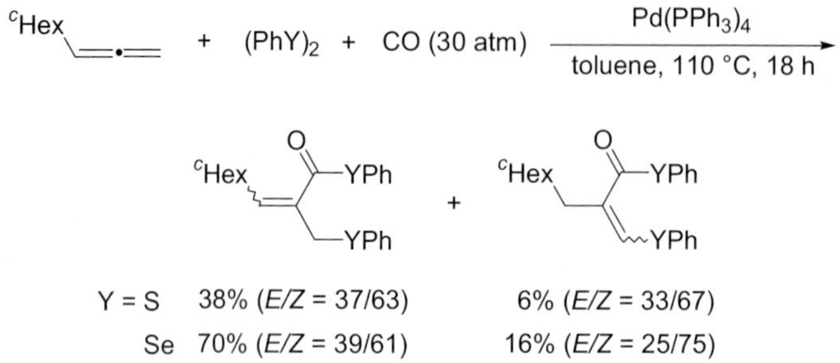

^{c}Hex-allene + $(PhY)_2$ + CO (30 atm) —$Pd(PPh_3)_4$, toluene, 110 °C, 18 h→ ^{c}Hex…C(O)YPh / YPh + ^{c}Hex…C(O)YPh / YPh

Y = S 38% (*E*/*Z* = 37/63) 6% (*E*/*Z* = 33/67)
Se 70% (*E*/*Z* = 39/61) 16% (*E*/*Z* = 25/75)

Scheme 11.7

addition of $(PhS)_2$ to allenes occurs to give the corresponding α,β-unsaturated thioesters as a regioisomeric mixture concerning the C–C double bond (Scheme 11.7) [10c]. Similarly, selenative carbonylation of allenes with CO and $(PhSe)_2$ successfully proceeds in the presence of $Pd(PPh_3)_4$ [10d]. Under higher pressure (60 atm) of carbon monoxide, the carbonylation occurs efficiently; however, the double-bond isomerization takes place competitively. The present carbonylation proceeds via (1) the formation of $Pd(YPh)_2L_n$ (Y = S, Se) by the oxidative addition process; (2) the subsequent thio- or selenopalladation of allenes; (3) CO insertion; and (4) reductive elimination of the product with the regeneration of the catalyst.

In the case of thiols (RSH), the corresponding hydrothiocarbonylation takes place successfully at the terminal double bond, as shown in Scheme 11.8 [11]. A possible pathway is as follows: (1) formation of an H-Pd-SR species via oxidative addition of H-SR to Pd(0); (2) coordination of the allene double bond and subsequent insertion into a Pd–H bond to form an allypalladium complex intermediate; (3) CO insertion resulting in the formation of an acylapalladium carbonyl species; and (4) reductive elimination affording the thioester derivates.

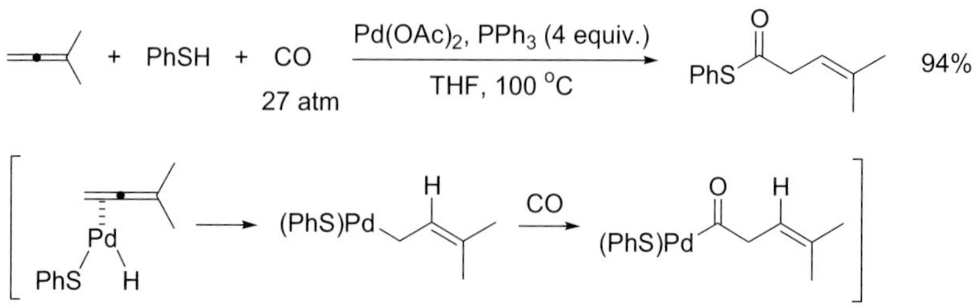

allene + PhSH + CO (27 atm) —$Pd(OAc)_2$, PPh_3 (4 equiv.), THF, 100 °C→ PhS-C(O)… 94%

[allene–Pd(PhS)(H) → (PhS)Pd…H —CO→ (PhS)Pd-C(O)…H]

Scheme 11.8

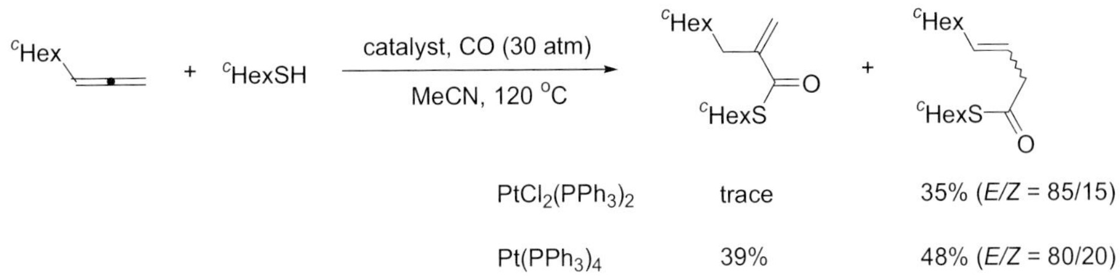

^{c}Hex-allene + ^{c}HexSH —catalyst, CO (30 atm), MeCN, 120 °C→ products

catalyst	terminal product	central product
$PtCl_2(PPh_3)_2$	trace	35% (*E/Z* = 85/15)
$Pt(PPh_3)_4$	39%	48% (*E/Z* = 80/20)

Scheme 11.9

Platinum complexes also exhibit catalytic activity toward the carbonylation of allene with CO and aliphatic thiol can react with allene regioselectively [12]. The $PtCl_2(PPh_3)_2$-catalyzed carbonylative thiolation of cyclohexylallene with (^{C} Hex-SH) as an aliphatic thiol carbon monoxide by the use of aliphatic thiol (^{c}Hex-CH) proceeds regioselectively at the terminal double bond of the allene. In the case of $Pt(PPh_3)_4$-catalyzed reaction, the thiolative carbonylation takes place at both terminal and central double bonds (Scheme 11.9).

Pt(cod)$(CH_3)_2$, which was expected to generate Pt(0) species *in situ*, could not catalyze this reaction; however, the addition of phosphine ligands causes the carbonylation reaction of allenes. This result suggests the carbonylation requires phosphine ligands on platinum.

The catalytic carbonylation of allenes can be employed for the synthesis of heterocyclic carbonyl compounds, as shown in Scheme 11.10 [13], which may proceed through oxidative addition of –SH (not Ar-I) to Pd and the following hydropalladation of allene.

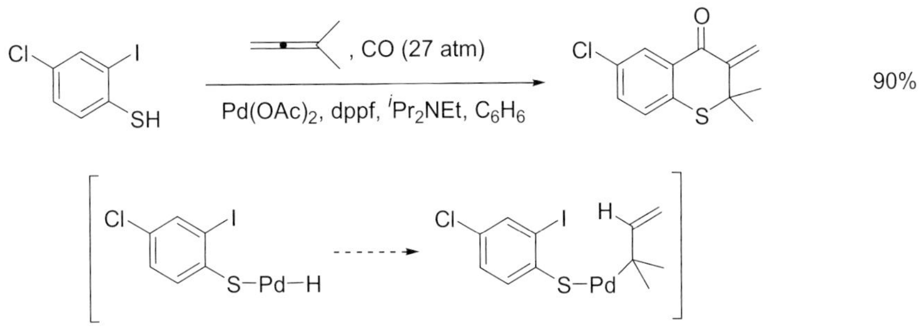

, CO (27 atm)

$Pd(OAc)_2$, dppf, $^{i}Pr_2NEt$, C_6H_6

90%

Scheme 11.10

+ CO + (Me, Ph, NH_2) —$Pd(PPh_3)_4$, AcOH, THF, 110 °C, 1 h→ 83%

[PdX]

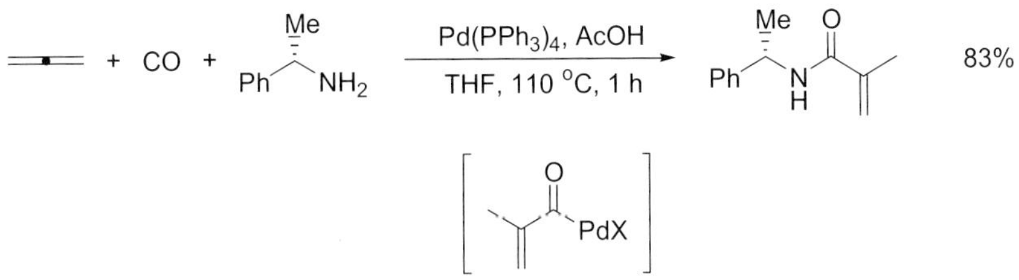

Scheme 11.11

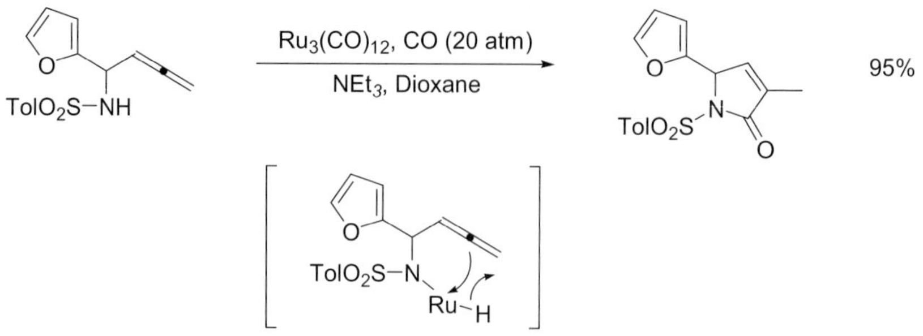

Scheme 11.12

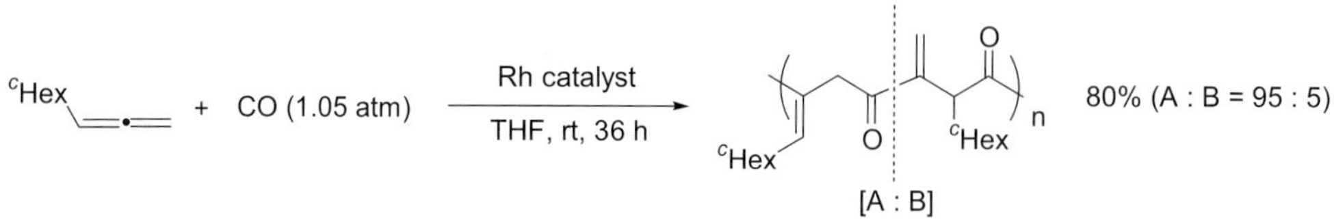

Scheme 11.13

Acrylamides can be synthesized conveniently by the palladium-catalyzed carbonylation of allenes with amines (Scheme 11.11) [14].

Development of new methods for the synthesis of cyclic amide compounds is important in view of medicinal chemistry. Scheme 11.12 indicates an example of ruthenium-catalyzed carbonylation of allenylamines with CO [15].

Allene carbonylation reactions is usefully applied to hydridometal-catalyzed copolymerization of allene and CO, which leads to polyketones [16]. The polymerization of allenes with CO is catalyzed by Rh catalyst, i.e. Rh[η^3-CH(Ar)C{C(=CHAr)CH_2C(=CHAr)CH_2CH_2CH=CHAr}CH_2]$(PPh_3)_2$ (Ar = C_6H_4-OCH_3-*p*), affording polyketones (Scheme 11.13). The molecular weight and the molecular weight distribution of the polymer is $M_n = 27\,700$ and $M_w/M_n = 1.09$. Wilkinson catalyst ($RhCl(PPh_3)_3$) or $RhH(CO)(PPh_3)_3$, in the presence of thiols, is a useful alternative catalyst for the copolymerization of allenes and CO (89 %, A:B=97:3) [12].

11.4 Carbometalation Process

Grigg and coworkers advocate useful sequential reactions involving a catalytic carbonylation process as the key step (Scheme 11.14) [17]: (1) oxidative addition of Ar-I to Pd(0) to form ArPdI; (2) insertion of CO and then allene to form π-allylpalladium intermediate; and (3) nucleophilic substitution.

These reactions proceed smoothly because the rate of insertion of CO is faster than the rate of insertion of allene into arylpalladium(II) species. Starting from the carbonylation products, a series of sulfur- and nitrogen-containing heterolytic compounds have been prepared conveniently (Scheme 11.15).

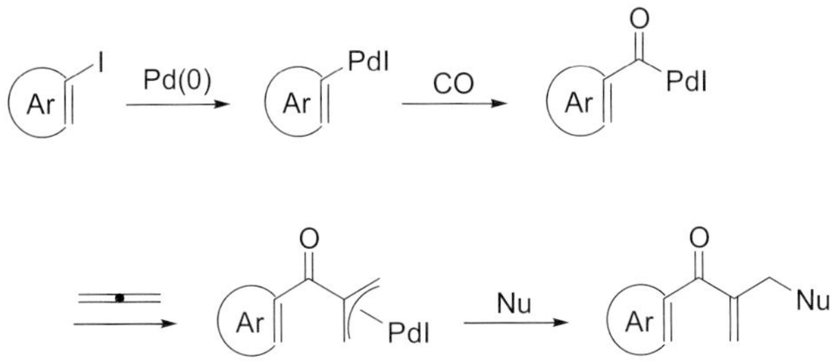

Scheme 11.14

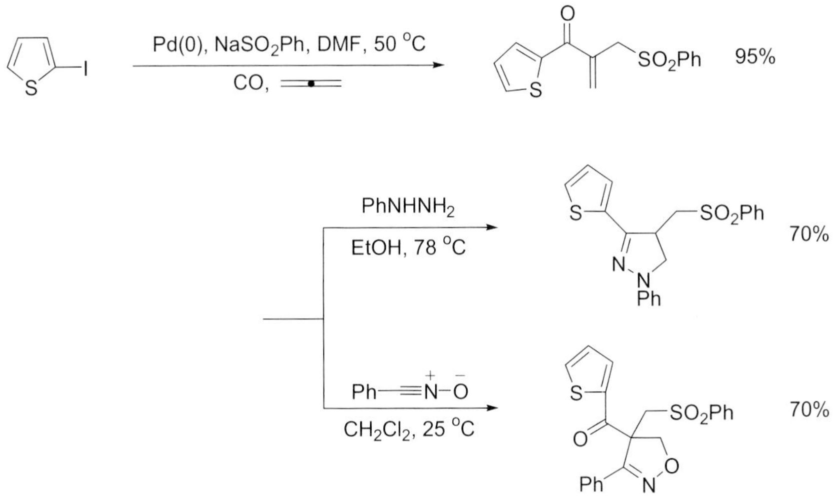

Scheme 11.15

Catalytic carbonylation of *o*-hydroxyaryl halides with allenes provides a useful tool to furanyl or pyranyl derivatives (Scheme 11.16) [18].

A three-component reaction of allenol, CO_2, and aryl or vinyl halide gives a cyclic carbonate in one pot in the presence of palladium catalyst (Scheme 11.17) [19].

The Ti(II)-mediated stereoselective reaction of allenyne with CO is reported to give bicyclic ketone (Scheme 11.18) [20]. The formation of the titanabicycle occurs from the less hindered side of the allene moiety, resulting in the production of bicyclic ketone stereoselectively. This reaction requires slightly excess amounts of the Ti reagent (not a catalytic reaction), but these results show the capability of stereoselective carbonylation of allenes.

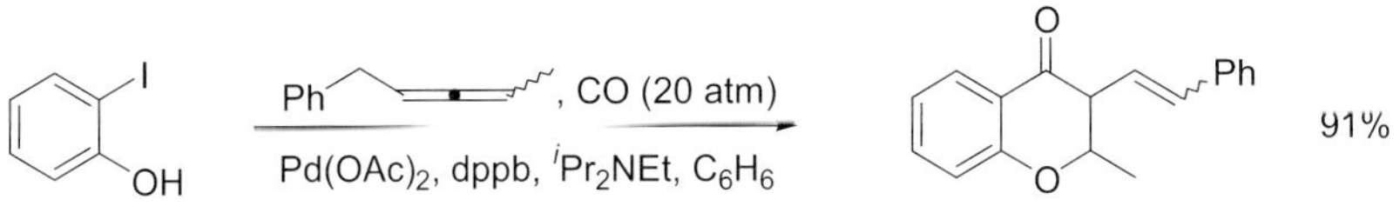

Scheme 11.16

Unsaturated bonds play important role for trapping palladium complex generated *in situ* if a reaction continuously progresses. By placing double bonds in suitable positions, inter- or intramolecular reaction occurs sequentially, affording the desired complex compounds (Scheme 11.19) [21].

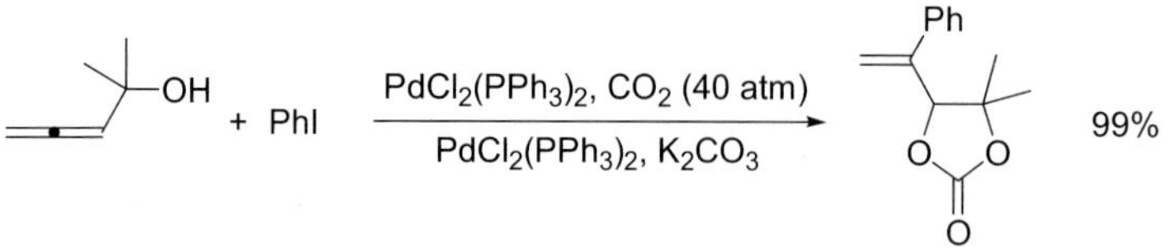

OH
+ PhI
PdCl2(PPh3)2, CO2 (40 atm)
PdCl2(PPh3)2, K2CO3
Ph
O
O
O
99%

Scheme 11.17

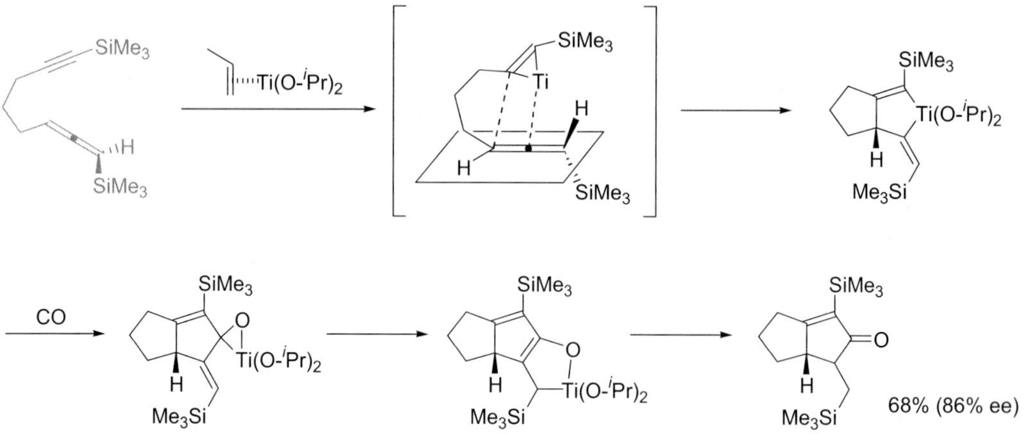

SiMe3
H
SiMe3
Ti(O-iPr)2
Ti
H
SiMe3
CO
O
Me3Si
68% (86% ee)

Scheme 11.18

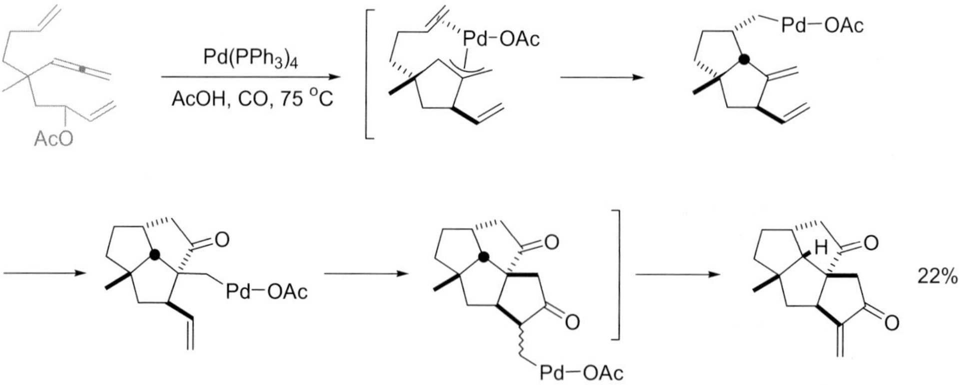

AcO
Pd(PPh3)4
AcOH, CO, 75 °C
Pd-OAc
Pd–OAc
O
H
22%

Scheme 11.19

In conclusion, the carbonylation of allenes provides useful methods for the synthesis of functionalized unsaturated carbonyl compounds and cyclic carbonyl compounds.

References

1 (a) N. Krause and S. Hashmi (eds) (2004) *Modern Allene Chemistry*, Wiley-VCH Verlag GmbH, New York. (b) E. Negishi (ed.) (2002) *Handbook of Organopalladium Chemistry for Organic Synthesis*, John Wiley & Sons, Inc., New York. (c) Brandsma, L.(2004) *Synthesis of Acetylenes, Allenes and Cumulenes*, Elsevier, Oxford.

2 For reviews, see (a) Zimmer, R., Dinesh, C.U., Nandanan, E. and Khan, F.A. ,(2000) *Chemical Reviews*, **100**, 3067. (b) Balme, G., Bossharth, E. and Monteiro, N. (2003) *Chemistry – A European Journal*, **9**, 4101. (c) Bates, R.W. and Satcharoen, V. (2002) *Chemical Society Reviews*, **31**, 12. (d) Ma, S. (2005) *Chemical Reviews*, **105**, 2829.

3 Alper, H., Hartstock, F.H. and Despeyroux, B. (1984) *Journal of the Chemical Society, Chemical Communications*, 905.

4 Gallagher, T., Davies, I.W., Jones, S.W., Lathbury, D., Mahon, M.F., Molloy, K.C., Shaw, R.W. and Vernon, P. (1992) *Journal of the Chemical Society, Perkin Transactions 1*, 433. (b) Lathbury, D., Vernon, P., and Gallagher, T. (1986) *Tetrahedron Letters*, **27**, 6009. (c) Fox, D.N.A. and Gallagher, T. (1990) *Tetrahedron*, **46**, 4697. (d) Fox, D.N.A., Lathbury, D., Mahon, M.F., Molloy, K.C. and Gallagher, T. (1991) *Journal of the American Chemical Society*, **113**, 2652.

5 Walkup, R.D. and Mosher, M.D. (1993) *Tetrahedron: Asymmetry*, **49**, 9285. (a) Walkup, R.D. and Park, G. (1987) *Tetrahedron Letters*, **28**, 1023. (b) Walkup, R.D. and Park, G. (1990) *Journal of the American Chemical Society*, **112**, 1597. (c) Walkup, R.D. and Mosher, M. D. (1993) *Tetrahedron*, **49**, 9285. (d) Walkup, R.D. and Mosher, M.D. (1994) *Tetrahedron Letters*, **35**, 8545. (e) Walkup, R.D., Guan, L., Kim, S.W. (1995) *Tetrahedron Letters*, **36**, 3805.

6 Nokami, J., Maihara, A. and Tsuji, J. (1990) *Tetrahedron Letters*, **31**, 5629.

7 (a) Piotti, M.E. and Alper, H. (1994) *The Journal of Organic Chemistry*, **59**, 1956. (b) Moriya, T., Miyaura, N. and Suzuki, A. (1994) *Synlett*, 149.

8 Murakami, M., Itami, K. and Ito, Y. (1998) *Angewandte Chemie-International Edition*, **37**, 3418.

9 Ogawa, A., Kawakami, J., Sonoda, N. and Hirao, T. (1996) *The Journal of Organic Chemistry*, **61**, 4161. (a) Kuniyasu, H., Ogawa, A., Miyazaki, S., Ryu, I., Kambe, N. and Sonoda, N. (1991) *Journal of the American Chemical Society*, **113**, 9796. (b) Kuniyasu, H., Ogawa, A., Sato, K., Ryu, I., Kambe, N. and Sonoda, N. (1992) *Journal of the American Chemical Society*, **114**, 5902. (c) Ogawa, A. (2000) *Journal of Organometallic Chemistry*, **611**, 463.

10 Kodama, S., Nishinaka, E., Nomoto, A., Sonoda, M. and Ogawa, A. Tetrahedron Letters, in press. (a) Ogawa, A., Kawakami, J., Sonoda, N. and Hirao, T. (1996) *Journal of Organic Chemistry*, **61**, 4161. (b) Ogawa, A., Kudo, A. and Hirao, T. (1998) *Tetrahedron Letters*, **39**, 5213. (c) Kodama, S., Nishinaka, E., Nomoto, A., Sonoda, M. and Ogawa, A. (2007) *Tetrahedron Letters*, **48**, 6312. (d) Kamiya, I., Nishinaka, E. and Ogawa, A. (2005) *Tetrahedron Leters*, **46**, 3649.

11 Xiao, W.-J., Vasapollo, G. and Alper, H. (1998) *The Journal of Organic Chemistry*, **63**, 2609.

12 Kajitani, M., Kamiya, I., Nomoto, A., Kihara, N. and Ogawa, A. (2006) *Tetrahedron*, **62**, 6355.

13 Xiao, W.-J. and Alper, H. (1999) *The Journal of Organic Chemistry*, **64**, 9646.

14 Grigg, R., Monteith, M., Sridharan, V. and Terrier, C. (1998) *Tetrahedron*, **54**, 3885.

15 Kang, S.-K., Kim, K.-J., Yu, C.-M., Hwang, J.-W. and Do, Y.-K. (2001) *Organic Letters*, **3**, 2851.

16 (a) Kacker, S. and Sen, A. (1997) *Journal of the American Chemical Society*, **119**, 10028. (b) Osakada, K., Choi, J.-C. and Yamamoto, T. (1997) *Journal of the American Chemical Society*, **119**, 12390.(c) Takenaka, Y. and Osakada, K. (2001) *Macromolecular Chemistry and Physics*, **202**, 3571.

17 Grigg, R., Brown, S., Sridharan, V. and Uttley, M.D. (1997) *Tetrahedron Letters*, **38**, 5031.

18 Okuro, K. and Alper, H. (1997) *The Journal of Organic Chemistry*, **62**, 1566.

19 Uemura, K., Shiraishi, D., Noziri, M. and Inoue, Y. (1999) *Bulletin of the Chemical Society of Japan*, **72**, 1063.

20 Urabe, H., Takeda, T., Hideura, D. and Sato, F. (1997) *Journal of the American Chemical Society*, **119**, 11295.

21 Doi, T., Yanagisawa, A., Nakanishi, S., Yamamoto, K. and Takahashi, T. (1996) *The Journal of Organic Chemistry*, **61**, 2602.

12
Homogeneous Carbonylation Reactions in the Synthesis of Compounds of Pharmaceutical Importance

Rita Skoda-Földes

12.1
Introduction

The production of pharmaceuticals usually involves multistep syntheses where the selectivity and yield of the individual steps are of utmost importance. As a consequence, development of highly efficient catalytic processes for the synthesis of these compounds becomes the central question in synthetic organic chemistry. Among these processes, carbonylation gained special attention as it involves both new carbon–carbon bond formation and the introduction of a synthetically useful functionality in the molecule.

The main goal of this chapter is to summarize the most important achievements concerning synthesis of pharmaceuticals and their important intermediates using carbonylation as the key step and to show how rather complex synthetic processes can be accomplished under carbonylation conditions. The content of the chapter is limited to the description of the developments published mainly in the past 10 years. The results related to the carbonylation of various model compounds are not included in this chapter.

12.2
Carbonylation of Alkenes (or Alkynes)

Along with other addition reactions over multiple bonds, this type of carbonylation perfectly fulfills today's need for processes with high atom economy. However, similar to other reactions, several problems concerning chemo-, regio-, and stereo-selectivity also emerge that should be solved by the proper choice of catalysts, additives, and reaction conditions.

Modern Carbonylation Methods. Edited by László Kollár

ISBN: 978-3-527-31896-4

12.2.1 Hydroformylation

Hydroformylation is one of the most widely used homogeneous catalytic reactions in the synthesis of pharmaceuticals. It offers the possibility of the introduction of a great variety of different moieties into organic compounds, as the formyl group can be converted into other functional groups via oxidation, reduction, nucleophilic addition reactions, and other such processes. In all of the reactions, achieving high selectivities for either the branched or the linear isomer, depending on the application, is critical.

Several catalytic systems have been developed for the selective synthesis of the branched aldehydes starting from vinylaromatics and 4-vinylazetidin-2-ones that can be converted into nonsteroidal anti-inflammatory agents or carbapenem antibiotics, respectively. The use of rhodium–phosphine complexes usually ensures excellent regioselectivity with branched/normal ratios up to 98.5/1.5 [1–9]. Moreover, for the synthesis of compounds with the desired pharmacological activity, hydroformylation has to be carried out in an enantioselective fashion leading to the *S*-enantiomers of aldehyde precursors of 2-arylpropionic acids such as ibuprofen or naproxen, and the *R*-species (i.e., β-form) of 4-(1′-formylethyl)azetidin-2-ones. Accordingly, several chiral ligands (**1**–**6**, Figure 12.1) have been tested in these reactions together with various rhodium(I) precursors [2–9]. It should be mentioned that enantioselective hydroformylation of vinylaromatics was attempted even in the presence of platinum complexes. Although these catalysts promoted high enantioselection in some cases, regioselectivity toward the branched aldehyde was too low for practical purposes [10].

Among the rhodium complexes, the catalyst containing the nonsymmetrical atropisomeric phosphine-phosphite chelating ligand (**1**) gave the best enantioselectivity (up to 92% ee) in the hydroformylation of the ibuprofen precursor 4-isobutylstyrene [2] whereas the use of the bulky diphosphite ligand **2** resulted in a remarkable regioselectivity with a branched/normal ratio of 98.5/1.5 and somewhat lower optical yield (82% ee) [3].

In the hydroformylation of 4-vinylazetidin-2-ones, the use of catalysts containing (*R*,*S*)-BINAPHOS (**1**) [6], (*S*,*S*)-BDPP (**3**) [7], or phosphine–phosphinite ligands, such as **5** [8] and **6** [6], resulted in excellent enantioselectivities (up to 97/3 β/α ratios). At the same time, regioselectivities toward the branched aldehyde either remained only fairly good (up to 76/24 branched/normal ratio) or depended greatly on the substrate/catalyst ratio as in the case of **3** [9].

In contrast to the previous examples, the preferred formation of linear aldehydes was the main target in some syntheses to construct a cyclic derivative with an appropriate ring size in the next step or for simply elongating a carbon chain. The linear aldehydes are the proper intermediates for the synthesis of indolizidine alkaloids [11], the tricyclic marine alkaloid lepadiformine [12], ACE inhibitors such as MDL 27210 and its analogues [13,14], and bryostatin, a remarkably potent anticancer agent [15]. Rhodium complexes of bisphosphite ligands provide one of the best known classes of linear-selective hydroformylation catalysts for simple α-olefins. Except for the lepadiformine intermediate, where hydroformylation was carried out in the presence of the $Rh(acac)(CO)_2/P(OPh)_3$ catalyst system, in other

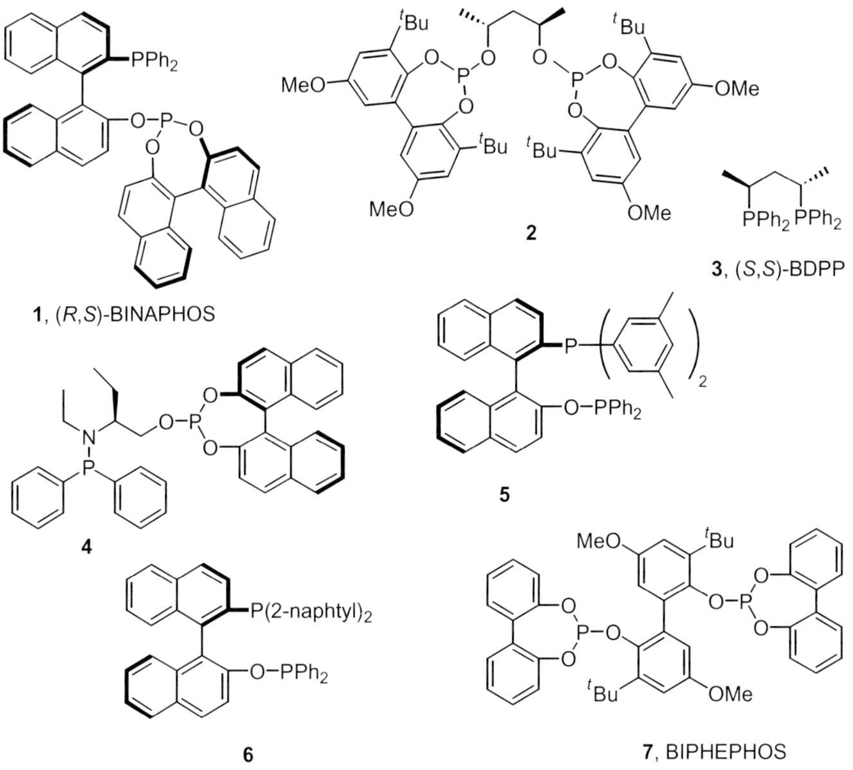
PPh2
O–P–O
tBu
MeO
OMe
1, (R,S)-BINAPHOS
2
3, (S,S)-BDPP
PPh2 PPh2
4
5
P
O–PPh2
P(2-naphtyl)2
6
7, BIPHEPHOS

Figure 12.1 Various ligands used in hydroformylation reactions.

reactions mentioned above, a rhodium/BIPHEPHOS (Figure 12.1, **7**) catalyst was used, and it indeed led to the desired linear aldehydes under mild reaction conditions and excellent selectivity. As an example, an intermediate of the ACE inhibitor MDL 27 210 was obtained in 80% yield at 65 °C and 4.8 bar pressure in 2 h with a linear/branched ratio of 99/1 [14].

In special cases, the desired regio- or stereoselectivity can be achieved with the help of the internal regio- or stereocontrol exerted by a moiety incorporated into a proper position of the substrate. During the synthesis of a key building block of the macrolide antibiotic bafilomycin A_1, the spatial shielding of one face of the double bond by the equatorial methyl group proved to result in the exclusive formation of the all-*anti* stereotriade aldehyde **9** [16].

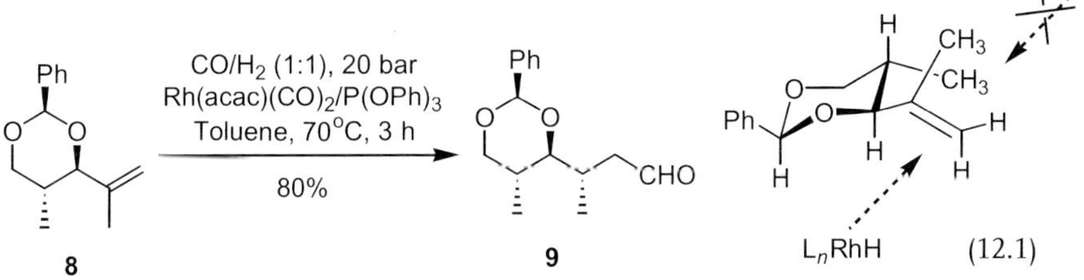
Ph
CO/H2 (1:1), 20 bar
Rh(acac)(CO)2/P(OPh)3
Toluene, 70°C, 3 h
80%
CHO
8
9
CH3
LnRhH

(12.1)

Hydroformylation can be incorporated efficiently into tandem reactions [17]. Various tryptamine-based pharmaceuticals were prepared by hydroformylation carried out in the presence of hydrazine derivatives. The aldehydes were generated *in situ* from the alkenes and trapped by the hydrazines to form the hydrazones that on cyclization gave the indole products. Application of water-soluble ligands such as TPPTS made it possible to execute hydroformylation and cyclization in one step in aqueous sulfuric acid [18].

Several synthetic approaches toward 3,3-diarylpropyl- or 4,4-diarylbutylamines (Figure 12.2) comprise hydroformylation as the key step. One possible route involves the reaction of the ω,ω-diarylalkylhalide, obtained in a few steps from the hydroformylation product linear aldehyde, with an appropriate amine [19]. On the contrary, the aldehydes can be converted directly into the amines via a transition metal catalyzed reductive amination reaction [20]. Although these methods have been employed efficiently for the synthesis of compounds **10**, **11** [19], and **12** [20], the most elegant solution is the direct hydroaminomethylation where the initial hydroformylation of the alkene is followed by the condensation of the intermediate aldehyde with the amine present in the reaction mixture and a final hydrogenation to give a saturated secondary or tertiary amine.

Initial attempts to prepare fenpiprane (**13**) by this method failed because of preferred substrate hydrogenation as a side reaction in the presence of the amine. This drawback was circumvented either by injecting the amine into the pressurized reaction vessel after completion of the hydroformylation step [21] or by using trialkyl

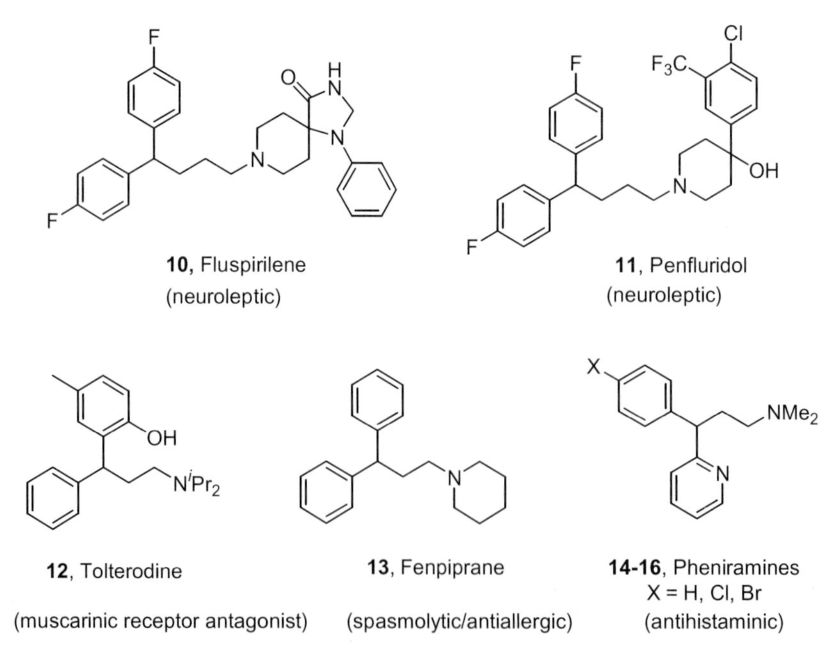

Figure 12.2 Pharmacologically active 3,3-diarylpropyl- or 4,4-diarylbutylamines.

phosphine ligands such as PBu_3 together with the rhodium catalyst during the tandem reaction [22]. The two different methods led to fenpiprane (**13**) in 78% [21] and 72% yield [22], respectively. Hydroaminomethylation was applied effectively for the synthesis of fluspirilene (**10**) and other pharmaceuticals with related structure; **10** was obtained in an overall 88% yield in four steps from a commercially available diaryl ketone [23]. Other examples of successful hydroaminomethylation involve the synthesis of pharmacologically active derivatives of phenothiazine, iminodibenzyl, carbazole, and pyrazole [24], as well as new aminomethyl- [25] and hydroxy-aminomethyl steroids [26]. At the same time it should be mentioned that the antihistaminic agent pheniramines (**14–16**) could not be synthesized by this method as regioselectivity was shifted to the undesired branched aldehydes because of the directing effect of the pyridine nitrogen [27].

Hydroformylation of substrates incorporating *O*- or *N*-nucleophilic moieties leads to cyclic hemiacetals, acetals, *O,N*-acetals, or enamines depending on the reaction conditions and the functional groups of the substrates. In the total synthesis of the anticancer agent, natural product leucascandrolide A (**24**), three different carbonylation steps were incorporated (Scheme 12.1). Alkene **20** underwent a cyclohydrocarbonylation reaction under hydroformylation conditions, resulting in the formation of hemiacetal **21**. The other two carbonylation steps involved formylation of **18** and intramolecular alkoxycarbonylation of alkene **22** [28]. Various tryptamine derivatives [29] and the framework of piperidine alkaloids [30] were also synthesized via cyclohydrocarbonylation starting from functionalized homoallylic amines or aniline derivatives, respectively.

There are some unique transformations of alkynes with adjacent reactive functional groups that take place under hydroformylation conditions. Such reactions were used in the synthesis of the thiazepinone [31] and the pyrrolinone [32] framework of pharmacologically active compounds. In the first case, the reaction of acetylenic thiazole **25** with carbon monoxide and hydrogen afforded thiazepinone **26** via a cyclohydrocarbonylative ring expansion (Equation 12.2), whereas pyrrolinones (e.g., **28**) were obtained by a tandem cyclohydrocarbonylation – CO insertion sequence (Equation 12.3).

25 → **26**: CO/H_2 (1:1), 21 bar; $(\eta^6\text{-}C_6H_5BPh_3)^-Rh^+(1,5\text{-COD})$; $P(OPh)_3$; CH_2Cl_2, 110 °C, 18 h; 90% (12.2)

27 → **28**: CO/H_2 (11:1), 42 bar; $(\eta^6\text{-}C_6H_5BPh_3)^-Rh^+(1,5\text{-COD})$; $P(OPh)_3$; CH_2Cl_2, 90 °C, 24 h; 78% (12.3)

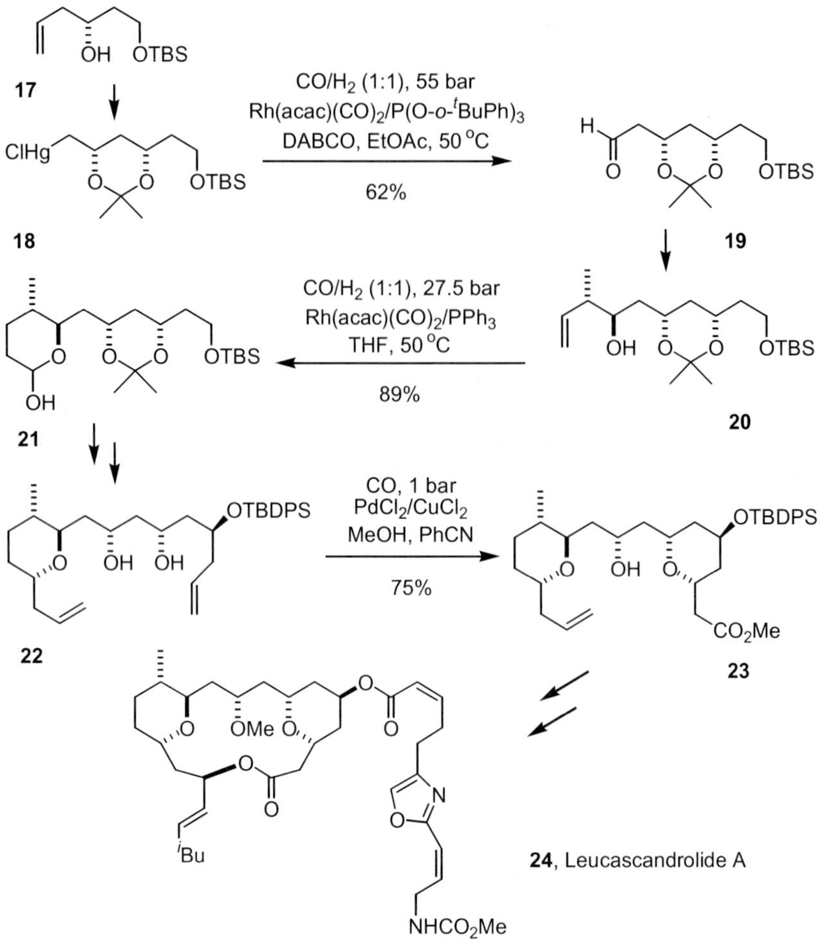

Scheme 12.1 Synthetic route to leucascandrolide A.

12.2.2
Hydrocarboxylation

One of the most important pharmaceutical applications of this reaction is the hydrocarboxylation of vinylaromatics, resulting in the direct formation of nonsteroidal analgesics such as ibuprofen [33] or ketoprofen [34]. Although the usual catalytic system consists of $PdCl_2$ and a tertiary phosphine together with HCl and $CuCl_2$ as promoters, the proper choice of ligands, additives, and solvents is essential to achieve good catalytic activity and selectivity [33]. The application of the TsOH/LiCl combination of additives instead of the commonly used $HCl/CuCl_2$ was a real breakthrough and led to a significant improvement in turnover frequencies (from 25 to $1240\,h^{-1}$) [35]. Novel cationic palladium complexes containing hemilabile N−O and N−N chelating ligands such as **29** showed both excellent catalytic activity (with turnover frequencies over 1300 h^{-1}) and selectivity toward the desired branched regioisomer (up to 99/1) [36].

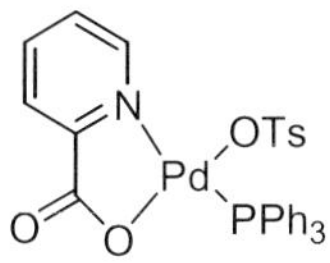

29 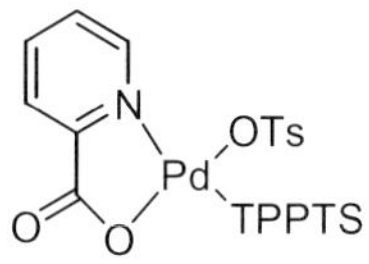 **30**

Even if the turnover frequency is very high for a homogeneous catalytic reaction, catalyst separation and reuse remain a problem to be solved. Sheldon and coworkers were the first to report on hydrocarboxylation of various olefins in the presence of the water-soluble $Pd(TPPTS)_3$ complex [37]. However, in the reaction of 4-isobutyl styrene, the activity of this catalyst was low, probably because of the very low water solubility of this particular substrate, and only moderate regioselectivity could be achieved. A significantly improved turnover frequency and selectivity was observed using another water-soluble complex **30** [38]. As another advantage, this latter catalyst could be recycled several times at 100–105 °C reaction temperatures with only negligible loss of catalytic activity.

12.2.3 Hydroesterification (Alkoxycarbonylation)

Hydroesterification followed by hydrolysis provides a third route to 2-arylpropionic acids starting from vinylaromatics besides the hydroformylation–oxidation or direct hydrocarboxylation routes discussed previously. Esters of ibuprofen [33] and ketoprofen [39] were synthesized in the presence of catalysts containing a Pd(II) precursor and tertiary phosphines. Immobilized catalysts such as $Pd(OAc)_2$–montmorillonite [40] and a silica-supported chitosan–palladium complex [41], resulting in the same regioselectivity as the homogeneous analogues, have also been used successfully. Some attempts were made to carry out the reaction in an enantioselective fashion but only moderate enantiomeric excesses were obtained together with low regio- and chemoselectivity and even poor catalytic activity [42].

16α-Substituted steroidal esters [43] and hydroxyesters [44] were obtained as main products in the reaction of androst-16-ene derivatives with alcohols and α,ω-diols, respectively, in the presence of a catalytic amount of $(Ph_3P)_2PdCl_2$ under 80–120 bar CO pressure. On the contrary, a Δ^5 double bond remained intact even under more severe conditions [43].

Hydroesterification of the C–C triple bond of a macrocycle has been used as the key step for the total synthesis of several members of the pseudopterane family possessing significant biological activity such as cytotoxicity or anti-inflammatory activity [45]. As any attempts to convert propargylic alcohol **31** directly into butenolide **34** by conventional methods failed, a new route was developed via atmospheric alkoxycarbonylation of mesylate **32** to allenic ester **33**.

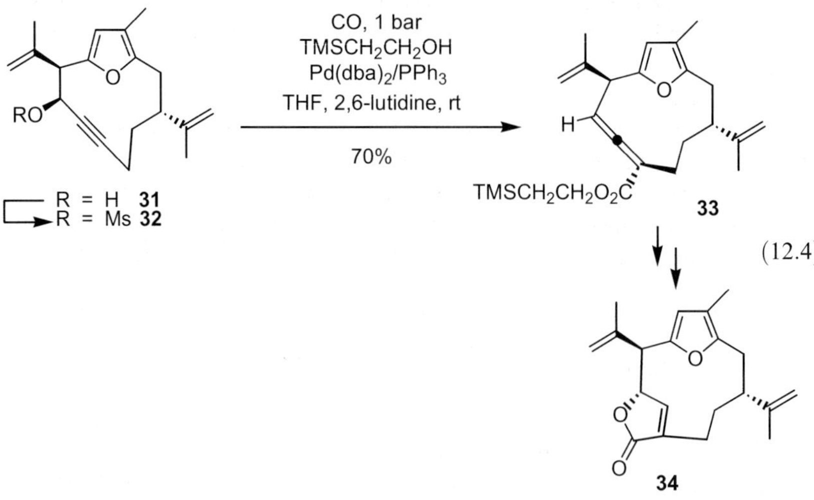

CO, 1 bar
$TMSCH_2CH_2OH$
$Pd(dba)_2/PPh_3$
THF, 2,6-lutidine, rt
70%

R = H **31**
R = Ms **32**

$TMSCH_2CH_2O_2C$ **33**

34

(12.4)

The intramolecular version of hydroesterification was used for the synthesis of steroidal 17-spirolactones such as the aldosterone antagonist eplerenone starting from 17-hydroxy-17-vinyl or 17-hydroxy-17-ethynyl derivatives [46].

In the presence of various functional groups in an appropriate distance from the alkenyl or alkynyl moieties, substituted alkenes or alkynes may undergo tandem reactions comprising hydroesterification as one of the individual steps. The bicylic *cis*-fused lactone core of plakortones, a novel class of activators of cardiac SR-Ca^{2+}-pumping ATPase, was efficiently accessed by a palladium(II)-mediated cyclization–intramolecular hydroesterification reaction sequence. Cyclization of the starting enediols was carried out using either a small excess of Pd(II) [47] or a catalytic amount of Pd(II) and an excess of Cu(II) as the reoxidant [48].

The (2*E*)-methoxymethylidene-1,6-dioxaspiro[4,5]-decane skeleton (**36**) was constructed in a highly stereoselective manner by a palladium(II)-catalyzed cyclization–alkoxycarbonylation sequence starting from the 3,4-dioxygenated-9-hydroxy-1-nonyn-5-one (**35**). The use of a catalytic amount of Pd(0) and a high excess of *p*-benzoquinone was essential to minimize the amount of the active palladium(II) catalyst in the reaction mixture and to prevent intramolecular acetal formation, leading to **37** as a side reaction [49].

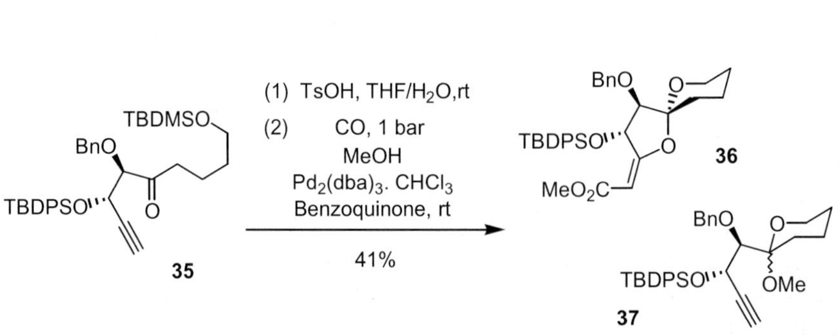

(1) TsOH, THF/H_2O,rt
(2) CO, 1 bar
MeOH
$Pd_2(dba)_3$. $CHCl_3$
Benzoquinone, rt
41%

35 TBDMSO, BnO, TBDPSO

36 BnO, TBDPSO, MeO_2C

37 BnO, TBDPSO, OMe

(12.5)

Another cyclization–carbonylation reaction sequence was used to form the benzofuran ring and to introduce the alkoxycarbonyl group in the 3-position in a single operation during the total synthesis of the natural product XH-14, a potent antagonist of the A_1 adenosine receptor [50]. The 8-azabicyclo-[3.2.1]octane and 9-azabicyclo-[4.2.1]nonane ring systems of ferruginine **42** [51] and anatoxin-a **43** [52] were obtained by aminocyclization and subsequent alkoxycarbonylation of aminocycloheptene and aminocyclooctene precursors, respectively. An intramolecular version of the reaction led to the key intermediates of the synthesis of 1-deoxynojirimycin and 1-deoxy-L-idonojirimycin starting from a highly substituted benzylaminohexene [53].

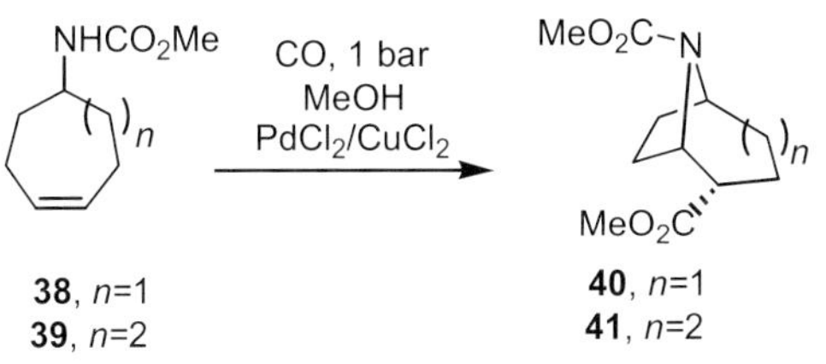

$NHCO_2Me$ — CO, 1 bar, MeOH, $PdCl_2/CuCl_2$ — MeO_2C–N … MeO_2C

38, *n*=1
39, *n*=2
40, *n*=1
41, *n*=2
42, Ferruginine **43**, Anatoxin-a

(12.6)

A tandem intramolecular Heck reaction/carbonylation was developed for the construction of the C20 quaternary center found in a precursor (**45**) of cytotoxic agent perophoramidine [54].

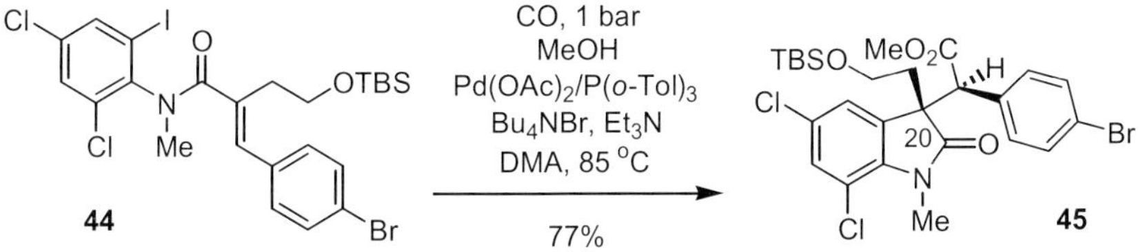

CO, 1 bar
MeOH
$Pd(OAc)_2/P(o\text{-}Tol)_3$
Bu_4NBr, Et_3N
DMA, 85 °C
77%

44 **45**

(12.7)

12.3 Carbonylation of Alcohols and Amines

This type of carbonylation reactions may lead to a great variety of products, such as carboxylic acids, esters, ureas, lactams, and so on, depending on the reaction conditions.

12.3.1 Hydrocarboxylation of Alcohols

As vinylaromatics, used as substrates in the synthetic routes toward 2-arylpropionic acids described in the previous chapters, are often obtained by the dehydration of the corresponding alcohols, direct carbonylation of 1-arylethanols provides a shorter access to ibuprofen or naproxen. However, in the presence of the $PdCl_2(PPh_3)_2$ catalyst, the use of strongly acidic medium is necessary to obtain ibuprofen from 1-(4-isobutylphenyl)ethanol with good (>98%) selectivity [55]. A significant improvement in catalytic activity (with turnover frequencies up to 850 h^{-1}) has been achieved by

Chaudhari's group by the use of either the $PdCl_2(PPh_3)_2$/TsOH/LiCl system [56] or catalyst **29** [36]. Some attempts were made to carry out hydrocarboxylation of 1-(4-isobutylphenyl)ethanol with polymer-supported catalysts, too [57]. The most effective catalytic system consisting of Pd/C, PPh_3, TsOH, and LiCl showed both excellent activity (with TOF up to 3375 h^{-1}) and selectivity toward ibuprofen (>99.2%) [58].

12.3.2
Alkoxycarbonylation of Alcohols

Reaction rates of hydrocarboxylation of 1-(6-methoxynaphtyl)ethanol usually fell behind that of 1-(4-isobutylphenyl)ethanol [56]. While seeking an efficient method to produce naproxen, some reports concerning alkoxycarbonylation of 1-(6-methoxynaphtyl)ethanol have been published. Good selectivities were achieved both with a homogeneous [59] and a polymer-supported $PdCl_2/PPh_3/CuCl_2$/TsOH system [60] but catalytic activities remained moderate. This reaction was also carried out in an enantioselective manner. By the use of the chiral DDPPI (1,4:3,6-dianhydro-2,5-dideoxy-2,5-bis(diphenylphosphino)-L-iditol) ligand the methyl ester of (*S*)-naproxen was obtained in a chemical yield of 90% with 81% ee [61].

12.3.3
Oxidative Carbonylation of Amines

Although many transition metals are known to promote oxidative carbonylation of amines to ureas, only palladium- and wolfram-catalyzed processes were used in the synthesis of compounds of pharmaceutical importance. Urea **46**, a potent antagonist of the neuropeptide Y5 receptor was obtained starting from 9-isopropyl-4-methyl-9*H*-carbazol-3-ylamine and morpholine, using the PdI_2/KI catalyst under 20 bar of 4/1 carbon monoxide/air mixture [62]. Intramolecular versions of this reaction were carried out with another system, using $W(CO)_6$ as the catalyst and I_2 as the oxidant during the synthesis of biotin methyl ester **47** [63], and the core structure of the HIV protease inhibitors **48** and **49** [64].

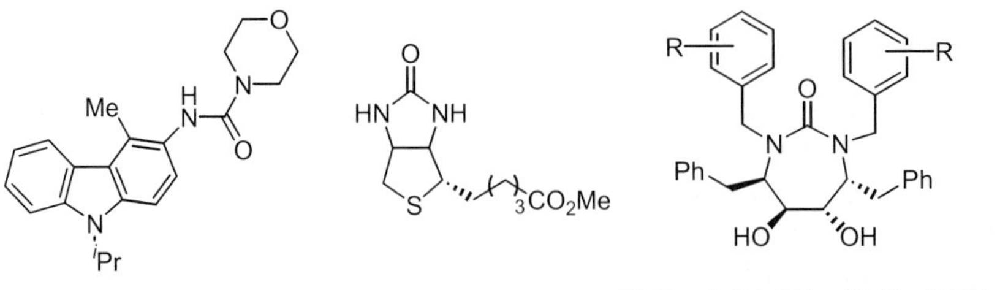

46 **47** **48**, R = 4-CH_2OH **49**, R = 3-NH_2

12.3.4
Carbonylation of Aziridines

Carbonylative ring expansion of aziridines in the presence of palladium or cobalt catalysts has proved to be a useful reaction for the construction of the β-lactam ring of carbapenem antibiotics. Complete but opposing stereoselectivities were observed with the two types of catalysts: in the presence of the $Pd_2(dba)_3 + PPh_3$ system, carbonylation of *trans*-aziridines led to the formation of *trans*-β-lactams [65], whereas using the $Co_2(CO)_8$ catalyst, *cis*-β-lactams from *trans*-aziridines and *trans*-β-lactams from *cis*-aziridines were obtained [66].

12.4 Carbonylation of Alkenyl/Aryl Halides or Triflates

Mild reaction conditions, high selectivities, and compatibility with other functional groups make this type of carbonylation a versatile tool in synthetic organic chemistry and especially in the synthesis of pharmaceuticals. They provide a straightforward route to the synthesis of various α,β-unsaturated carbonyl compounds and carboxylic acid derivatives starting from easily available substrates.

12.4.1 Hydroxycarbonylation

Atmospheric hydroxycarbonylation was applied as the key step in the synthesis of two 5α-reductase inhibitors: SB 209 963 (**50**) [67] and epristeride (**51**) [68]. In both cases a 3-bromo-3,5-diene moiety was converted into the 3-carboxylic acid.

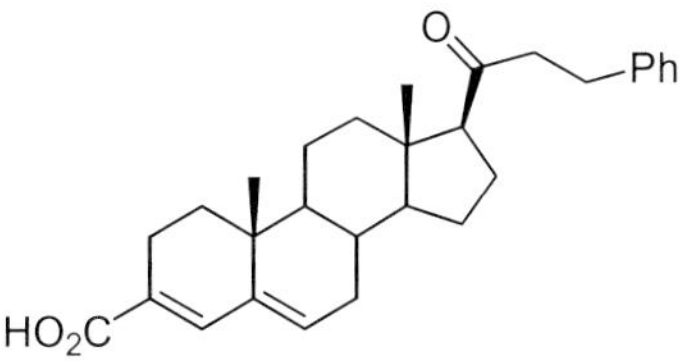

50, SB 209963

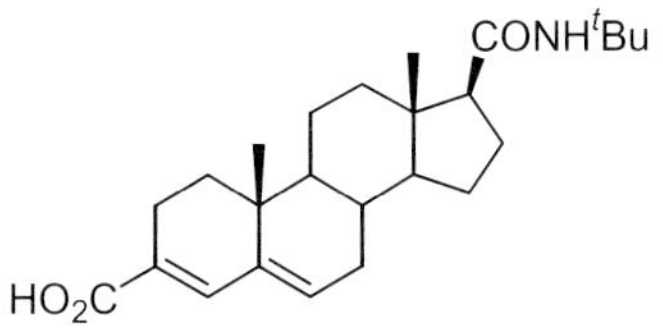

51, epristeride

The method developed by Cacchi and Lupi for the hydroxycarbonylation of various aryl triflates including steroidal substrates [69] was used effectively for the synthesis of an intermediate of the biologically active 8,10-di-*O*-methylbergenin [70]. The conversion of the rather unreactive aryl triflate substrate was achieved using DMSO as solvent that was shown to play a crucial role in the catalytic cycle by facilitating decomposition of the *in situ* formed acylpalladium complex and formation of the carboxylic acid product [69].

Although alkenyl/aryl halides or triflates are the usual substrates for such carbonylations, there are also a few examples for the conversion of benzyl derivatives. Among them, hydroxycarbonylation of 1-arylethyl chlorides or bromides leading to 2-arylpropionic acids can be of practical importance, but the use of acidic media, elevated temperature (110–125 °C), and high CO pressure (55 bar) is necessary to obtain satisfactory catalytic activity and chemoselectivity [71].

12.4.2
Alkoxycarbonylation

The main goal of alkoxycarbonylation is either the attachment of the alkoxycarbonyl moiety itself to an unsaturated carbon atom as in the case of new β-lactam antibiotics [72] or the incorporation of a functional group that can be converted into other moieties such as hydroxycarbonyl, hydroxymethylene, or formyl groups or lactones during the synthesis of the anti-inflammatory compound myrsinoic acid E (**52**) [73], and precursors of allocyathins (e.g., **53**) [74], clerocidine (**54**) [75], and 8-*epi*-xanthatin (**55**) [76]. (Figure 12.3).

As the ester group is often converted into other functional groups that do not retain the alkyl moiety of the alcohol reactant, methanol that usually shows higher reactivity compared to other alcohols is the nucleophile of choice in these reactions. Alkenyl and aryl iodides and enol triflates are the most reactive substrates that undergo alkoxycarbonylation at atmospheric CO pressure and relatively low temperature (50–60 °C). However, the reactivity may depend greatly on the presence of the nearby functional groups. As an example, the 6-iodo functionality of 6-iodo-4-aza-androst-5-en-3-one derivatives remained unchanged under carbonylation conditions, probably because of the coordination of the lactame nitrogen to palladium [77]. Alkoxycarbonylation of alkenyl bromides requires either the use of higher temperature and CO pressure, like during the synthesis of 2-arylacrylic esters from α-arylvinyl bromides [78], or that of chelating bisphosphine ligands [79]. In the latter case, the different reactivities of the iodo- and bromo-alkenyl moieties were exploited during the construction of the macrocycle (**56**) of macrosphelides in two different alkoxycarbonylation steps.

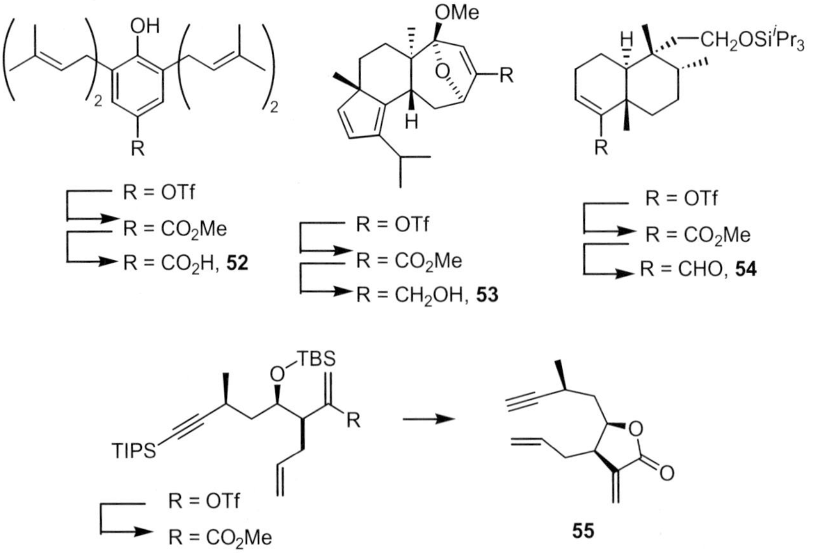

Figure 12.3 Different functional groups obtained through alkoxycarbonylation of enol- or aryl triflates.

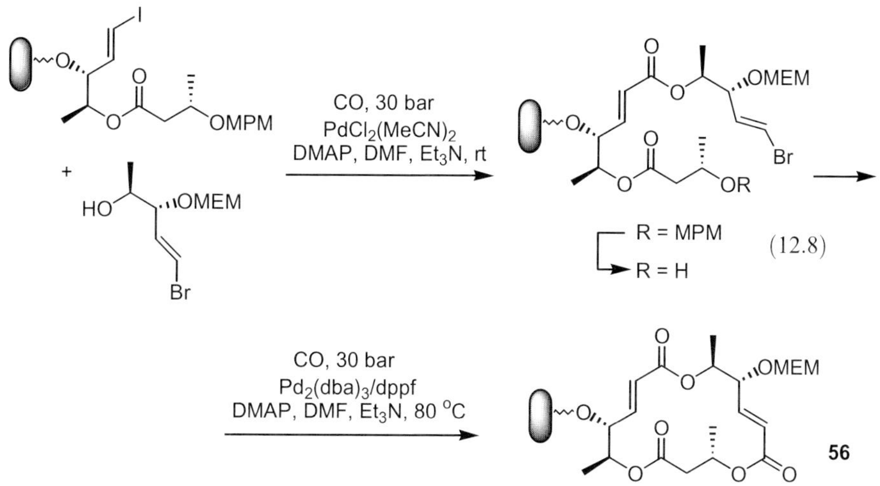

CO, 30 bar
$PdCl_2(MeCN)_2$
DMAP, DMF, Et_3N, rt

R = MPM
R = H

(12.8)

CO, 30 bar
$Pd_2(dba)_3$/dppf
DMAP, DMF, Et_3N, 80 °C

56

Although chloroaromatics generally display lower reactivity in palladium-catalyzed reactions, the presence of the heteroatom in heteroaryl chlorides can activate the C–Cl bond in C-2 and C-4 positions. Accordingly, alkoxycarbonylation of **57** in the presence of the $Pd(OAc)_2$/dppp catalyst followed by the annulation of the cyclopentanone ring was used for the construction of the tricyclic ring system (**59**) during the total synthesis of the anticancer drug irinotecan [80].

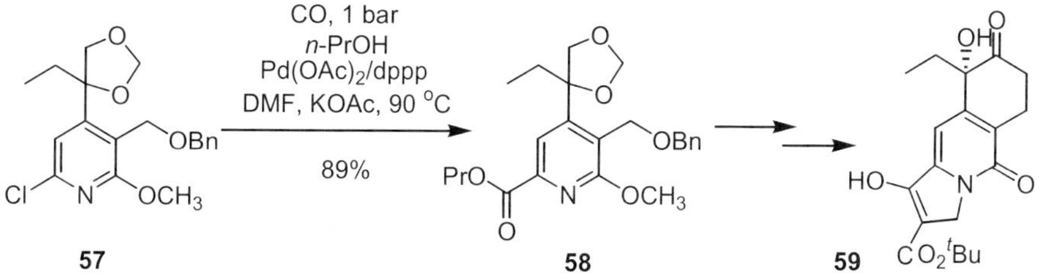

CO, 1 bar
n-PrOH
$Pd(OAc)_2$/dppp
DMF, KOAc, 90 °C
89%

57 **58** **59**

(12.9)

Even selective monocarbonylation of 2,5-dichloropyridine in the C-2 position could be achieved by the use of the $Pd(OAc)_2$/dppf catalyst and by limiting the amount of the Et_3N base to 1.01 equiv. to prevent overcarbonylation during the synthesis of a chloropyridine *N*-oxide building block of a potent phosphodiesterase IV inhibitor [81].

Alkoxycarbonylation of the relatively unreactive aryl triflates can be enhanced by the use of bidentate ligands such as dppp or dppf. The $Pd(OAc)_2$/dppp catalyst system turned out to be effective during the synthesis of tetrahydrocannabinol analogues [82] or protein–tyrosine phosphatase inhibitors, [83] whereas dppf was the ligand of choice in a palladium-catalyzed synthetic approach to huparzine A analogues [84] as well as during the development of new dihydrofolate reductase inhibitors involving the synthesis of a 4-propionyl benzoic acid ester as an important intermediate [85].

Intramolecular alkoxycarbonylation leading to lactone formation was the key step during the construction of the dihydropyranone ring of precursor **60** of marine anti-inflammatory sesterpenoid manoalide [86]. In the total synthesis of uncinine, the γ-alkylidene butenolide structure **61** was obtained in a single operation starting from a β-iodoenone through trapping the intermediate acylpalladium species with the carbonyl oxygen as the internal nucleophile [87].

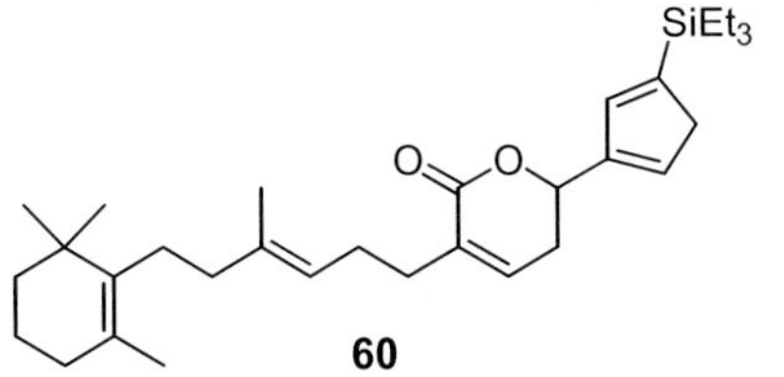

$SiEt_3$ THPO O O O

60 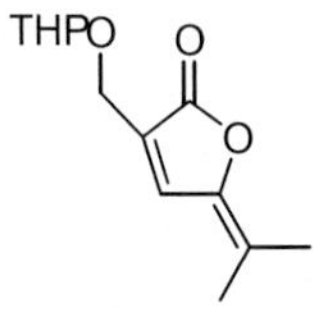 **61**

The phthalide structure of mycophenolic acid [88] and phthalideisoquinoline alkaloids [89] were constructed by intramolecular alkoxycarbonylation of substituted 2-bromobenzyl alcohols as substrates. Similar compounds were obtained by employing *in situ* generated carbon monoxide obtained from solid $Mo(CO)_6$ under microwave irradiation [90].

An interesting example for a tandem reaction integrating alkoxycarbonylation is the synthesis of the tetracyclic core of phomoidrides possessing protein farnesyl transferase inhibitory effect. The palladium–acyl intermediate, formed in the reaction of the palladium catalyst, the enol triflate **62** and carbon monoxide, has been trapped with the hemiketal formed from the hydroxy ketone in **62**, thus producing two rings in a single step. This was followed by a Cope rearrangement that was accomplished by simply raising the temperature to 110 °C after completion of carbonylation [91]. Another tandem reaction involving activation of enol triflate **64**, cyclization, and alkoxycarbonylation led to a useful A-ring precursor (**66**) of vitamin D analogues [92].

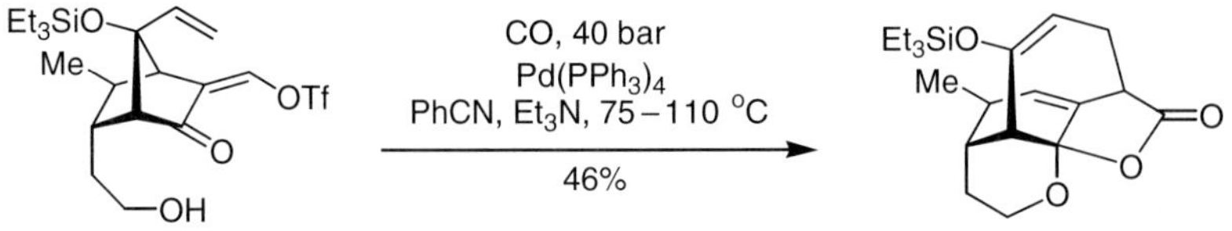

Et_3SiO Me OTf O OH **62** — CO, 40 bar; $Pd(PPh_3)_4$; PhCN, Et_3N, 75–110 °C; 46% → Et_3SiO Me O O O **63** (12.10)

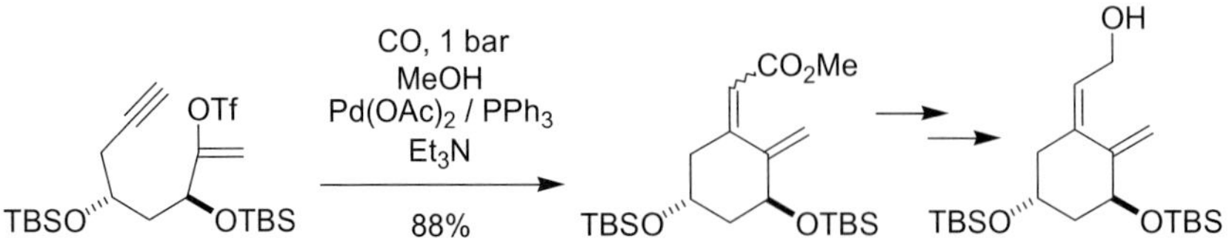

OTf TBSO OTBS **64** — CO, 1 bar; MeOH; $Pd(OAc)_2$ / PPh_3; Et_3N; 88% → CO_2Me TBSO OTBS **65** → → OH TBSO OTBS **66** (12.11)

Although alkyl iodides do not undergo the usual palladium-catalyzed carbonylation reactions, they can be converted into carboxylic acid derivatives by radical atom transfer carbonylation. The reaction of primary alkyl iodides was shown to be accelerated by $Pd(PPh_3)_4$ catalyst, and this method was effectively used for the construction of the lactone ring of (−)-hinokinin [93].

12.4.3
Aminocarbonylation

All substrates mentioned in the previous chapter easily undergo palladium-catalyzed aminocarbonylation also. It should be mentioned that even allylic carbonates instead of halides could be converted into carboxamides by this method, as it was demonstrated during the total synthesis of antillatoxin [94]. The use of aryl chlorides is somewhat restricted owing to the low reactivity of these type of substrates in oxidative addition to palladium(0), as well as the competitive amination observed, for example, in the reaction of 1-chloroisoquinoline [95]. Functionalization of a great number of pharmaceutically important skeletons, for example, steroids [96–99], tropenes [100], benzodiazepines [101], indoles [102], benzo[*c*]quinolizine derivatives [103], and quinolones [104], have been carried out.

The other advantage of aminocarbonylation involves the great variety of nucleophiles that can effectively be applied as reactants. Beside the commonly used primary and secondary amines, these include propargylamine in the synthesis of an intermediate of 8-*epi*-griseoviridin [105], arylamines, and heteroarylamines in the synthesis of gonadotropine-releasing hormone antagonists [104], substituted hydrazine derivatives [97], hydrazides [98], amino crown ethers [99], and amino acids [94,106]. Even sulfonamides were shown to be able to participate as nucleophiles in microwave-accelerated carbonylation in the presence of $Mo(CO)_6$ as the carbonyl source [107].

Intramolecular aminocarbonylation leading to the lactame structure has been used in the construction of the tetracyclic core of 8-oxoberbines (e.g., **68**) that were converted into protoberberine alkaloids (e.g., **69**) by reduction with lithium aluminum hydride [89].

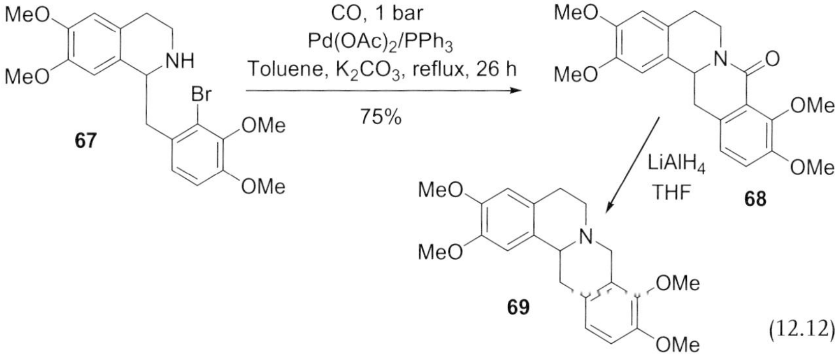

(12.12)

12.4.4
Carbonylative Coupling Reactions

Palladium-catalyzed coupling reactions of alkenyl/aryl halides or triflates with alkenes, alkynes, or organometallic reagents are readily used during the synthesis of pharmaceuticals because of the high yield and great selectivity that can be achieved under mild reaction conditions in these processes [108]. At the same time, there are only a few examples for the same reaction carried out in carbon monoxide atmosphere.

A carbonylative alkynylation–cyclocondensation reaction sequence starting from iodoindoles (e.g., **70**) was developed for the synthesis of meridianins, a group of natural products with protein kinase inhibition activity [109]. It is worth mentioning that carbonylation could be carried out with great selectivity, as in the example depicted in Equation 12.13, where the bromo substituent remained unchanged under the reaction conditions. A tandem carbonylative coupling–cyclization reaction of **73** with thiazolylacetylene **74** led to the formation of quionolone **75**, a key substructure of protease inhibitor BILN 2061 [110].

70 + TMS-acetylene → (CO, 1 bar; $PdCl_2(PPh_3)_2$/CuI; THF, Et_3N, rt, 48 h; 68%) → **71** → **72**, Meridianin C (12.13)

73 + **74** → (CO, 17 bar; $PdCl_2$(dppf); Et_2NH, 120 °C, 6h; 70%) → **75** (12.14)

Reaction of aryl or alkenyl triflates with arylboronic acid or $NaBPh_4$ in carbon monoxide atmosphere led to a mixture of products obtained by carbonylation and by direct coupling, as it was observed during the synthesis of grossularine intermediates [111] and steroidal ketones [112]. On the contrary, a similar reaction of alkenyl iodides led to the selective formation of the keto-compounds [112]. Synthesis of steroids possessing 17-formyl-16-ene moiety was carried out by the carbonylation of the corresponding 17-iodo-16-ene derivatives using tributyltin hydride as the hydrogen source [113]. The chemoselectivity of the reaction was greatly influenced by the reaction conditions. The application of flexible chelating phosphines such as dppb or dppp proved to be superior to the use of monodentate phosphines.

12.5 Concluding Remarks

The great number of publications cited in this chapter clearly shows that among other homogeneous catalytic methods, carbonylation reactions are applied more and more extensively during the synthesis of pharmaceuticals. In many cases, a target compound can be obtained by a single selective catalytic step instead of a stepwise synthesis, resulting in a large increase in the total yield.

References

1 Chen, A.C., Ren, L., Decken, A. and Crudden, C.M. (2000) *Organometallics*, **19**, 3459–3461.

2 Nozaki, K., Sakai, N., Nanno, T., Higashijima, T., Mano, S., Horiuchi, T. and Takaya, H. (1997) *Journal of the American Chemical Society*, **119**, 4413–4423.

3 Babin, J.E. and Whiteker, G.T. (1993) WO, 9303839, Union Carbide Chemicals.

4 Guimet, E., Parada, J., Diégez, M., Ruiz, A. and Claver, C. (2005) *Applied Catalysis A – General*, **282**, 215–220.

5 Chelucci, G., Marchetti, M. and Sechi, B. (1997) *Journal of Molecular Catalysis A: Chemical*, **122**, 111–114.

6 Nozaki, K., Li, W., Horiuchi, T. and Takaya, H. (1996) *Journal of Organic Chemistry*, **61**, 7658–7659.

7 Alper, H., Saito, T. and Miura, T. (2001) US 6169179 B1, Takasago Int. Corp.

8 Saito, T., Matsumura, K., Miura, T., Kumobayashi, H. and Yoshida, A. (1997) EP 0755937 A1, Takasago Int. Corp.

9 Cesarotti, E. and Rimoldi, I. (2004) *Tetrahedron: Asymmetry*, **15**, 3841–3845.

10 Stille, J.K., Su, H., Brechot, P., Parrinello, G. and Hegedus, L.S. (1991) *Organometallics*, **10**, 1183–1189.

11 Cuny, G.D. and Buchwald, S.L. (1995) *Synlett*, 519–522.

12 Sun, P., Sun, C. and Weinreb, S.M. (2002) *Journal of Organic Chemistry*, **67**, 4337–4345.

13 Daugs, E.D., Peng, W.-J. and Rand, C.L. (2005) WO 2005110986 A1, Dow Global Techn. Inc.

14 Horgan, S.W., Burkhouse, D.W., Cregge, R. J., Freund, D.W., LeTourneau, M., Margolin, A. and Webster, M.E. (1999) *Organic Process Research & Development*, **3**, 241–247.

15 Keck, G.E. and Truong, A.P. (2005) *Organic Letters*, **7**, 2149–2152.

16 Breit, B. and Zahn, S.K. (1998) *Tetrahedron Letters*, **39**, 1901–1904.

17 Eilbracht, P., Bärfacker, L., Buss, C., Hollmann, C., Kitsos-Rzychon, B.E., Kranemann, C.L., Rische, T., Roggenbuck, R. and Schmidt, A. (1999) *Chemical Reviews*, **99**, 3329–3365; Fogg, D.E. and dos Santos, E.N. (2004) *Coordination Chemistry Reviews*, **248**, 2365–2379.

18 Schmidt, A.M. and Eilbracht, P. (2005) *Journal of Organic Chemistry*, **70**, 5528–5535.

19 Botteghi, C., Marchetti, M., Paganelli, S. and Persi-Paoli, F. (2001) *Tetrahedron: Asymmetry*, **57**, 1631–1637.

20 Botteghi, C., Corrias, T., Marchetti, M., Paganelli, S. and Piccolo, O. (2002) *Organic Process Research & Development*, **6**, 379–383.

21 Botteghi, C., Cazzolato, L., Marchetti, M. and Paganelli, S. (1995) *Journal of Organic Chemistry*, **60**, 6612–6615.

22 Rische, T. and Eilbracht, P. (1999) *Tetrahedron: Asymmetry*, **55**, 1915–1920.

23 Schmidt, A., Marchetti, M. and Eilbracht, P. (2004) *Tetrahedron: Asymmetry*, **60**, 11487–11492.

24 Rische, T., Müller, K.-S. and Eilbracht, P. (1999) *Tetrahedron: Asymmetry*, **55**, 9801–9816.

25 Tőrös, S., Gémes-Pécsi, I., Heil, B., Mahó S. and Tuba, Z. (1992) *Journal of the Chemical Society, Chemical Communications*, 858–859.

26 Nagy, E., Heil, B. and Tőrös, S. (1999) *Journal of Organometallic Chemistry*, **586**, 101–105.

27 Botteghi, C., Ghelucci, G., Del Ponte, G., Marchetti, M. and Paganelli, S. (1994) *Journal of Organic Chemistry*, **59**, 7125–7127.

28 Hornberger, K.R., Hamblett, C.L. and Leighton, J.L. (2000) *Journal of the American Chemical Society*, **122**, 12894–12895.

29 Ojima, I. and Vidal, E.S. (1998) *Journal of Organic Chemistry*, **63**, 7999–8003.

30 Dong, Y. and Busacca, C.A. (1997) *Journal of Organic Chemistry*, **62**, 6464–6465.

31 Van den Hoven, B.G. and Alper, H. (2001) *Journal of the American Chemical Society*, **123**, 1017–1022.

32 Van den Hoven, B.G. and Alper, H. (2001) *Journal of the American Chemical Society*, **123**, 10214–10220.

33 Wu, T.-C. (1999) US, 5902898, Albemarle Corp.

34 Ramachandran, V., Wu, T.-C. and Berry, C. B. (1998) WO, 9837052, Albemarle Corp.

35 Seayad, A., Jayasree, S. and Chaudhari, R.V. (1999) *Organic Letters*, **1**, 459–461.

36 Jayasree, S., Seayad, A. and Chaudhari, R.V. (2000) *Organic Letters*, **2**, 203–206.

37 Papadogianakis, G., Verspui, G., Maat, L. and Sheldon, R.A. (1997) *Catalysis Letters*, **47**, 43–46.

38 Jayasree, S., Seayad, A. and Chaudhari, R.V. (2000) *Journal of the Chemical Society, Chemical Communications*, 1239–1240.

39 Ramminger, C., Zim, D., Lando, V.R., Fassina, V. and Monteiro, A.L. (2000) *Journal of the Brazilian Chemical Society*, **11**, 105–111.

40 Lee, C.W. and Alper, H. (1995) *Journal of Organic Chemistry*, **60**, 250–252.

41 Zhang, J. and Xia, C.-G. (2003) *Journal of Molecular Catalysis A: Chemical*, **206**, 59–65.

42 Oi, S., Nomura, M., Aiko, T. and Inoue, Y. (1997) *Journal of Molecular Catalysis A: Chemical*, **115**, 289–295.

43 Tőrös, S., Heil, B., Gálik, G. and Tuba, Z. (1992) *Tetrahedron Letters*, **33**, 3667–3668.

44 Nagy, E., Heil, B. and Tőrös, S. (1999) *Journal of Molecular Catalysis A: Chemical*, **143**, 229–232.

45 Marshall, J.A., Wallace, E.M. and Coan, P. S. (1995) *Journal of Organic Chemistry*, **60**, 796–797; Marshall, J.A., Bartley, G.S. and Wallace, E.M. (1996) *Journal of Organic Chemistry*, **61**, 5729–5735;Marshall, J.A. and Liao J. (1998) *Journal of Organic Chemistry*, **63**, 5962–5970.

46 Franczyk, T.S., Wagner, G.M., Pearlman, B. A., Padilla, A.G., Havens, J.L., Mackey, S.S. and Haifeng, W. (2004) WO 2004/085458, Pharmacia Corp.

47 Semmelhack, M.F. and Shanmugam, P. (2000) *Tetrahedron Letters*, **41**, 3567–3571.

48 Paddon-Jones, G.C., Hungerford, N.L., Hayes, P. and Kitching, W. (1999) *Organic Letters*, **1**, 1905–1907; Hayes, P.Y. and Kitching, W. (2002) *Journal of the American Chemical Society*, **124**, 9718–9719.

49 Miyakoshi, N., Aburano, D. and Mukai, C. (2005) *Journal of Organic Chemistry*, **70**, 6045–6052.

50 Lütjens, H. and Scammels, P.J. (1998) *Tetrahedron Letters*, **39**, 6581–6584.

51 Ham, W.-H., Jung, Y.H., Lee, K., Oh, C.-Y. and Lee, K.-Y. (1997) *Tetrahedron Letters*, **38**, 3247–3248.

52 Oh, C.-Y., Kim, K.-S. and Ham, W.-H. (1998) *Tetrahedron Letters*, **39**, 2133–2136.

53 Szolcsányi, P., Gracza, T., Koman, M., Prónayová N. and Liptaj, T. (2000) *Tetrahedron: Asymmetry*, **11**, 2579–2597.

54 Artman, G.D., III and Weinreb, S.M. (2003) *Organic Letters*, **5**, 1523–1526.

55 Sheldon, R.A., Maat, L. and Papadogianakis, G. (1996) US, 5536874, Hoechst Celanese Corp.;Jang, E.J., Lee, K. H., Lee, J.S. and Kim, Y.G. (1999) *Journal of Molecular Catalysis A: Chemical*, **138**, 25–36.

56 Seayad, A., Jayasree, S. and Chaudhari, R.V. (1999) *Catalysis Letters*, **61**, 99–103;Seayad, A., Jayasree, S. and Chaudhari, R.V. (2001) *Journal of Molecular Catalysis A: Chemical*, **172**, 151–164.

57 Jang, E.J., Lee, K.H., Lee, J.S. and Kim, Y.G. (1999) *Journal of Molecular Catalysis A: Chemical*, **144**, 431–440;Mukhopadhyay, K., Sarkar, B.R. and Chaudhari, R.V. (2002) *Journal of the American Chemical Society*, **124**, 9692–9693.

58 Jayasree, S., Seayad, A. and Chaudhari, R.V. (1999) *Journal of the Chemical Society, Chemical Communications*, 1067–1068.

59 Zhou, H., Cheng, J., Lu, S., Fu, H. and Wang, H. (1998) *Journal of Organometallic Chemistry*, **556**, 239–242.

60 Li, Y.-Y. and Xia, C.-G. (2001) *Applied Catalysis A – General*, **210**, 257–262.

61 Xie, B.H., Xia, C.-G., Lu, S.-J., Chen, K.-J., Kou, Y. and Yin, Y.-Q. (1998) *Tetrahedron Letters*, **39**, 7365–7368.

62 Gabriele, B., Salerno, G., Mancuso, R. and Costa, M. (2004) *Journal of Organic Chemistry*, **69**, 4741–4750.

63 Zhang, Y., Forinash, K., Phillios, C.R. and Mc-Elwee-White, L. (2005) *Green Chemistry*, **7**, 451–455.

64 Hylton, K.-G., Main, A.D. and Mc-Elwee-White, L. (2003) *Journal of Organic Chemistry*, **68**, 1615–1617.

65 Tanner, D. and Somfai, P. (1993) *Bioorganic & Medicinal Chemistry Letters*, **3**, 2415–2418.

66 Davoli, P., Moretti, I., Prati, F. and Alper, H. (1999) *Journal of Organic Chemistry*, **64**, 518–521; Davoli, P. and Prati, F. (2000) *Heterocycles*, **53**, 2379–2389.

67 Yu, M.S. and Baine, N.H. (1999) *Tetrahedron Letters*, **40**, 3123–3124.

68 McGuire, M.A., Sorenson, E., Klein, D.N. and Baine, N.H. (1998) *Synthetic Communications*, **28**, 1611–1615.

69 Cacchi, S. and Lupi, A. (1992) *Tetrahedron Letters*, **33**, 3939–3942.

70 Herzner, H., Palmacci, E.R. and Seeberger, P.H. (2002) *Organic Letters*, **4**, 2965–2967.

71 Elango, V. (2003) US 6555704 B1, BASF Corp.

72 Konaklieva, M.I., Shi, H. and Turos, E. (1997) *Tetrahedron Letters*, **38**, 8647–8650; Munroe, J.E. (1989) EP 0299728 A2, Eli Lilly and Company.

73 Makabe, H., Miyazaki, S., Kamo, T. and Hirota, M. (2003) *Bioscience, Biotechnology, and Biochemistry*, **67**, 2038–2041.

74 Ward, D.E., Gai, Y. and Qiao, Q. (2000) *Organic Letters*, **2**, 2125–2127; Snider, B.B., Vo, N.H. and O'Neil, S.V. (1998) *Journal of Organic Chemistry*, **63**, 4732–4740.

75 Almstead, J.-I.K., Demuth, T.P., Jr and Ledoussal, B. (1998) *Tetrahedron: Asymmetry*, **9**, 3179–3183.

76 Kummer, D.A., Brenneman, J.B. and Martin, S.F. (2005) *Organic Letters*, **7**, 4621–4623.

77 Skoda-Földes, R., Horváth, J., Tuba, Z. and Kollár, L. (1999) *Journal of Organometallic Chemistry*, **586**, 94–100.

78 Silveira, P.B. and Monteiro, A.L. (2006) *Journal of Molecular Catalysis A: Chemical*, **247**, 1–6.

79 Takahasi, T., Kusaka, S., Doi, T., Sunazuka, T. and Ōmura, S. (2003) *Angewandte Chemie – International Edition*, **42**, 5230–5234.

80 Henegar, K.E., Ashford, S.W., Baughman, T.A., Sih, J.C. and Gu, R.-L. (1997) *Journal of Organic Chemistry*, **62**, 6588–6597.

81 Albaneze-Walker, J., Murry, J.A., Soheili, A., Ceglia, S., Springfield, S.A., Bazaral, C., Dormer, P.G. and Hughes, D.L. (2005) *Tetrahedron: Asymmetry*, **61**, 6330–6336.

82 Sun, H., Mahadevan, A. and Razdan, R.K. (2004) *Tetrahedron Letters*, **45**, 615–617.

83 Ye, B. and Burke, T.R., Jr (1996) *Tetrahedron: Asymmetry*, **52**, 9963–9970.

84 Gemma, S., Butini, S., Fattorusso, C., Fiorini, I., Nacci, V., Bellebaum, K., McKissic, D., Saxena, A. and Campiani, G. (2003) *Tetrahedron: Asymmetry*, **59**, 87–93.

85 Gangjee, A., Zeng, Y., McGuire, J.J. and Kisliuk, R.L. (2002) *Journal of Medicinal Chemistry*, **45**, 1942–1948.

86 Pommier, A. and Kocieński, P.J. (1997) *Journal of the Chemical Society, Chemical Communications*, 1139–1140.

87 Fáková H., Pour, M., Kuneš J. and Šenel, P. (2005) *Tetrahedron Letters*, **46**, 8137–8140.

88 Lee, Y., Fujiwara, Y., Ujita, K., Nagatom, M., Ohta, H. and Shimizu, I. (2001) *Bulletin of the Chemical Society of Japan*, **74**, 1437–1443.

89 Orito, K., Miyazawa, M., Kanbayashi, R., Tokuda, M. and Suginome, H. (1999) *Journal of Organic Chemistry*, **64**, 6583–6596.

90 Wu, X., Mahalingam, A.K., Wan, Y. and Alterman, M. (2004) *Tetrahedron Letters*, **45**, 4635–4638.

91 Bio, M.M. and Leighton, J.L. (1999) *Journal of the American Chemical Society*, **121**, 890–891; Bio, M.M. and Leighton, J.L. (2000) *Organic Letters*, **2**, 2905–2907; Bio, M.M. and Leighton, J.L. (2003) *Journal of Organic Chemistry*, **68**, 1693–1700.

92 Mouriño, A., Torneiro, M., Vitale, C., Fernández, S., Pérez-Sestelo, J., Anné, S. and Gregorio, C. (1997) *Tetrahedron Letters*, **38**, 4713–4716.

93 Fukuyama, T., Nishitani, S., Inouye, T., Morimoto, K. and Ryu, I. (2006) *Organic Letters*, **8**, 1383–1386.

94 Loh, T.-P., Cao, G.-Q. and Yin, Z. (1999) *Tetrahedron Letters*, **40**, 2649–2652.

95 Kumar, K., Michalik, D., Garcia Castro, I., Tillack, A., Zapf, A., Arlt, M., Heinrich, T., Böttcher, H. and Beller, M. (2004) *Chemistry – A European Journal*, **10**, 746–757.

96 Tuba, Z., Horváth, J., Széles, J., Kollár, L. and Balogh, G. (1995) WO 9500531, Works of Gedeon Richter Ltd; Ács, P., Müller, E., Czira, G., Mahó S., Perreira, M. and Kollár L. (2006) *Steroids*, **71**, 875–879.

97 Skoda-Földes, R., Szarka, Z., Kollár, L., Dinya, Z., Horváth, J. and Tuba, Z. (1999) *Journal of Organic Chemistry*, **64**, 2134–2136.

98 Szarka, Z., Skoda-Földes, R., Horváth, J., Tuba, Z. and Kollár, L. (2002) *Steroids*, **67**, 581–586.

99 Petz, A., Gálik, Gy., Horváth, J., Tuba, Z., Berente, Z., Pintér, Z. and Kollár, L. (2001) *Synthetic Communications*, **31**, 335–341.

100 Horváth, L., Berente, Z. and Kollár, L. (2005) *Letters in Organic Chemistry*, **2**, 54–56.

101 Andrews, I.P., Atkins, R.J., Badham, N.F., Bellingham, R.K., Breen, G.F., Carey, J.S., Etridge, S.K., Hayes, J.F., Hussain, N., Morgan, D.O., Share, A.C., Smith, S.A.C., Walsgrove, T.C. and Wells, A.S. (2001) *Tetrahedron Letters*, **42**, 4915–4917.

102 Herbert, J.M. and McNeill, A.H. (1998) *Tetrahedron Letters*, **39**, 2421–2424.

103 Guarna, A. and Serio, M. (1997)WO 9729107 (to Appl. Research Syst.).

104 Walsh, T.F., Toupence, R.B., Young, J.R., Huang, S.X., Ujjainwalla, F., DeVita, R.J., Goulet, M.T., Wyvratt, M.J., Jr, Fisher, M.H., Lo, J.-L., Ren, N., Yudkovitz, J.B., Yang, Y.T., Cheng, K. and Smith, R.G. (2000) *Bioorganic & Medicinal Chemistry Letters*, **10**, 443–447.

105 Kuligowski, C., Bezzenine-Lafollée, S., Chaume, G., Mahuteau, J., Barrière, J.-C., Bacqué E., Pancrazi, A. and Ardisson, J. (2002) *Journal of Organic Chemistry*, **67**, 4565–4568.

106 Müller, E., Péczely, G., Skoda-Földes, R., Takács, E., Kokotos, G., Bellis, E. and Kollár, L. (2005) *Tetrahedron: Asymmetry*, **61**, 797–802; Takács, E., Skoda-Földes, R., Ács, P., Müller, E., Kokotos, G. and Kollár, L. (2006) *Letters in Organic Chemistry*, **3**, 62–67.

107 Wu, X., Rönn, R., Gossas, T. and Larhed, M. (2005) *Journal of Organic Chemistry*, **70**, 3094–3098.

108 Beller, M. and Bolm, C.(1998) *Transition Metals for Organic Synthesis*, vol. 1, Wiley-VCH Verlag GmbH, Weinheim, Germany.

109 Karpov, A.S., Merkul, E., Rominger, F. and Müller, T.J.J. (2005) *Angewandte Chemie – International Edition*, **44**, 6951–6956.

110 Haddad, N., Tan, J. and Farina, V. (2006) *Journal of Organic Chemistry*, **71**, 5031–5034.

111 Choshi, T., Yamada, S., Sugino, E., Kuwada, T. and Hibino, S. (1995) *Journal of Organic Chemistry*, **60**, 5899–5904.

112 Skoda-Földes, R., Székvölgyi, Z., Kollár, L., Berente, Z., Horváth, J. and Tuba, Z. (2000) *Tetrahedron: Asymmetry*, **56**, 3415–3418.

113 Petz, A., Horváth, J., Tuba, Z., Pintér, Z. and Kollár, L. (2002) *Steroids*, **67**, 777–781.

13
Palladium-Assisted Synthesis of Heterocycles via Carbonylation Reactions

Elisabetta Rossi

13.1
Introduction

The rationale of this chapter is to illustrate how palladium salts and complexes as catalysts are important in the recent development of carbonylation reactions that are devoted to the synthesis of heterocyclic compounds. Carbonylative metal-catalyzed processes have proven to be a very powerful tool for the synthesis of heterocycles, and a great number of reviews [1–16] and books [17–20] focusing or containing detailed descriptions on this argument have been published. *Inter alia*, cyclocarbonylation reactions giving rise to simple five-, six-, and seven-membered lactone and lactame rings have been recently reviewed [8,10] by several authors and are assessed in the chapter devoted to carbonylation reactions of alcohols and amines. Thus, this chapter does not present a complete historical coverage of the matter but presents the most significant developments of the past 10 years. The emphasis is on novel and synthetic transformations of genuine value to organic chemists. This review will especially focus on carbonylative reactions involving oxidative addition of Pd(0) to $C_{sp^2}-X$ bond (Section 13.2) and oxidative carbonylation reactions mediated by palladium(II) salts (Section 13.3).

13.2
Carbonylative Reactions Involving Oxidative Addition of Pd(0) to $C_{sp^2}-X$ Bond

The carbonylative palladium-catalyzed reactions discussed in this section proceed by oxidative addition of palladium(0) to the carbon–X bond of aryl/vinyl/acyl halides and pseudohalides followed by carbon monoxide insertion, giving rise to an acylpalladium intermediate. The acylpalladium intermediate can in turn react with various tethered nucleophiles (a), be involved in activation/hetero or carbopalladation steps with unsaturated carbon–carbon bonds (b), or participate in cascade reactions (c) (Scheme 13.1).

Modern Carbonylation Methods. Edited by László Kollár

ISBN: 978-3-527-31896-4

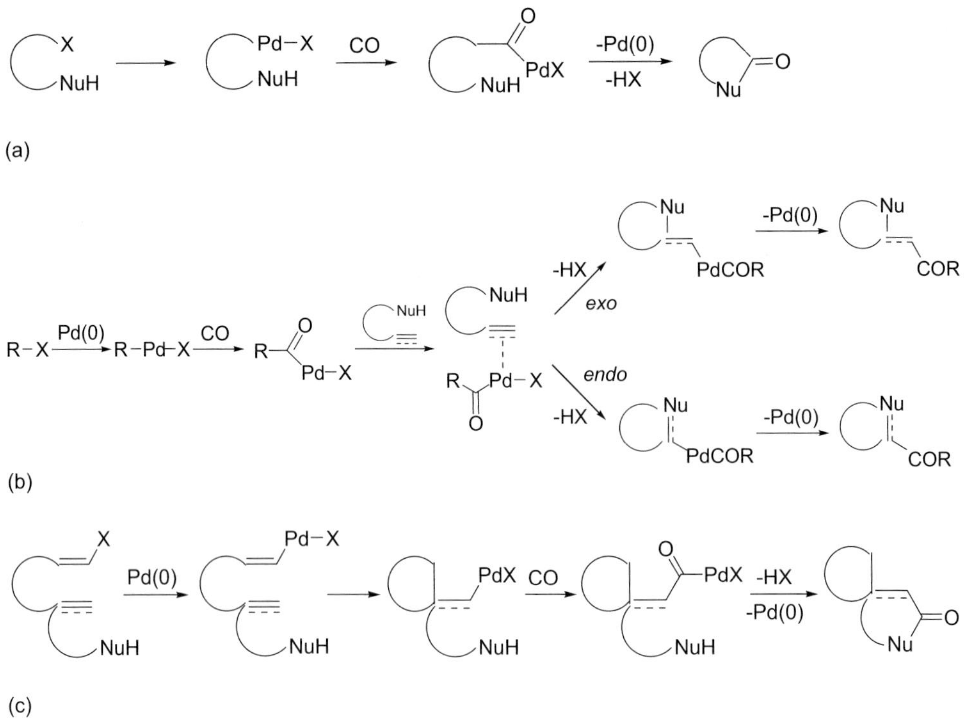

(a)

(b)

(c)

Scheme 13.1

All proposed reaction pathways end with reductive elimination of palladium(0) and formation of simple, condensed, or polyheterocyclic compounds.

13.2.1
Carbonylative Cyclizations Involving Heteronucleophilic Attack on an Acylpalladium Intermediate

Alper and coworkers successfully exploited the cyclocarbonylation reactions of *o*-iodophenols **1**, 2-hydroxy-3-iodopyridine **2**, and *o*-iodoanilines **3** with heterocumulenes, giving rise to benzo[*e*]-1,3-oxazin-4-one **4**, pyrido[3,2-*e*]-1,3-oxazin-4-one **5**, 4(3*H*)-quinazolinone **6**, and 4(3*H*)-quinazolindione **7** derivatives (Scheme 13.2) [21,22]. The reactions proceed in the presence of equimolar amount of palladium acetate and a bidentate phosphine ligand (dppb or dppf) and a base (*i*-Pr_2NEt or K_2CO_3), in benzene or THF at 80–100 °C under a pressure of 20 atm of carbon monoxide. Carbodiimides, isocyanate, and ketenimines were used as cumulenes, and good regioselectivities were achieved with unsymmetrical reactants. The reaction mechanism is believed to involve *in situ* formation of a carbamate ester or urea-type intermediate followed by palladium-catalyzed carbonylation and cyclization to yield the products.

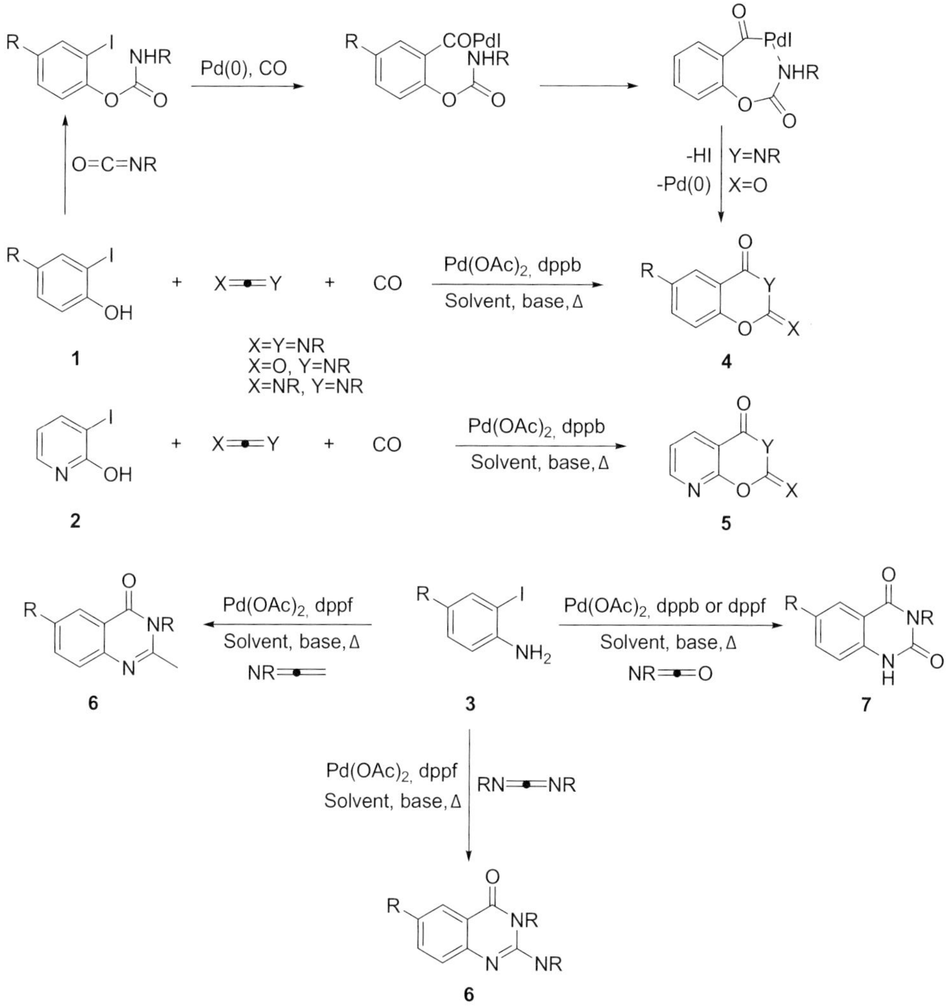

Scheme 13.2

When acid chlorides were used instead of cumulenes in the reaction with *o*-iodoanilines **3** and carbon monoxide (20 atm), 2-substituted-4*H*-3,1-benzoxazin-4-ones **8** were obtained in good to excellent yields, also in the absence of a phosphine ligand [23]. The reaction is believed to proceed via *in situ* amide formation from *o*-iodoaniline and acid chloride, followed by oxidative addition to Pd(0), CO insertion, and intramolecular cyclization (Scheme 13.3).

Protoberberine alkaloids have been prepared in a similar fashion by Orito *et al.* [24], who described the carbonylation of 1-(2′-bromo-3′,4′-dialkoxybenzyl)-tetrahydroisoquinolines **10** that were easily prepared by $NaBH_4$ reduction of the corresponding dihydroisoquinolines **9**, the Bischler–Napieralski cyclization products.

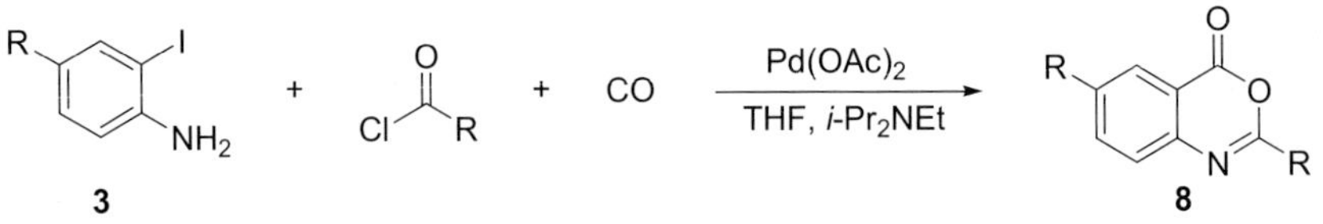

Scheme 13.3

By treatment with CO (1 atm) in the presence of $Pd(OAc)_2$, PPh_3, and excess K_2CO_3 in boiling toluene, 8-oxoberbines **11** were isolated in good yields. Further, by treating with excess $LiAlH_4$, these lactams were converted almost quantitatively into protoberberine alkaloids **12** (Scheme 13.4).

Starting from *o*-iodophenols **1** or *N*-substituted *o*-iodoanilines **3** and internal alkynes, Larock and coworkers described several annulation processes involving heteronucleophilic attack on the acylpalladium intermediate generated through intermolecular insertion of the internal alkyne into the σ-aryl palladium complex followed by CO insertion (Scheme 13.5). These annulation processes represent the first examples of intermolecular insertion of an alkyne on a σ-aryl palladium complex occurring in preference to CO insertion and allow the exclusive formation of coumarins **13** and 2-quinolones **14** [25–27]. Under optimized settings coumarins and 2-quinolones were obtained in moderate to good yields, whereas isomeric chromones and 4-quinolones, arising from reverse insertion sequence, were never isolated or detected. The selected reaction conditions utilize $Pd(OAc)_2$, 2 equiv. of pyridine, and n-Bu_4NCl, under CO (1 atm) in DMF at 100–120 °C. The only drawback encountered is related to the annulation of unsymmetrical alkynes that produces mixture of regioisomers arising from the two possible modes of alkyne insertion into the aryl–palladium bond. Both the regioselectivity and the yields of these annulations are affected by the steric bulk of the substituents on the alkyne moiety.

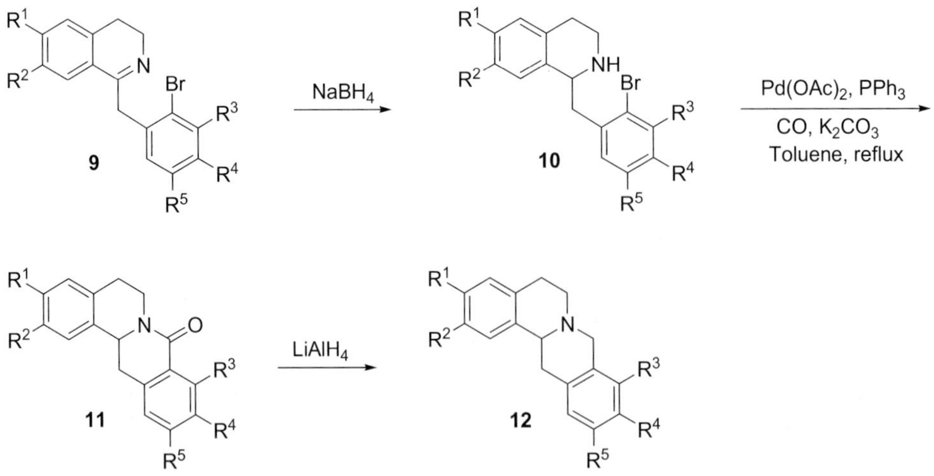

Scheme 13.4

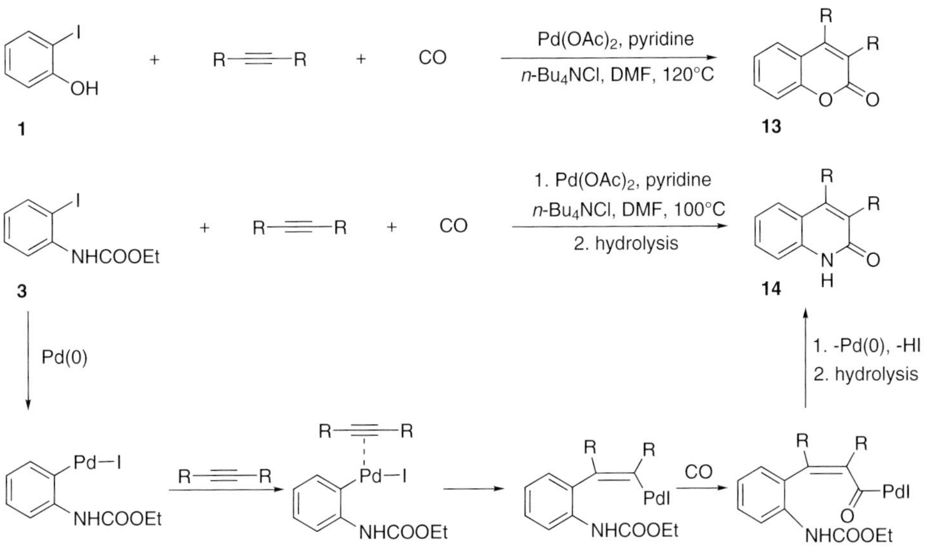

Scheme 13.5

Under the same reaction conditions, terminal alkynes afforded 3-substituted cumarines and 2-quinolones in moderate yields and regioselectivities (Scheme 13.6) [28].

However, when *o*-iodophenols **1** were reacted with terminal alkynes under Sonogashira conditions in the presence of carbon monoxide, the corresponding 4*H*-chromen-4-ones (flavones) were the sole reaction products. For example, ferrocene-containing compounds **15**, which are interesting both for academy and industry, have been synthesized from ethynylferrocene, *o*-iodophenols **1**, and CO (1 atm) under the conditions reported in Scheme 13.7 [29]. Probably, under these conditions, the alkyne acts as the nucleophile over the acylpalladium complex, and the reaction ends with intramolecular Michael addition of the phenolic oxygen over the activated triple bond.

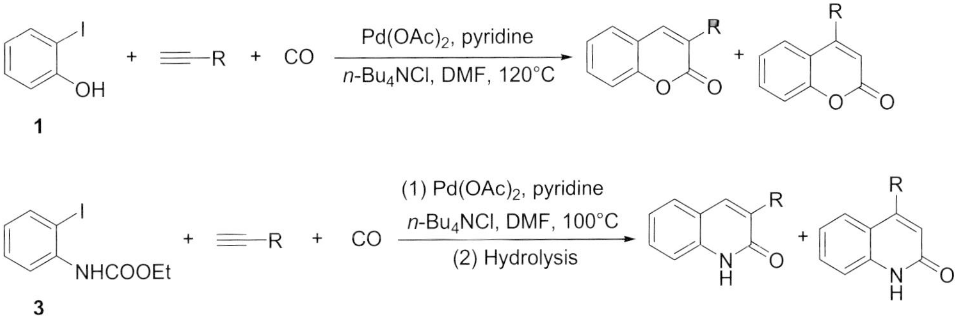

Scheme 13.6

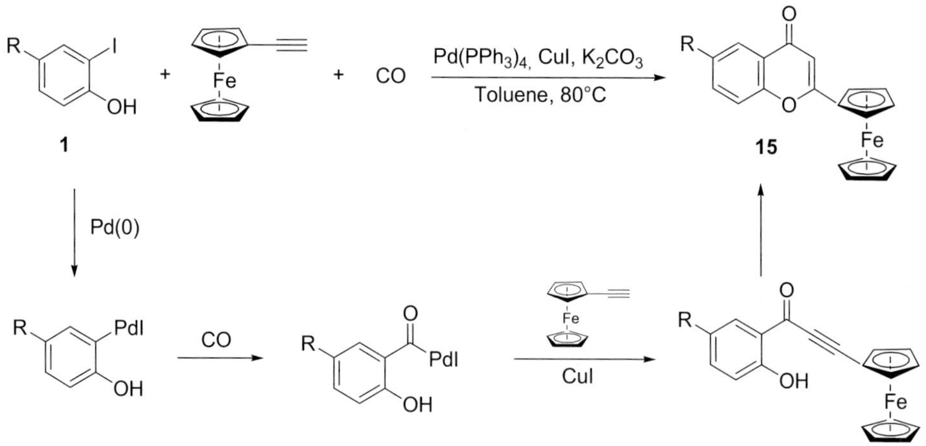

Scheme 13.7

Also, in the presence of $Pd(OAc)_2$(dppf) as the catalyst and DBU as the base in DMF at 60 °C under CO (1 atm), under copper-free conditions, these reactions involve the initial formation of an ynone arising from sequential CO and alkyne insertion. However, the reactions give a mixture of six-membered chromones **15**

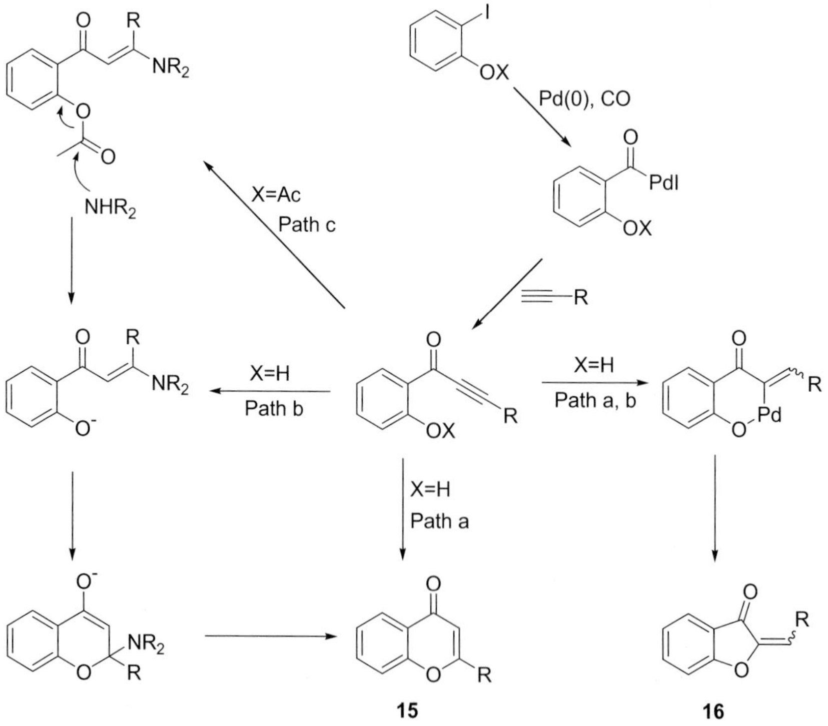

Scheme 13.8

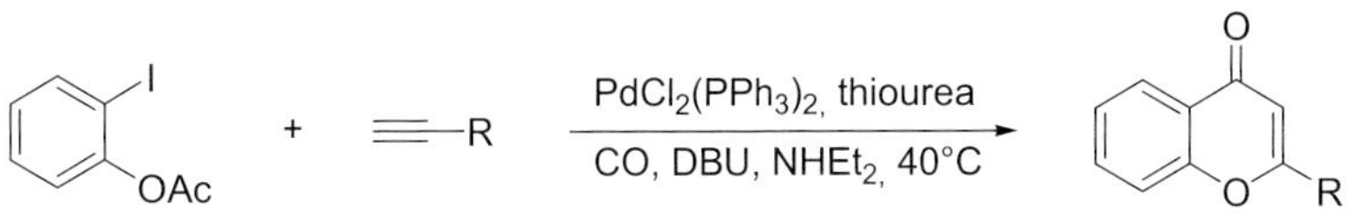

Scheme 13.9

and five-membered aurones **16** (Scheme 13.8, path a) [30], and this setback can be circumvented working with a large excess of diethylamine (Scheme 13.8, path b) [31,32].

More recently, an efficient strategy for these carbonylative annulations has been developed starting from *o*-iodophenol acetates, instead of *o*-iodophenols, and terminal acetylenes (Scheme 13.8, path c) [33]. The reaction, mediated by a palladium–thiourea complex under a balloon pressure of CO, generates diversified flavones **15** in high yields and in a regiospecific fashion (Scheme 13.9).

Several examples of lactonization processes involving carbonylative cyclization of oxygen nucleophiles with acylpalladium intermediates have been reported for the synthesis of natural occurring heterocyclic compounds. Orito *et al.* developed a general route for the synthesis of phthalideisoquinoline alkaloids bearing methoxy- and/or methylenedioxy substituents together with *nor*- and *iso*-types of the alkaloids (Scheme 13.10) [24]. Starting from the Bischler–Napieralski adducts **9**, reported in Scheme 13.4, the key steps are the stereoselective reduction of 1-(2′-bromobenzoyl)-3, 4-dihydroisoquinoline methiodide **17** with sodium borohydride or lithium aluminum hydride, followed by a palladium(0)-catalyzed carbonylation of the resultant *erythro*- and *threo*-amino alcohols **18** and **19**, assisted by chlorotrimethylsilane. In the lactonization step the isoquinoline moiety may act as an efficient internal trap for the evolved HBr molecule and can also generate a steric repulsion between the OH and Br (or Pd) groups, therefore providing a favorable approach for the lactonization step to the desired phthalideisoquinoline ring systems **20** and **21** and avoiding the drawbacks encountered in the lactonization process of 2-bromo-3, 4-dimethoxybenzyl alcohol, initially examined as the model compound. Indeed, under the same reaction conditions, 2-bromo-3,4-dimethoxybenzyl alcohol gives mixtures of desired phthalide together with the corresponding debrominated benzyl alcohol and benzaldehyde.

However, cyclocarbonylation of 2-bromo-3,5-dimethoxybenzyl alcohol **22** was successfully achieved by Shimizu and coworkers in the synthesis of mycophenolic acid **23**, a fermentation product produced by a number of penicillum species (Scheme 13.11) [34]. The synthesis was carried out using palladium-catalyzed carbonylation and Heck olefination. Thus, the reaction of 2-bromo-3, 5-dimethoxybenzyl alcohol in toluene under carbon monoxide (40 atm) at 180 °C in the presence of $Pd(OAc)_2/PPh_3$ using sodium carbonate as the base gave 5, 7-dimethoxyphthalide in 88% yield. The phthalide was converted into 6-iodo-5, 7-dimethoxy-4-methylphthalide and then reacted with isoprene and dimethyl malonate in the presence of palladium(0) catalyst, giving rise to a three-component

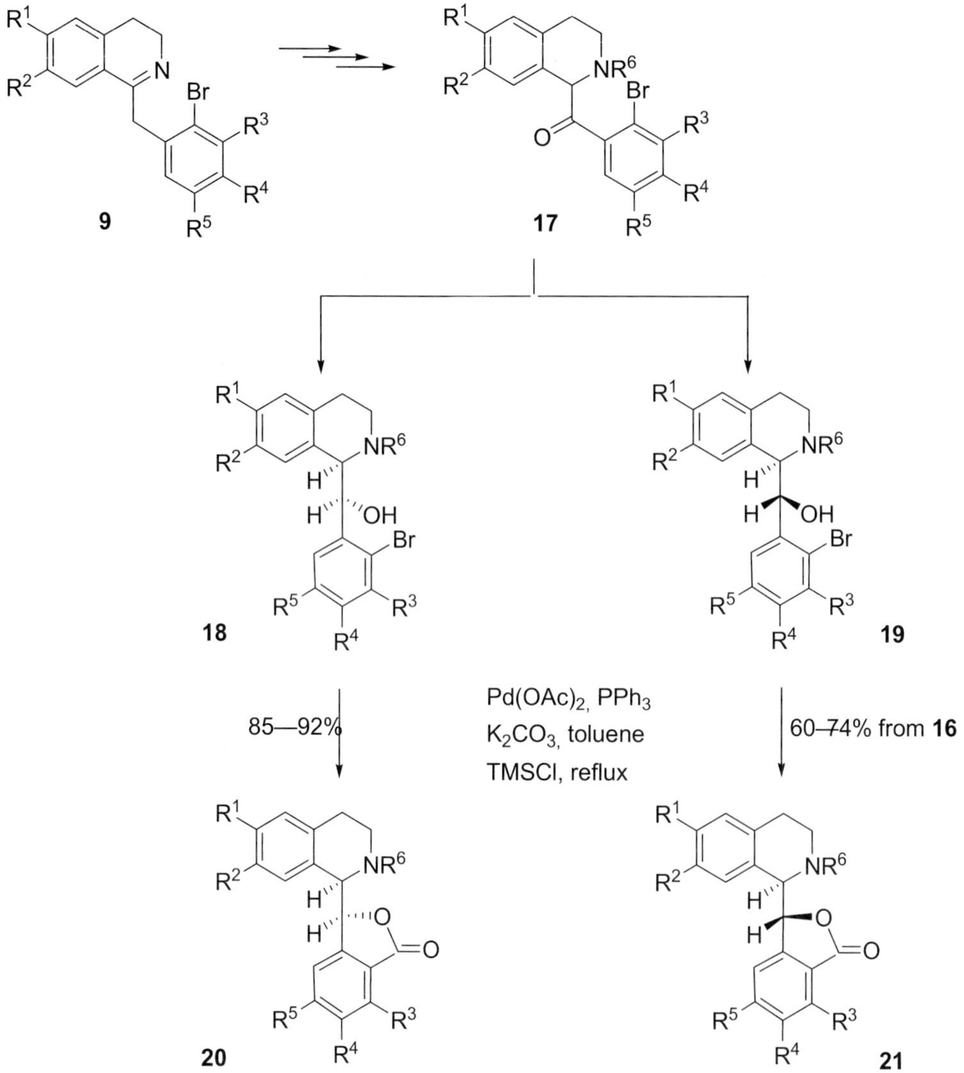

R1
R2
N
Br
R3
R4
R5
9
NR6
O
17
H
OH
18
19
85—92%
Pd(OAc)2, PPh3
K2CO3, toluene
TMSCl, reflux
60–74% from 16
20
21

Scheme 13.10

coupling product, which was finally converted into mycophenolic acid **23** in three steps. 4-NorMPA and 4-homoMPA were synthesized similarly.

A stereodefined lactone is the key intermediate in the synthesis of (+)-homopumiliotoxin 223G **25** proposed by Kibayashi and coworkers (Scheme 13.12) [35]. The essential feature leads in the carbonylative cyclization process involving a vinyliodide and a stereodefined tertiary alcohol, giving rise to lactone **24** in 99% yield under very straightforward and standard reaction conditions. Starting from **24**, (+)-homopumiliotoxin 223G **25** was obtained in three steps including cyclization to quinolizidine nucleus with a *Z*-alkylidene side chain.

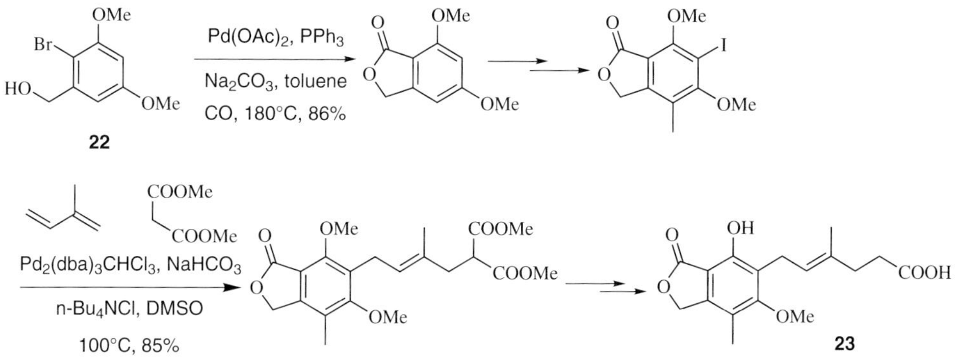

Scheme 13.11

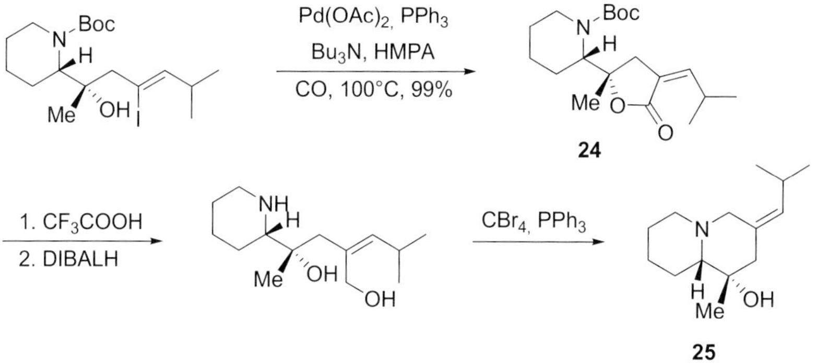

Scheme 13.12

An elegant approach to the synthesis of phomoidrides **26** was reported in several papers by Leighton. The phomoidrides are complex polycyclic natural compounds that posses some inhibitor activity against Ras farnesyl transferase and squalene synthase (Figure 13.1).

Leighton's work accounts for an efficient synthesis of the fully elaborated tetracyclic core of phomoidrides B and D. The synthesis involves, starting from triflate **27**, a novel carbonylation reaction that delivers a strained bicyclic pseudoester system **28**, whose strain in turn drives an *in situ* highly efficient silyloxy-Cope rearrangement that delivers the tetracyclic core of phomoidrides **29** (Scheme 13. 13, path a) [36–38]

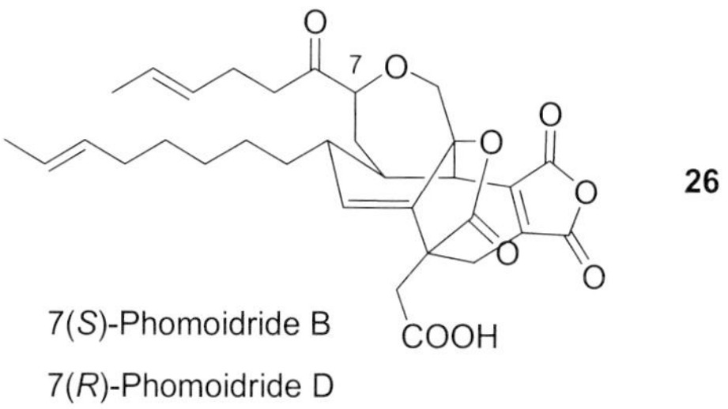

Figure 13.1

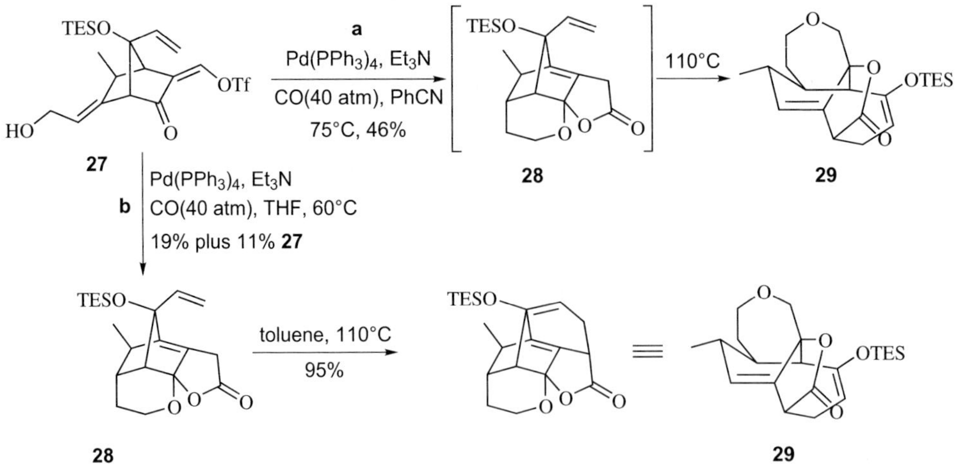

a: $Pd(PPh_3)_4$, Et_3N, CO(40 atm), PhCN, 75°C, 46%; 110°C

b: $Pd(PPh_3)_4$, Et_3N, CO(40 atm), THF, 60°C, 19% plus 11% **27**

toluene, 110°C, 95%

Scheme 13.13

As reported in the scheme, a careful choice of the reaction medium and temperature allowed the tandem process, giving rise to the desired compound **29** in 46% yield starting from **27**. Lowering of the reaction temperature to 60 °C and performing the reaction in THF instead of PhCN results in the isolation of the sole carbonylative cyclization product **28** in 19% yield, plus 11% recovered enol triflate **27** (Scheme 13.13, path b).

Moreover, in the past 10 years, a number of works involving a nucleophilic attack of a nitrogen nucleophile over an acylpalladium intermediate, generated *in situ* from simple and easily prepared starting compounds, has been reported to be dealing with the synthesis of heterocyclic compounds, that is, 2,3,4,5-tetrahydro-1*H*-2,4-benzodiazepine-1,3-dione derivatives **30** [39], tetrahydro-β-carboline/tetrahydriisoquinoline fused δ-lactam derivatives **31** and **32**, respectively [40], and substituted hydantoins **33** [41] (Scheme 13.14).

The synthesis of the hydantoins **33** was realized by Beller's group in the context of their work devoted to the synthesis of amino acid derivatives by palladium-catalyzed amidocarbonylation reactions. The acylpalladium intermediate is generated *in situ* by acid-promoted hydroxy–halogen exchange on the β-hydroxy-*N*-acyl derivative obtained from urea/aldehyde condensation, followed by oxidative palladium addition and CO insertion (Scheme 13.15) [42].

Some examples in which carbonylative and cyclization steps work separately in a one-pot procedure dealing with the synthesis of oxadiazoles **34** [43,44], quinolines, and naphthyridines **35** [45] have been reported. Oxadiazoles were prepared via palladium-catalyzed cyclocarbonylation of aryliodides or diaryliodonium salts with amidoximes (Scheme 13.16). The acylpalladium complex generated from oxidative addition–insertion reaction between aryliodide/iodonium salt, Pd(0), and carbon monoxide reacts with the amidoxime giving rise, after reductive elimination of Pd(0), to an *O*-acyl-amidoxime, which in turn undergoes cyclodehydratation to oxadiazole in moderate to good yields.

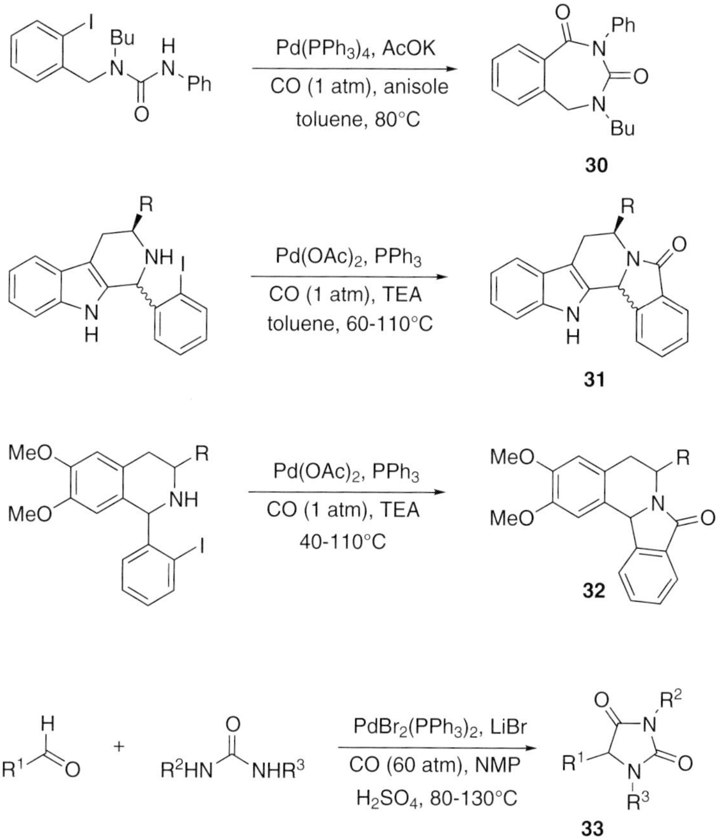

$Pd(PPh_3)_4$, AcOK
CO (1 atm), anisole
toluene, 80°C

30

$Pd(OAc)_2$, PPh_3
CO (1 atm), TEA
toluene, 60-110°C

31

$Pd(OAc)_2$, PPh_3
CO (1 atm), TEA
40-110°C

32

$PdBr_2(PPh_3)_2$, LiBr
CO (60 atm), NMP
H_2SO_4, 80-130°C

33

Scheme 13.14

Moreover, a new multicomponent domino reaction leading to 2-aryl-4-amino-quinolines and [1,8]naphthyridines starting from carbon monoxide, 2-ethynyl-arylamines, aryliodides, and primary amines was recently reported [45]. The palladium-mediated process (Scheme 13.17, cycle A) involves a triple-domino sequence, for example, carbonylative coupling and inter- and intramolecular

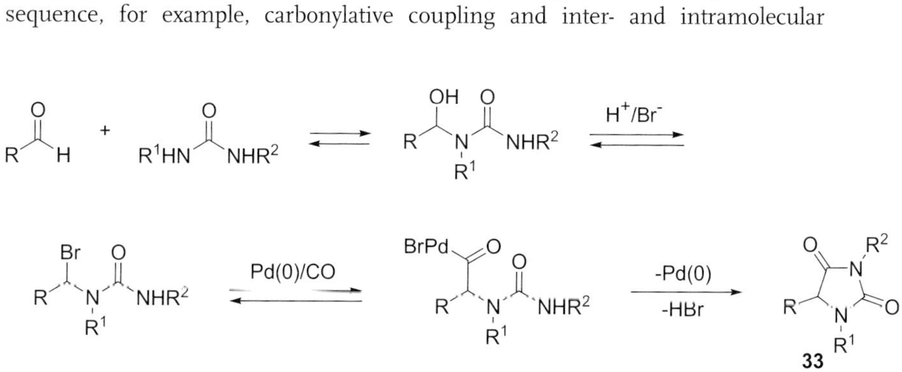

H^+/Br^-

Pd(0)/CO

-Pd(0)
-HBr

33

Scheme 13.15

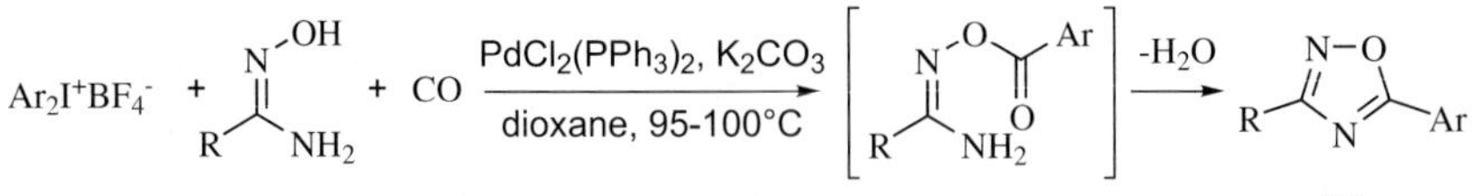

$Ar_2I^+BF_4^-$ + R–C(=N–OH)NH$_2$ + CO $\xrightarrow[\text{dioxane, 95-100°C}]{PdCl_2(PPh_3)_2,\ K_2CO_3}$ [R–C(NH$_2$)=N–O–C(=O)Ar] $\xrightarrow{-H_2O}$ **34**

Scheme 13.16

nucleophilic additions to triple carbon–carbon bond and to double carbon–oxygen bond, respectively. The success of the synthetic cycle, performed for the first time with primary amines, is related to the correct selection of the appropriate catalytic system with hidden competitive palladium-catalyzed carbonylative amidation reaction (Scheme 13.17, cycle B).

The use of 5% of $Pd(OAc)_2$ and 7% of $P(o\text{-tol})_3$ in THF at 100 °C under 6 atm of carbon monoxide allows the almost chemoselective synthesis of **35** with complete consumption of starting 2-ethynylphenylamine and restraint of amide side product formation (Scheme 13.18).

13.2.2 Carbonylative Cyclization Involving Activation/Hetero or Carbopalladation Steps with Unsaturated Carbon–Carbon Bonds

The carbonylative cyclization reactions of *o*-substituted arylalkynyl and arylethynyl derivatives represent a very useful strategy for the construction of simple and polysubstituted heterocyclic rings.

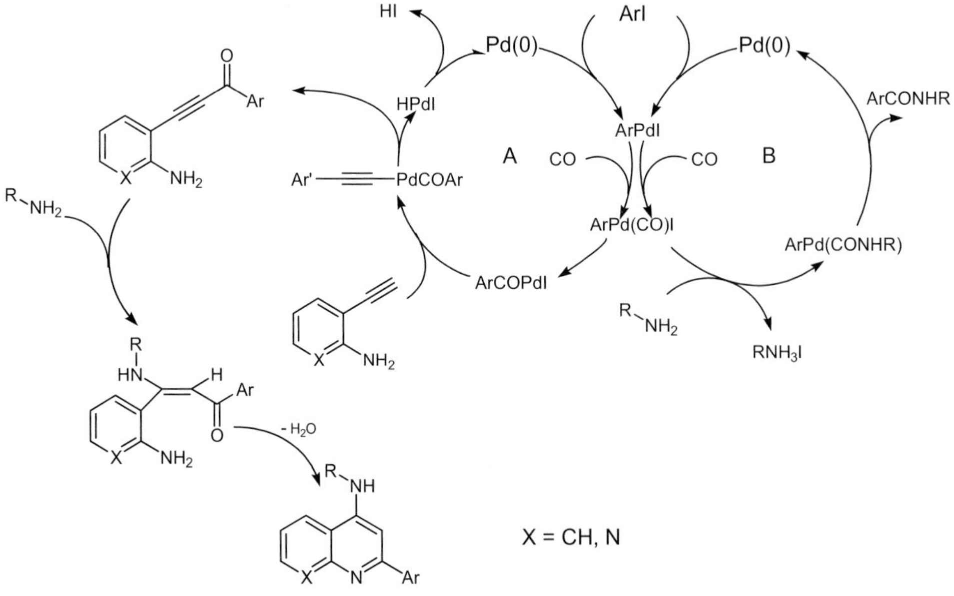

HI, Pd(0), ArI, Pd(0), ArCONHR, HPdI, ArPdI, A, CO, CO, B, Ar'—≡—PdCOAr, ArPd(CO)I, ArPd(CONHR), ArCOPdI, R-NH$_2$, RNH$_3$I, - H$_2$O, **35**, X = CH, N

Scheme 13.17

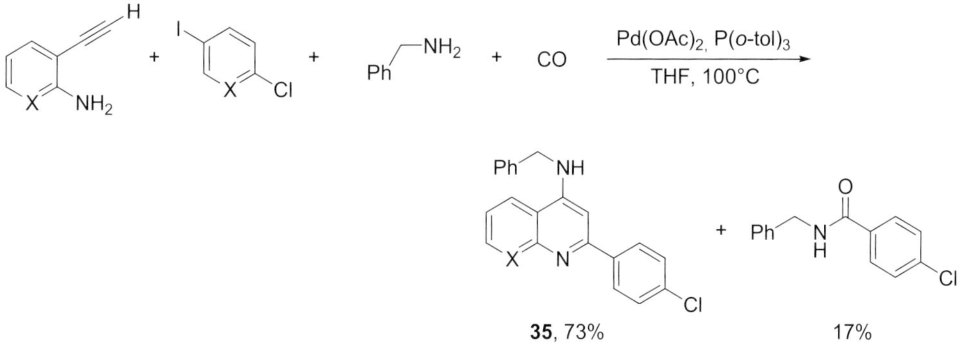

Scheme 13.18

Over the past 10 years, Cacchi and coworkers developed a highly efficient methodology for the construction of indole nucleus through a palladium-catalyzed reaction of *o*-alkynyltrifluoroacetanilides with aryl and vinyl iodides, bromides or triflates, alkyl halides, and allylestyers [1]. The proposed reaction mechanism involves aminopalladation/reductive elimination domino reaction (Scheme 13.19). The key step of the process is the intramolecular nucleophilic attack of a proximate nitrogen nucleophile across the C–C triple bond, activated through coordination to the palladium atom of an organopalladium complex generated *in situ*. The addition of carbon monoxide as a third component in the same reaction provides an easy route to indole derivatives containing an acyl group at C-3 (Scheme 13.19).

After some experimentations, it was found that the use of $Pd(PPh_3)_4$ or Pd $(OAc)_2(PPh_3)_2$ in acetonitrile under a balloon of carbon monoxide give good results with many aryl iodides. The use of a higher carbon monoxide pressure or, alternatively, $Pd_2(dba)_3$ and $P(o\text{-tol})_3$ under standard conditions was found necessary with

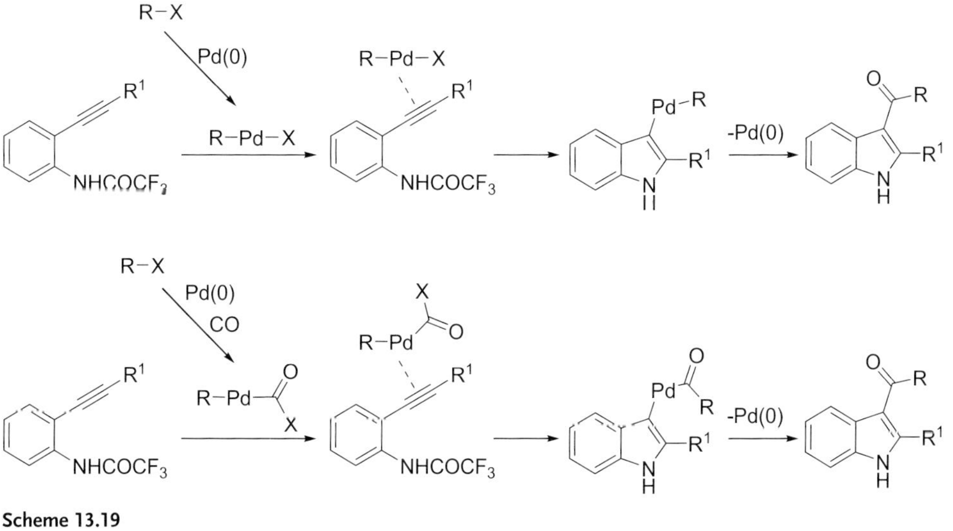

Scheme 13.19

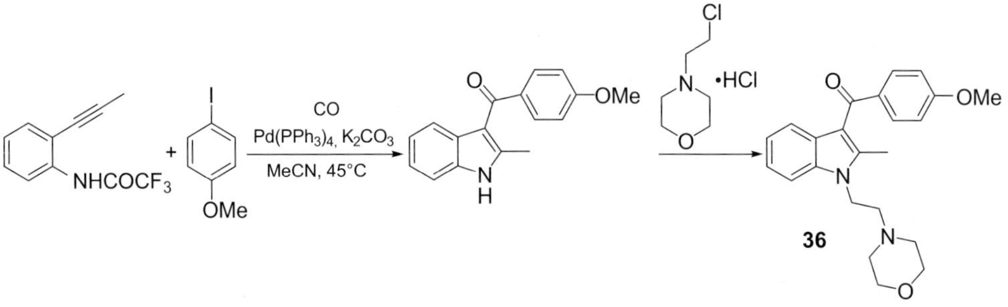

Scheme 13.20

aryl iodides containing electron-withdrawing substituents. With vinyl triflates, the use of anhydrous acetonitrile gave the best results [46]. The appointed standard methodology was applied to the synthesis of pravadoline **36**, an indole derivative designed as a nonacidic nonsteroidal anti-inflammatory drug (Scheme 13.20).

The authors next focused their attention on the potential use of 3-acylindoles bearing suitable functionality in a subsequent cyclization step [47]. Thus, 6-aryl-11*H*-indolo[3,2-*c*]quinolines **37** were prepared in one-pot process through a straightforward palladium-catalyzed carbonylative cyclization of *o*-(*o*′-aminophenylethynyl)trifluoroacetanilide with aryliodides followed by the cyclization of the resultant 3-acylindole derivative (Scheme 13.21).

Moreover, 6-trifluoromethyl-12-acylindolo[1,2-*c*]quinazolines **38** were prepared in high yield through the palladium-catalyzed reaction of bis(*o*-trifluoroacetamidophenyl) acetylene with aryl or vinyl halides and triflates, (Scheme 13.22) [48]. The reaction, which tolerates a variety of important functional groups, probably involves

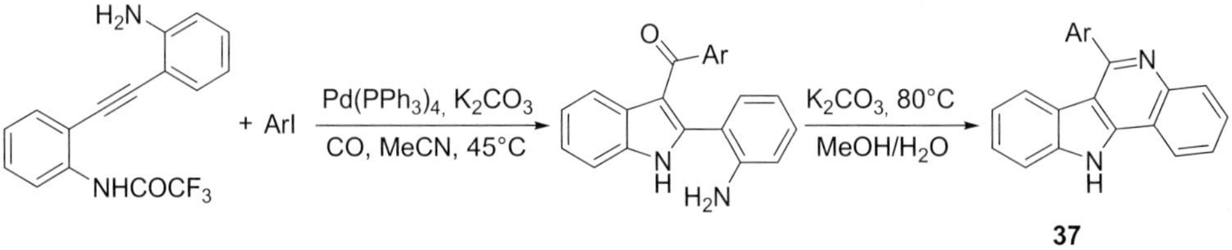

Scheme 13.21

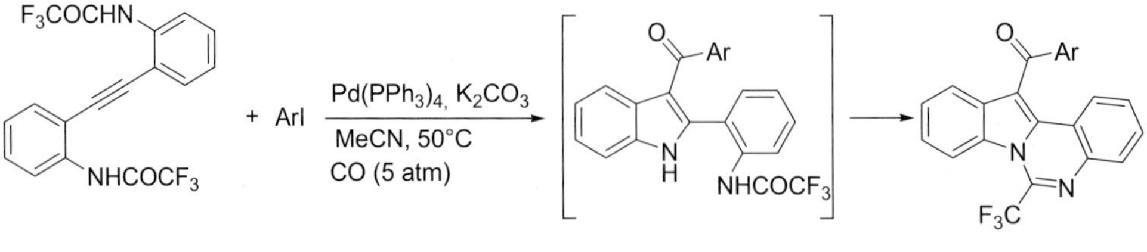

Scheme 13.22

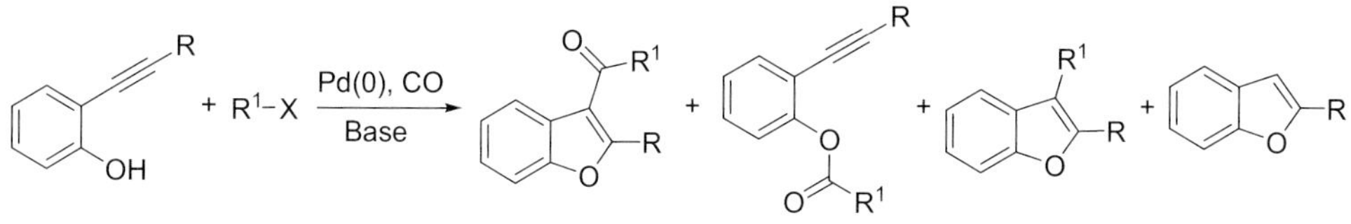

Scheme 13.23

the formation of a 3-acyl-2-(*o*-trifluoroacetamidophenyl)indole intermediate, followed by cyclization to the indoloquinazoline derivatives.

Taking into account the structural similarity between indoles and benzo[*b*]furans, an analogous approach has been used to generate the benzo[*b*]furan skeleton. In the presence of $Pd(PPh_3)_4$ and KOAc, under a balloon of carbon monoxide, *o*-alkynylphenol produced 2-substituted-3-acylbenzo[*b*]furans **39** in moderate yields with vinyl triflates (Scheme 13.23) [49]. Nevertheless, when vinyl triflates were substituted with aryl iodides, the acylated *o*-alkynylphenol was obtained as the sole product. Moreover, depending on the substitution pattern of the reagents, variable amounts of 2,3-disubstituted-benzo[*b*]furans and in some cases 2-substituted-benzo [*b*]furans have also been isolated.

More recently, aryl iodides have been involved in palladium-catalyzed carbonylative annulations with unsubstituted *o*-alkynylphenols by means of a cationic σ-acyl palladium complex with Lewis acid character and high reactivity toward coordination of alkynes (Scheme 13.24) [50]. Thus, using $Pd(PPh_3)_4$ (5%) with an equal amount of $AgBF_4$ under carbonylative annulation conditions in dibutyl ether, 2-aryl-3-acylbenzo [*b*]furans **39** have been obtained in 30% yield. In the same account, the authors report that substituted *o*-alkynylphenols are able to participate in palladium-catalyzed carbonylative annulations with aryl iodides under neutral reaction conditions, giving rise to the corresponding 2-aryl-3-acylbenzo[*b*]furans **39** in moderate to good yields, even if a rationale for the obtained results is not reported.

A combination of electronic and steric effects could account for a completely different course observed when *o*-ethynylphenols, instead of *o*-alkynylphenols, were used in the reactions with vinyl triflates under the same reaction conditions. In this case 3-alkylidene-2-coumaranones **40** were isolated in moderate yields beside variable amounts of carbonylative coupling derivatives (Scheme 13.25) [51].

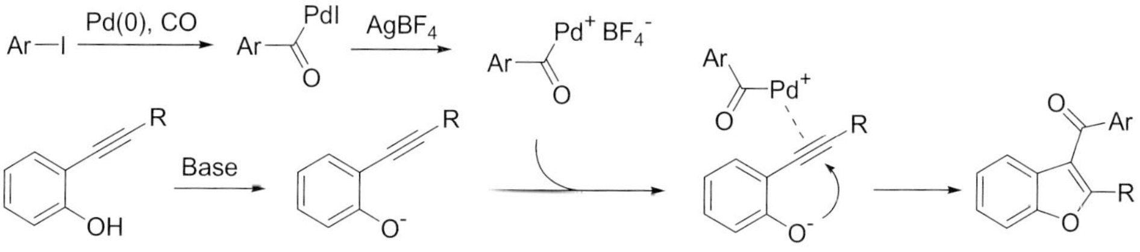

Scheme 13.24

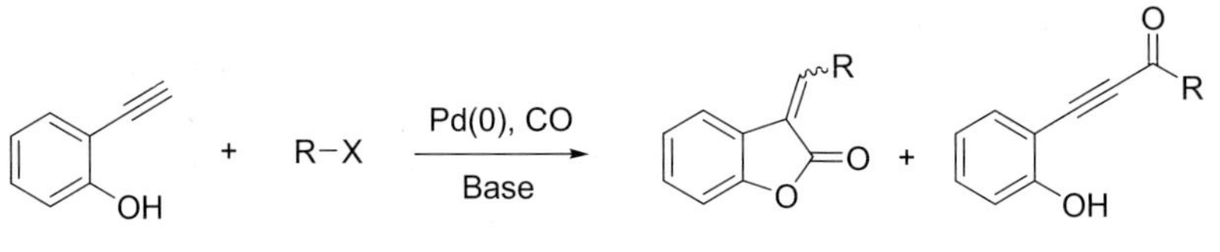

Scheme 13.25

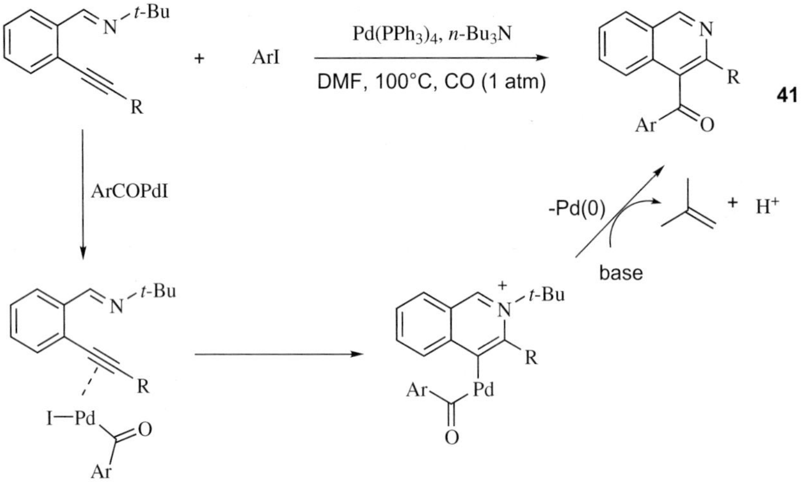

Scheme 13.26

An efficient synthetic approach to the 3-substituted 4-aroylisoquinolines **41** has been developed starting from *N-tert*-butyl-2-(1-alkynyl)benzaldimines and aryl halides by carbonylative cyclization reactions. The combination of 5 mol% of $Pd(PPh_3)_4$, and 5 equiv. of tri-*n*-butylamine in DMF at 100 °C under 1 atm of CO give the best results (Scheme 13.26) [52,53].

A palladium-catalyzed synthetic approach for the formation of a wide variety of isoindolinones was suggested to proceed via intra- and intermolecular carbopalladative addition across carbon–nitrogen double bond as an organometallic key step. For example, starting from 2-bromobenzaldehyde, 2-bromoacetophenone, and 2-iodobenzoyl chloride and differently substituted amines and amine derivatives, the synthesis of various isoindolin-1-ones **42**, **43**, **44** was accomplished in the presence of $PdCl_2(PPh_3)_2$ under carbon monoxide (13–20 atm) (Scheme 13.27) [54–57]. Buchwald *et al.* reported that haloarenes are coupled with primary and secondary amines in the presence of a palladium catalyst to give aminoarenes. However, no amination product was observed in the present reactions. Moreover, the reactions can lapse with intramolecular amination, β-hydride elimination, or alkoxycarbonylation steps, as reported in Scheme 13.27.

The Pd-catalyzed reactions of 2-propynyl-1,3-dicarbonyls with haloaryls and vinyl triflates at 60 °C in the presence of K_2CO_3 and, respectively, $Pd(OAc)_2/P(o\text{-tol})_3$ or Pd $(PPh_3)_4$ provided acyl furans **45** under 1 or 2 atm of CO (Scheme 13.28) [58,59].

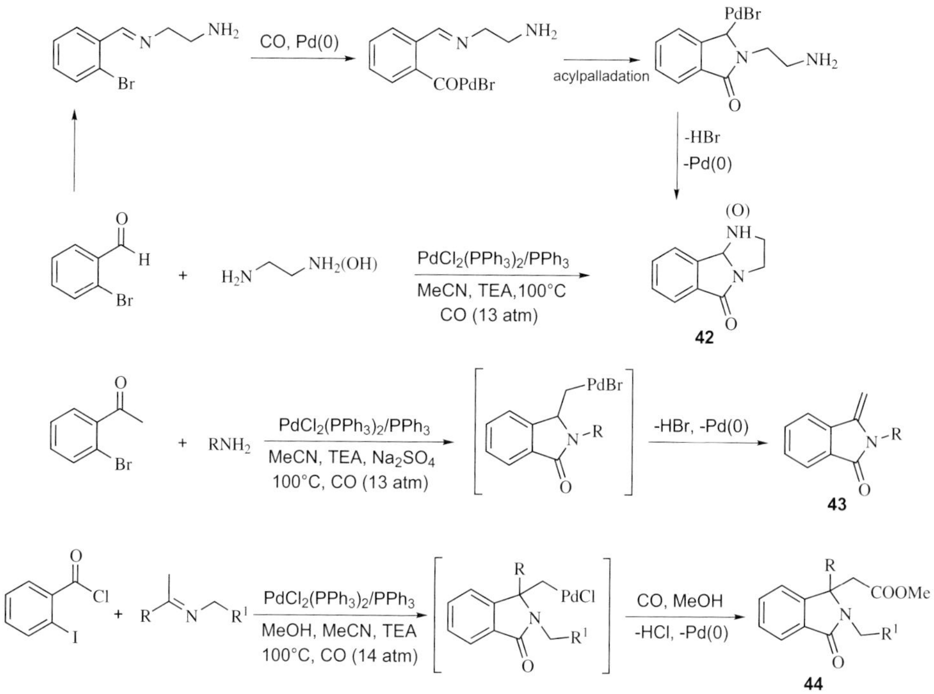

Scheme 13.27

A stereocontrolled synthetic approach to *Z*-substituted methylene-3,4-dihydro-2*H*-1-benzopyrans **46** is described from acyclic iodoalkyne using a regioselective palladium-catalyzed intramolecular carbopalladation reaction as the key step. The resulting vinyl–palladium(II) intermediates may undergo carbonylation reaction in the presence of CH_3OH to give the *Z*-ester. The optimized conditions employ Et_3N, $Pd(PPh_3)_4$, and AgOAc in a medium consisting of 1 : 1 : 0.2 CH_3OH–DMF–H_2O, CO (1 atm) at 100 °C (Scheme 13.29) [60].

Both palladium-catalyzed carbonylative ring-forming reactions of *o*-iodophenols and *o*-iodothiophenols with allenes and carbon monoxide involve a carbopalladation step and afford 1-benzopyran-4-one **47** and thiochroman-4-one **48** derivatives [61,62]. The reactions proceed in benzene at 100 °C, in autoclave at 20 atm of carbon monoxide and in the presence of $Pd(OAc)_2$, Hünig's base, dppb or dppf. However,

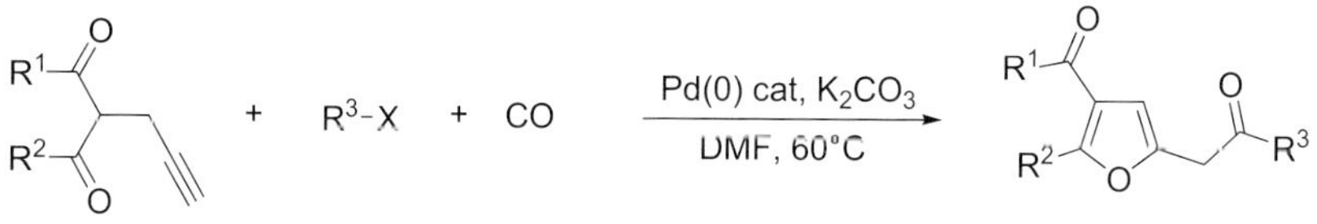

Scheme 13.28

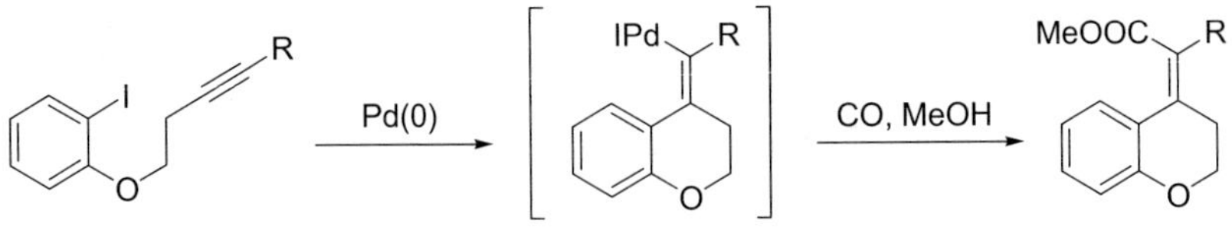
R
I
O
Pd(0)
IPd
R
O
CO, MeOH
MeOOC
R
O
46

Scheme 13.29

the proposed reaction mechanisms are quite different and involve addition of palladium(0) into carbon–iodine and carbon–sulfur bond. The reactions then proceed in the first case by regioselective addition of an initially formed aroylpalladium intermediate to the allenyl unit to produce a π-allylpalladium species; nucleophilic attack by the hydroxyl group at the π-allylpalladium intermediate would terminate the catalytic cycle, affording 1-benzopyran-4-one derivatives (Scheme 13.30). The reactions involving α- and α,α-substituted allenes are regioselective with nucleophilic attack of the hydroxyl group occurring exclusively at the more substituted carbon of the allene unit, whereas with unsymmetrical internal allenes the regioselectivity is regulated by electronic factors.

In the presence of *o*-iodothiophenol as the starting compound, it is conceivable that the reaction may proceed through oxidative addition of the *o*-iodothiophenol to palladium(0) followed by the insertion of allene to produce a σ-allylpalladium intermediate. Reductive elimination of Pd(0) would give an iodothioether. Oxidative addition of the latter to Pd(0) and subsequent CO insertion and intramolecular cyclization would afford thiochroman-4-ones **48** and regenerate the catalyst (Scheme 13.31).

Finally, α-, γ-, and δ-allene-substituted *p*-toluenesulfonamide bearing a tethered nitrogen nucleophile undergo a palladium(0)-catalyzed three-component carbonylation/coupling/cyclization reaction with aryl iodides and carbon monoxide to form pyrrolines **49**, pyrrolidine **50**, and/or piperidine **51** [63]. Also in this case, the reaction works at 20 atm of carbon monoxide in presence of $Pd(PPh_3)_4$, K_2CO_3 as a catalytic system and involves a π-allylpalladium intermediate generated by the

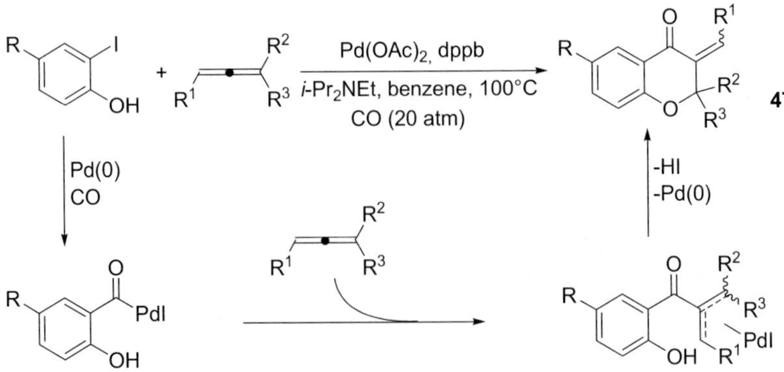
R
I
OH
+
R1
R2
R3
Pd(OAc)2, dppb
i-Pr2NEt, benzene, 100°C
CO (20 atm)
R
O
R1
R2
R3
O
47
Pd(0)
CO
-HI
-Pd(0)
R2
R1
R3
R
O
PdI
OH
R
O
R2
R3
PdI
OH
R1

Scheme 13.30

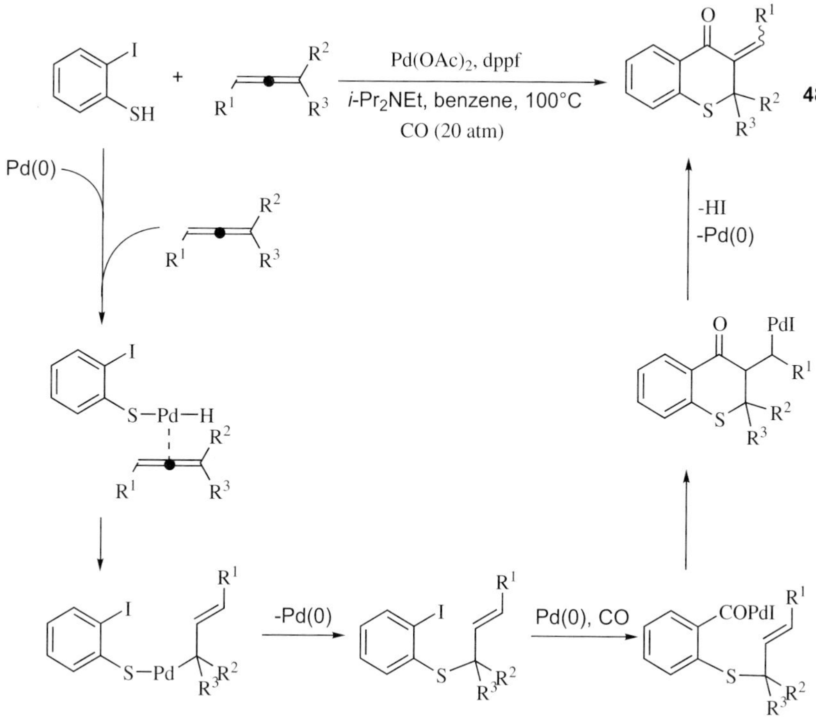

Scheme 13.31

addition of the acyl palladium complex to the central carbon atom of allene moiety (Scheme 13.32).

Mesoionic münchnones are usually prepared by cyclodehydratation of amino acid derivatives and are in equilibrium with the corresponding amido-substituted ketenes. These compounds are well known as substrates for 1,3-dipolar and [2 + 2] cycloaddition reactions giving rise, respectively, to five- and four-membered heterocycles. Arndtsen and coworkers recently developed a new and elegant way to

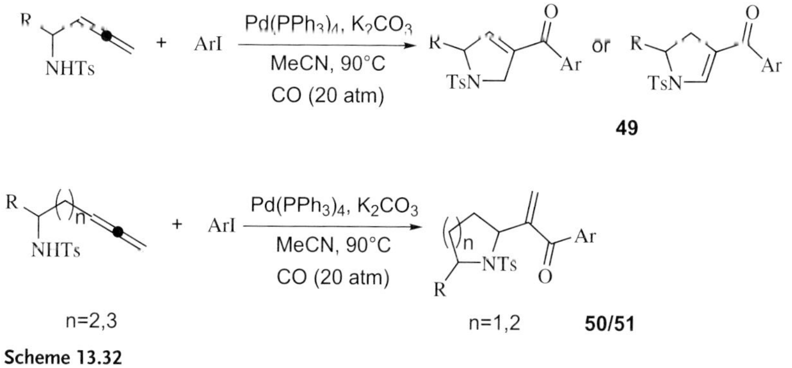

Scheme 13.32

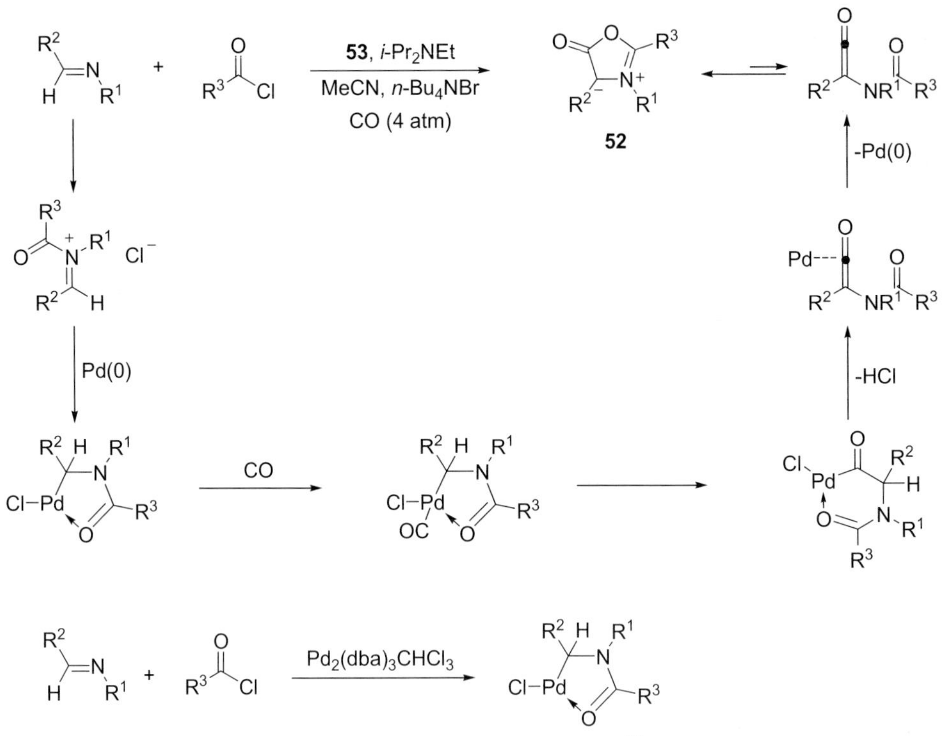

Scheme 13.33

generate *in situ* mesoionic münchnones starting from imines, acid chlorides, and carbon monoxide in the presence of Pd(0) (Scheme 13.33) [64].

The formation of münchnones **52** occur via oxidative addition of imine and acid chloride to Pd(0), CO coordination, insertion, and β-hydride elimination. The optimized reaction conditions (83% yield) for the synthesis of **52** involves CO (4 atm), Hünig's base, Bu_4NBr as additive, and the use of 5% [Pd-(Cl)[η^2-CH(*p*-tolyl)NBn(COPh)]$_2$ **53**, generated by pretreating $Pd_2(dba)_3{\cdot}CHCl_3$ with imine and acid chloride as a catalyst. This protocol provides a straightforward and high-yield catalytic route to prepare stable münchnones, moreover, by fine and slight modification of the catalytic systems, multicomponent route to heterocycles has been realized by adding to the reaction medium suitable cycloaddition partners (Scheme 13.34).

For example, in the presence of alkynes highly substituted pyrroles **54** were obtained in 56–88% yields [65], whereas working with a bimolecular amount of imine β-lactams **55** were obtained in 27–66 % isolated yields [66]. The reaction can also be performed in the presence of two differently substituted imines, that is, *N*-aryl/alkylimines and *N*-tosylimines [67], regioselectively yielding imidazoles **56**. It is worth noting that no products incorporating two of the same imine are observed. This selectivity is believed to result from the catalytic mechanism. In particular, the *N*-tosylimine is not sufficiently nucleophilic to interact with the acid chloride for

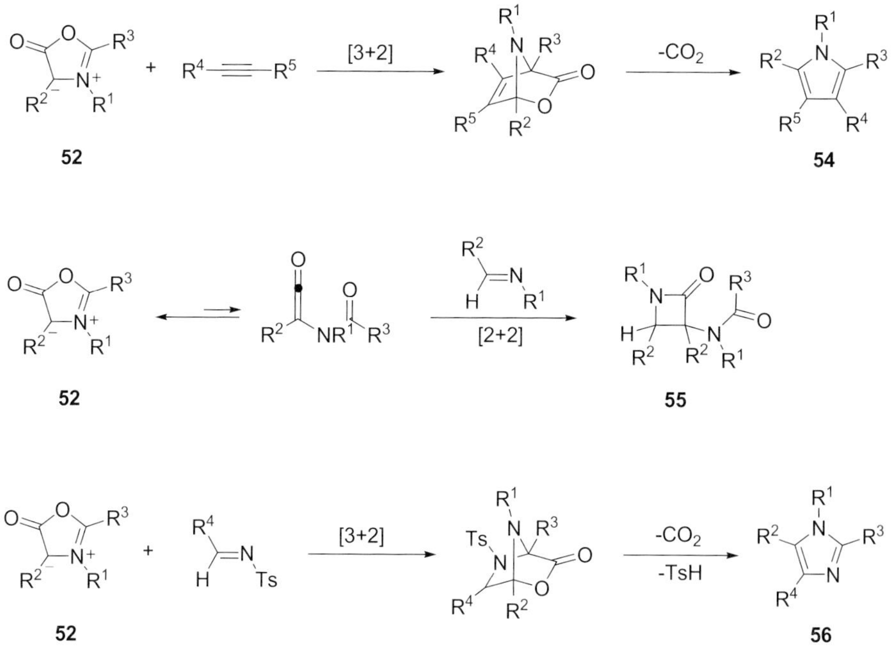

Scheme 13.34

the formation of münchnone **52** that in turn reacts exclusively with the more electron-poor imine via cycloaddition.

13.2.3
Cascade Reactions

Synthetic organic chemistry is concerned with developing highly efficient and environmentally tolerable synthetic procedures. The ideal synthesis should lead to the desired products in as few steps as possible, in overall good yields and by using environmentally compatible reagents. In recent years this aim was pursued using different strategies based on the optimization of the synthetic processes, and among these a useful approach consists in using, instead of traditional syntheses, cascade or domino processes. Indeed, it would be more efficient to form several bonds in one sequence without isolating the intermediate, changing the reaction conditions, or adding reagents. It is obvious that this type of reaction would allow the minimization of waste as, compared to stepwise reactions, the amount of solvents, reagents, adsorbents, and energy would be dramatically decreased. Cascade reactions involving a carbonylation step in combination with transition metal catalyzed reactions have been used by several research groups to approach the synthesis of different heterocyclic systems. In particular, the research group of Grigg developed a wide range of palladium-catalyzed cascade reactions that allow the incorporation of carbon monoxide in an heterocyclic ring through cascade cyclization, anion capture, relay switches, and molecular queues. These ring-forming reactions are characterized by

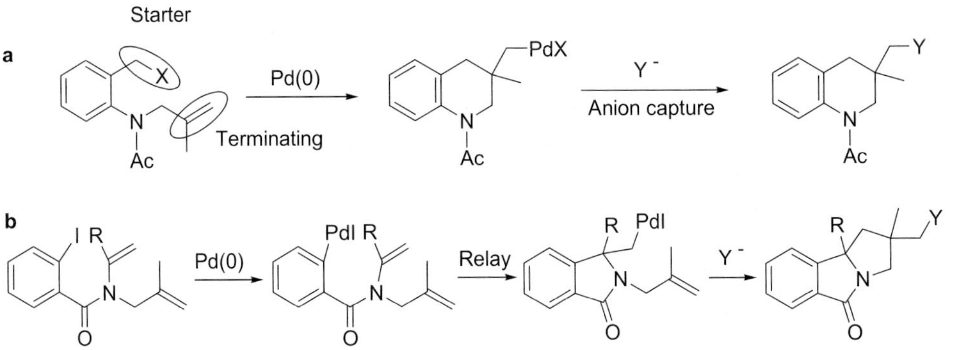

Starter

a

Pd(0)

Terminating

Y^-

Anion capture

b

Pd(0)

Relay

Y^-

Scheme 13.35

high tolerance of functionality, are effective on all types of C–C unsaturated bonds, and occur with excellent chemo-, regio-, and stereoselectivity. An example of palladium-catalyzed cascade cyclization–anion capture process is reported in Scheme 13.35a and involves a starter species, the alkyl halide, a terminating species, the alkene, and the anion capture agent, the hydride anion. Moreover, in the presence of suitable substrates, cyclization can proceed for two or more cycles in a relay phase before intercepting the anion capture agent (Scheme 13.35b).

These two-component processes can be transformed to multicomponent process if a relay phase is set up by incorporating intra- or intermolecular fragments. A representative three-component reaction is reported in Scheme 13.36.

These additional components, called by Grigg relay switches, are frequently carbon monoxide or allenes. Earlier works of this group has been carefully reviewed [68], thus only some examples of the most topical findings on cascade reactions are reported here. Recently, a termolecular cyclization–anion capture process employing carbon monoxide as a relay switch and hydride, organotin, or boron reagents as anion capture agents has been reported, which allowed the synthesis of five- and six-membered benzocondensed rings **57**. The reactions that involve cyclization and carboformylation steps were performed with *o*-iodosubstituted aryls containing a tethered carbon–carbon double bond, with carbon monoxide at atmospheric pressure and in the presence of diphenylmethylsilane as the hydride source. All reactions were carried out in toluene at 90 °C in the presence of $Pd(OAc)_2/PPh_3$ (Scheme 13.37) [69].

However, in the queuing processes terminating with stannanes or borane, as anion capture agents, the desired molecular queuing to **57** can be achieved only increasing

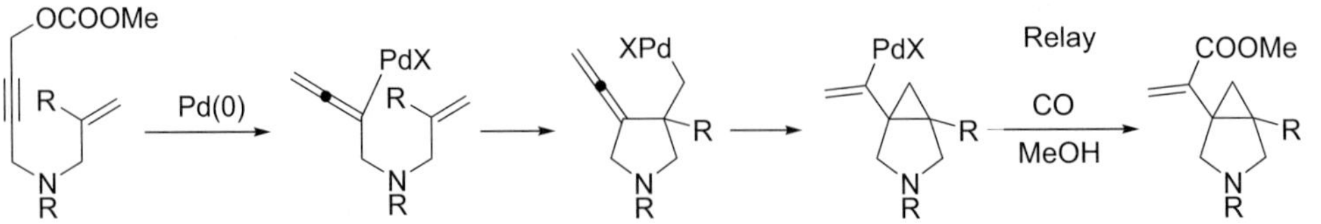

Pd(0)

Relay

CO

MeOH

Scheme 13.36

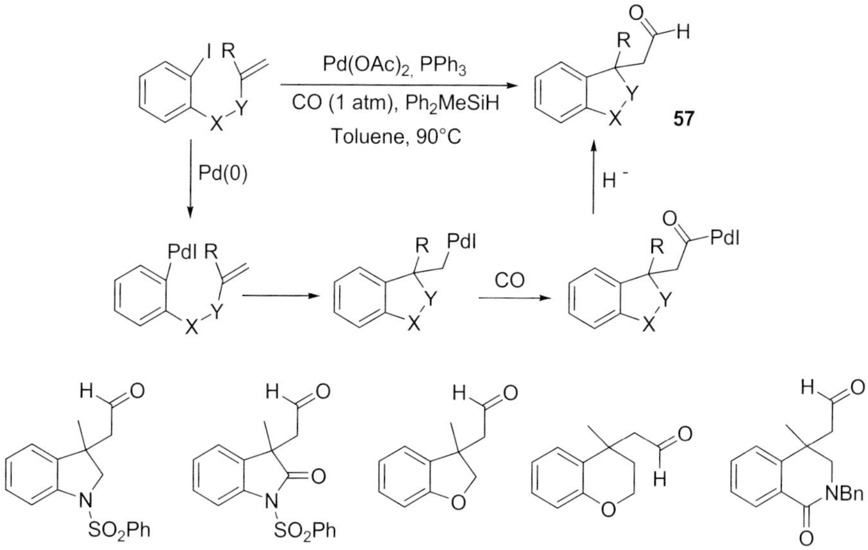

Scheme 13.37

the reaction temperature at 110 °C to avoid the competitive process, giving rise to direct capture compound **58** (Scheme 13.38).

Finally, when *in situ* generated vinylstannanes were used as anion capture reagents, an enone function can be introduced in the heterocyclic ring, providing new opportunities for further extended cascades by combination with Michael additions (Scheme 13.39) [70].

Conceptually similar palladium-catalyzed cascade reactions have been developed, involving molecular-queuing cycloaddition, cyclocondensation and Diels–Alder reactions [71], cyclization–anion-capture–olefin metathesis [72], carbonylation-allene insertion [73], carbonylation/amination/Michael addition [74], sequential Petasis reaction/palladium-catalyzed process [75], supported allenes as substrates [76], and palladium–ruthenium catalysts [77].

A very interesting three-component cascade process involving nucleophilic substitution/carbonylation/amination catalyzed by palladium nanoparticles affords

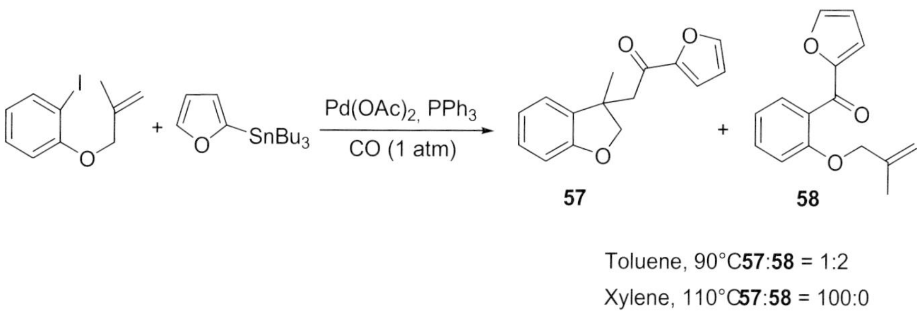

Toluene, 90°C **57**:**58** = 1:2

Xylene, 110°C **57**:**58** = 100:0

Scheme 13.38

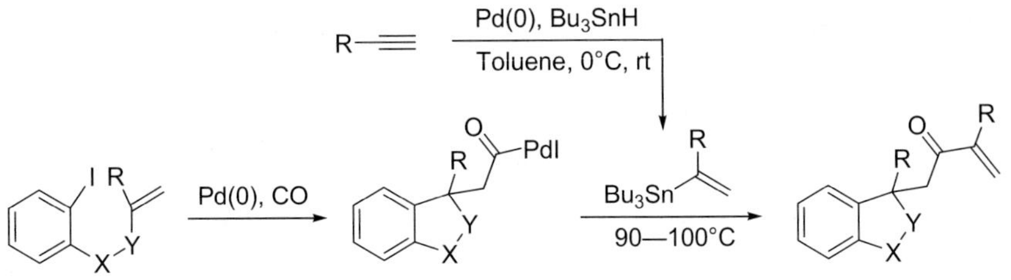

Scheme 13.39

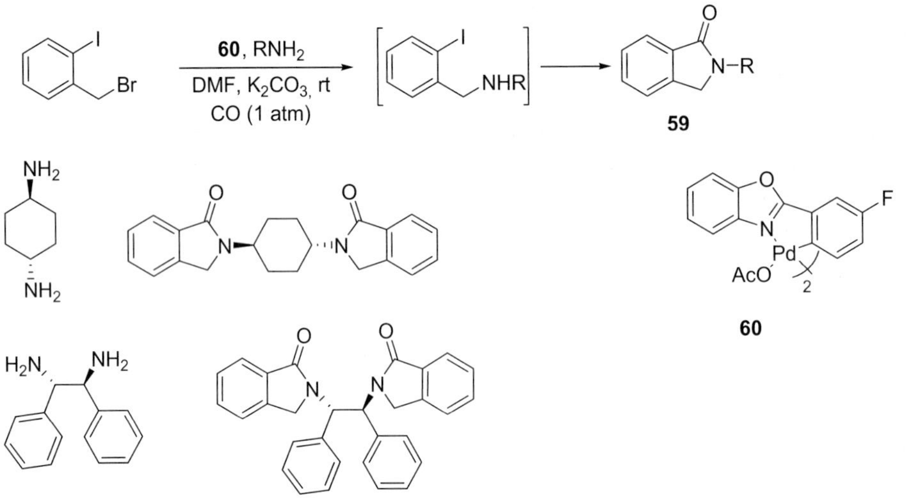

Scheme 13.40

isoindolinones **59** in good yields, starting from *o*-iodobenzyl bromide, primary amines, and carbon monoxide. In particular, the process involves reductive degradation of palladacycle **60** to palladium nanoparticles realized for the first time with carbon monoxide in DMF at room temperature (Scheme 13.40) [78]. Moreover, at room temperature, the reaction rate correlate to pK_a of the amine employed.

13.3 Carbonylative Reactions Involving Palladium(II) Salts

The carbonylation reactions reported in this section can be regarded as processes in which the carbon monoxide insertion into the organic substrate is mediated by a metal that at the same time undergoes a reduction of its oxidation state. When palladium(II) is used as a catalyst, the overall process is as described in Scheme 13.41.

The reaction requires a stoichiometric amount of palladium(II) salt or a reoxidant must be used to achieve a catalytic reaction (oxidative carbonylation) (Scheme 13.42). A great variety of external oxidants, either organic (i.e., benzoquinone) or inorganic (i.e., copper(II) salts), have been evaluated and described. Moreover, a particularly

Organic substrate + CO + Pd(II) ⟶ Carbonyl derivative + Pd(0)

Scheme 13.41

effective catalytic system consisting of PdI_2, an excess of iodide anions, and molecular oxygen has been developed and will be discussed in the last part of this section.

Many different substrates can undergo oxidative carbonylations, such as unsaturated and saturated hydrocarbons, aromatic and heteroaromatic derivatives, alcohols, phenols, and amines, giving a number of carbonyl compounds in a regio- and stereocontrolled fashion and with high degree of chemoselectivity. Oxidative carbonylation reactions have been recently and carefully reviewed [10–13]. In this section a general overview of the most recent developments in oxidative carbonylations mediated by catalytic palladium(II) compounds and directed toward the synthesis of heterocyclic compounds is presented. As already reported in the introduction, the cyclocarbonylation reactions ending with the synthesis of simple lactones and lactams are reported in the chapter that discusses the carbonylation of alcohols and amines and are recently been reviewed [7,9,79].

The palladium(II)-catalyzed alkoxy/aminocarbonylation reactions of alkenes and alkynes bearing a tethered nucleophile, usually oxygen or nitrogen nucleophiles, give rise to heterocyclic compounds and follow the mechanisms shown in Scheme 13.43.

Exo- or *endo*-cycloheteropalladation of double or triple, Pd(II)-coordinated carbon–carbon bond followed by CO insertion and displacement by nucleophiles ends with the elimination of Pd(0) and delivering of a new heterocyclic ring. Reactions involving an *exo*-cyclization route are most frequently observed.

For example (Scheme 13.44), treatment of γ-hydroxy olefins with a catalytic amount of $PdCl_2{-}(CH_3CN)_2$ and a stoichiometric amount of $CuCl_2$ in methanol under CO leads to intramolecular alkoxylation/carboalkoxylation sequence and formation of substituted tetrahydrofurans **61** [80–83]. Similarly, 4-pentenyl carbamates undergo palladium-catalyzed intramolecular amination/carboalkoxylation to form substituted pyrrolidine derivatives **62** [84,85].

Beside these pioneer works [80–85], several of the most recent reports in this field are summarized in Schemes 13.45–13.57.

Pd(II)-catalyzed intramolecular aminocarbonylation of olefins bearing many different nitrogen nucleophiles for the synthesis of nitrogen-containing heterocyclic

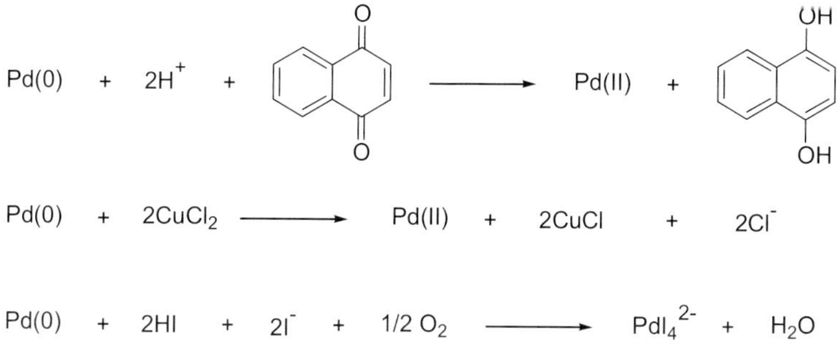

Pd(0) + $2H^+$ + [1,4-naphthoquinone] ⟶ Pd(II) + [1,4-dihydroxynaphthalene]

Pd(0) + $2CuCl_2$ ⟶ Pd(II) + 2CuCl + $2Cl^-$

Pd(0) + 2HI + $2I^-$ + 1/2 O_2 ⟶ PdI_4^{2-} + H_2O

Scheme 13.42

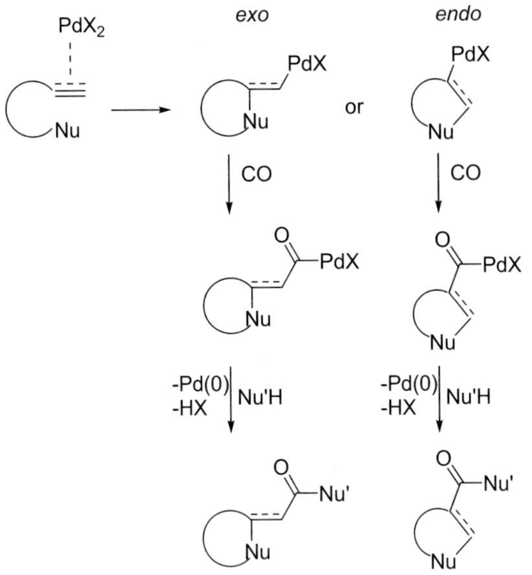

Scheme 13.43

derivatives (i.e., **63–68**, Scheme 13.45) has been recently examined under two typical conditions: acidic conditions (conditions a, $PdCl_2$ and $CuCl_2$ under 1 atm of CO at room temperature in methanol) and buffered conditions (conditions b, $PdCl_2$ and $CuCl_2$ under 1 atm of CO at 30 °C in trimethyl orthoacetate) [86]. Conditions a are strongly acidic, as 2 mol of HCl evolve during the catalytic cycle, whereas conditions b are buffered, as the HCl evolved is neutralized either by a base (i.e., NaOAc) or by a trimethyl orthoacetate.

Many nitrogen nucleophiles (*exo*-carbamates, *endo*-ureas, *exo*-ureas, and *exo*-tosylamides) smoothly undergo aminocarbonylation under conditions a, but not under conditions b. Endocarbamates, on the contrary, react under conditions b and not under conditions a. Probably, conditions b are sufficiently basic to generate the conjugate base of endocarbamates, a reactive species for aminocarbonylation. Thus,

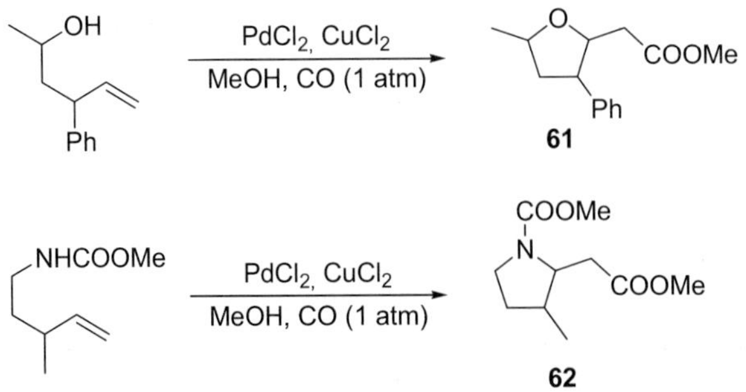

Scheme 13.44

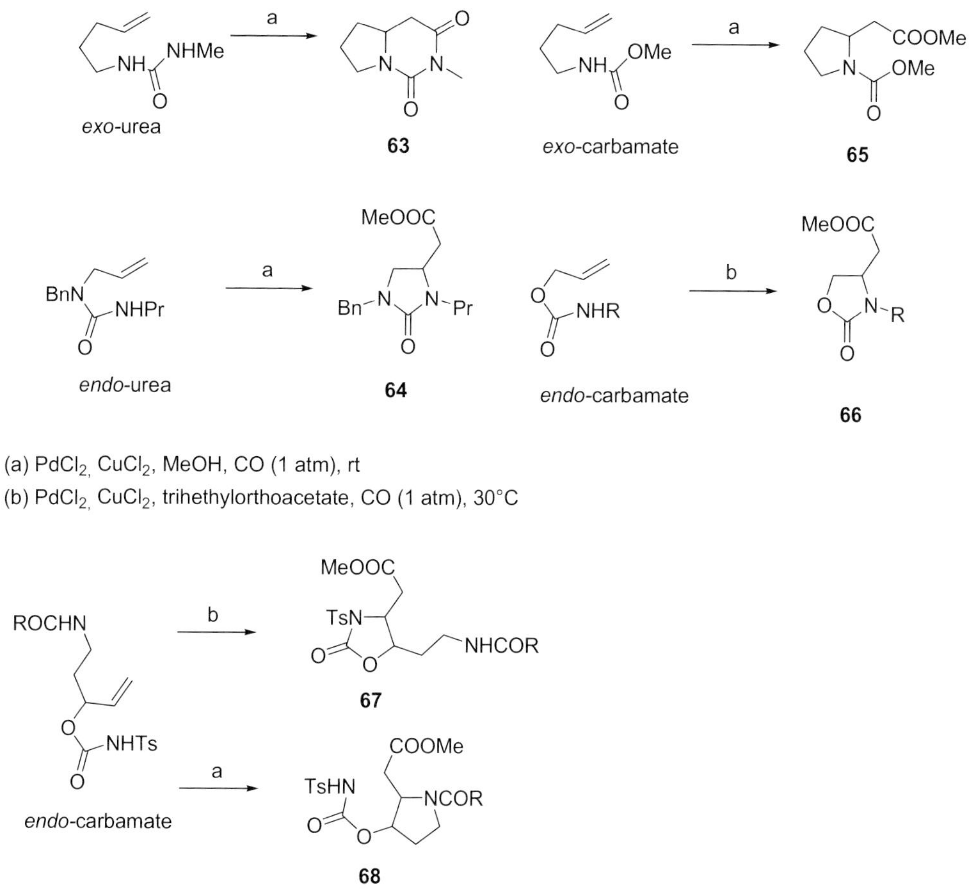

(a) $PdCl_2$, $CuCl_2$, MeOH, CO (1 atm), rt
(b) $PdCl_2$, $CuCl_2$, trihethylorthoacetate, CO (1 atm), 30°C

Scheme 13.45

completely chemoselective aminocarbonylation of endocarbamate derivatives bearing tethered nitrogen nucleophiles can be realized involving the endocarbamate moiety under conditions b or the other nitrogen nucleophiles under conditions a.

Interesting, when *O*-homoallylhydroxylamines were treated with palladium(II) and copper(II) acetate, the accurate selection of solvents and bases allows the construction of isooxazolidine ring **69** in good yields. The reaction occurs when hydroxylamines are substituted with an electron-withdrawing N-protecting group (Scheme 13.46) [87].

Using the original reaction conditions, reported by Tamaru *et al.* [84,85], with sodium acetate as the base, $PdCl_2$ as the catalyst, and $CuCl_2$ as a reoxidant in methanol, besides unreacted starting material isooxazolidine were isolated in low yields. However, working in the presence of a more soluble copper salt (copper acetate) and a good coordinating agent (acetonitrile), the use of stronger bases such as tetramethylguanidine or sodium methoxide resulted in the isolation of isooxazolidine **69** in good yields. Moreover, with carbamate group on nitrogen, a single

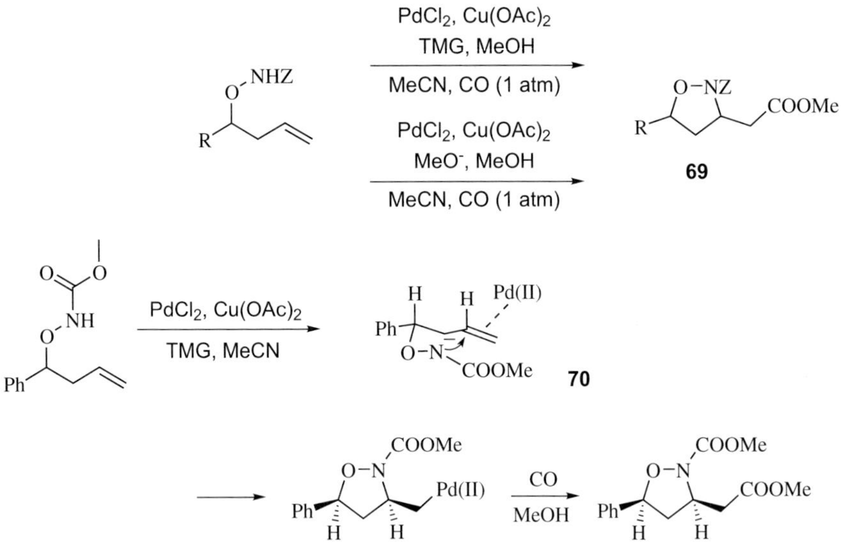

Scheme 13.46

diastereoisomer was obtained, and this is probably related to the formation of the less crowded envelopelike reactive intermediate **70**.

Following the original work of Tamaru on aminocarbonylation of alkenylamino derivatives [85], the first attempt to achieve the enantioselective version of these reactions was reported recently (Scheme 13.47) [88] for the synthesis of [4,4-dimethyl-1-(*p*-toluene-sulfonyl)-pyrrolidin-2-yl]-acetic acid methyl ester **71** in good yield with moderate enantioselectivity.

The reactions were performed in the presence of a series of different substituted chiral spiro bis(isoxazoline) ligands (SPRIX), Pd(II), and benzoquinone. The best results (53% ee), obtained under catalytic conditions, employ 30% mol Pd $(OCOCF_3)_2$/(M,*S*,*S*)-H-SPRIX **72**, 4 equiv. of benzoquinone at −20 °C under CO (1 atm) in methanol.

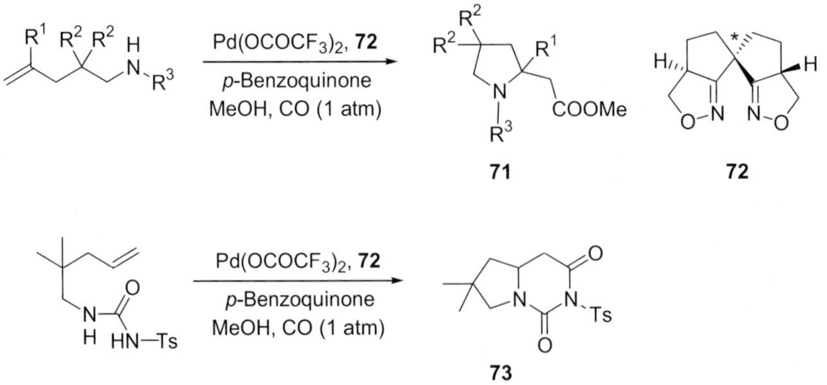

Scheme 13.47

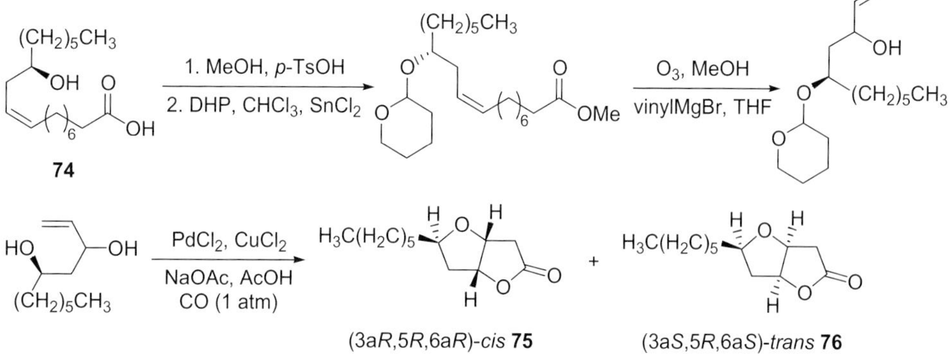

Scheme 13.48

When the aminocarbonylation reaction was carried out using 10 mol% of Pd $(OCOCF_3)_2$ and 12 mol% of (M,*S*,*S*)-H-SPRIX **72**, the reaction was completed in only 30 min giving rise to pirrolidine **71** in 70% yield, and with lower enantioselectivity (20% ee) along with the appearance of Pd black, which might indicate dissociation of the ligand from Pd under carbon monoxide atmosphere. In this type of reaction, interesting optically active bicyclic compounds, including two nitrogen atoms, are prepared in one step. Pd(II)/(M,*S*,*S*)-H-SPRIX catalyst **72** gave the bicyclic 5, 6-dihydrouracil derivative **73** as a single product in 54% ee. Product **73** would be derived by intramolecular nucleophilic attack of the tosylamide group instead of methanol toward acylpalladium intermediate.

Also with starting compounds containing two oxygen nucleophiles, both alkoxy/amino carbonylation and cyclocarbonylation (carbonylative cyclization to lactone) mechanisms can occur in sequence. Some bicyclic γ-lactones from parasitic wasps (Hymenoptera: Braconidae) were synthesized by hydrolytic kinetic resolution of epoxides and palladium(II)-catalyzed hydroxycyclization–carbonylation–lactonization of enediols. For example, *cis*- and *trans*-5-*n*-hexyl-tetrahydrofuro[3,2-b]furan-2(3*H*)-ones **75** and **76**, with the (3a*R*,5*R*,6a*R*) and (3a*S*,5*R*,6a*S*) configurations, respectively, were acquired in six steps from the inexpensive, readily available ricinoleic acid **74** (Scheme 13.48) [89].

The chemical profiling of parasitic wasps could be of great interest in their morphology and taxonomy definition. Moreover, certain species of these wasps in the family Braconidae have been assessed for a possible role in integrated pest management strategies, particularly in fruit fly control regimens.

Palladium(II)-mediated cyclization–alkoxycarbonylation reactions of alkynyl alcohols has proven to be a useful method for the synthesis of indole- and benzofuran-3-carboxylates via *endo*-cyclization pathway [90–92]. For example, Scammells realized the synthesis of naturally occurring benzofurane XH-14 **78** isolated from *Salvia miltiorrhiza* and found to be a potent antagonist of the A1 adenosine receptor. The synthesis of benzofurane skeleton **77** was accomplished in 68% yield starting from *o*-ethynylphenols in the presence of catalytic amount of $PdCl_2$, sodium acetate, and carbon monoxide in methanol and in the presence of $CuCl_2$ as reoxidant (Scheme 13.49) [90].

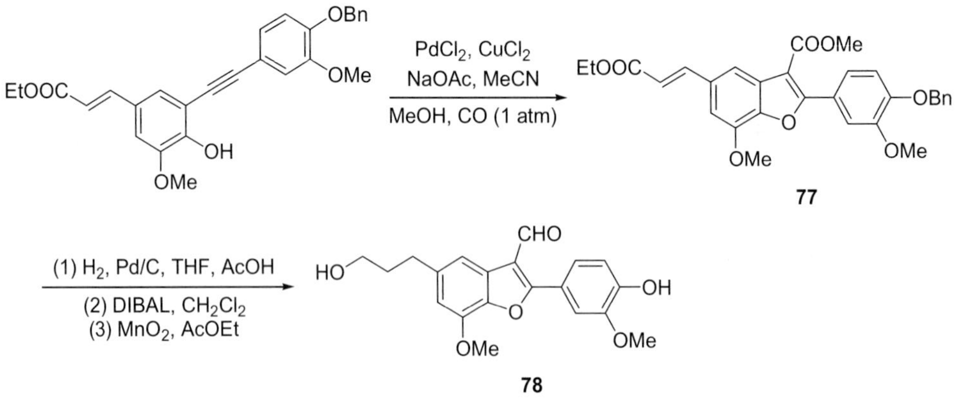

Scheme 13.49

More recently, a highly effective catalytic system was developed for the cyclization of both electron-rich and electron-poor *o*-ethynylphenols. Yang and coworkers introduced for the first time carbon tetrabromide as a reoxidant for the turnover of Pd(0) to Pd(II). A systematic study to identify the appropriate catalyst, base, and oxidative agent resulted in the choice of PdI_2–thiourea, Cs_2CO_3, and CBr_4, respectively. The reactions performed under CO (1 atm) in CH_3OH at 40 °C resulted in the isolation of benzofuran-3-carboxilates **79** in 78–85% yields (Scheme 13.50) [93]. The combinatorial synthesis of these substrates was also achieved by the same authors [94].

The synthesis of the corresponding furans has been achieved by palladium(II)-mediated cyclization–alkoxycarbonylation reactions of 4-yn-1-ols using $PdI_2/KI/O_2$ (the catalytic system will be discussed later), and the reactions give rise to a mixture of desired cyclic β-alkoxyacrylates along with some acetal products arising from acid-catalyzed addition of methanol (Scheme 13.51).

However, following the methodology developed by Kato and Akita, a great variety of β-alkoxyacrylates have been synthesized starting from 4-yn-1-ols [95], cyclic 2-propargyl-1-ols [96], or 2-propargyl-1-ones [97–99] (Scheme 13.52).

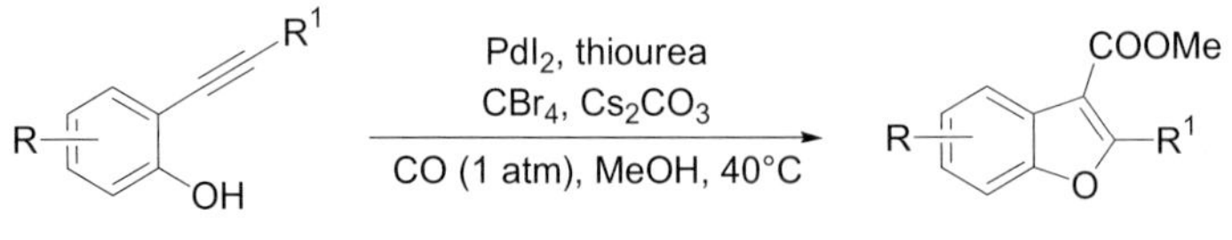

Scheme 13.50

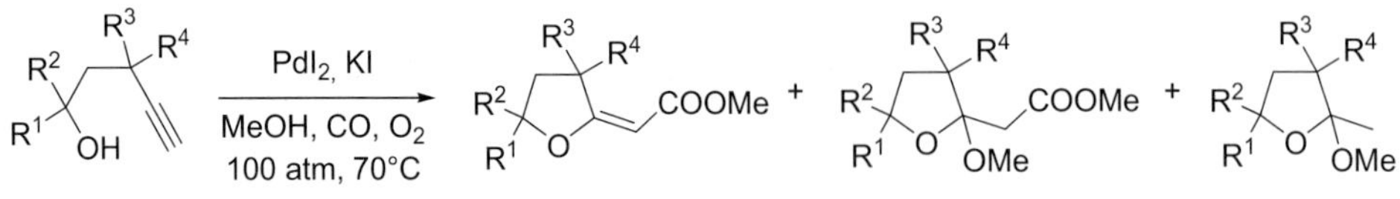

Scheme 13.51

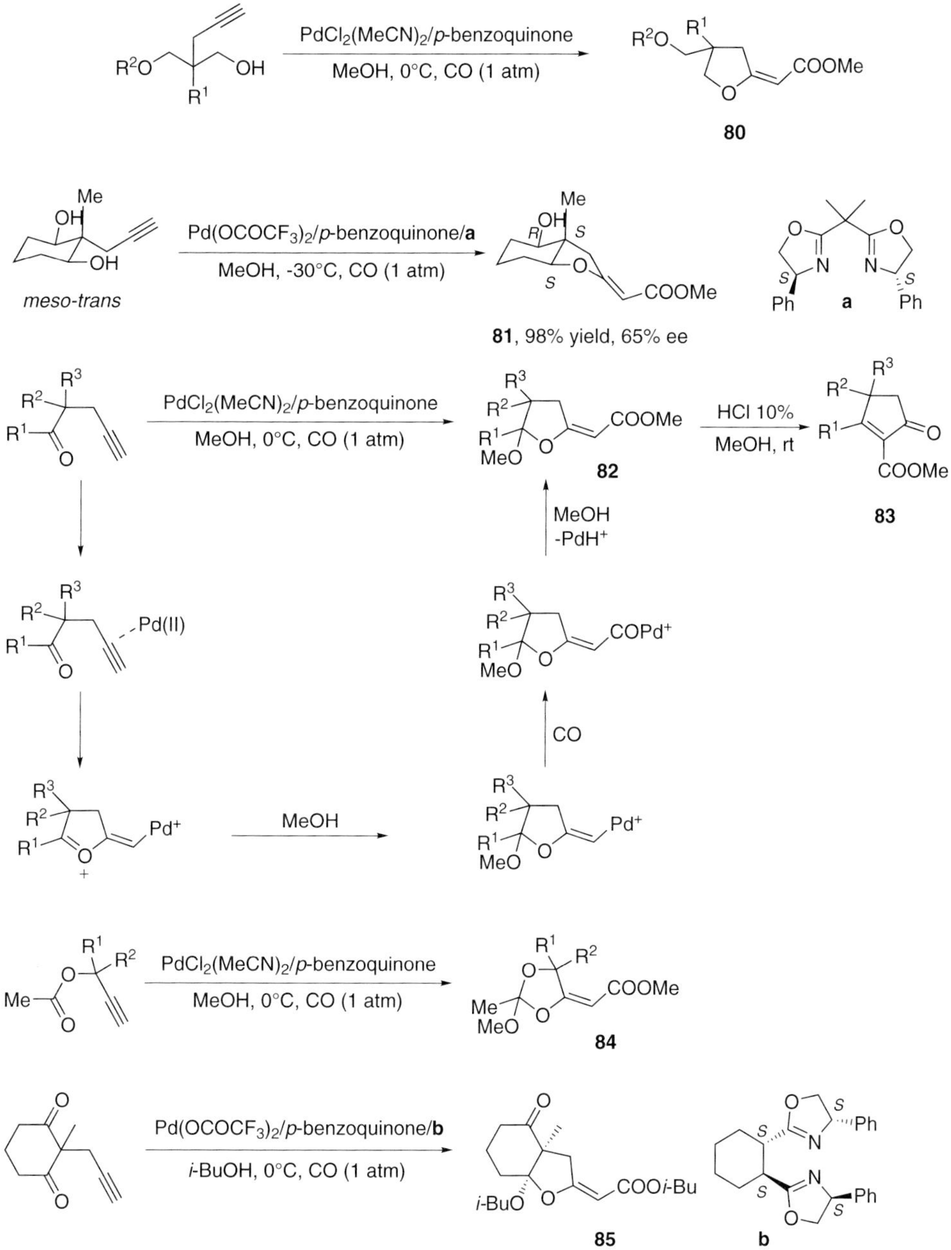

PdCl2(MeCN)2/p-benzoquinone
MeOH, 0°C, CO (1 atm)
COOMe
80
meso-trans
Pd(OCOCF3)2/p-benzoquinone/a
MeOH, -30°C, CO (1 atm)
COOMe
81, 98% yield, 65% ee
a
PdCl2(MeCN)2/p-benzoquinone
MeOH, 0°C, CO (1 atm)
82
HCl 10%
MeOH, rt
83
MeOH
-PdH+
Pd(II)
COPd+
CO
MeOH
Pd+
PdCl2(MeCN)2/p-benzoquinone
MeOH, 0°C, CO (1 atm)
84
Pd(OCOCF3)2/p-benzoquinone/b
i-BuOH, 0°C, CO (1 atm)
COOi-Bu
85
b

Scheme 13.52

The authors focused their attention on trapping the acid, which is coproduced from the reaction. *p*-Benzoquinone was found to be a very efficient reagent for trapping a proton of hydrochloric acid arising from the catalytic cycle and also from the reoxidation of Pd(0) species to Pd(II). Acyclic 4-yn-1-ols afforded (*E*)-cyclic-β-alkoxyacrylates **80** bearing functional groups such as acetate, hydroxyl, ketone, ester, terminal acetylene, and the acid-sensitive protecting groups (TBDPS, TBDMS, MOM, and THP). Cyclic 2-methyl-2-propargyl-1,3-diols in the presence of chiral bisoxazolines afforded (*E*)-bicyclic-β-alkoxyacrylates with moderate enantioselectivities as reported in Scheme 13.52 for the synthesis of **81**. Finally, when differently substituted 4-yn-1-ones were reacted under the same reaction conditions cyclic ketals **82** (easily converted into the corresponding 2-cyclopentenone carboxylate **83**), cyclic orthoesters **84**, and bicyclic ketals **85** had been achieved, also in a stereoselective manner, starting from α or α,α-disubstituted-4-yn-1-ones, propargylic acetates, and cyclic 2-propargyl-1,3-diones, respectively.

Recently, palladium(II)-catalyzed arylation/carboxylation reactions were reported for the synthesis of carbazoles starting from 2-alkenyl indoles. These reactions employ the electron-rich C-3 carbon atom of the indole nucleus as carbon nucleophile in cyclization/carboalkoxylation reactions. The conditions that proved effective for 2-(2,3 or 5-alkenyl)-3-unsubstituted indoles are similar to those reported for the cyclization of alkenes and alkynes bearing tethered heteronucleophiles. In particular, the best results has been obtained using $PdCl_2(CH_3CN)_2$, $CuCl_2$ in methanol under CO (1 atm). Under these conditions methyl (9-methyl-2,3,4,9-tetrahydro-4-carbazolyl)acetate **86** was isolated in 94% yield (Scheme 13.53) [100,101].

As reported at the beginning of this section, PdI_2 in the presence of an excess of iodide anions and molecular oxygen has been developed as an efficient, selective, and versatile catalytic system for promoting a variety of oxidative carbonylation processes, leading to heterocyclic derivatives [10]. These methodologies have been jointly developed principally by Gabriele and Salerno at the University of Arcavacata, Italy, and by Costa and Chiusoli at the University of Parma, Italy. The reason of the high activity of the PdI_4^{2-} effective catalyst lies in the efficient reoxidation of Pd(0) that involves oxidation of the *in situ* formed HI, by O_2, to I_2 that in turn undergoes oxidative addition to Pd(0) (Scheme 13.42). Under these conditions Pd(II) catalyst is restored without any external inorganic salt or organic reagent. For example, PdI_2/KI catalytic system promotes the oxidative carbonylation of prop-2-ynylamides to oxazolines **87** (Scheme 13.54) [102].

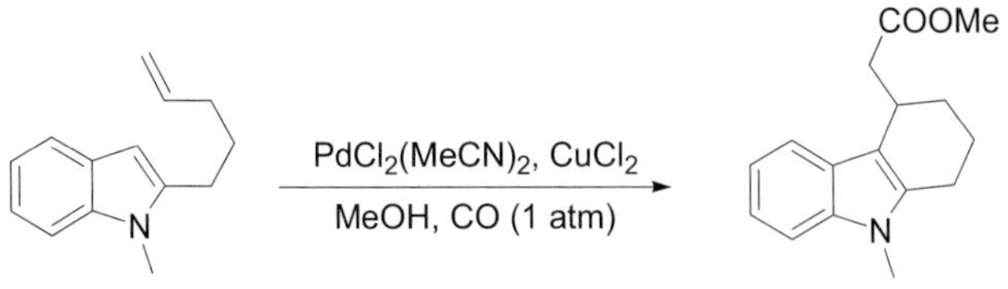

COOMe

N

$PdCl_2(MeCN)_2$, $CuCl_2$

MeOH, CO (1 atm)

N

86

Scheme 13.53

$R^1C(O)NH-C(R^2)(R^3)-C \equiv CH$ + CO + R^4OH + 1/2 O_2 → (PdI_2, KI; (20 atm)) → **87** (R^1, R^2, R^3, $COOR^4$; N, O)

Scheme 13.54

Under these or modified conditions, a plethora of five- and six-membered heterocyclic derivatives have been synthesized during the past 10 years, as partially summarized in the Schemes 13.55 and 13.56 [103–109].

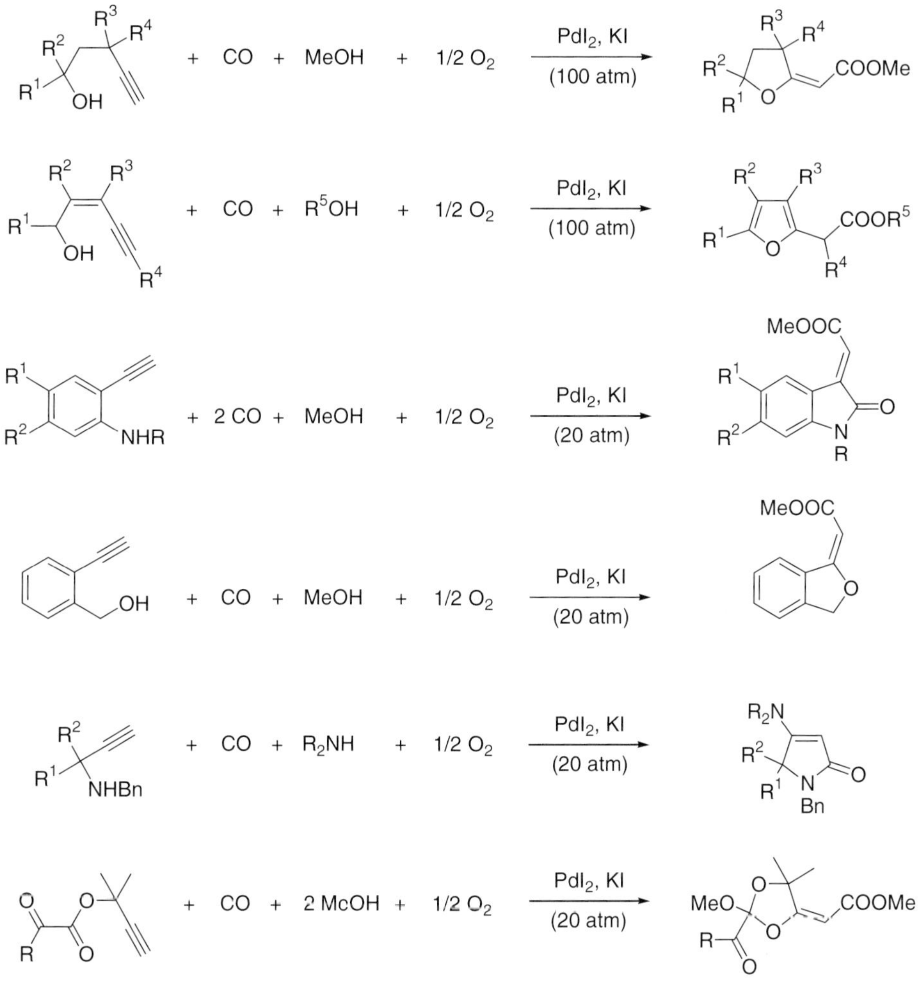

(R^1, R^2, R^3, R^4, OH) + CO + MeOH + 1/2 O_2 → (PdI_2, KI; (100 atm)) → (R^1, R^2, R^3, R^4, O, COOMe)

(R^1, R^2, R^3, R^4, OH) + CO + R^5OH + 1/2 O_2 → (PdI_2, KI; (100 atm)) → (R^1, R^2, R^3, R^4, O, $COOR^5$)

(R^1, R^2, NHR) + 2 CO + MeOH + 1/2 O_2 → (PdI_2, KI; (20 atm)) → (MeOOC, R^1, R^2, N, R, O)

(OH) + CO + MeOH + 1/2 O_2 → (PdI_2, KI; (20 atm)) → (MeOOC, O)

(R^1, R^2, NHBn) + CO + R_2NH + 1/2 O_2 → (PdI_2, KI; (20 atm)) → (R_2N, R^2, R^1, N, Bn, O)

(O, R, O, O) + CO + 2 MeOH + 1/2 O_2 → (PdI_2, KI; (20 atm)) → (MeO, O, O, R, O, COOMe)

Scheme 13.55

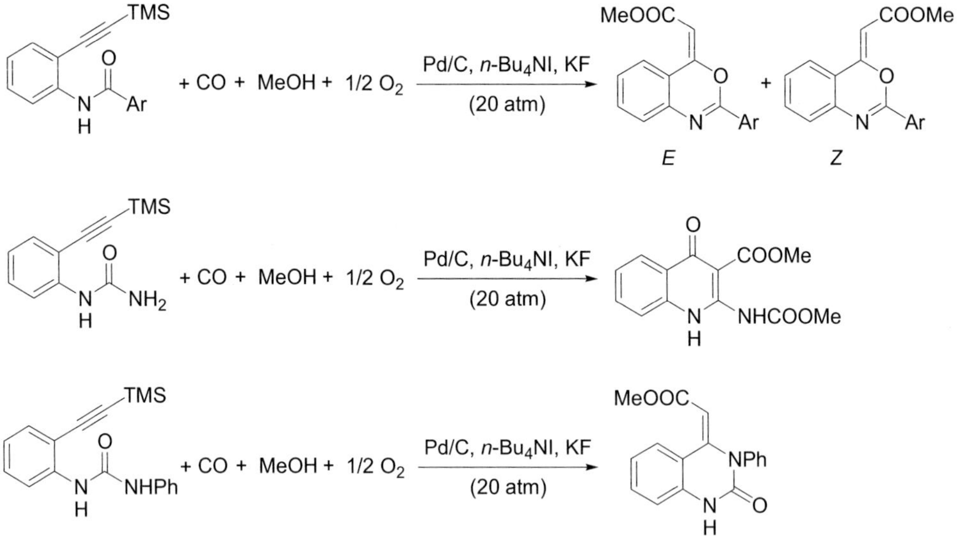

Scheme 13.56

Sometimes, nitrogen-containing substrates that are basic enough to be protonated by HI, evolved during the reactions, can inhibit the reoxidation of Pd(0) and therefore hinder the overall oxidative carbonylation process. When this occurs, a reactant able to reversibly bind the amino group (thus "freeing" the HI necessary for the reoxidation of Pd(0)) without hampering the cyclization–alkoxycarbonylation process was needed. Carbon dioxide effectively fulfills these requirements through the formation of a carbamate species with nitrogen functionalities present in the substrates. The nitrogen in the carbamate, although much less basic than in the substrate, may still act as nucleophile, because CO_2 can be eliminated during the cyclization process. An example of "buffered" process is reported in Scheme 13.57 [110,111].

Finally, working with slight different substrates, carbon dioxide and carbon monoxide can react in sequence as exemplified in the reaction of *N*-alkyl-substituted

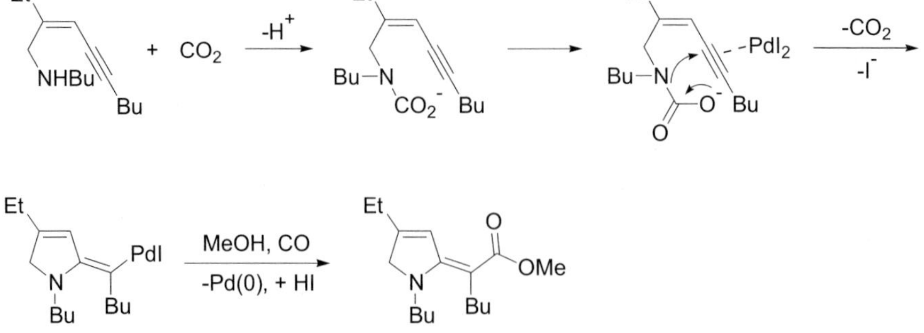

Scheme 13.57

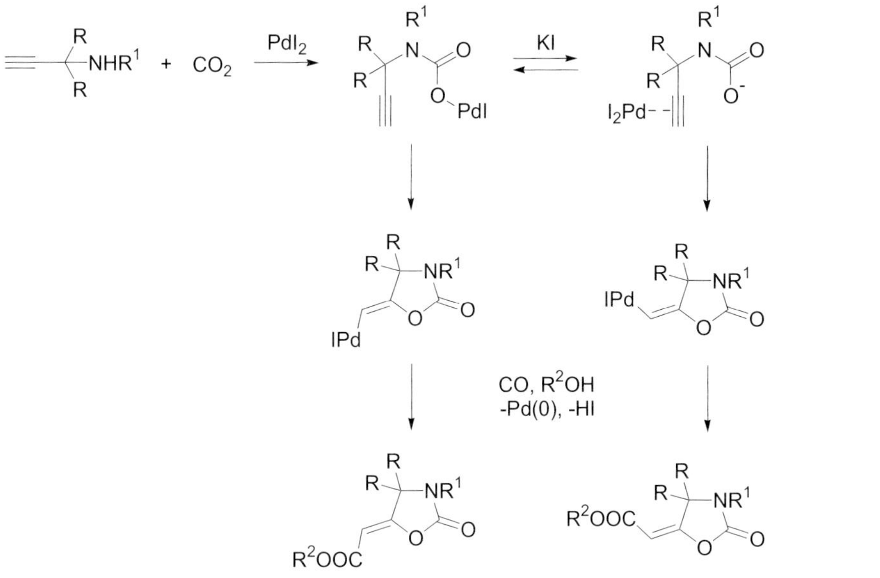

Scheme 13.58

dialkylpropynylamines giving rise, under catalytic conditions, to *Z* and *E* oxazolidin-2-ones **88** and **89** (Scheme 13.58) [112,113].

References

1 Battistuzzi, G., Cacchi, S. and Fabrizi, G. (2002) The aminopalladation/reductive elimination domino reaction in the construction of functionalized indole rings. *European Journal of Organic Chemistry*, 2671–2681.

2 Zeni, G. and Larock, R.C. (2006) Synthesis of heterocycles via palladium-catalyzed oxidative addition. *Chemical Reviews*, **106**, 4644–4680.

3 Kalck, P., Urrutigoïty, M. and Dechy-Cabaret, O. (2006) Hydroxy- and alkoxycarbonylation of alkenes and alkynes. *Topics in Organometallic Chemistry*, **18**, 97–123.

4 Muzart, J. (2005) Palladium-catalysed reactions of alcohols. Part D: Rearrangements, carbonylations, carboxylations and miscellaneous reactions. *Tetrahedron*, **61**, 9423–9463.

5 Skoda-Földes, R. and Kollár, L. (2002) Synthetic applications of palladium catalysed carbonylation of organic halides. *Current Organic Chemistry*, **6**, 1097–1119.

6 Varchi, G. and Ojima, I. (2006) Synthesis of heterocycles through hydrosilylation, silylformylation, silylcarbocyclization and cyclohydrocarbonylation reactions. *Current Organic Chemistry*, **10**, 1341–1362.

7 Vizer, S.A., Yerzhanov, K.B., Al Quntarb, A. A.A. and Dembitsky, V.M. (2004) Synthesis of heterocycles by carbonylation of acetylenic compounds. *Tetrahedron*, **60**, 5499–5538.

8 von Wangelin, A.J., Axel, J., Neumann, H. and Beller, M. (2006) Carbonylation of

aldehydes. *Topics in Organometallic Chemistry*, **18**, 207–221.

9 El Ali, B. and Alper, H. (2000) The application of transition metal catalysis for selective cyclocarbonylation reactions. synthesis of lactones and lactams. *Synlett*, 161–171.

10 Bartolo, G., Salerno, G. and Costa, M. (2006) Oxidative carbonylations. *Topics in Organometallic Chemistry*, **18**, 239–272.

11 Bartolo, G., Salerno, G. and Costa, M. (2004) PdI_2-catalyzed synthesis of heterocycles. *Synlett*, 2468–2483.

12 Bartolo, G., Salerno, G., Costa, M. and Chiusoli, G.P. (2004) Recent advances in the synthesis of carbonyl compounds by palladium-catalyzed oxidative carbonylation reactions of unsaturated substrates. *Current Organic Chemistry*, **8**, 919–946.

13 Bartolo, G., Salerno, G., Costa, M. and Chiusoli, G.P. (2003) Recent developments in the synthesis of heterocyclic derivatives by PdI_2-catalyzed oxidative carbonylation reactions. *Journal of Organometallic Chemistry*, **687**, 219–228.

14 Chatani, N. (2004) Selective carbonylation with ruthenium catalysts. *Topics in Organometallic Chemistry*, **11**, 173–195.

15 Morimoto, T. and Kakiuchi, K. (2003) Evolution of carbonylation catalysis: no need for carbon monoxide. *Angewandte Chemie – International Edition*, **42**, 5230–5234.

16 Tsuji, J. (1995) *Palladium Reagents and Catalysts: Innovations in Organic Synthesis*, John Wiley & Sons, Inc., New York.

17 Tsuji, J. (2000) *Transition Metals Reagents and Catalysts: Innovations in Organic Synthesis*, John Wiley & Sons, Inc., New York.

18 Li, J.J. and Gribble, G.W. (2000) *Palladium in Heterocyclic Chemistry*, Pergamon, New York.

19 Negishi, E. (2002) *Handbook of Organopalladium Chemistry for Organic Synthesis*, vols 1 and 2, John Wiley & Sons, Inc., New York.

20 Tsuji, J. (2003) *Palladium Reagents and Catalysts: New Perspectives for the 21st Century*, John Wiley & Sons, Inc., New York.

21 Larksarp, C. and Alper, H. (1999) Palladium-catalyzed cyclocarbonylation of *o*-iodophenols and 2-hydroxy-3-iodopyridine with heterocumulenes: regioselective synthesis of benzo[*e*]-1,3-oxazin-4-one and pyrido[3,2-*e*]-1,3-oxazin-4-one derivatives. *The Journal of Organic Chemistry*, **64**, 9194–9200.

22 Larksarp, C. and Alper, H. (2000) Palladium-catalyzed cyclocarbonylation of o-iodoanilines with heterocumulenes: regioselective preparation of 4 (3*H*)-quinazolinone derivatives. *The Journal of Organic Chemistry*, **65**, 2773–2777.

23 Larksarp, C. and Alper, H. (1999) A simple synthesis of 2-substituted-4*H*-3,1-benzoxazin-4-ones by palladium-catalyzed cyclocarbonylation of *o*-iodoanilines with acid chlorides. *Organic Letters*, **1**, 1619–1622.

24 Orito, K., Miyazawa, M., Kanbayashi, R., Tokuda, M. and Suginome, H. (1999) Synthesis of phthalideisoquinoline and protoberberine alkaloids and indolo[2,1-*a*] isoquinolines in a divergent route involving palladium(0)-catalyzed carbonylation. *The Journal of Organic Chemistry*, **64**, 6583–6596.

25 Kadnikov, D.V. and Larock, R.C. (2000) Synthesis of coumarins via palladium-catalyzed carbonylative annulation of internal alkynes by *o*-iodophenols. *Organic Letters*, **2**, 3643–3646.

26 Kadnikov, D.V. and Larock, R.C. (2003) Palladium-catalyzed carbonylative annulation of internal alkynes: synthesis of 3,4-disubstituted coumarins. *The Journal of Organic Chemistry*, **68**, 9423–9432.

27 Kadnikov, D.V. and Larock, R.C. (2004) Synthesis of 2-quinolones via palladium-catalyzed carbonylative annulation of internal alkynes by *N*-substituted *o*-iodoanilines. *The Journal of Organic Chemistry*, **69**, 6772–6780.

28 Kadnikov, D.V. and Larock, R.C. (2003) Palladium-catalyzed carbonylative annulation of terminal alkynes: synthesis of coumarins and 2-quinolones. *Journal of Organometallic Chemistry*, **687**, 425–435.

29 Ma, W., Li, X., Yang, J., Liu, Z., Chen, B. and Pana, X. (2006) Synthesis of aryl ferocenylethynyl ketones and 2-ferocenyl-4*H*-chromen-4-ones. *Synthesis*, 2489–2492.

30 Ciattini, P.G., Morera, E., Ortar, G. and Rossi, S.S. (1991) Preparative and regiochemical aspects of the palladium-catalyzed carbonylative coupling of 2-hydroxyaryl iodides with ethynylarenes. *Tetrahedron*, **47**, 6449–6456.

31 Kalinin, V.N., Shostakovskii, M.V. and Ponomarev, A.B. (1990) Palladium-catalyzed synthesis of flavones and chromones via carbonylative coupling of *o*-iodophenols with terminal acetylenes. *Tetrahedron Letters*, **31**, 4073–40766.

32 Torii, S., Okumoto, H., Xu, L.H., Sadakane, M., Shostakovsky, M.V., Ponomaryov, A.B. and Kalinin, V.N. (1993) Syntheses of chromones and quinolones via palladium-catalyzed carbonylation of *o*-iodophenols and anilines in the presence of acetylenes. *Tetrahedron*, **49**, 6773–6784.

33 Miao, H. and Yang, Z. (2000) Regiospecific carbonylative annulation of iodophenol acetates and acetylenes to construct the flavones by a new catalyst of palladium-thiourea-dppp complex. *Organic Letters*, **2**, 1765–1768.

34 Lee, Y., Fujiwara, Y., Ujita, K., Nagatomo, M., Ohta, H. and Shimizu, I. (2001) Syntheses of mycophenolic acid and its analogs by palladium methodology. *Bulletin of the Chemical Society of Japan*, **74**, 1437–1443.

35 Aoyagi, S., Hasegawa, Y., Hirashima, S. and Kibayashi, C. (1998) Total synthesis of (+)-homopumiliotoxin 223G. *Tetrahedron Letters*, **39**, 2149–2152.

36 Bio, M.M. and Leighton, J.L. (1999) An approach to the synthesis of CP-263,114: a remarkably facile silyloxy-Cope rearrangement. *Journal of the American Chemical Society*, **121**, 890–891.

37 Bio, M.M. and Leighton, J.L. (2000) Stereoconvergent palladium-catalyzed carbonylation of both *E* and Z isomers of a 2-trifloxy-1,3-butadiene. *Organic Letters*, **2**, 2905–2907.

38 Bio, M.M. and Leighton, J.L. (2003) An approach to the synthesis of the phomoidrides. *The Journal of Organic Chemistry*, **68**, 1693–1700.

39 Bocelli, G., Catellani, M., Cugini, F. and Ferraccioli, R. (1999) A new and efficient palladium-catalyzed synthesis of a 2,3,4,5-tetrahydro-1*H*-2,4-benzodiazepine-1,3-dione derivative. *Tetrahedron Letters*, **40**, 2623–2624.

40 Grigg, R., MacLachlan, W.S., MacPherson, D.T., Sridharan, V., Suganthan, S., Thornton-Petta, M. and Zhanga, J. (2000) Pictet-Spengler/palladium catalyzed allenylation and carbonylation processes. *Tetrahedron*, **56**, 6585–6594.

41 Beller, M., Eckert, M., Moradi, W.A. and Neumann, H. (1999) Palladium-catalyzed synthesis of substituted hydantoins – a new carbonylation reaction for the synthesis of amino acid derivatives. *Angewandte Chemie – International Edition*, **38**, 1454–1457.

42 Beller, M., Eckert, M. and Vollmüller, F. (1998) A new class of catalysts with superior activity and selectivity for amidocarbonylation reactions. *Journal of Molecular Catalysis A: Chemical*, **135**, 23–33.

43 Young, J.R. and De Vita, R.J. (1998) Novel synthesis of oxadiazoles via palladium catalysis. *Tetrahedron Letters*, **39**, 3931–3934.

44 Zhou, T. and Chen, Z. (2002) Hypervalent iodine in synthesis. 75. A convenient synthesis of oxadiazoles by palladium-catalyzed carbonylation of diaryliodonium salts and amidoximes. *Synthetic Communications*, **36**, 887–891.

45 Abbiati, G., Arcadi, A., Canevari, V., Capezzuto, L. and Rossi, E. (2005) Palladium-assisted multicomponent synthesis of 2-aryl-4 aminoquinolines and 2-aryl-4-amino[1,8]naphthyridines. *The Journal of Organic Chemistry*, **70**, 6454–6460.

46 Arcadi, A., Cacchi, S., Carnicelli, V. and Marinelli, F. (1994) 2-Substituted-3-acylindoles through the palladium-catalyzed carbonylative cyclization of 2-alkynyltrifluoroacetanilides with aryl halides and vinyl triflates. *Tetrahedron*, **50**, 437–452.

47 Cacchi, S., Fabrizi, G., Pace, P. and Marinelli, F. (1999) 6-Aryl-11*H*-indolo[3,2-*c*]quinolines through the palladium-catalyzed carbonylative cyclization of *o*-(*o*-aminophenyl)trifluoroacetanilides with aryl iodides. *Synlett*, 620–622.

48 Battistuzzi, G., Cacchi, S., Fabrizi, G., Marinelli, F. and Parisi, L.M. (2002) 12-Acylindolo[1,2-*c*]quinazolines by palladium-catalyzed cyclocarbonylation of *o*-alkynyltrifluoroacetanilides. *Organic Letters*, **4**, 1355–1358.

49 Arcadi, A., Cacchi, S., Del Rosario, M., Fabrizi, G. and Marinelli, F. (1996) Palladium-catalyzed reaction of *o*-ethynylphenols, *o*-((trimethylsilyl) ethynyl)phenyl acetates and *o*-alkynylphenols with unsaturated triflates or halides: a route to 2-substituted-, 2,3-disubstituted-, and 2-substituted-3-acylbenzo[*b*]furans. *The Journal of Organic Chemistry*, **61**, 9280–9288.

50 Hu, Y., Zhang, Y., Yang, Z. and Fathi, R. (2002) Palladium-catalyzed carbonylative annulation of *o*-alkynylphenols: syntheses of 2-substituted-3-aroyl-benzo[*b*]furans. *The Journal of Organic Chemistry*, **67**, 2365–2368.

51 Arcadi, A., Cacchi, S., Fabrizi, G. and Moro, L. (1999) Palladium-catalyzed cyclocarbonylation of *o*-ethynylphenols and vinyl triflates to form 3-alkylidene-2-coumaranones. *European Journal of Organic Chemistry*, 1137–1141.

52 Dai, G. and Larock, R.C. (2002) Synthesis of 3-substituted 4-aroylisoquinolines via Pd-catalyzed carbonylative cyclization of *o*-(1-alkynyl)benzaldimines. *Organic Letters*, **4**, 193–196.

53 Dai, G. and Larock, R.C. (2002) Synthesis of 3-substituted 4-aroylisoquinolines via Pd-catalyzed carbonylative cyclization of 2-(1-alkynyl)benzaldimines. *The Journal of Organic Chemistry*, **67**, 7042–7047.

54 Cho, C.S., Shim, H.S., Choi, H.-J., Kim, T.-J. and Shim, S.C. (2002) Palladium-catalyzed convenient synthesis of 3-methyleneisoindolin-1-ones. *Synthetic Communications*, **32**, 1821–1827.

55 Cho, C.S., Jiang, L.H. and Shim, S.C. (1998) Synthesis of isoindolinones by palladium-catalyzed carbonylative cyclization of 2-bromobenzaldehyde with diamines. *Synthetic Communications*, **28**, 849–857.

56 Cho, C.S., Chu, D.Y., Lee, D.Y., Shim, S.C., Kim, T.J., Lim, W.T. and Heo, N.H. (1997) Palladium-catalyzed diastereoselective synthesis of isoindolinones. *Synthetic Communications*, **27**, 4141–4158.

57 Cho, C.S., Shim, S.C., Choi, H.-J., Kim, T.-J., Shim, S.C. and Kim, M.C. (2000) Synthesis of isoindolin-1-ones via palladium-catalyzed intermolecular coupling and heteroannulation between 2-iodobenzoyl chloride and imines. *Tetrahedron Letters*, **41**, 3891–3893.

58 Arcadi, A. and Rossi, E. (1996) A Palladium-catalyzed domino reaction of 3-acetyl-5-hexyn-2-one with aryl iodides under carbon monoxide. *Tetrahedron Letters*, **37**, 6811–6814.

59 Arcadi, A., Cacchi, S., Fabrizi, G., Marinelli, F. and Parisi, L.M. (2003) Highly substituted furans from 2-propynyl-1,3-dicarbonyls and organic halides or triflates via the oxypalladation-reductive elimination domino reaction. *Tetrahedron*, **59**, 4661–4671.

60 Barberan, O., Alami, M. and Brion, J.-D. (2001) Synthesis of *E*- and *Z*-substituted methylene-3,4-dihydro-2*H*-1-benzopyrans by regio- and stereocontrolled palladium-catalyzed intramolecular cyclization. *Tetrahedron Letters*, **42**, 2657–2659.

61 Okuro, K. and Alper, H. (1997) Palladium-catalyzed carbonylation of *o*-iodophenols with allenes. *The Journal of Organic Chemistry*, **62**, 1566–1567.

62 Xiao, W.-J. and Alper, H. (1999) Regioselective carbonylative heteroannulation of *o*-iodothiophenols with allenes and carbon monoxide catalyzed by a palladium complex: a novel and efficient access to thiochroman-4-one derivatives. *The Journal of Organic Chemistry*, **64**, 9646–9652.

63 Kang, S.-K. and Kim, K.-J. (2001) Palladium (0)-catalyzed carbonylation-coupling-cyclization of allenic sulfonamides with aryl iodides and carbon monoxide. *Organic Letters*, **3**, 511–514.

64 Dhawan, R., Dghaym, R.D. and Arndtsen, B.A. (2003) The development of a catalytic synthesis of munchnones: a simple four-component coupling approach to α-amino acid derivatives. *Journal of the American Chemical Society*, **125**, 1474–1475.

65 Dhawan, R. and Arndtsen, B.A. (2004) Palladium-catalyzed multicomponent coupling of alkynes, imines, and acid chlorides: a direct and modular approach to pyrrole synthesis. *Journal of the American Chemical Society*, **126**, 468–469.

66 Dhawan, R., Dghaym, R.D., St Cyr, D.J. and Arndtsen, B.A. (2006) Direct, palladium-catalyzed, multicomponent synthesis of β-lactams from imines, acid chloride, and carbon monoxide. *Organic Letters*, **8**, 3927–3930.

67 Siamaki, A.R. and Arndtsen, B.A. (2006) A Direct, one step synthesis of imidazoles from imines and acid chlorides: a palladium catalyzed multicomponent coupling approach. *Journal of the American Chemical Society*, **128**, 6050–6051.

68 Grigg, R. and Sridharan, V. (1999) Palladium-catalysed cascade cyclisation anion capture, relay switches and molecular queues. *Journal of Organometallic Chemistry*, **576**, 65–87.

69 Brown, S., Clarkson, S., Grigg, R., Thomas, W.A., Sridharanb, V. and Wilson, D.M. (2001) Palladium catalysed queuing processes. Part 1: termolecular cyclization-anion capture employing carbon monoxide as a relay switch and hydride, organotin(IV) or boron reagents. *Tetrahedron*, **57**, 1347–1359.

70 Anwar, U., Casaschi, A., Griggp, R. and Sansano, J.Â.M. (2001) Palladium-catalysed queuing processes. Part 2: termolecular cyclization-anion capture employing carbon monoxide as a relay switch with *in situ* generated vinylstannanes. *Tetrahedron*, **57**, 1361–1367.

71 Grigg, R., Liu, A., Shaw, D., Suganthan, S., Washington, M.L., Woodall, D.E. and Yoganathan, G. (2000) Palladium-catalysed cascade molecular queuing-cycloaddition, cyclocondensation and Diels-Alder reactions. *Tetrahedron Letters*, **41**, 7129–7133.

72 Evans, P., Grigg, R., Ramzan, M.I., Sridharan, V. and York, M. (1999) Sequential and cascade palladium catalyzed cyclization-anion capture-olefin metathesis. *Tetrahedron Letters*, **40**, 3021–3024.

73 Grigg, R., Liu, A., Shaw, D., Suganthan, S., Washington, M.L., Woodall, D.E. and Yoganathan, G. (2000) Synthesis of quinol-4-ones and chroman-4-ones via a palladium-catalyzed cascade carbonylation-allene insertion. *Tetrahedron Letters*, **41**, 7125–7128.

74 Gai, X., Grigg, R., Khamnaen, T., Rajviroongit, S., Sridharan, V., Zhang, L., Collard, S. and Keep, A. (2003) Synthesis of 3-substituted isoindolin-1-ones via a palladium-catalysed 3-component carbonylation/amination/Michael addition process. *Tetrahedron Letters*, **44**, 7441–7443.

75 Grigg, R., Sridharan, V. and Thayaparan, A. (2003) Synthesis of novel cyclic α-amino acid derivatives via a one-pot sequential Petasis reaction/palladium catalysed process. *Tetrahedron Letters*, **44**, 9017–9019.

76 Grigg, R., MacLachlanb, W. and Rasparini, M. (2000) Palladium catalysed tetramolecular queuing cascades of aryl iodides, carbon monoxide, amines and a polymer supported allene. *Journal of the Chemical Society, Chemical Communications*, 2241–2242.

77 Grigg, R., Hodgson, A., Morris, J. and Sridharana, V. (2003) Sequential Pd/Ru-catalysed allenylation/olefin metathesis/ 1,3-dipolar cycloaddition route to novel heterocycles. *Tetrahedron Letters*, **44**, 1023–1026.

78 Grigg, R., Zhang, L., Collard, S. and Keep, A. (2003) Isoindolinones via a room temperature palladium nanoparticle-catalysed 3-component cyclative carbonylation-amination cascade. *Tetrahedron Letters*, **44**, 6979–6982.

79 Hegedus, L.S. (1999) *Transition Metals in the Synthesis of Complex Organic Molecules*, University Science Books, Sausalito, CA.

80 Hegedus, L.S., Allen, G.F. and Olsen, D.J. (1980) Palladium-assisted cyclization-insertion reactions, synthesis of functionalized heterocycles. *Journal of the American Chemical Society*, **102**, 3583–3587.

81 Semmelhack, M.F., Bozell, J.J., Sato, T., Wulff, W., Spiess, E. and Zask, A. (1982) Synthesis of nanaomycin a and deoxyfrenolicin by alkyne cycloaddition to chromium-carbene complexes. *Journal of the American Chemical Society*, **104**, 5850–5853.

82 Semmelhack, M.F. and Bodurow, C. (1984) Intramolecular alkoxypalladation/ carbonylation of alkenes. *Journal of the American Chemical Society*, **106**, 1496–1498.

83 White, J.D., Hong, J. and Robarge, L.A. (1999) Intramolecular palladium catalyzed alkoxy carbonylation of 6-hydroxy-1-octenes. Stereoselective synthesis of substituted tetrahydropyrans. *Tetrahedron Letters*, **40**, 1463–1466.

84 Tamaru, Y., Hojo, M., Higashimura, H. and Yoshida, Z-i. (1988) Urea as the most reactive and versatile nitrogen nucleophile for the palladium(2+)-catalyzed cyclization of unsaturated amines. *Journal of the American Chemical Society*, **110**, 3994–4002.

85 Tamaru, Y. Hojo, M. and Yoshida, Z.-i. (1988) Palladium(2+)-catalyzed intramolecular aminocarbonylation of 3-hydroxy-4-pentenylamines and 4-hydroxy-5-hexenylamine. *The Journal of Organic Chemistry*, **53**, 5731–5741.

86 Harayama, H., Abe, A., Sakado, T., Kimura, M., Fugami, K., Tanaka, S. and Tamaru, Y. (1997) Palladium(II)-catalyzed intramolecular aminocarbonylation of *endo*-carbamates under Wacker-type conditions. *The Journal of Organic Chemistry*, **62**, 2113–2122.

87 Bates, R.W. and Sa-Ei, K. (2002) *O*-Alkenyl hydroxylamines: a new concept for cyclofunctionalization. *Organic Letters*, **4**, 4225–4227.

88 Shinohara, T., Arai, M.A., Wakita, K., Arai, T. and Sasai, H. (2003) The first enantioselective intramolecular aminocarbonylation of alkenes promoted by Pd(II)-spiro bis(isoxazoline) catalyst. *Tetrahedron Letters*, **44**, 711–714.

89 Paddon-Jones, G.C., McErlean, C.S.P., Hayes, P., Moore, C.J., Konig, W.A. and Kitching, W. (2001) Synthesis and stereochemistry of some bicyclic γ-lactones from parasitic wasps (Hymenoptera: Braconidae). Utility of hydrolytic kinetic resolution of epoxides and palladium(II)-catalyzed hydroxycyclization-carbonylation-lactonization of ene-diols. *The Journal of Organic Chemistry*, **66**, 7487–7495.

90 Lütjens, H. and Scammells, P.J. (1998) Synthesis of natural products possessing a benzo[*b*]furan skeleton. *Tetrahedron Letters*, **39**, 6581–6584.

91 Lütjens, H. and Scammells, P.J. (1999) Synthesis of 2-substituted 3-acylbenzo[*b*] furans via the palladium catalysed carbonylative cyclisation of *ortho*-hydroxytolans. *Synlett*, 1079–1081.

92 Kondo, Y., Shiga, F., Murata, N., Sakamoto, T. and Yamanaka, H. (1994) Condensed heteroaromatic ring systems. XXIV. Palladium-catalyzed cyclization of 2-substituted phenylacetylenes in the presence of carbon monoxide. *Tetrahedron*, **50**, 11803–11812.

93 Nan, Y., Miao, H. and Yang, Z. (2000) A new complex of palladium-thiourea and carbon tetrabromide catalyzed carbonylative annulation of *o*-hydroxylarylacetylenes: efficient new synthetic technology for the synthesis of 2,3-disubstituted benzo[b]furans. *Organic Letters*, **3**, 297–299.

94 Liao, Y., Reitman, M., Zhang, Y., Fathi, R. and Yang, Z. (2002) Palladium(II)-mediated cascade carbonylative annulation of *o*-alkynyl-phenols on silyl linker-based macrobeads: a combinatorial synthesis of a 2,3-disubstituted benzo[b]furan library. *Organic Letters*, **4**, 2607–2609.

95 Kato, K., Nishimura, A., Yamamoto, Y. and Akita, H. (2001) Improved method for the synthesis of (*E*)-cyclic-β-alkoxyacrylates under mild conditions. *Tetrahedron Letters*, **42**, 4203–4205.

96 Kato, K., Tanaka, M., Yamamoto, Y. and Akita, H. (2002) Asymmetric cyclization–carbonylation of cyclic-2-methyl-2-propargyl-1,3-diols. *Tetrahedron Letters*, **43**, 1511–1513.

97 Kato, K., Yamamoto, Y. and Akita, H. (2002) Palladium(II)-mediated cyclization-carbonylation of 4-yn-1-ones: facile access to 2-cyclopentenone carboxylates. *Tetrahedron Letters*, **43**, 4915–4917.

98 Kato, K., Yamamoto, Y. and Akita, H. (2002) Unusual formation of cyclic-orthoesters by Pd(II)-mediated cyclization–carbonylation of propargylic acetates. *Tetrahedron Letters*, **43**, 6587–6590.

99 Kato, K., Tanaka, M., Yamamura, S., Yamamoto, Y. and Akita, H. (2003) Asymmetric cyclization–carbonylation of 2-propargyl-1,3-dione. *Tetrahedron Letters*, **44**, 3089–3092.

100 Liu, C. and Widenhoefer, R.A. (2004) Palladium-catalyzed cyclization/carboalkoxylation of alkenyl indoles. *Journal of the American Chemical Society*, **126**, 10250–10251.

101 Liu, C. and Widenhoefer, R.A. (2006) Scope and mechanism of the Pd^{II}-catalyzed arylation/carboalkoxylation of unactivated olefins with indoles. *Chemistry – A European Journal*, **12**, 2371–2382.

102 Bacchi, A., Costa, M., Gabriele, B., Pelizzi, G. and Salerno, G. (2002) Efficient and general synthesis of 5-(alkoxycarbonyl)methylene-3-oxazolines by palladium-catalyzed oxidative carbonylation of prop-2-ynylamides. *The Journal of Organic Chemistry*, **67**, 4450–4457.

103 Gabriele, B., Salerno, G., De Pascali, F., Costa, M. and Chiusoli, G.P. (2000) Palladium-catalyzed synthesis of 2*E*-(methoxycarbonyl)methylene] tetrahydrofurans: oxidative cyclization-methoxycarbonylation of 4-yn-1-ols versus cycloisomerization-hydromethoxylation. *Journal of Organometallic Chemistry*, **593–594**, 409–415.

104 Gabriele, B., Salerno, G., De Pascali, F., Costa, M. and Chiusoli, G.P. (1999) An efficient and general synthesis of furan-2-acetic esters by palladium-catalyzed oxidative carbonylation of (*Z*)-2-en-4-yn-1-ols. *The Journal of Organic Chemistry*, **64**, 7693–7699.

105 Gabriele, B., Salerno, G., Veltri, L., Costa, M. and Massera, C. (2001) Stereoselective synthesis of (*E*)-3-(methoxycarbonyl) methylene-1,3-dihydroindol-2-ones by palladium-catalyzed oxidative carbonylation of 2-ethynylanilines. *European Journal of Organic Chemistry*, 4607–4613.

106 Bacchi, A., Costa, M., Della Cà, N., Fabbricatore, M., Fazio, A., Gabriele, B., Nasi, C. and Salerno, G. (2004) Synthesis of 1-(alkoxycarbonyl)methylene-1,3-dihydroisobenzofurans and 4-(alkoxycarbonyl)benzo[c]pyrans by palladium-catalysed oxidative carbonylation of 2-alkynylbenzyl alcohols, 2-alkynylbenzaldehydes and 2-alkynylphenyl ketones. *European Journal of Organic Chemistry*, 574–585.

107 Costa, M., Della Cà, N., Gabriele, B., Massera, C., Salerno, G. and Soliani, M. (2004) Synthesis of 4*H*-3,1-benzoxazines, quinazolin-2-ones and quinoline-4-ones by palladium-catalyzed oxidative carbonylation of 2-ethynylaniline derivatives. *The Journal of Organic Chemistry*, **69**, 2469–2477.

108 Gabriele, B., Plastina, P., Salerno, G., Costa, M. (2005) A new synthesis of 4-dialkylamino-1,5-dihydropyrrol-2-ones by

Pd-catalyzed oxidative aminocarbonylation of 2-ynylamines. *Synlett*, 935–938.

109 Bacchi, A., Costa, M., Della Cà, N., Gabriele, B., Salerno, G. and Cassoni, S. (2005) Heterocyclic derivative syntheses by palladium-catalyzed oxidative cyclization-alkoxycarbonylation of substituted γ-oxoalkynes. *The Journal of Organic Chemistry*, **70**, 4971–4979.

110 Gabriele, B., Salerno, G., Fazio, A. and Campana, F.B. (2002) Unprecedented carbon dioxide effect on a Pd-catalysed oxidative carbonylation reaction: a new synthesis of pyrrole-2-acetic esters. *Journal of the Chemical Society, Chemical Communications*, 1408–1409.

111 Gabriele, B., Salerno, G., Fazio, A. and Veltri, L. (2006) Versatile synthesis of pyrrole-2-acetic esters and (pyridine-2-one)-3-acetic amides by palladium-catalyzed, carbon dioxide-promoted oxidative carbonylation of (*Z*)-(2-en-4-ynyl) amines. *Advanced Synthesis and Catalysis*, **348**, 2212–2222.

112 Bacchi, A., Chiusoli, G.P., Costa, M., Gabriele, B., Righi, C. and Salerno, G. (1997) Palladium-catalysed sequential carboxylation-alkoxycarbonylation of acetylenic amines. *Journal of the Chemical Society, Chemical Communications*, 1209–1210.

113 Chiusoli, G.P., Costa, M., Gabriele, B. and Salerno, G. (1999) Sequential reaction of carbon dioxide and carbon monoxide with acetylenic amines in the presence of a palladium catalyst. *Journal of Molecular Catalysis A: Chemical*, **143**, 297–310.

Index

Modern Carbonylation Methods. Edited by László Kollár

ISBN: 978-3-527-318964